The Law of Sines

In any $\triangle ABC$,

$$\frac{a}{\sin A} = \frac{b}{\sin B} = \frac{c}{\sin C}.$$

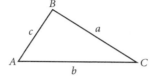

The Law of Cosines

In any $\triangle ABC$,

$$a^2 = b^2 + c^2 - 2bc \cos A,$$
$$b^2 = a^2 + c^2 - 2ac \cos B,$$
$$c^2 = a^2 + b^2 - 2ab \cos C.$$

Trigonometric Function Values of Special Angles

Graphs of Trigonometric Functions

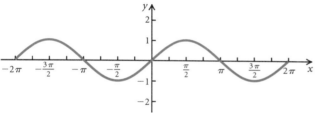

The sine function: $f(x) = \sin x$

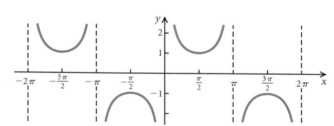

The cosecant function: $f(x) = \csc x$

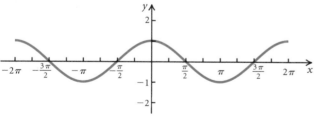

The cosine function: $f(x) = \cos x$

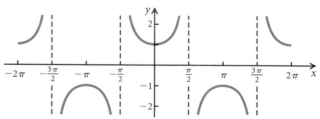

The secant function: $f(x) = \sec x$

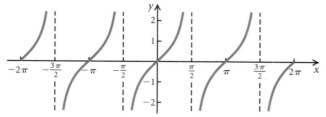

The tangent function: $f(x) = \tan x$

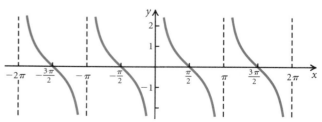

The cotangent function: $f(x) = \cot x$

Trigonometry

GRAPHS & MODELS

SECOND EDITION

Trigonometry

GRAPHS & MODELS

SECOND EDITION

Marvin L. Bittinger

Indiana University–Purdue University at Indianapolis

Judith A. Beecher

Indiana University–Purdue University at Indianapolis

David Ellenbogen

Community College of Vermont

Judith A. Penna

Indiana University–Purdue University at Indianapolis

Addison
Wesley

Boston   San Francisco   New York
London   Toronto   Sydney   Tokyo   Singapore   Madrid
Mexico City   Munich   Paris   Cape Town   Hong Kong   Montreal

Publisher	Jason A. Jordan
Project Manager	Kari Heen
Assistant Editor	Suzanne Alley
Managing Editor	Ron Hampton
Production Supervisor	Kathleen A. Manley
Editorial and Production Services	Martha K. Morong/Quadrata, Inc.
Art Editor	Geri Davis/The Davis Group, Inc.
Marketing Manager	Brenda Bravener
Illustrators	Scientific Illustrators and Barbara Harmon
Technical Art Consultants	Scott Silva and Joe Vetere
PrePress Buyer	Caroline Fell
Compositor	The Beacon Group
Text Designer	Geri Davis/The Davis Group, Inc.
Design Direction	Susan Carsten Raymond and Dennis Schaefer
Original watercolor on cover	*Portland Head Light, Maine* © 2001 Gina Hagen
Print Buyer	Evelyn Beaton

Photo Credits

62 (left), SuperStock   **62 (right),** M. Hyett/VIREO   **155,** Darryl Jones
159, courtesy of Goss Graphics Systems   **420,** UPI Newsphotos
446, SuperStock   **448,** The Image Bank   **451,** © 1985 Topps Chewing Gum,
Inc.   **492,** PhotoDisc   **500,** Ryan McVay, PhotoDisc

Library of Congress Cataloging-in-Publication Data

Trigonometry: graphs & models/Marvin L. Bittinger . . .
[et al.].—2nd ed.
      p. cm.
   Includes index.
    ISBN 0-201-61676-9
QA531.T75   2000
516.24—dc21

00-046899

2 3 4 5 6 7 8 9 10— QWT—0403

Contents

Appendixes

Preface

Note to Students

Our challenge and our goal when writing this edition of *Trigonometry: Graphs and Models* was to do everything possible to help you learn the concepts and skills contained between its covers. Every improvement we made and every feature we added was done with this objective in mind. We realize that your time is both valuable and limited, so we communicate in a highly visual way that allows you to focus easily and learn quickly and efficiently. Take advantage of the side-by-side algebraic and graphical solutions, the numerous graphing calculator windows, the Connecting the Concepts features, and the Study Tips. We included them to enable you to make the most of your study time and to be successful in this course.

Best wishes for a positive learning experience,

Marv Bittinger
Judy Beecher
Dave Ellenbogen
Judy Penna

Content Features

Trigonometry: Graphs and Models, Second Edition, covers college-level trigonometry and is appropriate for a one-term course in precalculus mathematics. Our approach is more interactive and more visual than that of most precalculus texts and our goal is to enhance the learning process through the use of technology. A course in intermediate algebra is a prerequisite for using this text.

• **Integrated Technology** In order to increase students' understanding of the course content through a visual means, we completely integrate graphing calculator technology throughout. In this text, the term "grapher" refers to all graphing calculator technology. The use of the grapher is woven throughout the text's exposition, exercise sets, and testing program without sacrificing algebraic skills. Grapher technology is included in order to enhance, not replace, students' mathematical skills and to alleviate the tedium associated with certain procedures. We assume that each student is required to have a grapher (or at least access to one) while enrolled in this course. (See pp. 91, 143, 217, 285, and 371.)

New!

• **Increased Visualization**   We use technology to make this edition much more visual than the first edition. Believing the adage "a picture is worth a thousand words," we include many new graphing calculator windows that illustrate algebraic concepts and skills. By displaying graphs, tables, or computations, we make it possible for students, particularly visual learners, to gain quick understanding of these concepts and skills. The additional windows also clearly illustrate how the graphing calculator can be used in combination with algebra to solve problems. (See pp. 18, 48, 240, and 349.)

• **Function Emphasis**   Since graphing calculator technology enables students to visualize a concept immediately, we believe it supports the early presentation of functions. We introduce graphing and functions in the first section of Chapter 1 and enhance and streamline the study of the family of functions (linear, quadratic, higher-degree polynomial, rational, exponential, logarithmic, and trigonometric) with the inclusion of the grapher. In addition, in order to amplify and add relevance to the study of functions, we incorporate applications with graphs throughout. (See pp. 14, 96, 161, and 169.)

New!

• **Connecting the Concepts**   Combining design and art, this feature highlights the importance of connecting concepts. When students are presented with concepts in a visual form (as in an outline or a chart) rather than merely in paragraphs of text, their *comprehension* is streamlined and their *retention* maximized. The visual aspect of this feature invites students to stop and check their understanding of how concepts work together in one section or in several sections. This concept check in turn enhances student performance on homework assignments. (See pp. 171, 343, and 411.)

• **Interactive Discoveries**   The grapher provides an exciting teaching opportunity and encourages students to discover and further investigate mathematical concepts. Our unique Interactive Discovery feature introduces new topics and provides a vehicle for students to see a concept in its visual form quickly. This feature reinforces the idea that grapher technology is an integral part of the course as well as an important learning tool. It invites students to develop analytic and reasoning skills while taking an active role in the learning process. (See pp. 31, 79, and 166.)

• **Side-by-Side Algebraic and Graphical Solutions**   When students are exposed to a variety of approaches to finding a solution, their skill in solving mathematical problems grows. To highlight both the algebraic- and graphical-solution approaches to solving equations, we use a two-column solution format in numerous examples (see pp. 243, 350, and 438) and we have increased the number of these two-column solutions in the Second Edition. In the algebraic/graphical side-by-side features, each method provides a complete solution. Side-by-side solutions emphasize that there is more than one way to obtain a result and illustrate the comparative efficiency and accuracy of the two methods.

• **Real-Data Applications** We encourage students to see and interpret the mathematics that appears every day in the world around them. Throughout the writing process, we conducted an energetic search for real-data applications and the result is a variety of examples and exercises that connect the mathematical content with the real world. Most of these applications feature source lines and frequently include charts and graphs. Many are drawn from the fields of health, business and economics, life and physical sciences, social science, and areas of general interest such as sports and daily life. (See pp. 34, 127, and 447–448.)

New!

• **Study Tips** Appearing in the text margins, Study Tips provide helpful study hints and briefly remind students to use the electronic and print supplements that accompany the text. (See pp. 49, 112, 189, and 257.)

New!

• **Review Icons** Placed next to the concept that a student is currently studying, a review icon references another text section or sections in which the student can find and review more content related to the concept at hand. (See pp. 282, 308, and 357.)

• **Technology Learning Aids** In order to minimize the amount of valuable class time needed to teach students how to use a grapher, we provide several aids that shorten the learning curve while increasing students' knowledge of graphing calculator fundamentals. The first learning aid is the "Introduction to Graphs and the Graphing Calculator" found at the beginning of the text. This section introduces students to the basic functions of the grapher. The other aids are the *Graphing Calculator Manual* packaged with each copy of the text (see p. xiv) and the series of Graphing Calculator Instructional Videos (see p. xv). All of these technology learning aids have been written and produced specifically for this text.

Pedagogical Features

• **Chapter Openers** Each chapter opens with an application relevant to the chapter content and is illustrated with both a technology window and situational art. The chapter openers also include a table of contents listing section titles. (See pp. 197 and 253.)

• **Section Objectives** Content objectives are listed at the beginning of each section. Together with subheadings throughout the section, these objectives provide a useful outline for both instructors and students. (See pp. 132, 211, and 254.)

• **Use of Color** This text uses full color in an extremely functional way, as seen in the design elements and artwork on nearly every page.

The choice of color has been carried out in a precise manner so that its use carries a consistent meaning, which enhances the readability of the text for both student and instructor. (See pp. 164, 186, and 226–227.)

- **Art Package**   The text contains over 1150 art pieces including a form of art called photorealism. Photorealism superimposes mathematics on a photograph and encourages students to see mathematics in familiar settings (see pp. 57 and 129). The exceptional situational art and statistical graphs throughout the text highlight the abundance of real-world applications while helping students visualize the mathematics (see pp. 34, 273, 396, and 451). The design and use of color with the grapher windows exemplifies the impact that technology has in today's mathematical curriculum (see pp. 69, 212, 238, and 438).

- **Annotated Examples**   Over 320 examples fully prepare the student for the exercise sets. Learning is carefully guided with numerous color-coded art pieces and step-by-step annotations, with substitutions and annotations highlighted in red. (See pp. 203 and 427.)

- **Variety of Exercises**   There are over 3300 exercises in this text. The exercise sets are enhanced with real-data applications and source lines, detailed art pieces, and technology windows that include both tables and graphs, and feature the following elements as well.

  Technology Exercises reflect the integration of the grapher throughout this text, and exercise sets include both grapher and nongrapher exercises. In some cases, detailed instruction lines indicate the approach that the student is expected to use. In others, we ask the student to choose the approach that seems best, thereby encouraging critical thinking. (See pp. 56, 75, 189, and 439.)

  Discussion and Writing Exercises encourage students to both consider and write about key mathematical ideas in each chapter. Many of these exercises are open-ended, making them particularly suitable for use in class discussions or as collaborative activities. (See pp. 77 and 274.)

  Skill Maintenance Exercises review concepts previously taught in the text and provide excellent review for a final examination. Answers to all Skill Maintenance exercises appear along with a section reference in the answer section at the back of the book. (See pp. 131 and 265.)

  Synthesis Exercises appear at the end of each exercise set and encourage critical thinking by requiring students to synthesize concepts from several sections or to take a concept a step further than in the general exercises. (See pp. 62, 191, and 422.)

- **Highlighted Information**   Certain material is presented and organized for efficient learning and review. Important definitions, properties, and rules are displayed in screened boxes, and summaries and procedures are listed in boxes outlined in color. (See pp. 184 and 280.)

- **Summary and Review**   The Summary and Review at the end of each chapter provides an extensive set of review exercises along with a list of important properties and formulas covered in that chapter. This feature provides excellent preparation for chapter tests and the final examination. Answers to all review exercises appear in the text along with section references that direct students to the appropriate material to reexamine if they have difficulty with a particular exercise. (See pp. 104 and 249.)

Supplements for the Instructor

For more information on these and other helpful instructor supplements, please contact your Addison Wesley Longman sales representative.

New!

Annotated Instructor's Edition *(ISBN 0-201-70409-9, bundled with the Graphing Calculator Manual)*
This specially bound version of the student edition contains answers to both even- and odd-numbered exercises at the back of the text.

Instructor's Solutions Manual *(ISBN 0-201-70395-5)*
The *Instructor's Solutions Manual* by Judith A. Penna contains worked-out solutions to all exercises in the exercise sets, including the Discussion and Writing exercises. It also includes a sample test with answers for each chapter and answers to the exercises in the appendixes. These sample tests are also included in the *Student's Solutions Manual*.

Printed Test Bank/Instructor's Resource Guide *(ISBN 0-201-70394-7)*
Prepared by Laurie Hurley, the *Printed Test Bank/Instructor's Resource Guide* contains the following:

- 4 free-response test forms for each chapter, following the format and with the same level of difficulty as the tests in the *Student's Solutions Manual*.
- 2 multiple-choice test forms for each chapter.
- 6 forms of a final examination, 4 with free-response questions and 2 with multiple-choice questions.
- Index to the Content Videos.
- Index to the Graphing Calculator Instructional Videos.
- Conversion Guide from the First Edition to the Second Edition.

TestGen-EQ with QuizMaster-EQ *(ISBN 0-201-70398-X)*
Available on a dual-platform Windows/Macintosh CD-ROM, this fully networkable software enables instructors to build, edit, print, and administer tests using a computerized test bank of questions organized according to the chapter content of the text. Tests can be printed or saved for on-line testing via a network or the Web, and the software can generate a variety of grading reports for tests and quizzes.

InterAct Math Plus *(Windows ISBN 0-201-63555-0;*
Macintosh ISBN 0-201-64805-9)
This networkable software provides course management and on-line administration for Addison Wesley Longman's InterAct Math Tutorial Software (see "Supplements for the Student"). InterAct Math Plus enables
instructors to create and administer on-line tests, summarize students'
results, and monitor students' progress in the tutorial software, providing an invaluable teaching and tracking resource.

Supplements for the Student

For more information on these and other helpful student supplements,
please contact your bookstore.

Graphing Calculator Manual *(ISBN 0-201-70396-3)*
The *Graphing Calculator Manual* by Judith A. Penna, with the assistance
of Daphne A. Bell, contains keystroke level instruction for the Texas
Instruments TI-82®, TI-83®, TI-83+®, TI-85®, TI-86®, and TI-89®.
Modules for other models are available on request. Contact your local
Addison Wesley Longman sales consultant for details. Bundled free with
every copy of the text, the *Graphing Calculator Manual* uses actual
examples and exercises from *Trigonometry: Graphs and Models,* Second
Edition, to help teach students to use the graphing calculator. The order
of topics in the *Graphing Calculator Manual* mirrors that of the text,
providing a just-in-time mode of instruction.

Student's Solutions Manual *(ISBN 0-201-34572-2)*
The *Student's Solutions Manual* by Judith A. Penna contains completely
worked-out solutions with step-by-step annotations for all the odd-
numbered exercises in the text, with the exception of the Discussion and
Writing exercises. It also includes a self-test with answers for each chapter and a final examination.

InterAct Math Tutorial CD-ROM *(ISBN 0-201-70399-8, stand-alone)*
This interactive tutorial software provides algorithmically generated
practice exercises that correlate at the objective level to the odd-numbered
exercises in the text. Each practice exercise is accompanied by an example and guided solution designed to involve students in the solution
process. The software recognizes common student errors and provides
appropriate feedback. Instructors can use InterAct Math Plus course
management software to create, administer, and track on-line tests and
monitor student performance during practice sessions.

New!

World Wide Web Supplement   www.aw.com/BBEP
This free Web site provides additional practice and learning resources
such as chapter quizzes, collaborative activities, and InterAct Math tutorial exercises.

New!

AWL Math Tutor Center *(ISBN 0-201-71049-8, stand-alone)*
The AWL Math Tutor Center is staffed by qualified mathematics instructors who provide students with tutoring on examples and exercises
from the textbook. Tutoring is available via toll-free telephone, fax, or
e-mail.

Videotapes *(ISBN 0-201-70362-9)*

Developed and produced especially for this text, these videotapes feature an engaging team of instructors presenting material and concepts from every section of the text in a format that stresses student interaction. The lecturers' presentations include examples and problems from the text and support an approach that emphasizes the use of technology, visualization, and problem solving.

Digital Video Tutor *(ISBN 0-201-70397-1, stand-alone)*

New!

The videos for this text are now available on CD-ROM, making it easy and convenient for students to watch video segments from a computer at home or on campus. The complete digitized video set, now affordable and portable for students, is ideal for distance learning or supplemental instruction.

InterAct MathXL *(12-month registration ISBN 0-201-71630-5, stand-alone)* www.mathxl.com

New!

This Web-based diagnostic testing and tutorial system allows students to take practice tests correlated to the textbook and receive customized study plans based on their results. Each time a student takes a practice test, the resulting study plan identifies areas for improvement and links to appropriate practice exercises and tutorials generated by InterAct Math. A course-management feature allows instructors to view students' test results, study plans, and practice work.

Graphing Calculator Instructional Videos *(ISBN 0-201-87352-4)*

Designed and produced specifically for *Trigonometry: Graphs and Models*, the Graphing Calculator Instructional Videos take students through procedures on the graphing calculator using content from the text. These videos include most topics covered in the *Graphing Calculator Manual*, as well as several additional topics. Every video section uses actual text examples or odd-numbered exercises from the text to help motivate students while they learn to use the graphing calculator. The videos are correlated to the sections of the text. A complete set of Graphing Calculator Instructional Videos is free to qualifying adopters.

Acknowledgments

We wish to express our genuine appreciation to a number of people who have contributed in special ways to the development of this textbook. Jason Jordan, Publisher, and Kari Heen, Executive Project Manager, at Addison Wesley Longman, shared our vision and provided encouragement and motivation. In addition, the production and marketing departments of Addison Wesley Longman brought to the project their unsurpassed commitment to excellence. The unwavering support from the Higher Education Group has been a continuing source of strength for this author team. For this we are most grateful. We are also indebted to Daphne Bell, Shane Goodwin, Laurie Hurley, and Barbara Johnson for their accuracy checking of the manuscript.

We would also like to thank the following reviewers for their invaluable contribution to the development of this text:

William Armstrong, *Lakeland Community College*
Jamie Ashby, *Texarkana College*
Brian Balman, *Johnson County Community College*
Ray Battee, *Cochise College*
Daphne Bell, *Motlow State Community College*
Susan Bradley, *Angelina College*
Beverly Broomell, *Suffolk County Community College*
Terry Cheng, *Irvine Valley College*
Wil Clarke, *La Sierra University*
Don Clayton, *Madisonville Community College*
John Collado, *South Suburban College*
Mary Ann Connors, *U.S. Military Academy*
Al Coons, *Pima Community College*
Cynthia Coulter, *Catawba Valley Community College*
Paul Cox, *Ricks College*
Lucio Della Vecchia, *Daytona Beach Community College — North*
Lisa DeLong Cuneo, *Pennsylvania State University — DuBois Campus*
Phil DeMarois, *Mt. Hood Community College*

Judy DeVoe, *Valencia Community College—East Campus*
Jo Dobbin, *Fayetteville Technical Community College*
Rob Farinelli, *Community College of Allegheny—Boyce Campus*
Joy Fett, *Moraine Valley Community College*
Edward Fey, *College of Lake County*
James Galloway, *Collin County Community College*
Rebecca Gehrke-Griswold, *Jefferson Community College—
   Southwest*
Shane Goodwin, *Ricks College*
William Grimes, *Central Missouri State University*
Lynn Gudmundsen, *Maui Community College*
Deborah Hanus, *Brookhaven College*
Daniel Harned, *Michigan State University*
Brian Hickey, *East Central College*
Robert Houston, *Rose State College*
Heidi Howard, *Florida Community College—Jacksonville*
Miles Hubbard, *St. Cloud State University*
Helen Kolman, *Central Piedmont Community College*
LuAnn Linton, *Arapahoe Community College*
Linda Long, *Ricks College*
William Martin, *Pima Community College—East Campus*
Judy Marwick, *Prairie State College*
Marsha May, *Midwestern State University*
Jean McArthur, *Joliet Junior College*
Timothy McLendon, *East Central College*
Sue Neal, *Wichita State University*
Linda O'Brien, *Johnson County Community College*
Tanya O'Keefe, *Darton College*
Nancy Olson, *Johnson County Community College*
Vic Perera, *Kent State University—Trumbull*
Kathy Rodgers, *University of Southern Indiana*
Mike Rosenborg, *Canyonville Christian Academy*
Randy Ross, *Morehead State University*
Keith Schwingendorf, *Purdue University North Central*
Joe Stickles, Jr., *University of Evansville*
Bob Sweetland, *Redlands Community College*
David Thielk, *Maui Community College*
Peter Uluave, *Utah Valley State College*
Bette Warren, *Eastern Michigan University*
Jon Weerts, *Triton College*
Brenda Wood, *Florida Community College—Jacksonville*

M.L.B.
J.A.B.
D.J.E.
J.A.P.

Feature Walkthrough

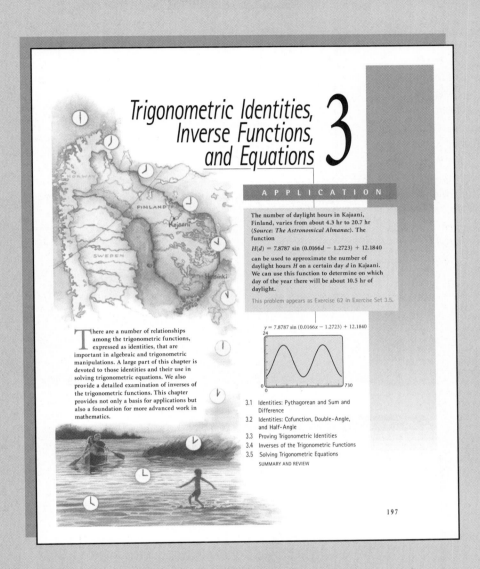

Trigonometric Identities, Inverse Functions, and Equations 3

APPLICATION

The number of daylight hours in Kajaani, Finland, varies from about 4.3 hr to 20.7 hr (*Source: The Astronomical Almanac*). The function

$$H(d) = 7.8787 \sin (0.0166d - 1.2723) + 12.1840$$

can be used to approximate the number of daylight hours H on a certain day d in Kajaani. We can use this function to determine on which day of the year there will be about 10.5 hr of daylight.

This problem appears as Exercise 62 in Exercise Set 3.5.

$y = 7.8787 \sin (0.0166x - 1.2723) + 12.1840$

There are a number of relationships among the trigonometric functions, expressed as identities, that are important in algebraic and trigonometric manipulations. A large part of this chapter is devoted to those identities and their use in solving trigonometric equations. We also provide a detailed examination of inverses of the trigonometric functions. This chapter provides not only a basis for applications but also a foundation for more advanced work in mathematics.

3.1 Identities: Pythagorean and Sum and Difference
3.2 Identities: Cofunction, Double-Angle, and Half-Angle
3.3 Proving Trigonometric Identities
3.4 Inverses of the Trigonometric Functions
3.5 Solving Trigonometric Equations
SUMMARY AND REVIEW

197

Applied Chapter Openers

Each chapter begins with an application showing students how chapter concepts are used in the real world. This application is modeled with a function and illustrated with a graph. The chapter openers are not only motivating to students but also help students develop their critical-thinking skills.

Connecting the Concepts

Comprehension is streamlined and retention is maximized when concepts are presented in a visual form. Combining design (usually in outline or chart form) and art, this feature emphasizes the importance of connecting concepts. Its visual aspect invites students to stop and check their understanding of how concepts work together in one or several sections. The increased understanding gained from this feature enhances student performance on homework assignments.

New!

CONNECTING THE CONCEPTS

COMPARING THE SINE AND COSINE FUNCTIONS

SINE FUNCTION

$y = \sin x$

COSINE FUNCTION

$y = \cos x$

SINE FUNCTION	COSINE FUNCTION
1. Continuous	1. Continuous
2. Period: 2π	2. Period: 2π
3. Domain: All real numbers	3. Domain: All real numbers
4. Range: $[-1, 1]$	4. Range: $[-1, 1]$
5. Amplitude: 1	5. Amplitude: 1
6. Odd: $\sin(-s) = -\sin s$	6. Even: $\cos(-s) = \cos s$

CONNECTING THE CONCEPTS

COMPARING EXPONENTIAL AND LOGARITHMIC FUNCTIONS

Usually, we use a number that is greater than 1 for the logarithmic base. In the following table, we compare exponential and logarithmic functions with bases a greater than 1. Similar statements would be made for a, where $0 < a < 1$. It is helpful to visualize the differences by carefully observing the graphs.

EXPONENTIAL FUNCTION

$y = a^x$
$f(x) = a^x$
$a > 1$
Continuous
One-to-one
Domain: All real numbers, $(-\infty, \infty)$
Range: All positive real numbers, $(0, \infty)$
Increasing
Horizontal asymptote is x-axis:
$(a^x \to 0 \text{ as } x \to -\infty)$
y-intercept: $(0, 1)$
There is no x-intercept.

LOGARITHMIC FUNCTION

$x = a^y$
$f^{-1}(x) = \log_a x$
$a > 1$
Continuous
One-to-one
Domain: All positive real numbers, $(0, \infty)$
Range: All real numbers, $(-\infty, \infty)$
Increasing
Vertical asymptote is y-axis:
$(\log_a x \to -\infty \text{ as } x \to 0^+)$
x-intercept: $(1, 0)$
There is no y-intercept.

Interactive
Discovery

We can graph the unit circle using a grapher. We use PARAMETRIC mode with the following window and let $X_{1T} = \cos T$ and $Y_{1T} = \sin T$. Here we use DEGREE mode.

WINDOW

Tmin = 0
Tmax = 360
Tstep = 15
Xmin = -1.5
Xmax = 1.5
Xscl = 1
Ymin = -1
Ymax = 1
Yscl = 1

Using the trace key and an arrow key to move the cursor around the unit circle, we see the T, X, and Y values appear on the screen. What do they represent? Repeat this exercise in RADIAN mode. What do the T, X, and Y values represent? (For more on parametric equations, see Section 5.8.)

EXAMPLE 9   Solve $\sin x + \cos x = 1$ in $[0, 2\pi)$.

Algebraic Solution

We have

$$\sin x + \cos x = 1$$
$$\sin^2 x + 2 \sin x \cos x + \cos^2 x = 1$$
Squaring both sides
$$2 \sin x \cos x + 1 = 1$$
Using $\sin^2 x + \cos^2 x = 1$
$$2 \sin x \cos x = 0$$
$$\sin 2x = 0.$$
Using $2 \sin x \cos x = \sin 2x$

We are looking for solutions x to the equation for which $0 \le x < 2\pi$. Multiplying by 2, we get $0 \le 2x < 4\pi$, which is the interval we consider to solve $\sin 2x = 0$. These values of $2x$ are 0, π, 2π, and 3π. Thus the desired values of x in $[0, 2\pi)$ satisfying this equation are 0, $\pi/2$, π, and $3\pi/2$. Now we check these in the original equation:

$$\sin 0 + \cos 0 = 0 + 1 = 1,$$
$$\sin \frac{\pi}{2} + \cos \frac{\pi}{2} = 1 + 0 = 1,$$
$$\sin \pi + \cos \pi = 0 + (-1) = -1,$$
$$\sin \frac{3\pi}{2} + \cos \frac{3\pi}{2} = (-1) + 0 = -1.$$

We find that π and $3\pi/2$ do not check, but the other values do. Thus the solutions in $[0, 2\pi)$ are

$$0 \quad \text{and} \quad \frac{\pi}{2}.$$

When the solution process involves squaring both sides, values are sometimes obtained that are not solutions of the original equation. As we saw in this example, it is important to check the possible solutions.

Graphical Solution

We can graph the left side and then the right side of the equation as seen in the first window below. Then we look for points of intersection. We could also rewrite the equation as $\sin x + \cos x - 1 = 0$, graph the left side, and look for the zeros of the function, as illustrated in the second window below. In each window, we see the solutions in $[0, 2\pi)$ as 0 and $\pi/2$.

$y_1 = \sin x + \cos x$, $y_2 = 1$

Xscl = $\frac{\pi}{2}$

$y = \sin x + \cos x - 1$

Xscl = $\frac{\pi}{2}$

This example illustrates a valuable advantage of the grapher—that is, with a grapher, extraneous solutions do not appear.

EXAMPLE 6 *Credit Card Volume.* The total credit card volume for Visa, MasterCard, American Express, and Discover has increased dramatically in recent years, as shown in the table on the next page.

a) Use a grapher to fit an exponential function to the data.
b) Graph the function with the scatterplot of the data.
c) Predict the total credit card volume for Visa, MasterCard, American Express, and Discover in 2003.

YEAR, x	CREDIT CARD VOLUME, y (IN BILLIONS)
1988, 0	$261.0
1989, 1	296.3
1990, 2	338.4
1991, 3	361.0
1992, 4	403.1
1993, 5	476.7
1994, 6	584.8
1995, 7	701.2
1996, 8	798.3
1997, 9	885.2

Source: CardWeb Inc.'s CardData

```
ExpReg
y=a*b^x
a=249.7267077
b=1.150881332
r²=.9877691696
r=.9938657704
```

Solution

a) We will fit an equation of the type $y = a \cdot b^x$ to the data, where x is the number of years since 1988. Entering the data into the grapher and carrying out the regression procedure, we find that the equation is

$$y = 249.7267077(1.150881332)^x.$$

The correlation coefficient is very close to 1. This gives us a good indication that the exponential function fits the data well.

b) The graph is shown at left.

$y = 249.7267077(1.150881332)^x$

c) We evaluate the function found in part (a) for $x = 15$ (2003 − 1988 = 15) and estimate that the total credit card volume in 2003 will be about $2056 billion, or $2,056,000,000,000.

```
2500
Y1=249.7267077463*1.150_

0  X = 15        Y = 2055.527   16
0
```

STUDY TIP

Success on the next exam can be planned. Include study time (even if only 30 minutes a day) in your daily schedule and commit to making this time a priority. Choose a time when you are most alert and a setting in which you can concentrate. You will be surprised how much more you can learn and retain if study time is included each day rather than in one long session before the exam.

The *lengths* of the sides of the triangle are used to define the six trigonometric ratios:

sine (sin),	cosecant (csc),
cosine (cos),	secant (sec),
tangent (tan),	cotangent (cot).

The **sine of** θ is the *length* of the side opposite θ divided by the *length* of the hypotenuse:

$$\sin \theta = \frac{\text{length of side opposite } \theta}{\text{length of hypotenuse}}.$$

The ratio depends on the measure of angle θ and thus is a function of θ. The notation $\sin \theta$ actually means $\sin (\theta)$, where sin, or sine, is the name of the function.

The **cosine of** θ is the *length* of the side adjacent to θ divided by the *length* of the hypotenuse:

$$\cos \theta = \frac{\text{length of side adjacent to } \theta}{\text{length of hypotenuse}}.$$

The six trigonometric ratios, or trigonometric functions, are defined as follows.

Review Icons

These icons prompt students to review topics necessary to understand a concept at hand, and the resulting review enhances students' comprehension. For easy navigation, the section number of the relevant review topic is given in the icon.

New!

Composition of Trigonometric Functions and Their Inverses

Various compositions of trigonometric functions and their inverses often occur in practice. For example, we might want to try to simplify an expression such as

$$\sin(\sin^{-1} x) \quad \text{or} \quad \sin\left(\text{arccot } \frac{x}{2}\right).$$

COMPOSITION OF FUNCTIONS
REVIEW SECTION 1.6.

In the expression on the left, we are finding "the sine of a number whose sine is x." Recall from Section 1.6 that if a function f has an inverse that is also a function, then

$$f(f^{-1}(x)) = x, \quad \text{for all } x \text{ in the domain of } f^{-1},$$

and

$$f^{-1}(f(x)) = x, \quad \text{for all } x \text{ in the domain of } f.$$

Thus, if $f(x) = \sin x$ and $f^{-1}(x) = \sin^{-1} x$, then

$\sin(\sin^{-1} x) = x$, for all x in the domain of $\sin^{-1}$,

which is any number on the interval $[-1, 1]$. Similar results hold for the other trigonometric functions.

---

distances between the points. However, to navigate they need to know the angle measurements. Calculate the angles for them.

31. *Airplanes.* Two airplanes leave an airport at the same time. The first flies 150 km/h in a direction of 320°. The second flies 200 km/h in a direction of 200°. After 3 hr, how far apart are the planes?

32. *Slow-pitch Softball.* A slow-pitch softball diamond is a square 65 ft on a side. The pitcher's mound is 46 ft from home. How far is it from the pitcher's mound to first base?

33. *Isosceles Trapezoid.* The longer base of an isosceles trapezoid measures 14 ft. The nonparallel sides measure 10 ft, and the base angles measure 80°.
  a) Find the length of a diagonal.
  b) Find the area.

34. *Area of Sail.* A sail that is in the shape of an isosceles triangle has a vertex angle of 38°. The angle is included by two sides, each measuring 20 ft. Find the area of the sail.

35. Three circles are arranged as shown in the figure below. Find the length PQ.

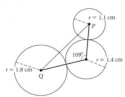

36. *Swimming Pool.* A triangular swimming pool measures 44 ft on one side and 32.8 ft on another side. These sides form an angle that measures 40.8°. How long is the other side?

Discussion and Writing

37. Try to solve this triangle using the law of cosines. Then explain why it is easier to solve it using the law of sines.

38. Explain why we cannot solve a triangle given SAS with the law of sines.

Skill Maintenance

Find the values.

39. $\sin 45°$

40. $\cos \dfrac{\pi}{6}$

41. $\cos\left(-\dfrac{2\pi}{3}\right)$

42. $\sin 300°$

Synthesis

43. *Canyon Depth.* A bridge is being built across a canyon. The length of the bridge is 5045 ft. From the deepest point in the canyon, the angles of elevation of the ends of the bridge are 78° and 72°. How deep is the canyon?

44. *Heron's Formula.* If a, b, and c are the lengths of the sides of a triangle, then the area K of the triangle is given by
$$K = \sqrt{s(s - a)(s - b)(s - c)},$$
where $s = \frac{1}{2}(a + b + c)$. The number s is called the *semiperimeter.* Prove Heron's formula. (*Hint:* Use the area formula $K = \frac{1}{2}bc \sin A$ developed in Section 4.1. Then use Heron's formula to find the area of the triangular swimming pool described in Exercise 36.)

Discussion and Writing Exercises

These innovative exercises encourage students both to consider and to write about key mathematical ideas in each section. Many of these exercises are open-ended, making them particularly suitable for use in class discussions or as collaborative activities.

Skill Maintenance Exercises

Skill Maintenance exercises review and reinforce skills previously presented in the text. These exercises provide excellent review for a final examination.

Synthesis Exercises

Synthesis exercises help build critical-thinking skills by requiring students to synthesize or combine learning objectives from the section being studied as well as from preceding sections in the text.

Chapter Summary and Review 6

Important Properties and Formulas

Exponential Function:	$f(x) = a^x$
The Number $e = 2.7182818284\ldots$	
Logarithmic Function:	$f(x) = \log_a x$
A Logarithm is an Exponent:	$\log_a x = y \longleftrightarrow x = a^y$
The Change-of-Base Formula:	$\log_b M = \dfrac{\log_a M}{\log_a b}$
The Product Rule:	$\log_a MN = \log_a M + \log_a N$
The Power Rule:	$\log_a M^p = p \log_a M$
The Quotient Rule:	$\log_a \dfrac{M}{N} = \log_a M - \log_a N$
Other Properties:	$\log_a a = 1, \qquad \log_a 1 = 0,$
	$\log_a a^x = x, \qquad a^{\log_a x} = x$
Base–Exponent Property:	$a^x = a^y \longleftrightarrow x = y$, for $a > 0$, $a \neq 1$
Exponential Growth Model:	$P(t) = P_0 e^{kt}$
Exponential Decay Model:	$P(t) = P_0 e^{-kt}$
Interest Compounded Continuously:	$P(t) = P_0 e^{kt}$
Limited Growth:	$P(t) = \dfrac{a}{1 + be^{-kt}}$

REVIEW EXERCISES

In Exercises 1–6, match the equation with one of figures (a)–(d), which follow. If needed, use a grapher.

a)

b)

c)

d)

1. $f(x) = e^{x-3}$
2. $f(x) = \log_3 x$
3. $y = -\log_3 (x + 1)$
4. $y = \left(\frac{1}{2}\right)^x$
5. $f(x) = 3(1 - e^{-x}),\ x \geq 0$
6. $f(x) = |\ln (x - 4)|$

End-of-Chapter Material

End-of-chapter material includes a summary and review of properties and formulas along with a complete set of Review Exercises. Review Exercises also include Discussion and Writing and Synthesis exercises. The answers to the Review Exercises, which appear at the back of the text, have text section references to further aid students.

Trigonometry

GRAPHS & MODELS

SECOND EDITION

Introduction to Graphs and the Graphing Calculator

- *Plot points by hand and using a grapher.*
- *Graph equations by hand and using a grapher.*
- *Find the point(s) of intersection of two graphs.*

Graphing calculators and computers equipped with graphing software are useful tools in applying and understanding mathematical concepts. **All such graphing utilities will be referred to as graphers in this text, although the emphasis will be on graphing calculators.**

When we think of the ways in which a grapher can be used, graphing equations might come to mind first. Although the grapher will be used extensively in this text to graph equations, many of its other uses will also be explored. These include performing calculations, evaluating expressions, solving equations, analyzing the graphs of equations, creating tables of data, finding mathematical models of real data, analyzing those models, and using them to make estimates and predictions.

Keystrokes and features vary among different brands and models of graphers. We will use features that are commonly found on many graphing calculators. **Specific keystrokes and instructions for using these features can be found in the *Graphing Calculator Manual* that accompanies this text.** You can also consult either your instructor or the user's manual for your particular grapher.

The goal of this text is to use the grapher as a tool to enhance the learning and understanding of mathematics and to relieve the tedium of some procedures. *Keep in mind that a grapher cannot be used effectively without a firm mathematical foundation upon which to build.* For example, expressions cannot be entered correctly nor can results be interpreted well if the relevant concepts are not understood and applied correctly.

The Use of the Grapher

A grapher is a tool that can be used in the process of learning and understanding mathematics. It should be used to enhance the understanding of concepts, not to replace the learning of skills.

1

Graphs

Graphs provide a means of displaying, interpreting, and analyzing data in a visual format. It is not uncommon to open a newspaper or magazine and encounter graphs. Shown below are examples of bar, circle, and line graphs.

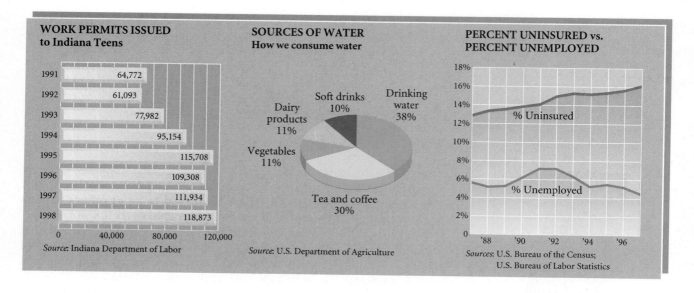

WORK PERMITS ISSUED
to Indiana Teens

Year	Permits
1991	64,772
1992	61,093
1993	77,982
1994	95,154
1995	115,708
1996	109,308
1997	111,934
1998	118,873

Source: Indiana Department of Labor

SOURCES OF WATER
How we consume water

Soft drinks 10%
Drinking water 38%
Dairy products 11%
Vegetables 11%
Tea and coffee 30%

Source: U.S. Department of Agriculture

PERCENT UNINSURED vs.
PERCENT UNEMPLOYED

% Uninsured
% Unemployed

Sources: U.S. Bureau of the Census;
U.S. Bureau of Labor Statistics

Many real-world situations can be modeled, or described mathematically, using equations in which two variables appear. We use a plane to graph a pair of numbers. To locate points on a plane, we use two perpendicular number lines, called **axes**, which intersect at (0, 0). We call this point the **origin**. The horizontal axis is called the **x-axis**, and the vertical axis is called the **y-axis**. (Other variables, such as a and b, can also be used.) The axes divide the plane into four regions, called **quadrants**, denoted by Roman numerals and numbered counterclockwise from the upper right. Arrows show the positive direction of each axis.

Each point (x, y) in the plane is called an **ordered pair.** The first number, x, indicates the point's horizontal location with respect to the y-axis, and the second number, y, indicates the point's vertical location with respect to the x-axis. We call x the **first coordinate**, **x-coordinate**, or **abscissa**. We call y the **second coordinate, y-coordinate,** or **ordinate**. Such a representation is called the **Cartesian coordinate system** in honor of the great French mathematician and philosopher René Descartes (1596–1650).

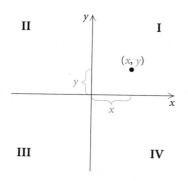

EXAMPLE 1 Graph and label the points $(-3, 5)$, $(4, 3)$, $(3, 4)$, $(-4, -2)$, $(3, -4)$, $(0, 4)$, $(-3, 0)$, and $(0, 0)$.

Solution To graph or **plot** $(-3, 5)$, we note that the x-coordinate tells us to move from the origin 3 units to the left of the y-axis. Then we move

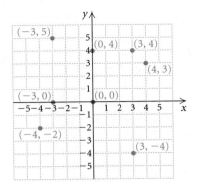

5 units up from the x-axis. To graph the other points, we proceed in a similar manner. (See the graph at left.) Note that the origin, $(0, 0)$, is located at the intersection of the axes and that the point $(4, 3)$ is different from the point $(3, 4)$.

A graph of a set of points like the one in Example 1 is called a **scatterplot**. A grapher can also be used to make a scatterplot. This involves, among other things, determining the portion of the xy-plane that will appear on the grapher's screen. That portion of the plane is called the **viewing window**.

The notation used in this text to denote a window setting consists of four numbers [L, R, B, T], which represent the **L**eft and **R**ight endpoints of the x-axis and the **B**ottom and **T**op endpoints of the y-axis, respectively. The window with the settings $[-10, 10, -10, 10]$ is the **standard viewing window,** shown in the following figure. On some graphers, the standard window can be selected quickly using the ZSTANDARD feature from the ZOOM menu.

Xmin and Xmax are used to set the left and right endpoints of the x-axis, respectively; Ymin and Ymax are used to set the bottom and top endpoints of the y-axis. The settings Xscl and Yscl give the scales for the axes. For example, Xscl = 1 and Yscl = 1 means that there is 1 unit between tick marks on each of the axes. Generally in this text, Xscl and Yscl are both assumed to be 1 unless different values are given. Some exceptions will occur when the scaling factor is different from 1 but readily apparent.

EXAMPLE 2 Use a grapher to graph the points $(-3, 5)$, $(4, 3)$, $(3, 4)$, $(-4, -2)$, $(3, -4)$, $(0, 4)$, $(-3, 0)$, and $(0, 0)$.

Solution We note that the x-coordinates of the given points range from -4 to 4 and the y-coordinates range from -4 to 5. Thus one good choice for the viewing window is the standard window, because x- and y-values both range from -10 to 10 in this window.

The coordinates of the points are entered into the grapher in lists using the EDIT operation from the STAT menu. Note that the $\boxed{(-)}$ key rather than the $\boxed{-}$ key *must* be used when we are entering negative numbers. We enter the first coordinates in one list, and then the corresponding second coordinates in the same order in a second list. Then we turn on and set up a STAT PLOT to draw a scatterplot using the data from these two lists.

Some graphers have a ZOOMSTAT feature that automatically selects a viewing window that displays all the data points in the lists for a given STAT PLOT. This feature is activated after the data have been entered.

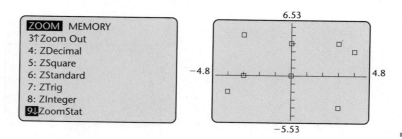

To turn off the plot, we return to the plot screen and select Off.

Solutions of Equations

Equations in two variables, like $2x + 3y = 18$, have solutions (x, y) that are ordered pairs such that when the first coordinate is substituted for x and the second coordinate is substituted for y, the result is a true equation.

EXAMPLE 3   Determine whether each ordered pair is a solution of $2x + 3y = 18$.

a) $(-5, 7)$ **b)** $(3, 4)$

Solution   We substitute the ordered pair into the equation and determine whether the resulting equation is true.

a) $\dfrac{2x + 3y = 18}{}$

$2(-5) + 3(7) \ ? \ 18$ We substitute -5 for x and 7 for y (alphabetical order).

$-10 + 21$

$11 \ | \ 18$ FALSE

The equation $11 = 18$ is false, so $(-5, 7)$ is not a solution.

b)

$$2x + 3y = 18$$

$$2(3) + 3(4) \ ? \ 18 \qquad \text{We substitute 3 for } x$$
$$\text{and 4 for } y.$$

$$6 + 12$$

$$18 \ \big| \ 18 \qquad \text{TRUE}$$

The equation $18 = 18$ is true, so $(3, 4)$ is a solution.

We can also perform these substitutions on a grapher. When we substitute -5 for x and 7 for y, we get 11. Since $11 \neq 18$, $(-5, 7)$ is not a solution of the equation. When we substitute 3 for x and 4 for y, we get 18, so $(3, 4)$ is a solution.

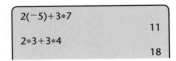

Graphs of Equations

The equation considered in Example 3 actually has an infinite number of solutions. Since we cannot list all the solutions, we will make a drawing, called a **graph**, that represents them.

> ### To Graph an Equation
>
> To **graph an equation** is to make a drawing that represents the solutions of that equation.

Shown at left are some suggestions for making hand-drawn graphs.

EXAMPLE 4   Graph: $2x + 3y = 18$.

Solution   To find ordered pairs that are solutions of this equation, we can replace either x or y with any number and then solve for the other variable. For instance, if x is replaced with 0, then

$$2 \cdot 0 + 3y = 18$$
$$3y = 18$$
$$y = 6. \qquad \text{Dividing by 3}$$

Thus $(0, 6)$ is a solution. If x is replaced with 5, then

$$2 \cdot 5 + 3y = 18$$
$$10 + 3y = 18$$
$$3y = 8 \qquad \text{Subtracting 10}$$
$$y = \tfrac{8}{3}. \qquad \text{Dividing by 3}$$

Suggestions for Hand-Drawn Graphs

1. Use graph paper.
2. Draw axes and label them with the variables.
3. Use arrows on the axes to indicate positive directions.
4. Scale the axes; that is, mark numbers on the axes.
5. Calculate solutions and list the ordered pairs in a table.
6. Plot the ordered pairs, look for patterns, and complete the graph. Label the graph with the equation being graphed.

Thus $\left(5, \frac{8}{3}\right)$ is a solution. If y is replaced with 0, then

$$2x + 3 \cdot 0 = 18$$
$$2x = 18$$
$$x = 9. \qquad \text{Dividing by 2}$$

Thus (9, 0) is a solution.

We continue choosing values for one variable and finding the corresponding values of the other. We list the solutions in a table, and then plot the points. Note that the points appear to lie on a straight line.

x	y	(x, y)
0	6	$(0, 6)$
5	$\frac{8}{3}$	$\left(5, \frac{8}{3}\right)$
9	0	$(9, 0)$
-1	$\frac{20}{3}$	$\left(-1, \frac{20}{3}\right)$

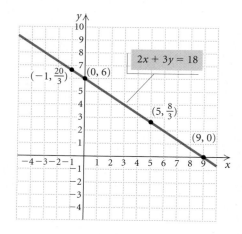

In fact, were we to graph additional solutions of $2x + 3y = 18$, they would be on the same straight line. Thus, to complete the graph, we use a straightedge to draw a line as shown in the figure. That line represents all solutions of the equation.

EXAMPLE 5   Use a grapher to create a table of ordered pairs that are solutions of the equation $y = \frac{1}{2}x + 1$. Then graph the equation by hand.

Solution   We use the TABLE feature on a grapher to create the table of ordered pairs. We must first enter the equation on the *equation-editor,* or "$y =$", screen. Many graphers require an equation to be entered in the form "$y =$" as this one is written. If an equation is not initially given in this form, it must be solved for y before it is entered in the grapher.

```
Plot1   Plot2   Plot3
\Y1■(1/2)X+1
\Y2=
\Y3=
\Y4=
\Y5=
\Y6=
\Y7=
```

On a grapher, we enter $y = (1/2)x + 1$. Since $1/2x$ is interpreted as $(1/2)x$ by some graphers and as $1/(2x)$ by others, we use parentheses to ensure that the equation is interpreted correctly. Then we set up a table in AUTO mode by designating a value for TBLSTART and a value for ΔTBL. The grapher will produce a table starting with the value of TBLSTART and continuing by adding ΔTBL to supply succeeding x-values. For the equation $y = \frac{1}{2}x + 1$, we let TBLSTART $= -3$ and ΔTBL $= 1$.

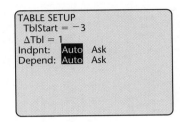

Next, we plot some of the points given in the table and draw the graph.

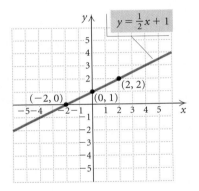

In the equation $y = \frac{1}{2}x + 1$, the value of y *depends* on the value chosen for x, so x is said to be the **independent variable** and y the **dependent variable.**

We can also graph an equation on a grapher. Be sure that the stat plots are turned off, as described on p. 4, before graphing equations like those below. Failure to do this could affect the viewing window and prevent the desired graph from being seen.

EXAMPLE 6 Graph using a grapher: $y = \frac{1}{2}x + 1$.

Solution We enter $y = \frac{1}{2}x + 1$ on the equation-editor screen in the form $y = (1/2)x + 1$, select a viewing window, and draw the graph. The graph is shown in the standard window.

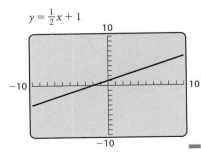

EXAMPLE 7   Graph: $y = x^2 - 5$.

Solution   We select values for x and find the corresponding y-values. Then we plot the points and connect them with a smooth curve.

X	Y₁
-3	4
-2	-1
-1	-4
0	-5
1	-4
2	-1
3	4

X = -3

x	y	(x, y)
-3	4	$(-3, 4)$
-2	-1	$(-2, -1)$
-1	-4	$(-1, -4)$
0	-5	$(0, -5)$
1	-4	$(1, -4)$
2	-1	$(2, -1)$
3	4	$(3, 4)$

① Select values for x.

② Compute values for y.

We can also graph this equation using a grapher, as shown below.

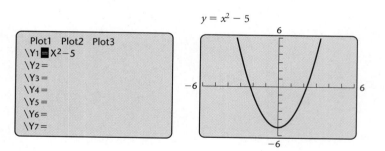

$y = x^2 - 5$

Plot1 Plot2 Plot3
\Y₁ = X²−5
\Y₂ =
\Y₃ =
\Y₄ =
\Y₅ =
\Y₆ =
\Y₇ =

EXAMPLE 8   Graph: $y = x^2 - 9x - 12$.

Solution   We make a table of values, plot enough points to obtain an idea of the shape of the curve, and connect them with a smooth curve.

x	y
-3	24
-1	-2
0	-12
2	-26
4	-32
5	-32
10	-2
12	24

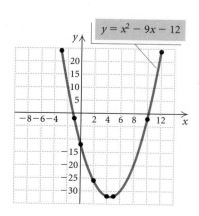

$y = x^2 - 9x - 12$

We can also graph the equation using a grapher. We enter $y = x^2 - 9x - 12$ on the equation-editor screen and graph the equation in the standard window.

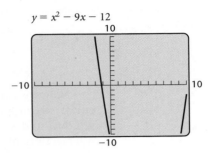

$y = x^2 - 9x - 12$

Note that this window does not give a good picture of the graph. Because the graph is cut off to the right of $x = 10$ and below $y = -10$, it appears as though Xmax should be larger than 10 and Ymin should be less than -10. Since the graph rises steeply in the second quadrant, we could also let Xmin be -5 rather than -10. We try $[-5, 15, -15, 10]$, with Yscl $= 2$.

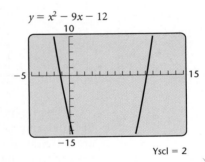

$y = x^2 - 9x - 12$

Yscl = 2

The graph is still cut off below Ymin, or -15. We try other settings for Ymin until we find one that shows the lower portion of the graph. One good window is $[-5, 15, -35, 10]$, with Yscl $= 5$.

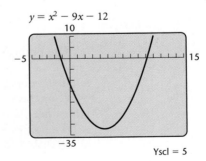

$y = x^2 - 9x - 12$

Yscl = 5

Finding Points of Intersection

There are many situations in which we want to determine the point(s) of intersection of two graphs.

EXAMPLE 9 Use a grapher to find the point of intersection of the graphs of $x - y = -5$ and $y = 4x + 10$.

Solution Since equations must be entered in the form "$y =$" on many graphers, we solve the first equation for y:

$$x - y = -5$$
$$x = y - 5 \qquad \text{Adding } y \text{ on both sides}$$
$$x + 5 = y. \qquad \text{Adding 5 on both sides}$$

Now we can enter $y_1 = x + 5$ and $y_2 = 4x + 10$ on the equation-editor screen and graph the equations. We begin by using the standard window and see that it is a good choice because it shows the point of intersection of the graphs. Next, we use the INTERSECT feature from the CALC menu to find the coordinates of the point of intersection. (See the *Graphing Calculator Manual* that accompanies this text for the keystrokes.)

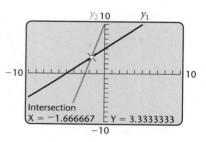

The graphs intersect at the point $(-1.666667, 3.3333333)$, which is a decimal approximation for the point of intersection. If the coordinates are rational numbers, their exact values can be found using the ▶FRAC feature from the MATH menu. (The keystrokes for doing these conversions are given in the *Graphing Calculator Manual* that accompanies this text.)

The point of intersection is $\left(-\frac{5}{3}, \frac{10}{3}\right)$.

If the coordinates in Example 9 had not been rational numbers, the ▶FRAC operation would have returned the original decimal approximation rather than a fraction.

If a pair of equations has more than one point of intersection, we use the INTERSECT feature repeatedly to find the coordinates of all the points.

FIGURE 1

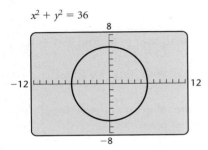

FIGURE 2

Squaring a Viewing Window

When we graph a circle, we select a viewing window in which the distance between units is visually the same on both axes. This procedure is called **squaring the viewing window.** We do this so that the graph will not be distorted. A graph of the circle $x^2 + y^2 = 36$ in a nonsquared window is shown in Fig. 1.

On many graphers, the distance between units on the y-axis is about $\frac{2}{3}$ the distance between units on the x-axis. When we choose a window in which the length of the y-axis is $\frac{2}{3}$ the length of the x-axis, the window will be squared. The windows with dimensions $[-6, 6, -4, 4]$, $[-9, 9, -6, 6]$, and $[-12, 12, -8, 8]$ are examples of squared windows. A graph of the circle $x^2 + y^2 = 36$ in a squared window is shown in Fig. 2. Methods of graphing circles will be discussed in Section 5.2 and Appendix B.

Many graphers have an option on the ZOOM menu that squares the window automatically.

Exercise Set

Graph and label the given points by hand.

1. $(4, 0)$, $(-3, -5)$, $(-1, 4)$, $(0, 2)$, $(2, -2)$
2. $(1, 4)$, $(-4, -2)$, $(-5, 0)$, $(2, -4)$, $(4, 0)$
3. $(-5, 1)$, $(5, 1)$, $(2, 3)$, $(2, -1)$, $(0, 1)$
4. $(4, 0)$, $(4, -3)$, $(-5, 2)$, $(-5, 0)$, $(-1, -5)$

Use a grapher to make a scatterplot of the given points.

5. $(-3, 2)$, $(-6, -5)$, $(0, -1)$, $(4, 1)$, $(3, 0)$
6. $(0, 0)$, $(4, -6)$, $(-6, 4)$, $(3, 1)$, $(-5, -3)$
7. $(4.3, -6.7)$, $(-5.1, 3.6)$, $(2.8, 0)$ $(-4.5, -8.4)$, $(7.3, 2.4)$
8. $(-3.2, -4.4)$, $(0, -8.3)$, $(1.6, 7.8)$, $(5.4, -2.9)$, $(-6.7, 1.8)$

Use substitution to determine whether the given ordered pairs are solutions of the given equation.

9. $(1, -1)$, $(0, 3)$; $y = 2x - 3$
10. $(2, 5)$, $(-2, -5)$; $y = 3x - 1$
11. $\left(\frac{2}{3}, \frac{3}{4}\right)$, $\left(1, \frac{3}{2}\right)$; $6x - 4y = 1$
12. $(1.5, 2.6)$, $(-3, 0)$; $x^2 + y^2 = 9$
13. $\left(-\frac{1}{2}, -\frac{4}{5}\right)$, $\left(0, \frac{3}{5}\right)$; $2a + 5b = 3$
14. $\left(0, \frac{3}{2}\right)$, $\left(\frac{2}{3}, 1\right)$; $3m + 4n = 6$
15. $(-0.75, 2.75)$, $(2, -1)$; $x^2 - y^2 = 3$
16. $(2, -4)$, $(4, -5)$; $5x + 2y^2 = 70$

Use a grapher to create a table of values with TBLSTART $= -3$ *and* ΔTBL $= 1$. *Then graph the equation by hand.*

17. $y = 3x + 5$
18. $y = -2x - 1$
19. $x - y = 3$
20. $x + y = 4$
21. $2x + y = 4$
22. $3x - y = 6$
23. $x - 4y = 5$
24. $x + 3y = 6$
25. $3x - 4y = 12$
26. $2x + 3y = -6$
27. $2x + 5y = -10$
28. $4x - 3y = 12$
29. $y = -x^2$
30. $y = x^2$
31. $y = x^2 - 3$
32. $y = 4 - x^2$
33. $y = x^2 - 2x - 3$
34. $y = x^2 + 2x - 1$

In Exercises 35–38, use a grapher to match the equation with one of the graphs (a)–(d), which follow.

a)

b)

c)

d)

35. $y = 3 - x$ **36.** $2x - y = 6$

37. $y = x^2 + 2x + 1$ **38.** $y = 4 - x^2$

Use a grapher to graph the equation in the standard window.

39. $y = 2x + 1$ **40.** $y = 3x - 4$

41. $4x + y = 7$ **42.** $5x + y = -8$

43. $y = \frac{1}{3}x + 2$ **44.** $y = \frac{3}{2}x - 4$

45. $2x + 3y = -5$ **46.** $3x + 4y = 1$

47. $y = x^2 + 6$ **48.** $y = x^2 - 8$

49. $y = 2 - x^2$ **50.** $y = 5 - x^2$

51. $y = x^2 + 4x - 2$

52. $y = x^2 - 5x + 3$

Graph the equation in the standard window and in the given window. Determine which window better shows the shape of the graph and where it crosses the x- and y-axes.

53. $y = x - 20$
$[-25, 25, -25, 25]$, with Xscl = 5 and Yscl = 5

54. $y = -2x + 24$
$[-15, 15, -10, 30]$, with Xscl = 3 and Yscl = 5

55. $y = 3x^2 - 6$
$[-4, 4, -4, 4]$

56. $y = 8 - x^2$
$[-3, 3, -3, 3]$

57. $y = -\frac{1}{6}x^2 + \frac{1}{12}$
$[-1, 1, -0.3, 0.3]$, with Xscl = 0.1 and Yscl = 0.1

58. $y = \frac{1}{2}x^2 - \frac{1}{4}x - \frac{1}{8}$
$[-0.8, 1.2, -0.5, 0.5]$, with Xscl = 0.1 and Yscl = 0.1

Use a grapher to find the point(s) of intersection of the graphs of the given pair of equations.

59. $y = 2x + 3, \ y = x + 4$

60. $y = 3x - 1, \ y = x + 5$

61. $y = 4 - x, \ y = 2x - 1$

62. $y = 5 - 2x, \ y = x + 7$

63. $y = 4x - 7, \ y = 4 - 2x$

64. $y = 5x - 3, \ y = 8 - x$

65. $y = x^2, \ y = 4$

66. $y = -x^2, \ y = -1$

67. $y = x^2 - 3x, \ y = x + 4$

68. $y = x^2 - 5x, \ y = x - 1$

Graphs, Functions, and Models 1

People in the United States create more garbage than people in any other nation. The graph below presents data regarding the total amount of solid waste for certain years (*Source: Parade Magazine*, June 13, 1999). Using linear regression, we can model the data with the linear function

$$y = f(x) = 3.65x - 7069.37,$$

where x is the year and y is the amount of solid waste, in millions of tons. We can use this function to predict amounts of solid waste in future years.

This problem appears as Example 2 in Appendix C.

This environmental application is just one example of the widespread use of graphs in today's society. In this chapter, we will focus attention on functions and equations with graphs that can be used to model real-life situations mathematically. Such models are invaluable when doing analyses and making predictions.

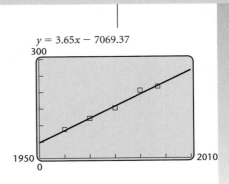

$y = 3.65x - 7069.37$

Functions, Graphs, and Graphers

1.1

- *Determine whether a correspondence or a relation is a function.*
- *Find function values, or outputs, using a formula.*
- *Find the domain and the range of a function.*
- *Determine whether a graph is that of a function.*
- *Solve applied problems using functions.*

We now focus attention on a concept that is fundamental to many areas of mathematics—the idea of a *function*.

Functions

We first consider an application.

Thunder Time Related to Lightning Distance. During a thunderstorm, it is possible to calculate how far away, y (in miles), lightning is when the sound of thunder arrives x seconds after the lightning has been sighted. It is known that the distance, in miles, is $\frac{1}{5}$ of the time, in seconds. If we hear the sound of thunder 15 seconds after we've seen the lightning, we know that the lightning is $\frac{1}{5} \cdot 15$, or 3 miles away. Similarly, 5 sec corresponds to 1 mi, 2 sec to $\frac{2}{5}$ mi, and so on. We can express this relationship with a set of ordered pairs, a graph, and an equation.

x	y	ORDERED PAIRS: (x, y)	CORRESPONDENCE
0	0	$(0, 0)$	$0 \longrightarrow 0$
1	$\frac{1}{5}$	$\left(1, \frac{1}{5}\right)$	$1 \longrightarrow \frac{1}{5}$
2	$\frac{2}{5}$	$\left(2, \frac{2}{5}\right)$	$2 \longrightarrow \frac{2}{5}$
5	1	$(5, 1)$	$5 \longrightarrow 1$
10	2	$(10, 2)$	$10 \longrightarrow 2$
15	3	$(15, 3)$	$15 \longrightarrow 3$

The ordered pairs express a relationship, or correspondence, between the first and second coordinates. We can see this relationship in the graph as well. The equation that describes the correspondence is

$$y = \frac{1}{5}x.$$

This is an example of a *function*. In this case, distance is a function of time.

Let's consider some other correspondences before giving the definition of a function.

DOMAIN		RANGE
To each registered student	there corresponds	an I. D. number.
To each mountain bike sold	there corresponds	its price.
To each number between −3 and 3	there corresponds	the square of that number.

In each example, the first set is called the **domain** and the second set is called the **range**. For each member, or **element**, in the domain, there is *exactly one* member of the range to which it corresponds. Thus each registered student has exactly *one* I. D. number, each mountain bike has exactly *one* price, and each number between −3 and 3 has exactly *one* square. Each correspondence is a *function*.

Function

A **function** is a correspondence between a first set, called the **domain**, and a second set, called the **range**, such that each member of the domain corresponds to *exactly one* member of the range.

It is important to note that not every correspondence between two sets is a function.

EXAMPLE 1 Determine whether each of the following correspondences is a function.

a)

b) Juan ⟶ Casandra
    Boris ⟵ Rebecca
    Nelson ⟶ Helga
    Bernie ⟶ Natasha

Solution

a) This correspondence *is* a function because each member of the domain corresponds to exactly one member of the range. Note that the definition of a function does allow more than one member of the domain to correspond to a member of the range.

b) This correspondence *is not* a function because there is a member of the domain (Boris) that is paired with two different members of the range (Rebecca and Helga).

EXAMPLE 2   Determine whether each of the following correspondences is a function.

DOMAIN	CORRESPONDENCE	RANGE
a) Years in which a presidential election occurs	The person elected	A set of presidents
b) The integers	Each integer's cube root	A set of real numbers
c) All states in the United States	A senator from that state	The set of all U.S. senators

Solution

a) This correspondence *is* a function, because in each presidential election *exactly one* president is elected.

b) This correspondence *is* a function, because each integer has *exactly one* cube root.

c) This correspondence *is not* a function, because each state can be paired with *two* different senators.

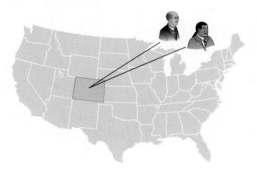

When a correspondence between two sets is not a function, it is still an example of a **relation**.

Relation

A **relation** is a correspondence between a first set, called the **domain**, and a second set, called the **range**, such that each member of the domain corresponds to *at least one* member of the range.

All the correspondences in Examples 1 and 2 are relations, but, as we have seen, not all are functions. Relations are sometimes written as sets of ordered pairs (as we saw earlier in the example on lightning) in which elements of the domain are the first coordinates of the ordered pairs and elements of the range are the second coordinates. For example, instead of writing $-3 \longrightarrow 9$, as we did in Example 1(a), we could write the ordered pair $(-3, 9)$.

EXAMPLE 3 Determine whether each of the following relations is a function. Identify the domain and the range.

a) $\{(-2, 5), (5, 7), (0, 1), (4, -2)\}$

b) $\{(9, -5), (9, 5), (2, 4)\}$

c) $\{(-5, 3), (0, 3), (6, 3)\}$

Solution

FIGURE 1

a) The relation *is a function* because *no* two ordered pairs have the same first coordinate and different second coordinates (see Fig. 1).

  The domain is the set of all first coordinates: $\{-2, 5, 0, 4\}$.

  The range is the set of all second coordinates: $\{5, 7, 1, -2\}$.

FIGURE 2

b) The relation *is not* a function because the ordered pairs $(9, -5)$ and $(9, 5)$ have the same first coordinate and different second coordinates (see Fig. 2).

  The domain is the set of all first coordinates: $\{9, 2\}$.

  The range is the set of all second coordinates: $\{-5, 5, 4\}$.

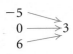

FIGURE 3

c) The relation *is a function* because *no* two ordered pairs have the same first coordinate and different second coordinates (see Fig. 3).

  The domain is $\{-5, 0, 6\}$.

  The range is $\{3\}$.

Notation for Functions

Functions used in mathematics are often given by equations. They generally require that certain calculations be performed in order to determine which member of the range is paired with each member of the domain. For example, in the "Introduction to Graphs and the Graphing Calculator," we graphed the function $y = x^2 - 5$ by doing calculations like the following:

$$\text{for } x = 3, y = 3^2 - 5 = 4,$$
$$\text{for } x = 2, y = 2^2 - 5 = -1, \quad \text{and}$$
$$\text{for } x = 1, y = 1^2 - 5 = -4.$$

A more concise notation is often used. For $y = x^2 - 5$, the **inputs** (members of the domain) are values of x substituted into the equation. The **outputs** (members of the range) are the resulting values of y. If we call the function f, we can use x to represent an arbitrary *input* and

$f(x)$—read "f of x", or "f at x", or "the value of f at x"—to represent the corresponding *output*. In this notation, the function given by $y = x^2 - 5$ is written as $f(x) = x^2 - 5$ and the above calculations would be

$$f(3) = 3^2 - 5 = 4,$$
$$f(2) = 2^2 - 5 = -1,$$
$$f(1) = 1^2 - 5 = -4. \qquad \text{Keep in mind that } f(x) \text{ does not mean } f \cdot x.$$

Thus, instead of writing "when $x = 3$, the value of y is 4," we can simply write "$f(3) = 4$," which can also be read as "f of 3 is 4" or "for the input 3, the output of f is 4." The letters g and h are also often used to name functions.

EXAMPLE 4 A function f is given by $f(x) = 2x^2 - x + 3$. Find each of the following.

a) $f(0)$ 　　　　　　　　　　　　　**b)** $f(-7)$

c) $f(5a)$ 　　　　　　　　　　　　　**d)** $f(a - 4)$

Solution We can think of this formula as follows:

$$f(\blacksquare) = 2(\blacksquare)^2 - (\blacksquare) + 3.$$

Then to find an output for a given input we think: "Whatever goes in the blank on the left goes in the blank(s) on the right." This gives us a "recipe" for finding outputs.

a) $f(0) = 2(0)^2 - 0 + 3 = 0 - 0 + 3 = 3$

b) $f(-7) = 2(-7)^2 - (-7) + 3 = 2 \cdot 49 + 7 + 3 = 108$

c) $f(5a) = 2(5a)^2 - 5a + 3 = 2 \cdot 25a^2 - 5a + 3 = 50a^2 - 5a + 3$

d) $f(a - 4) = 2(a - 4)^2 - (a - 4) + 3 = 2(a^2 - 8a + 16) - a + 4 + 3$

$$= 2a^2 - 16a + 32 - a + 4 + 3$$
$$= 2a^2 - 17a + 39$$

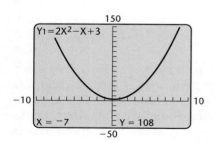

A grapher can be used to find function values. At left, we illustrate finding $f(-7)$ from part (b), first with the TABLE feature set in ASK mode and then with the VALUE feature from the CALC menu. In both screens, we see that $f(-7) = 108$. Consult either the manual for your particular grapher or the *Graphing Calculator Manual* that accompanies this text for further information on finding function values. ▬

Graphs of Functions

We graph functions the same way we graph equations. We find ordered pairs (x, y) or $(x, f(x))$, plot points, and complete the graph.

EXAMPLE 5 Graph each of the following functions using a grapher.

a) $f(x) = x^2 - 5$

b) $f(x) = x^3 - x$

c) $f(x) = \sqrt{x + 4}$

Solution   Most graphers do not use function notation "$f(x) = \ldots$" to enter a function formula. Instead, we must enter the function using "$y = \ldots$". The graphs follow.

a) $f(x) = x^2 - 5$

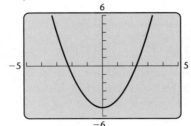

b) $f(x) = x^3 - x$

c) $f(x) = \sqrt{x + 4}$

To find a function value, like $f(3)$, from a graph, we locate the input 3 on the horizontal axis, move vertically to the graph of the function, and then move horizontally to find the output on the vertical axis. For the function $f(x) = x^2 - 5$, we see that $f(3) = 4$. We can check our work using a grapher.

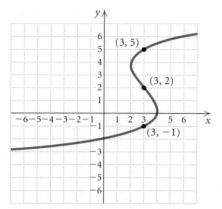

Since 3 is paired with more than one member of the range, the graph does not represent a function.

We know that when one member of the domain is paired with two or more different members of the range, the correspondence *is not* a function. Thus, when a graph contains two or more different points with the same first coordinate, the graph cannot represent a function (see the graph at left). Points sharing a common first coordinate are vertically above or below each other. This leads us to the *vertical-line test*.

The Vertical-Line Test

If it is possible for a vertical line to cross a graph more than once, then the graph is *not* the graph of a function.

To apply the vertical-line test, we try to find a vertical line that crosses the graph more than once. If we succeed, then the graph is not that of a function. If we do not, then the graph is that of a function.

EXAMPLE 6   Which of graphs (a) through (f) (in red) are graphs of functions?

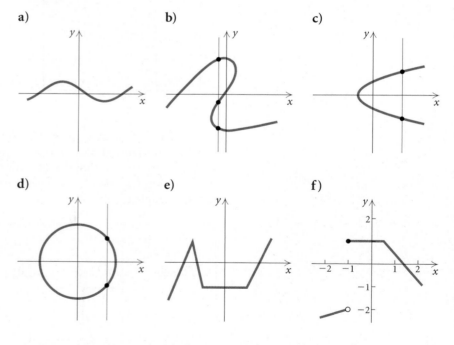

In graph (f), the solid dot shows that $(-1, 1)$ belongs to the graph. The open circle shows that $(-1, -2)$ does *not* belong to the graph.

Solution   Graphs (a), (e), and (f) are graphs of functions because we cannot find a vertical line that crosses any of them more than once. In (b), the vertical line crosses the graph in three points and so it is not that of a function. Also, in (c) and (d), we can find a vertical line that crosses the graph more than once. ▬

Finding Domains of Functions

When a function f, whose inputs and outputs are real numbers, is given by a formula, the *domain* is understood to be the set of all inputs for which the expression is defined as a real number. When a substitution results in an expression that is not defined as a real number, we say that the function value *does not exist* and that the number being substituted *is not* in the domain of the function.

EXAMPLE 7   Find the indicated function values. Simplify, if possible.

a) $f(1)$ and $f(3)$, for $f(x) = \dfrac{1}{x-3}$

b) $g(16)$ and $g(-7)$, for $g(x) = \sqrt{x} + 5$

Solution

a) $f(1) = \dfrac{1}{1-3} = \dfrac{1}{-2} = -\dfrac{1}{2}$;

$f(3) = \dfrac{1}{3-3} = \dfrac{1}{0}$

Since division by 0 is not defined, the number 3 is not in the domain of f. In a table from a grapher, this is indicated with an ERROR message. Thus, $f(3)$ does not exist.

b) $g(16) = \sqrt{16} + 5 = 4 + 5 = 9$;

$g(-7) = \sqrt{-7} + 5$

Since $\sqrt{-7}$ is not defined as a real number, the number -7 is not in the domain of g. Note the ERROR message in the table at left. Thus, $g(-7)$ does not exist. ▬

   Inputs that make a denominator 0 or a square-root radicand negative are not in the domain of a function.

EXAMPLE 8   Find the domain of each of the following functions.

a) $f(x) = \dfrac{1}{x-3}$ **b)** $g(x) = \sqrt{x} + 5$

c) $h(x) = \dfrac{3x^2 - x + 7}{x^2 + 2x - 3}$ **d)** $F(x) = x^3 + |x|$

Solution

a) The input 3 results in a denominator of 0. The domain is $\{x \,|\, x \neq 3\}$. We can also write the solution using interval notation and the symbol ∪ for the **union** or inclusion of both sets: $(-\infty, 3) \cup (3, \infty)$.

b) We can substitute any number for which the radicand is nonnegative, that is, for which $x \geq 0$. Thus the domain is $\{x \mid x \geq 0\}$, or the interval $[0, \infty)$.

c) Although we can substitute any real number in the numerator, we must avoid inputs that make the denominator 0. To find those inputs, we solve $x^2 + 2x - 3 = 0$, or $(x + 3)(x - 1) = 0$. Thus the domain consists of the set of all real numbers except -3 and 1, or $\{x \mid x \neq -3$ and $x \neq 1\}$, or $(-\infty, -3) \cup (-3, 1) \cup (1, \infty)$.

d) All substitutions are suitable. The domain is the set of all real numbers, $\mathbb{R}$, or $(-\infty, \infty)$.   ▬

Domain and Range Using a Grapher

Keep the following in mind regarding the *graph* of a function:

Domain = the set of a function's inputs, found on the horizontal axis;

Range = the set of a function's outputs, found on the vertical axis.

By carefully examining the graph of a function, we may be able to determine the function's domain as well as its range. Consider the graph of $f(x) = \sqrt{4 - (x - 1)^2}$, shown below. We look for the inputs on the x-axis that correspond to a point on the graph. We see that they extend from -1 to 3, inclusive. Thus the domain is $\{x \mid -1 \leq x \leq 3\}$, or the interval $[-1, 3]$.

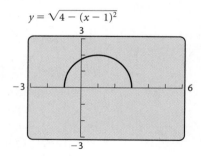

$$y = \sqrt{4 - (x - 1)^2}$$

To find the range, we look for the outputs on the y-axis. We see that they extend from 0 to 2, inclusive. Thus the range of this function is $\{y \mid 0 \leq y \leq 2\}$, or the interval $[0, 2]$. We can confirm our results using the TRACE feature, moving the cursor from left to right along the curve. We can also confirm our results using the TABLE feature.

EXAMPLE 9   Use a grapher to graph each of the following functions. Then estimate the domain and the range of each.

a) $f(x) = \sqrt{x + 4}$ **b)** $f(x) = x^3 - x$

c) $f(x) = \dfrac{12}{x}$ **d)** $f(x) = x^4 - 2x^2 - 3$

Solution

a)

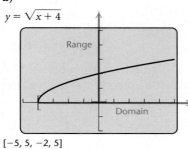

$y = \sqrt{x + 4}$

$[-5, 5, -2, 5]$

Domain $= [-4, \infty)$;
range $= [0, \infty)$

b)

$y = x^3 - x$

$[-3, 3, -4, 4]$

Domain $=$ all real numbers,
$(-\infty, \infty)$; range $=$ all real
numbers, $(-\infty, \infty)$

c)

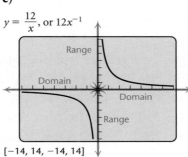

$y = \dfrac{12}{x}$, or $12x^{-1}$

$[-14, 14, -14, 14]$

Since the graph does not touch or
cross either axis, 0 is excluded
both as an input and as an output.
Domain $= (-\infty, 0) \cup (0, \infty)$;
range $= (-\infty, 0) \cup (0, \infty)$

d)

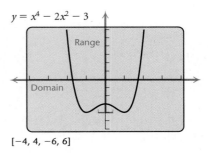

$y = x^4 - 2x^2 - 3$

$[-4, 4, -6, 6]$

Domain $=$ all real numbers,
$(-\infty, \infty)$; range $= [-4, \infty)$

Always consider adding the reasoning of Example 8 to a graphical analysis. Think, "What can I substitute?" to find the domain. Think, "What do I get out?" to find the range. Thus, in Example 9(d), it might not look like the domain is all real numbers because the graph rises steeply, but by reexamining the equation we see that we can indeed substitute any real number for *x*.

Applications of Functions

EXAMPLE 10 *Speed of Sound in Air.* The speed *S* of sound in air is a function of the temperature *t*, in degrees Fahrenheit, and is given by

$$S(t) = 1087.7 \sqrt{\frac{5t + 2457}{2457}},$$

where *S* is in feet per second.

a) Graph the function using the viewing window $[-1000, 2000, -1000, 3000]$, with Xscl = 500 and Yscl = 500.

b) Find the speed of sound in air when the temperature is 0°, 32°, 70°, and −10° Fahrenheit.

Solution

a) The graph is shown below. Note that $S(t)$ must be changed to y and t must be changed to x when the function is entered in a grapher.

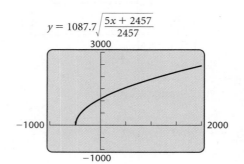

$$y = 1087.7 \sqrt{\dfrac{5x + 2457}{2457}}$$

X	Y₁
0	1087.7
32	1122.6
70	1162.6
−10	1076.6

X =

b) We use a grapher with the TABLE feature set in ASK mode to compute the function values. We find that

$$S(0) = 1087.7 \text{ ft/sec,}$$
$$S(32) \approx 1122.6 \text{ ft/sec,}$$
$$S(70) \approx 1162.6 \text{ ft/sec,} \quad \text{and}$$
$$S(-10) \approx 1076.6 \text{ ft/sec.}$$

CONNECTING THE CONCEPTS

FUNCTION CONCEPTS

Formula for f: $f(x) = 5 + 2x^2 - x^4$.

For every input, there is exactly one output.

$(1, 6)$ is on the graph.

For the input 1, the output is 6.

$f(1) = 6$

Domain: set of all inputs $= (-\infty, \infty)$

Range: set of all outputs $= (-\infty, 6]$

GRAPH

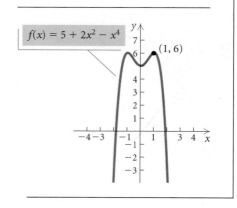

$f(x) = 5 + 2x^2 - x^4$

Exercise Set 1.1

In Exercises 1–14, determine whether the correspondence is a function.

1. $a \longrightarrow w$
$b \longrightarrow y$
$c \longrightarrow z$

2. $m \longrightarrow q$
$n \longrightarrow r$
$o \longrightarrow s$

3. $-6 \longrightarrow 36$
$-2 \longrightarrow 4$
2

4. $-3 \longrightarrow 2$
$1 \longrightarrow 4$
$5 \longrightarrow 6$
$9 \longrightarrow 8$

5. $m \longrightarrow A$
$n \longrightarrow B$
$r \longrightarrow C$
$s \longrightarrow D$

6. $a \longrightarrow r$
$b \longrightarrow s$
$c \longrightarrow t$
d

7. WORLD'S TEN LARGEST EARTHQUAKES (1900–1999)

LOCATION AND DATE	MAGNITUDE
Chile (May 22, 1960)	9.5
Alaska (March 28, 1964)	9.2
Russia (Nov. 4, 1952)	9.0
Ecuador (Jan. 31, 1906)	8.8
Alaska (March 9, 1957)	
Kuril Islands (Nov. 6, 1958)	8.7
Alaska (Feb. 4, 1965)	
India (Aug. 15, 1950)	8.6
Argentina (Nov. 11, 1922)	8.5
Indonesia (Feb. 1, 1938)	

(*Source*: National Earthquake Information Center, U.S. Geological Survey)

8.

ANIMAL	SPEED ON THE GROUND (IN MILES PER HOUR)
Cheetah	70
Lion	50
Reindeer	32
Giraffe	
Elephant	25
Giant tortoise	0.17

(*Source*: *Time Almanac*, 1999, p. 581)

DOMAIN	CORRESPONDENCE	RANGE
9. A set of cars in a parking lot	Each car's license number	A set of numbers
10. A set of people in a town	A doctor a person uses	A set of doctors
11. A set of members of a family	Each person's eye color	A set of colors
12. A set of members of a rock band	An instrument each person plays	A set of instruments
13. A set of students in a class	A student sitting in a neighboring seat	A set of students
14. A set of bags of chips on a shelf	Each bag's weight	A set of weights

Determine whether the relation is a function. Identify the domain and the range.

15. $\{(2, 10), (3, 15), (4, 20)\}$

16. $\{(3, 1), (5, 1), (7, 1)\}$

17. $\{(-7, 3), (-2, 1), (-2, 4), (0, 7)\}$

18. $\{(1, 3), (1, 5), (1, 7), (1, 9)\}$

19. $\{(-2, 1), (0, 1), (2, 1), (4, 1), (-3, 1)\}$

20. $\{(5, 0), (3, -1), (0, 0), (5, -1), (3, -2)\}$

21. A graph of a function f is shown below. Find $f(-1)$, $f(0)$, and $f(1)$.

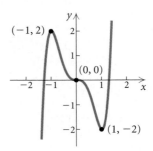

22. A graph of a function g is shown here. Find $g(-2)$, $g(0)$, and $g(2.4)$.

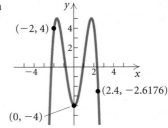

23. Given that $g(x) = 3x^2 - 2x + 1$, find each of the following.

  a) $g(0)$ **b)** $g(-1)$
  c) $g(3)$ **d)** $g(-x)$
  e) $g(1 - t)$

24. Given that $f(x) = 5x^2 + 4x$, find each of the following.

  a) $f(0)$ **b)** $f(-1)$
  c) $f(3)$ **d)** $f(t)$
  e) $f(t - 1)$

25. Given that $g(x) = x^3$, find each of the following.

  a) $g(2)$ **b)** $g(-2)$
  c) $g(-x)$ **d)** $g(3y)$
  e) $g(2 + h)$

26. Given that $f(x) = 2|x| + 3x$, find each of the following.

  a) $f(1)$ **b)** $f(-2)$
  c) $f(-x)$ **d)** $f(2y)$
  e) $f(2 - h)$

27. Given that $g(x) = \dfrac{x - 4}{x + 3}$, find each of the following.

  a) $g(5)$ **b)** $g(4)$
  c) $g(-3)$ **d)** $g(-16.25)$
  e) $g(x + h)$

28. Given that $f(x) = \dfrac{x}{2 - x}$, find each of the following.

  a) $f(2)$ **b)** $f(1)$
  c) $f(-16)$ **d)** $f(-x)$
  e) $f\left(-\dfrac{2}{3}\right)$

29. Find $g(0)$, $g(-1)$, $g(5)$, and $g\left(\frac{1}{2}\right)$ for

$$g(x) = \frac{x}{\sqrt{1 - x^2}}.$$

30. Find $h(0)$, $h(2)$, and $h(-x)$ for
$$h(x) = x + \sqrt{x^2 - 1}.$$

Find the domain of the function. Do not use a grapher.

31. $f(x) = 7x + 4$

32. $f(x) = |3x - 2|$

33. $f(x) = 4 - \dfrac{2}{x}$

34. $f(x) = \dfrac{1}{x^4}$

35. $f(x) = \dfrac{x + 5}{2 - x}$

36. $f(x) = \dfrac{8}{x + 4}$

37. $f(x) = \dfrac{1}{x^2 - 4x - 5}$

38. $f(x) = \dfrac{x^4 - 2x^3 + 7}{3x^2 - 10x - 8}$

Use a grapher to graph the function. Then visually estimate the domain and the range.

39. $f(x) = |x|$ **40.** $f(x) = |x| - 10.3$

41. $f(x) = \sqrt{9 - x^2}$ **42.** $f(x) = -\sqrt{25 - x^2}$

43. $f(x) = (x - 1)^3 + 2$ **44.** $f(x) = (x - 2)^4 + 1$

45. $f(x) = \sqrt{7 - x}$ **46.** $f(x) = \sqrt{x + 8}$

47. $f(x) = -x^2 + 4x - 1$ **48.** $f(x) = 2x^2 - x^4 + 5$

In Exercises 49–56, determine whether the graph is that of a function. An open dot indicates that the point does not belong to the graph.

49.

50.

51.

52.

53.

54.

55.

56.

In Exercises 57–60, determine whether the graph is that of a function. Find the domain and the range.

57.

58.

59.

60.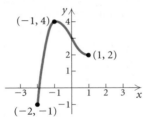

61. *Boiling Point and Elevation.* The elevation E, in meters, above sea level at which the boiling point of water is t degrees Celsius is given by the function

$$E(t) = 1000(100 - t) + 580(100 - t)^2.$$

At what elevation is the boiling point 99.5°? 100°?

62. *Average Price of a Movie Ticket.* The average price of a movie ticket, in dollars, can be estimated by the function P given by

$$P(x) = 0.1522x - 298.592$$

where x is the year. Thus, $P(2000)$ is the average price of a movie ticket in 2000. The price is lower than what might be expected due to lower prices for matinees, senior citizens' discounts, and so on.

a) Use the function to predict the average price in 2003 and 2010.

b) When will the average price be $8.00?

63. *Territorial Area of an Animal.* The territorial area of an animal is defined to be its defended, or exclusive region. For example, a lion has a certain region over which it is considered ruler. It has been shown that the territorial area I, in acres, of predatory animals is a function of body weight w, in pounds, and is given by the function

$$T(w) = w^{1.31}.$$

Find the territorial area of animals whose body weights are 0.5 lb, 10 lb, 20 lb, 100 lb, and 200 lb.

Discussion and Writing

To the student and the instructor: The discussion and writing exercises are meant to be answered with one or more sentences. They can be discussed and answered collaboratively by the entire class or by small groups. Because of their open-ended nature, the answers to these exercises do not appear at the back of the book. They are denoted by the words "Discussion and Writing."

64. Explain in your own words what a function is.

65. Explain in your own words the difference between the domain of a function and the range of a function.

Synthesis

To the student and the instructor: The synthesis exercises found at the end of every exercise set challenge students to combine concepts or skills studied in that section or in preceding parts of the text.

66. Make a hand-drawn graph of a function for which the domain is $[-4, 4]$ and the range is $[1, 2] \cup [3, 5]$. Answers may vary.

67. Give an example of two different functions that have the same domain and the same range, but have no pairs in common. Answers may vary.

68. Make a hand-drawn graph of a function for which the domain is $[-3, -1] \cup [1, 5]$ and the range is $\{1, 2, 3, 4\}$. Answers may vary.

69. Suppose that for some function f, $f(x - 1) = 5x$. Find $f(6)$.

Linear Functions, Slope, and Applications

1.2

- • *Graph linear functions and equations, finding the slope and the y-intercept.*
- • *Determine equations of lines.*
- • *Solve applied problems involving linear functions.*

In real-life situations, we often need to make decisions on the basis of limited information. When the given information is used to formulate an equation or inequality that at least approximates the situation mathematically, we have created a **model**. One of the most frequently used mathematical models is *linear*—the graph of a linear model is a straight line.

Linear Functions

Let's begin to examine the connections among equations, functions, and graphs that are straight lines.

Interactive Discovery

Graph each of the following equations using a grapher. (Some graphers are not able to graph equations of the type $x = a$. Consult your manual.) Which of these equations have graphs that are lines? Which of the equations whose graphs are lines are also functions? Look for patterns.

$$y = x,$$ $$y = x^2,$$
$$4x + 5y = 20,$$ $$y = 5 - x^2,$$

$$y = \frac{x}{2} + 12,$$ $$y = 1.5x - 0.05,$$

$$y = -2x - 5,$$ $$y = \frac{1}{x},$$

$$x = -3,$$ $$y = 6.2,$$
$$y = \sqrt{x},$$ $$x = 4.9.$$

We have the following results and related terminology.

Linear Functions

A function f is a **linear function** if it can be written as

$$f(x) = mx + b,$$

where m and b are constants.

(If $m = 0$, the function is the **constant function** $f(x) = b$. If $m = 1$ and $b = 0$, the function is the **identity function** $f(x) = x$.)

Horizontal and Vertical Lines

Horizontal lines are given by equations of the type $y = b$ or $f(x) = b$. (They are functions.)
Vertical lines are given by equations of the type $x = a$. (They are *not* functions.)

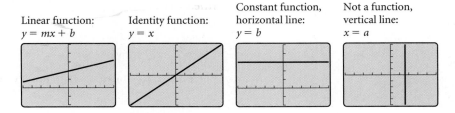

Linear function: $y = mx + b$	Identity function: $y = x$	Constant function, horizontal line: $y = b$	Not a function, vertical line: $x = a$

The Linear Function $f(x) = mx + b$ and Slope

To attach meaning to the constant m in the equation $f(x) = mx + b$, we first consider an application. FaxMax is an office machine business that currently has two stores in locations A and B in the same city. Their total costs for the same time period are given by two functions shown in the tables and graphs that follow. The variable x represents time, in months. The variable y represents total costs, in thousands of dollars, over that amount of time. Look for a pattern.

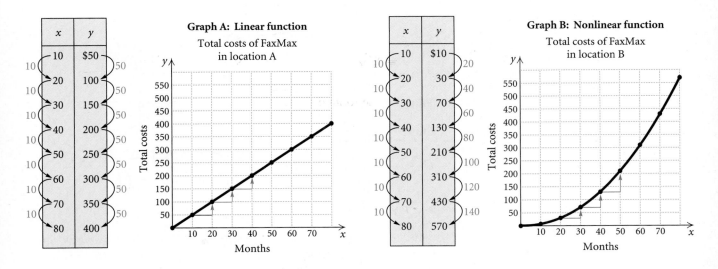

x	y
10	$50
20	100
30	150
40	200
50	250
60	300
70	350
80	400

Graph A: Linear function

Total costs of FaxMax in location A

x	y
10	$10
20	30
30	70
40	130
50	210
60	310
70	430
80	570

Graph B: Nonlinear function

Total costs of FaxMax in location B

We see in graph A that *every* change of 10 months results in a $50 thousand change in total costs. But in graph B, changes of 10 months do *not* result in constant changes in total costs. This is a way to distinguish linear from nonlinear functions. The rate at which a linear function changes, or the steepness of its graph, is constant.

Mathematically, we define a line's steepness, or **slope**, as the ratio of its vertical change (rise) to the corresponding horizontal change (run).

Slope

The **slope _m_** of a line containing points (x_1, y_1) and (x_2, y_2) is given by

$$m = \frac{\text{rise}}{\text{run}}$$

$$= \frac{\text{the change in } y}{\text{the change in } x}$$

$$= \frac{y_2 - y_1}{x_2 - x_1} = \frac{y_1 - y_2}{x_1 - x_2}.$$

EXAMPLE 1 Make a hand-drawn graph of the function $f(x) = -\frac{2}{3}x + 1$ and determine its slope.

Solution Since the equation for f is in the form $f(x) = mx + b$, we know it is a linear function. We can graph it by connecting two points on the graph with a straight line. We calculate two ordered pairs, plot the points, graph the function, and determine the slope:

$$f(3) = -\frac{2}{3} \cdot 3 + 1 = -1;$$

$$f(9) = -\frac{2}{3} \cdot 9 + 1 = -5;$$

Pairs: $(3, -1)$, $(9, -5)$;

$$\text{Slope} = m = \frac{y_2 - y_1}{x_2 - x_1}$$

$$= \frac{-5 - (-1)}{9 - 3}$$

$$= \frac{-4}{6} = -\frac{2}{3}.$$

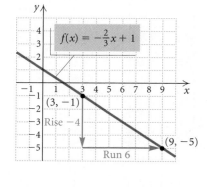

The slope is the same for any two points on a line. Thus, to check our work, note that $f(6) = -\frac{2}{3} \cdot 6 + 1 = -3$. Using the points $(6, -3)$ and $(3, -1)$, we have

$$m = \frac{-1 - (-3)}{3 - 6} = \frac{2}{-3} = -\frac{2}{3}.$$

We can also use the points in the opposite order when computing slope, so long as we are consistent:

$$m = \frac{-3 - (-1)}{6 - 3} = \frac{-2}{3} = -\frac{2}{3}.$$

Note also that the slope of the line is the number m in the equation for the function $f(x) = -\frac{2}{3}x + 1$.   ▬

The slope of the line given by $y = mx + b$ is m.

Let's explore the effect of the slope m in linear equations of the type $f(x) = mx$.

Interactive
Discovery

Graph the following equations using a grapher:

$$y_1 = x, \qquad y_2 = 2x, \qquad y_3 = 5x, \quad \text{and} \quad y_4 = 10x.$$

Try entering these equations as $y_1 = \{1, 2, 5, 10\}x$. What do you think the graph of $y = 128x$ will look like?

    Clear the screen and graph the following equations:

$$y_1 = x, \qquad y_2 = 0.75x, \qquad y_3 = 0.48x, \quad \text{and} \quad y_4 = 0.12x.$$

What do you think the graph of $y = 0.000029x$ will look like?

    Again clear the screen and graph each set of equations:

$$y_1 = -x, \qquad y_2 = -2x, \qquad y_3 = -4x, \quad \text{and} \quad y_4 = -10x$$

and $y_1 = -x, \qquad y_2 = -\frac{2}{3}x, \qquad y_3 = -\frac{7}{20}x, \quad \text{and} \quad y_4 = -\frac{1}{10}x.$

From your observations, what do you think the graphs of $y = -200x$ and $y = -\frac{17}{100,000}x$ will look like?

If a line slants up from left to right, the change in x and the change in y have the same sign, so the line has a positive slope. The larger the slope is, the steeper the line. If a line slants down from left to right, the change in x and the change in y are of opposite signs, so the line has a negative slope. The larger the absolute value of the slope, the steeper the line. When $m = 0$, $y = 0x$, or $y = 0$. The graph of this equation is shown on the right below. Note that this horizontal line is the x-axis.

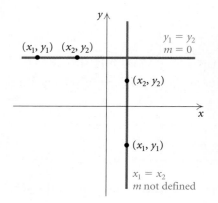

Horizontal and Vertical Lines

If a line is horizontal, the change in y for any two points is 0 and the change in x is nonzero. Thus a horizontal line has slope 0.

If a line is vertical, the change in y for any two points is nonzero and the change in x is 0. Thus the slope is *not defined* because we cannot divide by 0.

Note that zero slope and an undefined slope are two very different concepts.

EXAMPLE 2 Graph each linear equation and determine its slope.

a) $x = -2$

b) $y = \dfrac{5}{2}$

Solution

a) Since y is missing in $x = -2$, any value for y will do.

FIGURE 1

x	y
-2	0
-2	3
-2	-4

Choose any number for y; x must be -2.

The graph (see Fig. 1) is a *vertical line* 2 units to the left of the y-axis. The slope is not defined. The graph is not the graph of a function.

b) Since x is missing in $y = \frac{5}{2}$, any value for x will do.

FIGURE 2

x	y
0	$\frac{5}{2}$
-3	$\frac{5}{2}$
1	$\frac{5}{2}$

Choose any number for x; y must be $\frac{5}{2}$.

The graph (see Fig. 2) is a *horizontal line* $\frac{5}{2}$, or $2\frac{1}{2}$, units above the x-axis. The slope is 0. The graph is the graph of a function.

Applications of Slope

Slope has many real-world applications. Numbers like 2%, 4%, and 7% are often used to represent the **grade** of a road. Such a number is meant to tell how steep a road is on a hill or mountain. For example, a 4% grade means that the road rises 4 ft for every horizontal distance of 100 ft.

The concept of grade is also used with a treadmill. During a treadmill test, a cardiologist might change the slope, or grade, of the treadmill to measure its effect on heart rate. Another example occurs in hydrology. The strength or force of a river depends on how far the river falls vertically compared to how far it flows horizontally.

EXAMPLE 3   *Ramp for the Handicapped.*   Construction laws regarding access ramps for the handicapped state that every vertical rise of 1 ft requires a horizontal run of 12 ft. What is the grade, or slope, of such a ramp?

Solution   The grade, or slope, is given by

$$m = \tfrac{1}{12} \approx 0.083 \approx 8.3\%.$$

Slope can also be considered as an **average rate of change.** To find the average rate of change between any two data points on a graph, we determine the slope of the line that passes through the two points.

EXAMPLE 4 *Newspaper Circulation.* On September 30, 1999, the last edition of *The Indianapolis News*, an afternoon daily paper, rolled off the press. Circulation of the paper had dropped approximately 70% in the 10 preceding years, as shown in the following graph. Find the average rate of change in the circulation from 1988 to 1998.

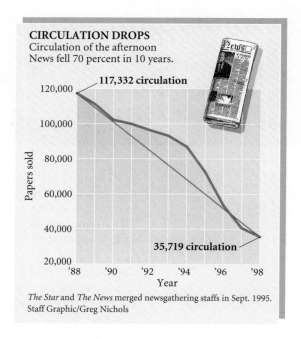

CIRCULATION DROPS
Circulation of the afternoon News fell 70 percent in 10 years.

The Star and *The News* merged newsgathering staffs in Sept. 1995.
Staff Graphic/Greg Nichols

Solution   We determine the coordinates of two points on the graph. In this case, we will use (1988, 117,332) and (1998, 35,719). Then we compute the slope, or average rate of change, as follows:

$$\text{Slope} = \text{Average rate of change} = \frac{\text{change in } y}{\text{change in } x}$$

$$= \frac{35,719 - 117,332}{1998 - 1988} = \frac{-81,613}{10} = -8161.3.$$

This result tells us that, on average, each year the circulation of the paper decreased by approximately 8161 copies. The *average rate of change* over the 10-year period was −8161 copies per year.

Slope–Intercept Equations of Lines

Let's explore the effect of the constant b in linear equations of the type $f(x) = mx + b$.

*Interactive
Discovery*

Begin with the graph of $y = x$. Now graph the lines $y = x + 3$ and $y = x - 4$ in the same viewing window. Try entering these equations as $y = x + \{0, 3, -4\}$ and compare the graphs. How do the lines differ from $y = x$? What do you think the line $y = x - 6$ will look like?

Try graphing $y = -0.5x$, $y = -0.5x - 4$, and $y = -0.5x + 3$ in the same viewing window. Describe what happens to the graph of $y = -0.5x$ when a number b is added.

Compare the graphs of the equations

$$y = 3x$$

and

$$y = 3x - 2.$$

Note that the graph of $y = 3x - 2$ is a shift down of the graph of $y = 3x$, and that $y = 3x - 2$ has y-intercept $(0, -2)$. That is, the graph is parallel to $y = 3x$ and it crosses the y-axis at $(0, -2)$. The point $(0, -2)$ is called the **y-intercept** of the graph.

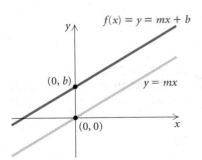

The Slope – Intercept Equation

The linear function f given by

$$f(x) = mx + b$$

has a graph that is a straight line parallel to $y = mx$. The constant m is called the slope, and the y-intercept is $(0, b)$.

We know that a nonvertical line has slope m and y-intercept $(0, b)$ and that its equation can be written as $y = mx + b$. The advantage of writing the equation in the form $y = mx + b$ is that we can read the slope m and the y-intercept $(0, b)$ directly from the equation.

EXAMPLE 5   Find the slope and the y-intercept of the line with equation $y = -0.25x - 3.8$.

$y = -0.25x - 3.8$

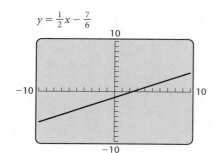

$y = \frac{1}{2}x - \frac{7}{6}$

Solution

$$y = \underbrace{-0.25\,x}_{} \; \underbrace{-\; 3.8}_{}$$

$$\text{Slope} = -0.25; \quad y\text{-intercept} = (0, -3.8)$$ —

Any equation whose graph is a straight line is a **linear equation.** To find the slope and the y-intercept of the graph of a linear equation, we can solve for y, and then read the information from the equation.

EXAMPLE 6   Find the slope and the y-intercept of the line with equation $3x - 6y - 7 = 0$.

Solution   We solve for y:

$$3x - 6y - 7 = 0$$
$$-6y = -3x + 7 \qquad \text{Adding } -3x \text{ and } 7 \text{ on both sides}$$
$$-\tfrac{1}{6}(-6y) = -\tfrac{1}{6}(-3x + 7) \qquad \text{Multiplying by } -\tfrac{1}{6}$$
$$y = \tfrac{1}{2}x - \tfrac{7}{6}.$$

Thus the slope is $\frac{1}{2}$, and the y-intercept is $\left(0, -\frac{7}{6}\right)$. —

EXAMPLE 7   A line has slope $-\frac{7}{9}$ and y-intercept $(0, 16)$. Find an equation of the line.

Solution   We use the slope–intercept equation and substitute $-\frac{7}{9}$ for m and 16 for b:

$$y = mx + b$$
$$y = -\tfrac{7}{9}x + 16.$$ —

Point–Slope Equations of Lines

Suppose that we have a nonvertical line and that the coordinates of point P_1 are (x_1, y_1). We can think of P_1 as fixed and imagine a movable point P on the line with coordinates (x, y). Thus the slope is given by

$$\frac{y - y_1}{x - x_1} = m.$$

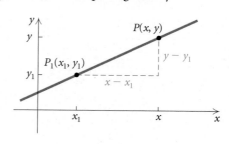

Multiplying on both sides by $x - x_1$, we get the *point-slope equation* of the line:

$$(x - x_1) \cdot \frac{y - y_1}{x - x_1} = m \cdot (x - x_1)$$
$$y - y_1 = m(x - x_1).$$

Point-Slope Equation

The **point-slope equation** of the line with slope m passing through (x_1, y_1) is

$$y - y_1 = m(x - x_1).$$

Thus if we know the slope of a line and the coordinates of one point on the line, we can find an equation of the line.

EXAMPLE 8 Find an equation of the line containing the point $\left(\frac{1}{2}, -1\right)$ and with slope 5.

Solution If we substitute in $y - y_1 = m(x - x_1)$, we get

$$y - (-1) = 5\left(x - \frac{1}{2}\right),$$

which simplifies as follows:

$$y + 1 = 5x - \frac{5}{2}$$
$$y = 5x - \frac{5}{2} - 1$$
$$y = 5x - \frac{7}{2}. \quad \text{Slope-intercept equation} \quad \blacksquare$$

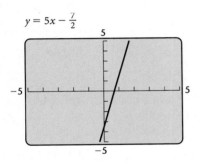

$y = 5x - \frac{7}{2}$

EXAMPLE 9 Find an equation of the line containing the points $(2, 3)$ and $(1, -4)$.

Solution We first determine the slope:

$$m = \frac{-4 - 3}{1 - 2}$$
$$= \frac{-7}{-1}$$
$$= 7.$$

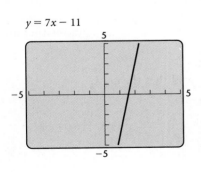

$y = 7x - 11$

Then we substitute 7 for m and either of the points $(2, 3)$ or $(1, -4)$ for (x_1, y_1) in the point-slope equation. In this case, we use $(2, 3)$. We get

$$y - 3 = 7(x - 2),$$

which simplifies to

$$y - 3 = 7x - 14$$
$$y = 7x - 11. \quad \blacksquare$$

Parallel Lines

How can we examine a pair of equations to determine whether their graphs are parallel—that is, they do not intersect? We can explore this using a grapher.

Interactive Discovery

Graph each of the following pairs of lines. Try to determine whether they are parallel. You will probably need to change viewing windows and either zoom in or zoom out to decide for sure. Complete the table and look for a pattern.

PAIRS OF EQUATIONS	SLOPES		PARALLEL?
	m_1	m_2	
$y = -0.38x + 4.2$, $y = -0.37x - 5.1$			
$y = -0.38x + 4.2$, $50y + 19x = 21$			
$2x + 1 = y$, $y + 3x = 4$			
$y = 8.02x - 11.3$, $y = 7.98x - 11.3$			
$y = -3$, $y = 4.2$			
$x = -5$, $x = 2$ (Use the DRAW menu to graph $x = -5$.)			

Can you find a way to determine whether two lines are parallel without graphing?

If two different lines are vertical, then they are parallel. Thus two equations such as $x = a_1$, $x = a_2$, where $a_1 \neq a_2$, have graphs that are *parallel lines*. Two nonvertical lines such as $y = mx + b_1$, $y = mx + b_2$, where the slopes are the same and $b_1 \neq b_2$, also have graphs that are *parallel lines*.

Parallel Lines
Vertical lines are **parallel**. Nonvertical lines are **parallel** if and only if they have the same slope and different y-intercepts.

Perpendicular Lines

How can we examine a pair of equations to determine whether they are perpendicular? Let's explore this using a grapher.

Interactive Discovery

Graph each of the following pairs of lines using the window $[-12, 12, -8, 8]$. This is an example of a **square** window. (See "Introduction to Graphs and the Graphing Calculator" and Appendix B.) Try to determine whether the lines are perpendicular. *You must use square viewing windows so that you can see whether angles between lines seem to be 90°.* Complete the table and look for a pattern.

PAIRS OF EQUATIONS	SLOPES m_1	m_2	PRODUCT OF SLOPES $m_1 \cdot m_2$	PERPENDICULAR?
$y = -\frac{2}{5}x + 3, \ y = \frac{5}{2}x + 3$				
$y = 0.1875x - 2, \ y = -\frac{16}{3}x - 5.1$				
$3y + 4x = -21, \ 4y - 3x = 8$				
$y = -3x + 4, \ y = 2x - 7$				
$x = -5, \ y = 2$				

Can you find a way to determine whether two lines are perpendicular without graphing?

Perpendicular Lines

Two lines with slopes m_1 and m_2 are **perpendicular** if and only if the product of their slopes is -1:

$$m_1 m_2 = -1.$$

Lines are also **perpendicular** if one is vertical ($x = a$) and the other is horizontal ($y = b$).

If a line has slope m_1, the slope m_2 of a line perpendicular to it is $-1/m_1$ (the slope of one line is the opposite of the reciprocal of the other).

EXAMPLE 10   Determine whether each of the following pairs of lines is parallel, perpendicular, or neither.

a) $y + 2 = 5x, \ 5y + x = -15$
b) $2y + 4x = 8, \ 5 + 2x = -y$
c) $2x + 1 = y, \ y + 3x = 4$

Solution   We use the slopes of the lines to determine whether the lines are parallel or perpendicular. We can create graphs on a grapher as a check.

a) We solve each equation for y:

$$y = 5x - 2, \qquad y = -\tfrac{1}{5}x - 3.$$

The slopes are 5 and $-\tfrac{1}{5}$. Their product is -1, so the lines are perpendicular (see Fig. 1).

b) Solving each equation for y, we get

$$y = -2x + 4, \qquad y = -2x - 5.$$

We see that $m_1 = -2$ and $m_2 = -2$. Since the slopes are the same and the y-intercepts, 4 and -5, are different, the lines are parallel (see Fig. 2).

c) Solving each equation for y, we get

$$y = 2x + 1, \qquad y = -3x + 4.$$

Since $m_1 = 2$ and $m_2 = -3$, we know that $m_1 \neq m_2$ and that $m_1 m_2 \neq -1$. It follows that the lines are neither parallel nor perpendicular (see Fig. 3).

$y_1 = 5x - 2, \quad y_2 = -\tfrac{1}{5}x - 3$

FIGURE 1

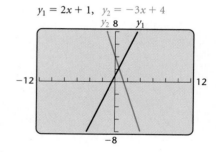

$y_1 = -2x + 4, \quad y_2 = -2x - 5$

FIGURE 2

$y_1 = 2x + 1, \quad y_2 = -3x + 4$

FIGURE 3

EXAMPLE 11   Write equations of the lines (a) parallel and (b) perpendicular to the graph of the line $4y - x = 20$ and containing the point $(2, -3)$.

Solution   We first solve $4y - x = 20$ for y to get $y = \tfrac{1}{4}x + 5$. Thus the slope is $\tfrac{1}{4}$.

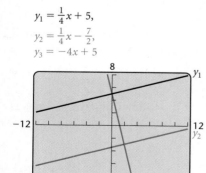

$y_1 = \tfrac{1}{4}x + 5,$

$y_2 = \tfrac{1}{4}x - \tfrac{7}{2},$

$y_3 = -4x + 5$

a) The line parallel to the given line will have slope $\tfrac{1}{4}$. We use the point–slope equation for a line with slope $\tfrac{1}{4}$ and containing the point $(2, -3)$:

$$y - y_1 = m(x - x_1)$$
$$y - (-3) = \tfrac{1}{4}(x - 2)$$
$$y + 3 = \tfrac{1}{4}x - \tfrac{1}{2}$$
$$y = \tfrac{1}{4}x - \tfrac{7}{2}.$$

b) The slope of the perpendicular line is -4. Now we use the point–slope equation to write an equation for a line with slope -4 and containing the point $(2, -3)$:

$$y - y_1 = m(x - x_1)$$
$$y - (-3) = -4(x - 2)$$
$$y + 3 = -4x + 8$$
$$y = -4x + 5.$$

We can visualize our work using a grapher.

Summary of Terminology About Lines

TERMINOLOGY	MATHEMATICAL INTERPRETATION
Slope	$m = \dfrac{y_2 - y_1}{x_2 - x_1}$, or $\dfrac{y_1 - y_2}{x_1 - x_2}$
Slope–intercept equation	$y = mx + b$
Point–slope equation	$y - y_1 = m(x - x_1)$
Horizontal lines	$y = b$
Vertical lines	$x = a$
Parallel lines	$m_1 = m_2,\ b_1 \neq b_2$
Perpendicular lines	$m_1 m_2 = -1$

Applications of Linear Functions

We now consider an application of linear functions.

Humerus

EXAMPLE 12 *Height Estimates.* An anthropologist can use linear functions to estimate the height of a male or a female, given the length of the humerus, the bone from the elbow to the shoulder. The height, in centimeters, of an adult male with a humerus of length x, in centimeters, is given by the function

$$M(x) = 2.89x + 70.64.$$

The height, in centimeters, of an adult female with a humerus of length x is given by the function

$$F(x) = 2.75x + 71.48.$$

A 26-cm humerus was uncovered in a ruins. Assuming it was from a female, how tall was she?

Algebraic Solution

We substitute into the function:

$$F(26) = 2.75(26) + 71.48$$
$$= 142.98.$$

Thus the female was 142.98 cm tall.

Graphical Solution

We graph $F(x) = 2.75x + 71.48$ and use the VALUE feature to find $F(26)$.

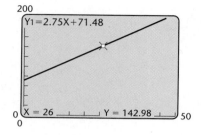

The female was 142.98 cm tall.

Theoretically, the domain of the function is the set of all real numbers. However, the context of the problem dictates a different domain. One could not find a bone with a length of 0 or less. Thus the domain consists of positive real numbers, that is, the interval $(0, \infty)$. (A more realistic domain might be 20 cm to 60 cm, the interval $[20, 60]$.) ▬

Exercise Set 1.2

In Exercises 1–4, the table of data contains input–output values for a function. Answer the following questions for each table.

a) *Is the change in the inputs x the same?*
b) *Is the change in the outputs y the same?*
c) *Is the function linear?*

1.

x	y
−3	7
−2	10
−1	13
0	16
1	19
2	22
3	25

2.

x	y
20	12.4
30	24.8
40	49.6
50	99.2
60	198.4
70	396.8
80	793.6

3.

x	y
11	3.2
26	5.7
41	8.2
56	9.3
71	11.3
86	13.7
101	19.1

4.

x	y
2	−8
4	−12
6	−16
8	−20
10	−24
12	−28
14	−36

Find the slope of the line containing the given points.

5.

6.

7.

8.

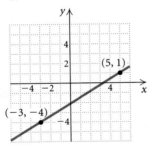

9. (9, 4) and (−1, 2)

10. (−3, 7) and (5, −1)

11. (4, −9) and (−5, 6)

12. (−6, −1) and (2, −13)

13. (π, −3) and (π, 2)

14. ($\sqrt{2}$, −4) and (0.56, −4)

15. (a, a^2) and ($a + h$, $(a + h)^2$)

16. (a, $3a + 1$) and ($a + h$, $3(a + h) + 1$)

17. *Road Grade.* Using the following figure, find the

road grade and an equation giving the height y as a function of the horizontal distance x.

18. *Grade of Treadmills.* A treadmill is 5 ft long and is set at an 8% grade. How high is the end of the treadmill?

19. Find the rate of change of the tuition and fees at public two-year colleges.

Source: Statistical Abstract of the United States

20. Find the rate of change of the cost of a formal wedding.

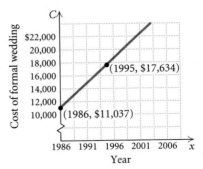

Source: Modern Bride Magazine

21. *Running Rate.* An Olympic marathoner passes the 10-km point of a race after 50 min and arrives at the 25-km point $1\frac{1}{2}$ hr later. Find the speed (average rate of change) of the runner.

22. *Work Rate.* As a typist resumes work on a project, $\frac{1}{6}$ of the paper has already been typed. Six hours later, the paper is $\frac{3}{4}$ done. Calculate the worker's typing rate.

Find the slope and the y-intercept of the equation.

23. $y = \frac{3}{5}x - 7$ **24.** $f(x) = -2x + 3$

25. $f(x) = 5 - \frac{1}{2}x$ **26.** $y = 2 + \frac{3}{7}x$

27. $3x + 2y = 10$ **28.** $2x - 3y = 12$

29. $4x - 3f(x) - 15 = 6$ **30.** $9 = 3 + 5x - 2f(x)$

Write a slope–intercept equation for a line with the given characteristics.

31. $m = \frac{2}{9}$, y-intercept $(0, 4)$

32. $m = -\frac{3}{8}$, y-intercept $(0, 5)$

33. $m = -4$, y-intercept $(0, -7)$

34. $m = \frac{2}{7}$, y-intercept $(0, -6)$

35. $m = -4.2$, y-intercept $\left(0, \frac{3}{4}\right)$

36. $m = -4$, y-intercept $\left(0, -\frac{3}{2}\right)$

37. $m = \frac{2}{9}$, passes through $(3, 7)$

38. $m = -\frac{3}{8}$, passes through $(5, 6)$

39. $m = 3$, passes through $(1, -2)$

40. $m = -2$, passes through $(-5, 1)$

41. $m = -\frac{2}{5}$, passes through $(-4, -1)$

42. $m = \frac{2}{3}$, passes through $(-4, -5)$

43. Passes through $(-1, 5)$ and $(2, -4)$

44. Passes through $(2, -1)$ and $(7, -11)$

45. Passes through $(7, 0)$ and $(-1, 4)$

46. Passes through $(-3, 7)$ and $(-1, -5)$

Determine whether the pair of lines is parallel, perpendicular, or neither.

47. $x + 2y = 5$,
    $2x + 4y = 8$

48. $2x - 5y = -3$,
    $2x + 5y = 4$

49. $y = 4x - 5$,
    $4y = 8 - x$

50. $y = 7 - x$,
    $y = x + 3$

Write a slope–intercept equation for a line passing through the given point that is parallel to the given line. Then write a second equation for a line passing through the given point that is perpendicular to the given line.

51. $(3, 5)$, $y = \frac{2}{7}x + 1$

52. $(-1, 6)$, $f(x) = 2x + 9$

53. $(-7, 0)$, $y = -0.3x + 4.3$

54. $(-4, -5)$, $2x + y = -4$

55. $(3, -2)$, $3x + 4y = 5$

56. $(8, -2)$, $y = 4.2(x - 3) + 1$

57. $(3, -3)$, $x = -1$

58. $(4, -5)$, $y = -1$

59. *Ideal Weight.*   One way to estimate the ideal weight of a woman is to multiply her height by 3.5 and subtract 110. Let $W =$ the ideal weight, in pounds, and $h =$ height, in inches.

a) Express W as a linear function of h.
b) Graph W.
c) Find the ideal weight of a woman whose height is 62 in.
d) Find the domain of the function.

60. *Pressure at Sea Depth.*   The function P, given by

$$P(d) = \frac{1}{33}d + 1,$$

gives the pressure, in atmospheres (atm), at a depth d, in feet, under the sea.

a) Graph P.
b) Find $P(0)$, $P(5)$, $P(10)$, $P(33)$, and $P(200)$.
c) Find the domain of the function.

61. *Stopping Distance on Glare Ice.*   The stopping distance (at some fixed speed) of regular tires on glare ice is a function of the air temperature F, in degrees Fahrenheit. This function is estimated by

$$D(F) = 2F + 115,$$

where $D(F)$ is the stopping distance, in feet, when the air temperature is F, in degrees Fahrenheit.

a) Graph D.
b) Find $D(0°)$, $D(-20°)$, $D(10°)$, and $D(32°)$.
c) Explain why the domain should be restricted to $[-57.5°, 32°]$.

62. *Anthropology Estimates.*   Consider Example 12 and the function

$$M(x) = 2.89x + 70.64$$

for estimating the height of a male.

a) Find the height of a male if a 26-cm humerus is found in an archeological dig.
b) What is the domain of M?

63. *Reaction Time.*   Suppose that while driving a car, you suddenly see a school crossing guard standing in the road. Your brain registers the information and sends a signal to your foot to hit the brake. The car travels a distance D, in feet, during this time, where D is a function of the speed r, in miles per hour, that the car is traveling when you see the crossing guard. That reaction distance is a linear function given by

$$D(r) = \frac{11r + 5}{10}.$$

a) Find the slope of this line and interpret its meaning in this application.
b) Graph D.
c) Find $D(5)$, $D(10)$, $D(20)$, $D(50)$, and $D(65)$.
d) What is the domain of this function? Explain.

64. *Straight-line Depreciation.*   A company buys a new color printer for $5200 to print banners for a sales campaign. The printer is purchased on January 1 and is expected to last 8 yr, at the end of which time its *trade-in*, or *salvage value*, will be $1100. If the company figures the decline or depreciation in value to be the same each year, then the salvage value V, after t years, is given by the linear function

$$V(t) = \$5200 - \$512.50t, \quad \text{for } 0 \le t \le 8.$$

a) Graph V.
b) Find $V(0)$, $V(1)$, $V(2)$, $V(3)$, and $V(8)$.
c) Find the domain and the range of this function.

In Exercises 65 and 66, express the slope as a ratio of two quantities.

65.

Number of weeks
after planting

66.

Number of hours
spent walking

67. *Total Cost.*   The Cellular Connection charges $60 for a phone and $40 per month under its economy plan. Write and graph an equation that can be used to determine the total cost, $C(t)$, of operating a

Cellular Connection phone for t months. Then find the total cost for 6 months.

68. *Total Cost.*   Twin Cities Cable Television charges a $35 installation fee and $30 per month for "deluxe" service. Write and graph an equation that can be used to determine the total cost, $C(t)$, for t months of deluxe cable television service. Then find the total cost for 8 months of service.

*In Exercises 69 and 70, the term **fixed costs** refers to the start-up costs of operating a business. This includes machinery and building costs. The term **variable costs** refers to what it costs a business to produce or service one item.*

69. Kara's Custom Tees experienced fixed costs of $800 and variable costs of $3 per shirt. Write an equation that can be used to determine the total expenses encountered by Kara's Custom Tees. Then graph the equation and determine the total cost of producing 75 shirts.

70. It's My Racquet experienced fixed costs of $500 and variable costs of $2 for each tennis racquet that is restrung. Write an equation that can be used to determine the total expenses encountered by It's My Racquet. Then graph the equation and determine the total cost of restringing 150 tennis racquets.

Discussion and Writing

71. Discuss why the graph of a vertical line $x = a$ cannot represent y as a function of x.

72. Explain as you would to a fellow student how the numerical value of slope can be used to describe the slant and the steepness of a line.

Skill Maintenance

To the student and the instructor: The skill maintenance exercises review skills covered previously in the text. You can expect such exercises in every exercise set. They provide excellent review for a final examination. Answers to all skill maintenance exercises appear, along with section references, in the answer section at the back of the book.

If $f(x) = x^2 - 3x$, find each of the following.

73. $f(-5)$

74. $f(5)$

75. $f(-a)$

76. $f(a + h)$

Synthesis

77. Find k so that the line containing the points $(-3, k)$ and $(4, 8)$ is parallel to the line containing the points $(5, 3)$ and $(1, -6)$.

78. Find an equation of the line passing through the point $(4, 5)$ perpendicular to the line passing through the points $(-1, 3)$ and $(2, 9)$.

79. *Fahrenheit and Celsius Temperatures.*   Fahrenheit temperature F is a linear function of Celsius temperature C. When C is 0, F is 32; and when C is 100, F is 212. Use these data to express F as a function of C and to express C as a function of F.

Suppose that f is a linear function. Then $f(x) = mx + b$. Determine whether each of the following is true or false.

80. $f(c + d) = f(c) + f(d)$

81. $f(cd) = f(c)f(d)$

82. $f(kx) = kf(x)$

83. $f(c - d) = f(c) - f(d)$

Let $f(x) = mx + b$. Find a formula for $f(x)$ given each of the following.

84. $f(3x) = 3f(x)$

85. $f(x + 2) = f(x) + 2$

More on Functions

1.3

- *Graph functions, looking for intervals on which the function is increasing, decreasing, or constant, and determine relative maxima and minima.*
- *Given an application, find a function formula that models the application; find the domain of the function and function values, and then graph the function.*
- *Graph functions defined piecewise.*
- *Find the sum, the difference, the product, and the quotient of two functions, and determine their domains.*

Because functions occur in so many real-world situations, it is important to be able to analyze them carefully.

Increasing, Decreasing, and Constant Functions

On a given interval, if the graph of a function rises from left to right, it is said to be **increasing** on that interval. If the graph drops from left to right, it is said to be **decreasing**. If the function stays the same from left to right, it is said to be **constant**.

Interactive Discovery

Graph each of the following functions on a grapher using the given viewing window. Then use the TRACE feature, moving from left to right along each graph. As you move the cursor from left to right, note that the x-coordinate always increases. Observe what happens to the y-coordinate. Where does it increase, decrease, or remain constant? Confirm the results with a table.

$$y = 5 - (x + 2)^2, \quad [-2, 4, -10, 6];$$
$$y = 4, \quad [-20, 20, -3, 6];$$
$$y = |x + 1| + |x - 2| - 5, \quad [-6, 6, -6, 6].$$

We are led to the following definitions.

Increasing, Decreasing, and Constant Functions

A function f is said to be **increasing** on an open interval I, if for all a and b in that interval, $a < b$ implies $f(a) < f(b)$.

A function f is said to be **decreasing** on an open interval I, if for all a and b in that interval, $a < b$ implies $f(a) > f(b)$.

A function f is said to be **constant** on an open interval I, if for all a and b in that interval, $f(a) = f(b)$.

EXAMPLE 1   Determine the intervals on which the function in the figure below is **(a)** increasing; **(b)** decreasing; and **(c)** constant.

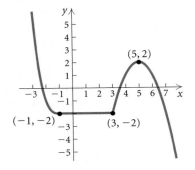

Solution

a) The function is increasing on the interval $(3, 5)$.

b) The function is decreasing on the intervals $(-\infty, -1)$ and $(5, \infty)$.

c) The function is constant on the interval $(-1, 3)$.

Relative Maximum and Minimum Values

Consider the graph shown below. Note the "peaks" and "valleys" at the points c_1, c_2, and c_3. The function value $f(c_2)$ is called a **relative maximum** (plural, **maxima**). Each of the function values $f(c_1)$ and $f(c_3)$ is called a **relative minimum** (plural, **minima**).

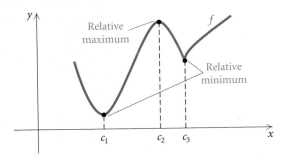

> ### Relative Maxima and Minima
>
> Suppose that f is a function for which $f(c)$ exists for some c in the domain of f. Then:
>
> $f(c)$ is a **relative maximum** if there exists an open interval I containing c such that $f(c) > f(x)$, for all x in I where $x \neq c$; and
>
> $f(c)$ is a **relative minimum** if there exists an open interval I containing c such that $f(c) < f(x)$, for all x in I where $x \neq c$.

Simply stated, $f(c)$ is a relative maximum if $f(c)$ is the highest point in some open interval, and $f(c)$ is a relative minimum if $f(c)$ is the lowest point in some open interval.

We can use the MAXIMUM or MINIMUM features on a grapher to approximate relative maxima or minima. If you take a calculus course, you will learn an algebraic method for determining exact values of relative maxima and minima.

EXAMPLE 2 Use a grapher to determine any relative maxima or minima of the function $f(x) = 0.1x^3 - 0.6x^2 - 0.1x + 2$ and to determine intervals on which the function is increasing or decreasing.

Solution We first graph the function, experimenting with the window as needed. The curvature is seen fairly well with $[-4, 6, -3, 3]$. Using the MAXIMUM and MINIMUM features, we determine the relative maximum value and the relative minimum value of the function.

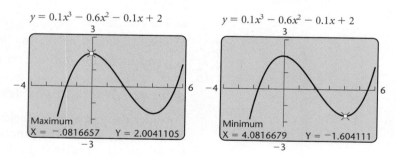

We see that the relative maximum value of the function is about 2.004. It occurs when $x \approx -0.082$. We also approximate the relative minimum: -1.604 at $x \approx 4.082$.

We note that the graph starts rising, or increasing, from the left and stops increasing at the relative maximum. From this point, the graph decreases to the relative minimum and then begins to rise again. The function is *increasing* on the intervals

$$(-\infty, -0.082) \quad \text{and} \quad (4.082, \infty)$$

and *decreasing* on the interval

$$(-0.082, 4.082).$$

STUDY TIP

Have reasonable expectations about the time you need to study math. Make an hour-by-hour schedule of your typical week. Include work, school, home, sleep, study, and leisure times. Try to schedule time for study when you are most alert. Choose a setting that will enable you to maximize your powers of concentration. Plan for success, and it will happen.

Applications of Functions

Many real-world situations can be modeled by functions.

EXAMPLE 3   *Car Distance.* Jamie and Ruben drive away from a restaurant at right angles to each other. Jamie's speed is 65 mph and Ruben's is 55 mph.

a) Express the distance between the cars as a function of time.

b) Find the domain of the function.

c) Graph the function.

Solution

a) Suppose 1 hr goes by. At that time, Jamie has traveled 65 mi and Ruben has traveled 55 mi. We can use the Pythagorean theorem then to find the distance between them. This distance would be the length of the hypotenuse of a triangle with legs measuring 65 mi and 55 mi. After 2 hr, the triangle's legs would measure 130 mi and 110 mi. Observing that the distances will always be changing, we make a drawing and let t = the time, in hours, that Jamie and Ruben have been driving since leaving the restaurant.

After t hours, Jamie has traveled $65t$ miles and Ruben $55t$ miles. We can use the Pythagorean theorem:

$$[d(t)]^2 = (65t)^2 + (55t)^2.$$

Because distance must be nonnegative, we need only consider the positive square root when solving for $d(t)$:

$$d(t) = \sqrt{(65t)^2 + (55t)^2} = \sqrt{4225t^2 + 3025t^2}$$
$$= \sqrt{7250t^2} = \sqrt{7250}\sqrt{t^2}$$
$$\approx 85.15|t| \qquad \text{Approximating the root to two decimal places}$$
$$\approx 85.15t. \qquad \text{Since } t \geq 0, |t| = t.$$

Thus, $d(t) = 85.15t$, $t \geq 0$.

b) Since the time traveled must be nonnegative, the domain is the set of nonnegative real numbers $[0, \infty)$.

c) Because of the ease of using a grapher, we can almost always visualize a problem by making a graph. Here we observe that the function is increasing.

$y = 85.15x$

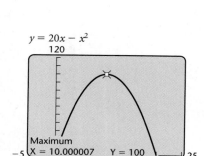

$y = 20x - x^2$

EXAMPLE 4 *Storage Area.* The Sound Shop has 20 ft of dividers with which to set off a rectangular area for the storage of overstock. If a corner of the store is used, the partition need only form two sides of a rectangle.

a) Express the floor area of the storage space as a function of the length of the partition.

b) Find the domain of the function.

c) Graph the function.

d) Find the dimensions that maximize the floor area.

Solution

a) Note that the dividers will form two sides of a rectangle. If, for example, 14 ft of dividers are used for the length of the rectangle, that would leave $20 - 14$, or 6 ft of dividers for the width. Thus if $x =$ the length, in feet, of the rectangle, then $20 - x =$ the width. We represent this information in a sketch, as shown at left.

The area, $A(x)$, is given by

$$A(x) = x(20 - x) \qquad \text{Area = length · width}$$
$$= 20x - x^2.$$

The function $A(x) = 20x - x^2$ can be used to express the rectangle's area as a function of the length.

b) Because the rectangle's length must be positive and less than 20 ft (why?), we restrict the domain of A to $\{x \mid 0 < x < 20\}$, that is, the interval $(0, 20)$.

c) The graph is shown at left in the viewing window $[-5, 25, 0, 120]$.

d) We use the MAXIMUM feature as in Example 2. The maximum value of the area function on $(0, 20)$ is 100 when $x = 10$. Thus the dimensions that maximize the area are

$$\text{Length} = x = 10 \text{ ft} \quad \text{and}$$
$$\text{Width} = 20 - x = 20 - 10 = 10 \text{ ft.}$$

Functions Defined Piecewise

Sometimes functions are defined **piecewise** using different output formulas for different parts of the domain.

EXAMPLE 5 Make a hand-drawn graph of the function defined as

$$f(x) = \begin{cases} 4, & \text{for } x \leq 0, \\ 4 - x^2, & \text{for } 0 < x \leq 2, \\ 2x - 6, & \text{for } x > 2. \end{cases}$$

Solution We create the graph in three parts, as shown and described at the top of the following page.

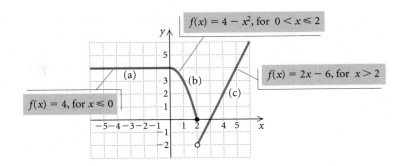

a) We graph $f(x) = 4$ *only* for inputs x less than or equal to 0 (that is, on the interval $(-\infty, 0]$).

b) We graph $f(x) = 4 - x^2$ *only* for inputs x greater than 0 and less than or equal to 2 (that is, on the interval $(0, 2]$).

c) We graph $f(x) = 2x - 6$ *only* for inputs x greater than 2 (that is, on the interval $(2, \infty)$).

To graph a function defined piecewise using a grapher, consult your user's manual or the *Graphing Calculator Manual* that accompanies this text. You might try the following way to enter the function formula for Example 5. It incorporates parenthetic descriptions of the intervals. With this method, you need to graph in DOT mode.

$$f(x) = 4 \ (x \le 0) \ + (4 - x^2) \ (x > 0)(x \le 2) \ + (2x - 6) \ (x > 2)$$

EXAMPLE 6   Make a hand-drawn graph of the function defined as

$$f(x) = \begin{cases} \dfrac{x^2 - 4}{x + 2}, & \text{for } x \ne -2, \\ 3, & \text{for } x = -2. \end{cases}$$

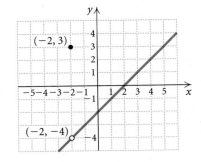

Solution   When $x \ne -2$, the denominator of $(x^2 - 4)/(x + 2)$ is nonzero, so we can simplify:

$$\frac{x^2 - 4}{x + 2} = \frac{(x + 2)(x - 2)}{x + 2} = x - 2.$$

Thus,

$$f(x) = x - 2, \quad \text{for } x \ne -2.$$

The graph of this part of the function consists of a line with a "hole" at the point $(-2, -4)$, indicated by the open dot. At $x = -2$, we have $f(-2) = 3$, so the point $(-2, 3)$ is plotted above the open dot.

$y = (x^2 - 4)/(x + 2)$

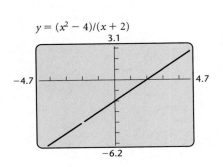

The hand-drawn graph of $f(x) = (x^2 - 4)/(x + 2)$ for $x \ne -2$, in Example 6, can be checked using a grapher as shown at left. When $y = (x^2 - 4)/(x + 2)$ is graphed, the hole may or may not be visible, depending on the window dimensions chosen. First, graph using the dimensions of your choosing. Then use the ZDECIMAL feature from the

---

X	Y1	
−2.3	−4.3	
−2.2	−4.2	
−2.1	−4.1	
−2	ERROR	
−1.9	−3.9	
−1.8	−3.8	
−1.7	−3.7	
X = −2		

ZOOM menu. Note the hole will appear. (See the *Graphing Calculator Manual* that accompanies this text for further details on selecting window dimensions.) If we use the TABLE feature set in ASK mode, as shown in the table at left, the ERROR message indicates that −2 is *not* in the domain of the function g given by $g(x) = (x^2 - 4)/(x + 2)$. However, −2 *is* in the domain of the piecewise function f because $f(-2)$ is defined to be 3.

A function with importance in calculus and computer programming is the *greatest integer function, f,* denoted as $f(x) = [\![x]\!]$, or int(x).

Greatest Integer Function

$f(x) = [\![x]\!]$ = the greatest integer less than or equal to x.

```
int(1.9)
                    1
int(2)
                    2
int(−3.06)
                   −4
```

The greatest integer function pairs each input with the greatest integer less than or equal to that input. Thus, 1, 1.2, and 1.9 are all paired with the y-value 1. Similarly, 2, 2.3, and 2.8 are all paired with 2, and −3.4, −3.06, and −3.99027 are all paired with −4. These values can be checked with a grapher using the int(feature from the NUM submenu in the MATH menu.

EXAMPLE 7   Graph $f(x) = [\![x]\!]$ and determine its domain and range.

Solution   The greatest integer function can also be defined by a piecewise function with an infinite number of statements. When plotting points by hand, it can be helpful to use the TABLE feature on a grapher.

X	Y1	
−2.5	−3	
−2	−2	
−1.45	−2	
−1	−1	
−.75	−1	
0	0	
1.8	1	
X = 1.8		

$$f(x) = [\![x]\!] = \begin{cases} \vdots & \\ -3 & \text{for } -3 \le x < -2 \\ -2 & \text{for } -2 \le x < -1 \\ -1 & \text{for } -1 \le x < 0 \\ 0 & \text{for } 0 \le x < 1 \\ 1 & \text{for } 1 \le x < 2 \\ 2 & \text{for } 2 \le x < 3 \\ 3 & \text{for } 3 \le x < 4 \\ \vdots & \end{cases}$$

We see that the domain of $f(x)$ is the set of all real numbers, $(-\infty, \infty)$, and the range is the set of all integers, $\{\ldots, -3, -2, -1, 0, 1, 2, 3, \ldots\}$.

If we used a grapher set in CONNECTED mode for Example 7, we would see the graph on the left below. CONNECTED mode connects points (or rectangles called **pixels** on a grapher) with line segments. We would see the graph on the right if we used DOT mode. The DOT mode is preferable, though even it does not show the open dots at the endpoints of the segments.

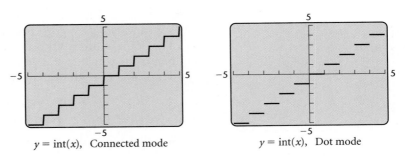

$y = \text{int}(x)$, Connected mode $y = \text{int}(x)$, Dot mode

Thus we must be careful when analyzing graphs of functions on graphers. The use of technology must be combined with an understanding of the basic concepts of functions. This is why we balance technology with a discussion of hand-drawn graphs.

The Algebra of Functions: Sums, Differences, Products, and Quotients

We now use addition, subtraction, multiplication, and division to combine functions and obtain new functions.

Consider the following two functions f and g:

$$f(x) = x + 2 \quad \text{and} \quad g(x) = x^2 + 1.$$

Since $f(3) = 3 + 2 = 5$ and $g(3) = 3^2 + 1 = 10$, we have

$$f(3) + g(3) = 5 + 10 = 15,$$
$$f(3) - g(3) = 5 - 10 = -5,$$
$$f(3) \cdot g(3) = 5 \cdot 10 = 50, \quad \text{and} \quad \frac{f(3)}{g(3)} = \frac{5}{10} = \frac{1}{2}.$$

In fact, so long as x is in the domain of both f and g, we can easily compute $f(x) + g(x)$, $f(x) - g(x)$, $f(x) \cdot g(x)$, and, assuming $g(x) \neq 0$, $f(x)/g(x)$. Notation has been developed to facilitate this work.

Sums, Differences, Products, and Quotients of Functions

If f and g are functions and x is in the domain of each function, then

$(f + g)(x) = f(x) + g(x), \qquad (f - g)(x) = f(x) - g(x),$

$(fg)(x) = f(x) \cdot g(x), \qquad (f/g)(x) = f(x)/g(x),$

$\qquad\qquad\qquad\qquad\qquad$ provided $g(x) \neq 0.$

EXAMPLE 8 Given that $f(x) = x + 1$ and $g(x) = \sqrt{x + 3}$, find each of the following.

a) $(f + g)(x)$ **b)** $(f + g)(6)$ **c)** $(f + g)(-4)$

Solution

a) $(f + g)(x) = f(x) + g(x)$
$= x + 1 + \sqrt{x + 3}$ This cannot be simplified.

b) We can find $(f + g)(6)$ provided 6 is in the domain of each function. The domain of f is all real numbers. The domain of g is all real numbers x for which $x + 3 \geq 0$, or $x \geq -3$. This is the interval $[-3, \infty)$. Thus, 6 is in both domains, so we have

$$f(6) = 6 + 1 = 7, \qquad g(6) = \sqrt{6 + 3} = \sqrt{9} = 3,$$
$$(f + g)(6) = f(6) + g(6) = 7 + 3 = 10.$$

Another method is to use the formula found in part (a):

$$(f + g)(6) = 6 + 1 + \sqrt{6 + 3} = 7 + \sqrt{9} = 7 + 3 = 10.$$

We can check our work using a grapher by entering $y_1 = x + 1$, $y_2 = \sqrt{x + 3}$, and $y_3 = y_1 + y_2$ in the $y =$ screen. Then on the home screen, we find Y3(6).

c) To find $(f + g)(-4)$, we must first determine whether -4 is in the domain of each function. We note that -4 is not in the domain of g, $[-3, \infty)$. That is, $\sqrt{-4 + 3}$ is not a real number. Thus, $(f + g)(-4)$ does not exist.

It is useful to view the concept of the sum of two functions graphically. In the graph below, we see the graphs of two functions f and g and their sum, $f + g$. Consider finding $(f + g)(4) = f(4) + g(4)$. We can locate $g(4)$ on the graph of g and use a compass to measure it. Then we move that length on top of $f(4)$ and add. The sum gives us $(f + g)(4)$.

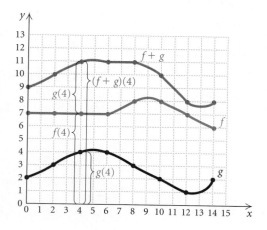

With this in mind, let's view Example 8 from a graphical perspec-

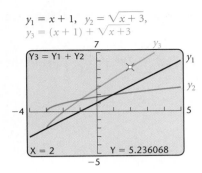

$y_1 = x + 1,\ \ y_2 = \sqrt{x+3},$
$y_3 = (x+1) + \sqrt{x+3}$

Y3 = Y1 + Y2

X = 2   Y = 5.236068

tive. Let's use a grapher to graph

$$y_1 = x + 1, \qquad y_2 = \sqrt{x+3}, \quad \text{and} \quad y_3 = x + 1 + \sqrt{x+3}.$$

See the graph at left. Note that because the domain of y_2 is $[-3, \infty)$, the graph of the sum exists only on $[-3, \infty)$. By using the VALUE feature and jumping the cursor vertically among the three graphs, we can confirm that the y-coordinates of the graph of $(f + g)(x)$ are the sums of the corresponding y-coordinates of the graphs of $f(x)$ and $g(x)$.

EXAMPLE 9   Given that $f(x) = x^2 - 4$ and $g(x) = x + 2$, find each of the following.

a) $(f + g)(x)$ **b)** $(f - g)(x)$

c) $(fg)(x)$ **d)** $(f/g)(x)$

e) $(gg)(x)$

f) The domain of $f + g$, $f - g$, fg, and f/g

Solution

a) $(f + g)(x) = f(x) + g(x) = (x^2 - 4) + (x + 2) = x^2 + x - 2$

b) $(f - g)(x) = f(x) - g(x) = (x^2 - 4) - (x + 2) = x^2 - x - 6$

c) $(fg)(x) = f(x) \cdot g(x) = (x^2 - 4)(x + 2) = x^3 + 2x^2 - 4x - 8$

d) $(f/g)(x) = \dfrac{f(x)}{g(x)}$

$$= \dfrac{x^2 - 4}{x + 2} \qquad \qquad \begin{array}{l}\text{Note that } (f/g)(x) \text{ is} \\ \text{undefined when } x = -2.\end{array}$$

$$= \dfrac{(x + 2)(x - 2)}{x + 2} \qquad \text{Factoring}$$

$$= x - 2 \qquad\qquad\quad \text{Removing a factor of 1: } \dfrac{x + 2}{x + 2} = 1$$

Thus, $(f/g)(x) = x - 2$ with the added stipulation that $x \neq -2$.

e) $(gg)(x) = g(x) \cdot g(x) = [g(x)]^2 = (x + 2)^2 = x^2 + 4x + 4$

f) The domain of f is the set of all real numbers. The domain of g is also the set of all real numbers. The domain of $f + g$, $f - g$, and fg is the set of numbers in the intersection of the domains—that is, the set of numbers in both domains, which is again the set of real numbers. For f/g, we must exclude -2, since $g(-2) = 0$. Thus the domain of f/g is the set of real numbers excluding -2, or $(-\infty, -2) \cup (-2, \infty)$. ▬

Difference Quotients

It is important for students preparing for calculus to be able to simplify rational expressions like

$$\dfrac{f(x + h) - f(x)}{h}.$$

In calculus, this process will be referred to as finding the **difference quotient.**

EXAMPLE 10   For the function f given by $f(x) = 2x^2 - x - 3$, find the difference quotient

$$\frac{f(x + h) - f(x)}{h}.$$

SOLUTION   We first find $f(x + h)$:

$$f(x + h) = 2(x + h)^2 - (x + h) - 3$$
$$= 2[x^2 + 2xh + h^2] - x - h - 3$$
$$= 2x^2 + 4xh + 2h^2 - x - h - 3.$$

Then

$$\frac{f(x + h) - f(x)}{h}$$

$$= \frac{[2x^2 + 4xh + 2h^2 - x - h - 3] - [2x^2 - x - 3]}{h}$$

$$= \frac{2x^2 + 4xh + 2h^2 - x - h - 3 - 2x^2 + x + 3}{h}$$

$$= \frac{4xh + 2h^2 - h}{h}$$

$$= \frac{h(4x + 2h - 1)}{h \cdot 1} = \frac{h}{h} \cdot \frac{4x + 2h - 1}{1} = 4x + 2h - 1.$$

Exercise Set **1.3**

Determine intervals on which the function is (a)
increasing, (b) decreasing, and (c) constant.

1.

2.

3.

4.

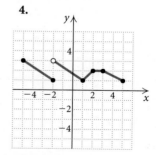

Graph the function using the given viewing window.
Find the intervals on which the function is increasing or
decreasing and find any relative maxima or minima.
Change viewing windows if it seems appropriate for
further analysis.

5. $f(x) = x^2$,
   $[-4, 4, -1, 10]$

6. $f(x) = 4 - x^2$,
   $[-5, 5, -5, 5]$

7. $f(x) = 5 - |x|$,
[−10, 10, −10, 10]

8. $f(x) = |x + 3| - 5$,
[−10, 10, −10, 10]

9. $f(x) = x^2 - 6x + 10$,
[−4, 10, −1, 20]

10. $f(x) = -x^2 - 8x - 9$,
[−10, 2, −10, 10]

11. $f(x) = -x^3 + 6x^2 - 9x - 4$,
[−3, 7, −20, 15]

12. $f(x) = 0.2x^3 - 0.2x^2 - 5x - 4$,
[−10, 10, −30, 20]

13. $f(x) = 1.1x^4 - 5.3x^2 + 4.07$,
[−4, 4, −4, 8]

14. $f(x) = 1.2(x + 3)^4 + 10.3(x + 3)^2 + 9.78$,
[−9, 3, −40, 100]

15. *Temperature During an Illness.* The temperature of a patient during an illness is given by the function

$$T(t) = -0.1t^2 + 1.2t + 98.6, \quad 0 \le t \le 12,$$

where T is the temperature, in degrees Fahrenheit, at time t, in days, after the onset of the illness.

a) Graph the function using a grapher.
b) At what time was the patient's temperature the highest? What was the highest temperature?

16. *Advertising Effect.* A software firm estimates that it will sell N units of a new CD-ROM video game after spending a dollars on advertising, where

$$N(a) = -a^2 + 300a + 6, \quad 0 \le a \le 300,$$

and a is measured in thousands of dollars.

a) Graph the function using a grapher.
b) Find the relative maximum.
c) For what advertising expenditure will the greatest number of games be sold? How many games will be sold for that amount?

Use a grapher. Find the intervals on which each function is increasing or decreasing. Consider the entire set of real numbers if no domain is given.

17. $f(x) = \dfrac{8x}{x^2 + 1}$ **18.** $f(x) = \dfrac{-4}{x^2 + 1}$

19. $f(x) = x\sqrt{4 - x^2}$, for $-2 \le x \le 2$

20. $f(x) = -0.8x\sqrt{9 - x^2}$, for $-3 \le x \le 3$

21. *Rising Balloon.* A hot-air balloon rises straight up from the ground at a rate of 120 ft/min. The balloon is tracked from a rangefinder on the ground at point P, which is 400 ft from the release point Q of the balloon. Let d = the distance from the balloon to the rangefinder at point P and t = the

time, in minutes, since the balloon was released. Express d as a function of t.

22. *Airplane Distance.* An airplane is flying at an altitude of 3700 ft. The slanted distance directly to the airport is d feet. Express the horizontal distance h as a function of d.

23. *Garden Area.* Yardbird Landscaping has 48 m of fencing with which to enclose a rectangular garden. If the garden is x meters long, express the garden's area as a function of the length.

24. *Triangular Flag.* A scout troop is designing a triangular flag so that the length of its base, in inches, is 7 less than twice the height, h. Express the area of the flag as a function of the height.

25. *Inscribed Rhombus.* A rhombus is inscribed in a rectangle that is w meters wide with a perimeter of 40 m. Each vertex of the rhombus is a midpoint of a side of the rectangle. Express the area of the rhombus as a function of the rectangle's width. (*Hint*: Consider the area of the rectangle.)

26. *Tablecloth Area.*　A tailor uses 16 ft of lace to trim the edges of a rectangular tablecloth. If the tablecloth is *w* feet wide, express its area as a function of the width.

27. *Golf Distance Finder.*　A device used in golf to estimate the distance *d*, in yards, to a hole measures the size *s*, in inches, that the 7-ft pin appears to be in a viewfinder. Express the distance *d* as a function of *s*.

28. *Gas Tank Volume.*　A gas tank has ends that are hemispheres of radius *r* feet. The cylindrical midsection is 6 ft long. Express the volume of the tank as a function of *r*.

29. *Play Space.*　A daycare center has 30 ft of dividers with which to enclose a rectangular play space in a corner of a large room. The sides against the wall require no partition. Suppose the play space is *x* feet long.

　a) Express the area of the play space as a function of *x*.
　b) Find the domain of the function.
　c) Graph the function.
　d) What dimensions yield the maximum area?

30. *Corral Design.*　A rancher has 360 yd of fencing with which to enclose two adjacent rectangular corrals, one for horses and one for cattle. A river forms one side of the corrals. Suppose the width of each corral is *x* yards.

　a) Express the total area of the two corrals as a function of *x*.
　b) Find the domain of the function.
　c) Graph the function.
　d) What dimensions yield the maximum area?

31. *Volume of a Box.*　From a 12-cm by 12-cm piece of cardboard, square corners are cut out so that the sides can be folded up to make a box.

　a) Express the volume of the box as a function of the length *x*, in centimeters, of a cut-out square.
　b) Find the domain of the function.
　c) Graph the function.
　d) What dimensions yield the maximum volume?

32. *Molding Plastics.*　Perfect Plastics plans to produce a one-component vertical file by bending the long side of an 8-in. by 14-in. sheet of plastic along two lines to form a U shape.

　a) Express the volume of the file as a function of the height *x*, in inches, of the file.
　b) Find the domain of the function.
　c) Graph the function.
　d) How tall should the file be in order to maximize the volume that the file can hold?

33. *Area of an Inscribed Rectangle.*   A rectangle that is x feet wide is inscribed in a circle of radius 8 ft.

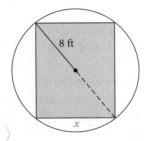

8 ft

x

a) Express the area of the rectangle as a function of x.
b) Find the domain of the function.
c) Graph the function.
d) What dimensions maximize the area of the rectangle?

34. *Cost of Material.*   A rectangular box with volume 320 ft^3 is built with a square base and top. The cost is \$1.50/ft^2 for the bottom, \$2.50/ft^2 for the sides, and \$1/ft^2 for the top. Let x = the length of the base, in feet.

x

x

x

x

a) Express the cost of the box as a function of x.
b) Find the domain of the function.
c) Graph the function.
d) What dimensions minimize the cost of the box?

Make a hand-drawn graph of each of the following. Check your results using a grapher.

35. $f(x) = \begin{cases} \frac{1}{2}x, & \text{for } x < 0, \\ x + 3, & \text{for } x \geq 0 \end{cases}$

36. $f(x) = \begin{cases} -\frac{1}{3}x + 2, & \text{for } x \leq 0, \\ x - 5, & \text{for } x > 0 \end{cases}$

37. $f(x) = \begin{cases} -\frac{3}{4}x + 2, & \text{for } x < 4, \\ -1, & \text{for } x \geq 4 \end{cases}$

38. $f(x) = \begin{cases} 4, & \text{for } x \leq -2, \\ x + 1, & \text{for } -2 < x < 3, \\ -x, & \text{for } x \geq 3 \end{cases}$

39. $f(x) = \begin{cases} x + 1, & \text{for } x \leq -3, \\ -1, & \text{for } -3 < x < 4, \\ \frac{1}{2}x, & \text{for } x \geq 4 \end{cases}$

40. $f(x) = \begin{cases} \dfrac{x^2 - 9}{x + 3}, & \text{for } x \neq -3, \\ 5, & \text{for } x = -3 \end{cases}$

41. $f(x) = \begin{cases} 2, & \text{for } x = 5, \\ \dfrac{x^2 - 25}{x - 5}, & \text{for } x \neq 5 \end{cases}$

42. $f(x) = \begin{cases} \dfrac{x^2 + 3x + 2}{x + 1}, & \text{for } x \neq -1, \\ 7, & \text{for } x = -1 \end{cases}$

43. $f(x) = [\![x]\!]$
44. $f(x) = 2[\![x]\!]$
45. $g(x) = 1 + [\![x]\!]$
46. $h(x) = \frac{1}{2}[\![x]\!] - 2$

Graph using a grapher.

47. $f(x) = \begin{cases} \sqrt[3]{x}, & \text{for } x \leq -1, \\ x^2 - 3x, & \text{for } -1 < x < 4, \\ \sqrt{x - 4}, & \text{for } x \geq 4 \end{cases}$

48. $f(x) = \begin{cases} \left| \dfrac{x}{3} + 2 \right|, & \text{for } x < 5, \\ \sqrt[3]{x - 8}, & \text{for } x \geq 5 \end{cases}$

Given that $f(x) = x^2 - 3$ and $g(x) = 2x + 1$, find each of the following, if it exists.

49. $(f + g)(5)$
50. $(f - g)(3)$
51. $(f - g)(-1)$
52. $(fg)(0)$
53. $(fg)(2)$
54. $(fg)(-2)$
55. $(f/g)(-1)$
56. $(f/g)\left(-\frac{1}{2}\right)$
57. $(fg)\left(-\frac{1}{2}\right)$
58. $(f/g)(-\sqrt{3})$
59. $(f/g)(\sqrt{3})$
60. $(f - g)(0)$

For each pair of functions in Exercises 61–64:

a) Find the domain of f, g, $f + g$, $f - g$, fg, ff, f/g, and g/f.

b) Find $(f + g)(x)$, $(f - g)(x)$, $(fg)(x)$, $(ff)(x)$, $(f/g)(x)$, and $(g/f)(x)$.

61. $f(x) = x - 3$, $g(x) = \sqrt{x + 4}$

62. $f(x) = x^2 - 1$, $g(x) = 2x + 5$

63. $f(x) = x^3$, $g(x) = 2x^2 + 5x - 3$

64. $f(x) = x^2$, $g(x) = \sqrt{x}$

For each pair of functions in Exercises 65–68, find $(f + g)(x)$, $(f - g)(x)$, $(fg)(x)$, *and* $(f/g)(x)$.

65. $f(x) = x^2 - 4$, $g(x) = x^2 + 2$

66. $f(x) = x^2 + 2$, $g(x) = 4x - 7$

67. $f(x) = \sqrt{x - 7}$, $g(x) = x^2 - 25$

68. $f(x) = \sqrt{x - 1}$, $g(x) = \sqrt{3x - 8}$

In Exercises 69–74, consider the functions F and G as shown in the following graph.

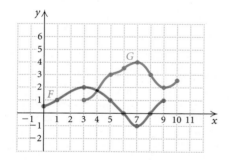

69. Find the domain of F, the domain of G, and the domain of $F + G$.

70. Find the domain of $F - G$, FG, and F/G.

71. Find the domain of G/F.

72. Graph $F + G$.

73. Graph $G - F$.

74. Graph $F - G$.

75. *Total Cost, Revenue, and Profit.* In economics, functions that involve revenue, cost, and profit are used. For example, suppose that $R(x)$ and $C(x)$ denote the total revenue and the total cost, respectively, of producing a new kind of tool for King Hardware Stores. Then the difference

$$P(x) = R(x) - C(x)$$

represents the total profit for producing x tools.

Given

$$R(x) = 60x - 0.4x^2 \quad \text{and} \quad C(x) = 3x + 13,$$

find each of the following.

a) $P(x)$

b) $R(100)$, $C(100)$, and $P(100)$

c) Using a grapher, graph the three functions in the viewing window [0, 160, 0, 3000].

76. *Total Cost, Revenue, and Profit.* Given that

$$R(x) = 200x - x^2 \quad \text{and} \quad C(x) = 5000 + 8x,$$

for a new radio produced by Clear Communication, find each of the following. (See Exercise 75.)

a) $P(x)$

b) $R(175)$, $C(175)$, and $P(175)$

c) Using a grapher, graph the three functions in the viewing window [0, 200, 0, 10,000].

For each function f, construct and simplify the difference quotient

$$\frac{f(x + h) - f(x)}{h}.$$

77. $f(x) = 3x^2 - 2x + 1$

78. $f(x) = 5x^2 + 4x$

79. $f(x) = x^3$

80. $f(x) = 2|x| + 3x$

81. $f(x) = \dfrac{x - 4}{x + 3}$

82. $f(x) = \dfrac{x}{2 - x}$

Discussion and Writing

83. Describe a real-world situation that could be modeled by a function that is, in turn, increasing, then constant, and finally decreasing.

84. Simply stated, a *continuous function* is a function whose graph can be drawn without lifting the pencil from the paper. Examine several functions in this exercise set to see if they are continuous. Then explore the continuous functions to find the relative maxima and minima. For continuous functions, how can you connect the ideas of increasing and decreasing on an interval to relative maxima and minima?

Skill Maintenance

Find the domain of the function.

85. $f(x) = -2x - 3$

86. $g(x) = \dfrac{1}{(x - 3)^2}$

87. $h(x) = \dfrac{x}{|x + 5|}$

88. $f(x) = -5$

Synthesis

Using a grapher, estimate the interval on which each function is increasing or decreasing and any relative maxima or minima.

89. $f(x) = x^4 + 4x^3 - 36x^2 - 160x + 400$

90. $f(x) = 3.22x^5 - 5.208x^3 - 11$

91. *Parking Costs.* A parking garage charges $2 for up to (but not including) 1 hr of parking, $4 for up to 2 hr of parking, $6 for up to 3 hr of parking, and so on. Let $C(t) =$ the cost of parking for t hours.

   a) Graph the function.
   b) Write an equation for $C(t)$ using the greatest integer notation $[\![t]\!]$.

92. If $[\![x + 2]\!] = -3$, what are the possible inputs for x?

93. If $([\![x]\!])^2 = 25$, what are the possible inputs for x?

94. *Minimizing Power Line Costs.* A power line is constructed from a power station at point A to an island at point C, which is 1 mi directly out in the water from a point B on the shore. Point B is 4 mi downshore from the power station at A. It costs

$5000 per mile to lay the power line under water and $3000 per mile to lay the power line under ground. The line comes to the shore at point S downshore from A. Let $x =$ the distance from B to S.

   a) Express the cost C of laying the line as a function of x.
   b) At what distance x should the line come to shore in order to minimize cost?

95. *Volume of an Inscribed Cylinder.* A right circular cylinder of height h and radius r is inscribed in a right circular cone with a height of 10 ft and a base with radius 6 ft.

   a) Express the height h of the cylinder as a function of r.
   b) Express the volume V of the cylinder as a function of r.
   c) Express the volume V of the cylinder as a function of h.

Symmetry and Transformations
1.4

- *Determine whether a graph is symmetric with respect to the x-axis, the y-axis, and the origin.*
- *Determine whether a function is even, odd, or neither even nor odd.*
- *Given the graph of a function, graph its transformation under translations, reflections, stretchings, and shrinkings.*

Symmetry

Symmetry occurs often in nature and in art. For example, when viewed from the front, the bodies of most animals are at least approximately symmetric. This means that each eye is the same distance from the center of the bridge of the nose, each shoulder is the same distance from the center of the chest, and so on. Architects have used symmetry for thousands of years to enhance the beauty of buildings.

A knowledge of symmetry in mathematics helps us graph and analyze equations and functions.

Consider the points $(4, 2)$ and $(4, -2)$ that appear in the graph of $x = y^2$ (see Fig. 1 at the top of the following page). Points like these have the same x-value but opposite y-values and are **reflections** of each other across the x-axis. If for any point (x, y) on a graph the point $(x, -y)$ is also on the graph, then the graph is said to be **symmetric with respect to the x-axis.** If we fold the graph on the x-axis, the parts above and below the x-axis will coincide.

Consider the points $(3, 4)$ and $(-3, 4)$ that appear in the graph of $y = x^2 - 5$ (see Fig. 2 at the top of the following page). Points like these have the same y-value but opposite x-values and are **reflections** of each other across the y-axis. If for any point (x, y) on a graph the point $(-x, y)$ is also on the graph, then the graph is said to be **symmetric**

with respect to the y-axis. If we fold the graph on the y-axis, the parts to the left and right of the y-axis will coincide.

FIGURE 1

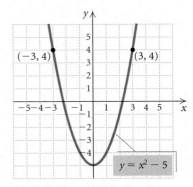

FIGURE 2

Consider the points $(3, \sqrt{7})$ and $(-3, -\sqrt{7})$ that appear in the graph of $x^2 = y^2 + 2$ (see Fig. 3). Note that if we take the opposites of the coordinates of one pair, we get the other pair. If for any point (x, y) on a graph the point $(-x, -y)$ is also on the graph, then the graph is said to be **symmetric with respect to the origin.** Visually, if we rotate the graph 180° about the origin, the resulting figure coincides with the original.

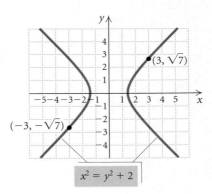

FIGURE 3

Algebraic Tests of Symmetry

x-axis: If replacing y with $-y$ produces an equivalent equation, then the graph is *symmetric with respect to the x-axis.*

y-axis: If replacing x with $-x$ produces an equivalent equation, then the graph is *symmetric with respect to the y-axis.*

Origin: If replacing x with $-x$ and y with $-y$ produces an equivalent equation, then the graph is *symmetric with respect to the origin.*

EXAMPLE 1 Test $y = x^2 + 2$ for symmetry with respect to the x-axis, the y-axis, and the origin.

Algebraic Solution

x-AXIS

We replace y with $-y$:

$$y = x^2 + 2$$
$$\downarrow$$
$$-y = x^2 + 2$$
$$y = -x^2 - 2. \qquad \text{Multiplying by } -1 \\ \text{on both sides}$$

The resulting equation is not equivalent to the original equation, so the graph *is not* symmetric with respect to the x-axis.

y-AXIS

We replace x with $-x$:

$$y = x^2 + 2$$
$$y = (-x)^2 + 2$$
$$y = x^2 + 2. \qquad \text{Simplifying}$$

The resulting equation is equivalent to the original equation, so the graph *is* symmetric with respect to the y-axis.

ORIGIN

We replace x with $-x$ and y with $-y$:

$$y = x^2 + 2$$
$$\downarrow \qquad \downarrow$$
$$-y = (-x)^2 + 2$$
$$-y = x^2 + 2 \qquad \text{Simplifying}$$
$$y = -x^2 - 2.$$

The resulting equation is not equivalent to the original equation, so the graph *is not* symmetric with respect to the origin.

Graphical Solution

We use a grapher to graph the equation.

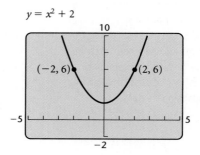

$y = x^2 + 2$

Note that if the graph were folded on the x-axis, the parts above and below the x-axis would not coincide. If it were folded on the y-axis, the parts to the left and right of the y-axis would coincide. If we rotated it 180°, the resulting graph would not coincide with the original graph.

Thus we see that the graph *is not* symmetric with respect to the x-axis or the origin. The graph *is* symmetric with respect to the y-axis.

The algebraic method is often easier to apply than the graphical, especially with equations that we may not be able to graph easily. It is also often more precise.

EXAMPLE 2 Test $x^2 + y^4 = 5$ for symmetry with respect to the x-axis, the y-axis, and the origin.

Algebraic Solution

***x*-AXIS**

We replace y with $-y$:

$$x^2 + y^4 = 5$$

$$x^2 + (-y)^4 = 5$$

$$x^2 + y^4 = 5.$$

The resulting equation is equivalent to the original equation. Thus the graph is symmetric with respect to the x-axis.

We leave it to the student to verify algebraically that the graph is also symmetric with respect to the y-axis and the origin.

Graphical Solution

To graph $x^2 + y^4 = 5$ using a grapher, we first solve the equation for y:

$$y = \pm\sqrt[4]{5 - x^2}.$$

Then we enter on the $y=$ screen the equations

$$y_1 = \sqrt[4]{5 - x^2} \quad \text{and} \quad y_2 = -\sqrt[4]{5 - x^2}.$$

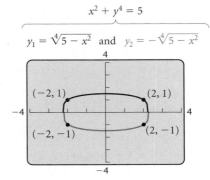

We see symmetry with respect to both axes and with respect to the origin.

Even and Odd Functions

Now we relate symmetry to graphs of functions.

Even and Odd Functions

If the graph of a function f is symmetric with respect to the y-axis, we say that it is an **even function.** That is, for each x in the domain of f, $f(x) = f(-x)$.

If the graph of a function f is symmetric with respect to the origin, we say that it is an **odd function.** That is, for each x in the domain of f, $f(-x) = -f(x)$.

Algebraic Procedure for Determining Even and Odd Functions

Given the function $f(x)$:

1. Find $f(-x)$ and simplify. If $f(x) = f(-x)$, then f is even.

2. Find $-f(x)$, simplify, and compare with $f(-x)$ from step (1). If $f(-x) = -f(x)$, then f is odd.

Except for the function $f(x) = 0$, a function cannot be *both* even and odd. Thus if we see in step (1) that $f(x) = f(-x)$ (that is, f is even), we need not continue.

EXAMPLE 3   Determine whether each of the following functions is even, odd, or neither.

a) $f(x) = 5x^7 - 6x^3 - 2x$ **b)** $h(x) = 5x^6 - 3x^2 - 7$

a)

Algebraic Solution

$f(x) = 5x^7 - 6x^3 - 2x$

1. $f(-x) = 5(-x)^7 - 6(-x)^3 - 2(-x)$
$= 5(-x^7) - 6(-x^3) + 2x$

       $(-x)^7 = (-1 \cdot x)^7 = (-1)^7 x^7 = -x^7$

$= -5x^7 + 6x^3 + 2x$

We see that $f(x) \neq f(-x)$. Thus, f is *not* even.

2. $-f(x) = -(5x^7 - 6x^3 - 2x)$
$= -5x^7 + 6x^3 + 2x$

We see that $f(-x) = -f(x)$. Thus, f is odd.

Graphical Solution

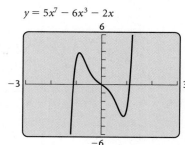

$y = 5x^7 - 6x^3 - 2x$

We see that the graph appears to be symmetric with respect to the origin. The function is odd.

b)

Algebraic Solution

$h(x) = 5x^6 - 3x^2 - 7$

1. $h(-x) = 5(-x)^6 - 3(-x)^2 - 7$
$= 5x^6 - 3x^2 - 7$

We see that $h(x) = h(-x)$. Thus the function is even.

Graphical Solution

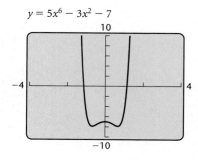

$y = 5x^6 - 3x^2 - 7$

We see that the graph appears to be symmetric with respect to the y-axis. The function is even.

Transformations of Functions

The graphs of some basic functions are shown below. Others can be seen on the inside back cover.

Squaring function:
$y = x^2$

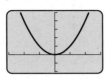

Square root function:
$y = \sqrt{x}$

Cubing function:
$y = x^3$

Cube root function:
$y = \sqrt[3]{x}$

Reciprocal function:
$y = \dfrac{1}{x}$

Absolute value function:
$y = |x|$

These functions can be considered building blocks for many other functions. We can create graphs of new functions by shifting them horizontally or vertically, stretching or shrinking them, and reflecting them across an axis. We now consider these *transformations*.

Vertical and Horizontal Translations

Suppose that we have a function given by $y_1 = f(x)$. Let's explore the graphs of new functions $y_2 = f(x) + b$ and $y_3 = f(x) - b$, for $b > 0$.

Interactive
Discovery

Consider the function $y_1 = \frac{1}{5}x^4$. Graph it and the functions $y_2 = \frac{1}{5}x^4 + 5$ and $y_3 = \frac{1}{5}x^4 - 3$ using different graph styles and the same viewing window, $[-5, 5, -8, 10]$. What pattern do you see? Test it with some other graphs.

The effect of adding a constant to or subtracting a constant from $f(x)$ in $y = f(x)$ is a shift of $f(x)$ up or down. Such a shift is called a **vertical translation.**

Vertical Translation

For $b > 0$,

the graph of $y = f(x) + b$ is the graph of $y = f(x)$ shifted *up b* units;

the graph of $y = f(x) - b$ is the graph of $y = f(x)$ shifted *down b* units.

Suppose that we have a function given by $y_1 = f(x)$. Let's explore the graphs of the new functions $y_2 = f(x + d)$ and $y_3 = f(x - d)$, for $d > 0$.

Consider the function $y_1 = \frac{1}{5}x^4$. Graph it and the functions $y_2 = \frac{1}{5}(x - 3)^4$ and $y_3 = \frac{1}{5}(x + 7)^4$ using the viewing window $[-10, 10, -2, 10]$. What pattern do you see? Test it with some other graphs.

The effect of adding a constant to the x-value or subtracting a constant from the x-value in $y = f(x)$ is a shift of $f(x)$ to the right or left. Such a shift is called a **horizontal translation**.

> *Horizontal Translation*
>
> For $d > 0$:
>
> the graph of $y = f(x - d)$ is the graph of $y = f(x)$ shifted *right d* units;
>
> the graph of $y = f(x + d)$ is the graph of $y = f(x)$ shifted *left d* units.

EXAMPLE 4   Graph each of the following. Before doing so, describe how each graph can be obtained from one of the basic graphs shown on the preceding page.

a) $g(x) = x^2 - 6$ **b)** $g(x) = \sqrt{x + 2}$
c) $g(x) = |x - 4|$ **d)** $h(x) = \sqrt{x + 2} - 3$

Solution

a) To graph $g(x) = x^2 - 6$, think of the graph of $f(x) = x^2$. Since $g(x) = f(x) - 6$, the graph of $g(x) = x^2 - 6$ is the graph of $f(x) = x^2$, shifted, or translated, *down* 6 units (see Fig. 1).

b) To graph $g(x) = \sqrt{x + 2}$, think of the graph of $f(x) = \sqrt{x}$. Since $g(x) = f(x + 2)$, the graph of $g(x) = \sqrt{x + 2}$ is the graph of $f(x) = \sqrt{x}$, shifted *left* 2 units (see Fig. 2).

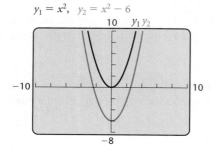

$y_1 = x^2, \ y_2 = x^2 - 6$

FIGURE 1

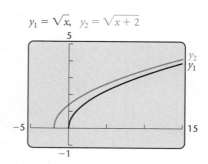

$y_1 = \sqrt{x}, \ y_2 = \sqrt{x + 2}$

FIGURE 2

c) To graph $g(x) = |x - 4|$, think of the graph of $f(x) = |x|$. Since $g(x) = f(x - 4)$, the graph of $g(x) = |x - 4|$ is the graph of $f(x) = |x|$ shifted *right* 4 units (see Fig. 3).

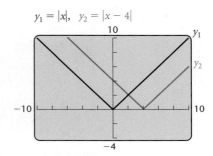

FIGURE 3

d) To graph $h(x) = \sqrt{x + 2} - 3$, think of the graph of $f(x) = \sqrt{x}$. In part (b), we found that the graph of $g(x) = \sqrt{x + 2}$ is the graph of $f(x) = \sqrt{x}$ shifted left 2 units. Since $h(x) = g(x) - 3$, we shift the graph of $g(x) = \sqrt{x + 2}$ *down* 3 units. Together, the graph of $f(x) = \sqrt{x}$ is shifted *left* 2 units and *down* 3 units (see Fig. 4).

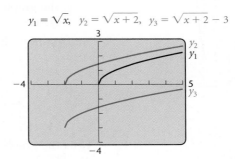

FIGURE 4

Reflections

Suppose that we have a function given by $y = f(x)$. Let's explore the graphs of the new functions $g(x) = -f(x)$ and $g(x) = f(-x)$.

Interactive

Discovery

Consider the functions $y_1 = \frac{1}{5}x^4$ and $y_2 = -\frac{1}{5}x^4$. (Here $y_1 = f(x)$ and $y_2 = -f(x)$.) Graph y_1 using the viewing window $[-10, 10, -10, 10]$. Then graph y_2 using the same viewing window. What do you observe? Test your observation with some other functions in which $y_2 = -y_1$.

Consider the functions $y_1 = 2x^3 - x^4 + 5$ and $y_2 = 2(-x)^3 - (-x)^4 + 5$. (Here $y_1 = f(x)$ and $y_2 = f(-x)$.) Graph y_1 using the viewing window $[-4, 4, -10, 10]$. Then graph y_2 in the same viewing window. What do you observe? Test your observation with some other functions in which x is replaced with $-x$.

Given the graph of $y_1 = f(x)$, we can reflect each point across the x-axis to obtain the graph of $y_2 = -f(x)$. We can reflect each point of y_1 across the y-axis to obtain the graph of $y_2 = f(-x)$. The new graphs are called **reflections** of $f(x)$.

Reflections

The graph of $y = -f(x)$ is the **reflection** of the graph of $y = f(x)$ across the x-axis.

The graph of $y = f(-x)$ is the **reflection** of the graph of $y = f(x)$ across the y-axis.

EXAMPLE 5 Graph each of the following. Before doing so, describe how each graph can be obtained from the graph of $f(x) = x^3 - 4x^2$.

a) $g(x) = (-x)^3 - 4(-x)^2$ **b)** $h(x) = 4x^2 - x^3$

Solution

a) We first note that

$$f(-x) = (-x)^3 - 4(-x)^2 = g(x).$$

Thus the graph of g is a reflection of the graph of f across the y-axis (see Fig. 1).

b) We first note that

$$\begin{aligned} -f(x) &= -(x^3 - 4x^2) \\ &= -x^3 + 4x^2 \\ &= 4x^2 - x^3 \\ &= h(x). \end{aligned}$$

Thus the graph of h is a reflection of the graph of f across the x-axis (see Fig. 2).

$y_1 = x^3 - 4x^2, \quad y_2 = (-x)^3 - 4(-x)^2$

FIGURE 1

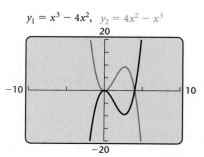

$y_1 = x^3 - 4x^2, \quad y_2 = 4x^2 - x^3$

FIGURE 2

Vertical and Horizontal Stretchings and Shrinkings

Suppose that we have a function given by $y_1 = f(x)$. Let's explore the graphs of the new functions $y_2 = af(x)$ and $y_3 = f(cx)$.

Interactive Discovery

Graph the function $y_1 = x^3 - x$ using the viewing window $[-3, 3, -1, 1]$. Then graph $y_2 = \frac{1}{10}(x^3 - x)$ using the same viewing window. Turn off y_2 and graph $y_3 = 2(x^3 - x)$. Then turn off y_3 and graph $y_4 = -2(x^3 - x)$. What pattern do you see? Test it with some other graphs.

Now graph y_1 and $y_2 = (2x)^3 - (2x)$ using the same viewing window. Turn off y_2 and graph $y_3 = \left(\frac{1}{2}x\right)^3 - \left(\frac{1}{2}x\right)$. Then turn off y_3 and graph $y_4 = \left(-\frac{1}{2}x\right)^3 - \left(-\frac{1}{2}x\right)$. What pattern do you see? Test it with some other graphs.

Consider any function f given by $y = f(x)$. Multiplying on the right by any constant a, where $|a| > 1$, to obtain $g(x) = af(x)$ will *stretch* the graph vertically away from the x-axis. If $0 < |a| < 1$, then the graph will be flattened or *shrunk* vertically toward the x-axis. If $a < 0$, the graph is also reflected across the x-axis.

> *Vertical Stretching and Shrinking*
>
> The graph of $y = af(x)$ can be obtained from the graph of $y = f(x)$ by
>
> > stretching vertically for $|a| > 1$, or
> > shrinking vertically for $0 < |a| < 1$.
>
> For $a < 0$, the graph is also reflected across the x-axis.

The constant c in the equation $g(x) = f(cx)$ will *stretch* the graph of $y = f(x)$ horizontally away from the y-axis if $0 < |c| < 1$. If $|c| > 1$, the graph will be *shrunk* horizontally toward the y-axis. If $c < 0$, the graph is also reflected across the y-axis.

> *Horizontal Stretching and Shrinking*
>
> The graph of $y = f(cx)$ can be obtained from the graph of $y = f(x)$ by
>
> > shrinking horizontally for $|c| > 1$, or
> > stretching horizontally for $0 < |c| < 1$.
>
> For $c < 0$, the graph is also reflected across the y-axis.

It is instructive to use these concepts now to create hand-drawn graphs from a given graph.

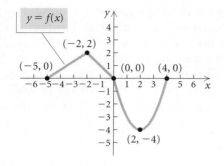

EXAMPLE 6   Shown at left is a graph of $y = f(x)$ for some function f. No formula for f is given. Make a hand-drawn graph of each of the following.

a) $g(x) = f(2x)$ **b)** $h(x) = f\left(\frac{1}{2}x\right)$ **c)** $t(x) = f\left(-\frac{1}{2}x\right)$

Solution

a) Since $|2| > 1$, the graph of $g(x) = f(2x)$ is a horizontal shrinking of the graph of $y = f(x)$. We can consider the key points $(-5, 0)$, $(-2, 2)$, $(0, 0)$, $(2, -4)$, and $(4, 0)$ on the graph of $y = f(x)$. The transformation divides each x-coordinate by 2 to obtain the key points $(-2.5, 0)$, $(-1, 2)$, $(0, 0)$, $(1, -4)$, and $(2, 0)$ of the graph of $g(x) = f(2x)$. The graph is shown below.

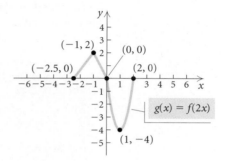

b) Since $\left|\frac{1}{2}\right| < 1$, the graph of $h(x) = f\left(\frac{1}{2}x\right)$ is a horizontal stretching of the graph of $y = f(x)$. We again consider the key points $(-5, 0)$, $(-2, 2)$, $(0, 0)$, $(2, -4)$, and $(4, 0)$ on the graph of $y = f(x)$. The transformation divides each x-coordinate by $\frac{1}{2}$ (which is the same as multiplying by 2) to obtain the key points $(-10, 0)$, $(-4, 2)$, $(0, 0)$, $(4, -4)$, and $(8, 0)$ of the graph of $h(x) = f\left(\frac{1}{2}x\right)$. The graph is shown below.

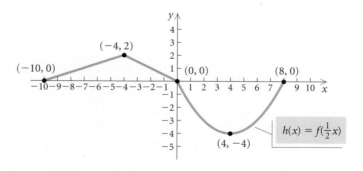

c) The graph of $t(x) = f\left(-\frac{1}{2}x\right)$ can be obtained by reflecting the graph in part (b) across the y-axis. It is shown at the top of the following page.

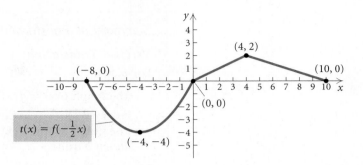

EXAMPLE 7 Use the graph of $y = f(x)$ given in Example 6. Make a hand-drawn graph of

$$y = -2f(x - 3) + 1.$$

Solution

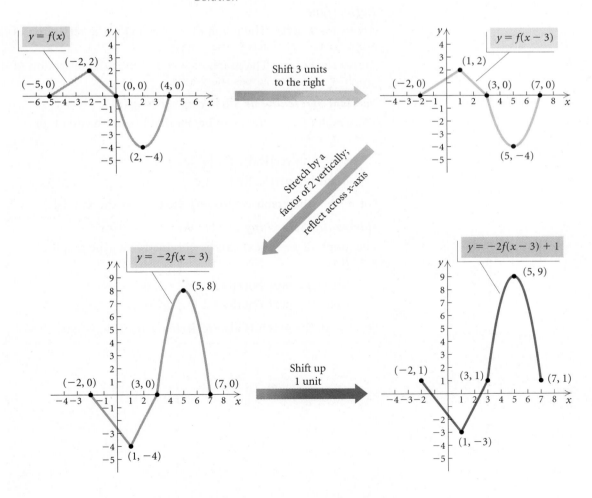

Summary of Transformations of y = f(x)

Vertical Translation: y = f(x) ± b

For $b > 0$,

> the graph of $y = f(x) + b$ is the graph of $y = f(x)$ shifted *up b* units;
>
> the graph of $y = f(x) - b$ is the graph of $y = f(x)$ shifted *down b* units.

Horizontal Translation: y = f(x ∓ d)

For $d > 0$,

> the graph of $y = f(x - d)$ is the graph of $y = f(x)$ shifted *right d* units;
>
> the graph of $y = f(x + d)$ is the graph of $y = f(x)$ shifted *left d* units.

Reflections

Across the x-axis: The graph of $y = -f(x)$ is the reflection of the graph of $y = f(x)$ across the x-axis.

Across the y-axis: The graph of $y = f(-x)$ is the reflection of the graph of $y = f(x)$ across the y-axis.

Vertical Stretching or Shrinking: y = af(x)

The graph of $y = af(x)$ can be obtained from the graph of $y = f(x)$ by

> stretching vertically for $|a| > 1$, or
>
> shrinking vertically for $0 < |a| < 1$.

For $a < 0$, the graph is also reflected across the x-axis.

Horizontal Stretching or Shrinking: y = f(cx)

The graph of $y = f(cx)$ can be obtained from the graph of $y = f(x)$ by

> shrinking horizontally for $|c| > 1$, or
>
> stretching horizontally for $0 < |c| < 1$.

For $c < 0$, the graph is also reflected across the y-axis.

Exercise Set **1.4**

Determine visually whether the graph is symmetric with
respect to the x-axis, the y-axis, and the origin.

1.

2.

3.

4.

5.

6.

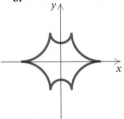

First, graph the equation and determine visually whether
it is symmetric with respect to the x-axis, the y-axis,
and the origin. Then verify your assertion algebraically.

7. $y = |x| - 2$

8. $y = |x + 5|$

9. $5y = 4x + 5$

10. $2x - 5 = 3y$

11. $5y = 2x^2 - 3$

12. $x^2 + 4 = 3y$

13. $y = \dfrac{1}{x}$

14. $y = -\dfrac{4}{x}$

Test algebraically whether the graph is symmetric with
respect to the x-axis, the y-axis, and the origin. Then
check your work graphically, if possible, using a grapher.

15. $5x - 5y = 0$

16. $6x + 7y = 0$

17. $3x^2 - 2y^2 = 3$

18. $5y = 7x^2 - 2x$

19. $y = |2x|$

20. $y^3 = 2x^2$

21. $2x^4 + 3 = y^2$

22. $2y^2 = 5x^2 + 12$

23. $3y^3 = 4x^3 + 2$

24. $3x = |y|$

25. $xy = 12$

26. $xy - x^2 = 3$

Determine visually whether the function is even, odd, or
neither even nor odd.

27.

28.

29.

30.

31.

32.

Test algebraically whether the function is even, odd, or
neither even nor odd. Then check your work graphically,
where possible, using a grapher.

33. $f(x) = -3x^3 + 2x$

34. $f(x) = 7x^3 + 4x - 2$

35. $f(x) = 5x^2 + 2x^4 - 1$

36. $f(x) = x + \dfrac{1}{x}$

37. $f(x) = x^{17}$

38. $f(x) = \sqrt[3]{x}$

39. $f(x) = \dfrac{1}{x^2}$

40. $f(x) = x - |x|$

41. $f(x) = 8$

42. $f(x) = \sqrt{x^2 + 1}$

Graph each of the following using a grapher. Before
doing so, describe how the graph can be obtained from
one of the basic graphs at the beginning of this section.

43. $f(x) = x^2 + 1$

44. $g(x) = |3x|$

45. $g(x) = (x + 5)^3$

46. $f(x) = \frac{1}{2}\sqrt[3]{x}$

47. $f(x) = -x^2$

48. $f(x) = |x - 3| - 4$

49. $f(x) = 3\sqrt{x} - 5$

50. $f(x) = 5 - \dfrac{1}{x}$

51. $g(x) = \left|\frac{1}{3}x\right| - 4$

52. $f(x) = \frac{2}{3}x^3 - 4$

53. $f(x) = (x + 5)^2 - 4$

54. $f(x) = (-x)^3 - 5$

55. $f(x) = -\frac{1}{4}(x - 5)^2$

56. $g(x) = \sqrt{-x} + 5$

57. $f(x) = \dfrac{1}{x + 3} + 2$

58. $f(x) = 3(x + 4)^2 - 3$

Write an equation for a function that has a graph with the given characteristics. Check your answer using a grapher.

59. The shape of $y = x^2$, but upside-down and shifted right 8 units

60. The shape of $y = \sqrt{x}$, but shifted left 6 units and down 5 units

61. The shape of $y = |x|$, but shifted left 7 units and up 2 units

62. The shape of $y = x^3$, but upside-down and shifted right 5 units

63. The shape of $y = 1/x$, but shrunk vertically by a factor of $\frac{1}{2}$ and shifted down 3 units

64. The shape of $y = x^2$, but shifted right 6 units and up 2 units

65. The shape of $y = x^2$, but upside-down and shifted right 3 units and up 4 units

66. The shape of $y = |x|$, but stretched horizontally by a factor of 2 and shifted down 5 units

67. The shape of $y = \sqrt{x}$, but reflected across the y-axis and shifted left 2 units and down 1 unit

68. The shape of $y = 1/x$, but reflected across the x-axis and shifted up 1 unit

69. The shape of $y = x^3$, but shifted left 4 units and shrunk vertically by a factor of 0.83

A graph of $y = f(x)$ follows. No formula for f is given. In Exercises 70–75, make a hand-drawn graph of the equation.

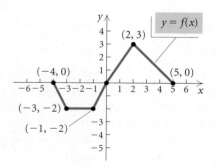

70. $g(x) = \frac{1}{2}f(x)$

71. $g(x) = -2f(x)$

72. $g(x) = f(2x)$

73. $g(x) = f\left(-\frac{1}{2}x\right)$

74. $g(x) = -3f(x + 1) - 4$

75. $g(x) = -\frac{1}{2}f(x - 1) + 3$

The graph of the function f is shown in figure (a). In Exercises 76–83, match the function g with one of the graphs (a)–(h), which follow. Some graphs may be used more than once.

a)

b)

c)

d)

e)

f)

g)

h)

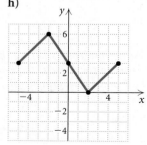

76. $g(x) = f(x) + 3$

77. $g(x) = f(-x) + 3$

78. $g(x) = -f(-x)$

79. $g(x) = -f(x) + 3$

80. $g(x) = \frac{1}{3}f(x) - 3$

81. $g(x) = \frac{1}{3}f(x - 2)$

82. $g(x) = -f(x + 2)$

83. $g(x) = \frac{1}{3}f(x + 2)$

For each pair of functions, determine algebraically if $g(x) = f(-x)$. Then, using the TABLE *feature on a grapher, check your answers.*

84. $f(x) = \frac{1}{4}x^4 + \frac{1}{5}x^3 - 81x^2 - 17$,
$g(x) = \frac{1}{4}x^4 + \frac{1}{5}x^3 + 81x^2 - 17$

85. $f(x) = 2x^4 - 35x^3 + 3x - 5$,
$g(x) = 2x^4 + 35x^3 - 3x - 5$

A graph of the function $f(x) = x^3 - 3x^2$ is shown below. Exercises 86–89 show graphs of functions transformed from this one. Find a formula for each function.

86.

87.

88.

89.

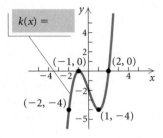

Discussion and Writing

90. Describe conditions under which you would know whether a polynomial function $f(x) = a_n x^n + a_{n-1}x^{n-1} + \cdots + a_2 x^2 + a_1 x + a_0$ is even or odd without using an algebraic or graphical procedure. Explain.

91. Consider the constant function $f(x) = 0$. Determine whether the graph of this function is symmetric with respect to the x-axis, the y-axis, and the origin. Determine whether this function is even or odd. In general, can a function be symmetric with respect to the x-axis? Explain.

92. Without drawing the graph, describe what the graph of $f(x) = |x^2 - 9|$ looks like.

93. Explain in your own words why the graph of $y = f(-x)$ is a reflection of the graph of $y = f(x)$ across the y-axis.

Skill Maintenance

94. Given $f(x) = 4x^3 - 5x$, find each of the following.
a) $f(2)$ **b)** $f(-2)$ **c)** $f(a)$ **d)** $f(-a)$

95. Given $f(x) = 5x^2 - 7$, find each of the following.
a) $f(-3)$ **b)** $f(3)$ **c)** $f(a)$ **d)** $f(-a)$

96. Find the slope and the y-intercept of the line with equation $2x - 9y + 1 = 0$.

97. Write an equation of the line perpendicular to the graph of the line $8x - y = 10$ and containing the point $(-1, 1)$.

Synthesis

Determine whether the function is even, odd, or neither even nor odd. Use any method.

98. $f(x) = \dfrac{x^2 + 1}{x^3 - 1}$

99. $f(x) = x\sqrt{10 - x^2}$

Determine whether the graph is symmetric with respect to the x-axis, the y-axis, and the origin. Use any method.

100.

$x^3 = y^2(2 - x)$

101.

$x^4 = y^2 + x^6$

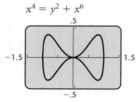

102.

$(x^2 + y^2)^2 = 2xy$

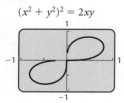

103.

$y^2 + 4xy^2 - y^4 = x^4 - 4x^3 + 3x^2 + 2x^2y^2$

104. The graph of $f(x) = |x|$ passes through the points $(-3, 3)$, $(0, 0)$, and $(3, 3)$. Transform this function to one whose graph passes through the points $(5, 1)$, $(8, 4)$, and $(11, 1)$. Check your answer using a grapher.

Graph each of the following using a grapher. Before doing so, describe how the graph can be obtained from a more basic graph. Give the domain and the range of the function.

105. $f(x) = \left[\!\left[x - \tfrac{1}{2} \right]\!\right]$

106. $f(x) = |\sqrt{x} - 1|$

107. If $(3, 4)$ is a point on the graph of $y = f(x)$, what point do you know is on the graph of $y = 2f(x)$? of $y = 2 + f(x)$? of $y = f(2x)$?

108. Find the zeros of $f(x) = 3x^5 - 20x^3$. Then without using a grapher, state the zeros of $f(x - 3)$ and $f(x + 8)$.

109. If $(-1, 5)$ is a point on the graph of $y = f(x)$, find b such that $(2, b)$ is on the graph of $y = f(x - 3)$.

State whether each of the following is true or false.

110. The product of two odd functions is odd.

111. The sum of two even functions is even.

112. The product of an even function and an odd function is odd.

113. Show that if f is *any* function, then the function E defined by

$$E(x) = \frac{f(x) + f(-x)}{2}$$

is even.

114. Show that if f is *any* function, then the function O defined by

$$O(x) = \frac{f(x) - f(-x)}{2}$$

is odd.

115. Consider the functions E and O of Exercises 113 and 114.

a) Show that $f(x) = E(x) + O(x)$. This means that every function can be expressed as the sum of an even and an odd function.

b) Let $f(x) = 4x^3 - 11x^2 + \sqrt{x} - 10$. Express f as a sum of an even function and an odd function.

Quadratic Functions and Applications

1.5

- *Find the vertex, the line of symmetry, and the maximum or minimum value of a quadratic function using the method of completing the square.*
- *Graph quadratic functions.*
- *Solve applied problems involving maximum and minimum function values.*

We define quadratic functions as follows.

Quadratic Functions

A **quadratic function** f is a second-degree polynomial function

$$f(x) = ax^2 + bx + c, \quad a \neq 0,$$

where a, b, and c are real numbers.

We have already graphed some quadratic functions in this chapter. Now we develop methods for analyzing the graph of any quadratic function. These techniques will allow us to find the exact value of a maximum or a minimum.

Graphing Quadratic Functions of the Type $f(x) = a(x - h)^2 + k$

The graph of a quadratic function is called a **parabola**. We will see that the graph of every parabola evolves from the graph of the squaring function $f(x) = x^2$ using transformations.

Interactive Discovery

Think of transformations and look for patterns. Consider the following functions:

$$y_1 = x^2, \quad y_2 = -0.4x^2,$$
$$y_3 = -0.4(x - 2)^2, \quad y_4 = -0.4(x - 2)^2 + 3.$$

Graph y_1 and y_2. How do you get from the graph of y_1 to y_2?
Graph y_2 and y_3. How do you get from the graph of y_2 to y_3?
Graph y_3 and y_4. How do you get from the graph of y_3 to y_4?
  Consider the following functions:

$$y_1 = x^2, \quad y_2 = 2x^2,$$
$$y_3 = 2(x + 3)^2, \quad y_4 = 2(x + 3)^2 - 5.$$

Graph y_1 and y_2. How do you get from the graph of y_1 to y_2?
Graph y_2 and y_3. How do you get from the graph of y_2 to y_3?
Graph y_3 and y_4. How do you get from the graph of y_3 to y_4?

TRANSFORMATIONS

REVIEW SECTION 1.4.

We get the graph of $f(x) = a(x - h)^2 + k$ from the graph of $f(x) = x^2$ as follows:

$$f(x) = x^2$$

$$\downarrow$$

$$f(x) = ax^2 \qquad \text{Vertical stretching or shrinking with a reflection across the } x\text{-axis if } a < 0$$

$$\downarrow$$

$$f(x) = a(x - h)^2 \qquad \text{Horizontal translation}$$

$$\downarrow$$

$$f(x) = a(x - h)^2 + k. \qquad \text{Vertical translation}$$

Consider the following graphs of the form $f(x) = a(x - h)^2 + k$. The point (h, k) at which the graph turns is called the **vertex**. The maximum or minimum value of $f(x)$ occurs at the vertex. Each graph has a line $x = h$ that is called the **line of symmetry.**

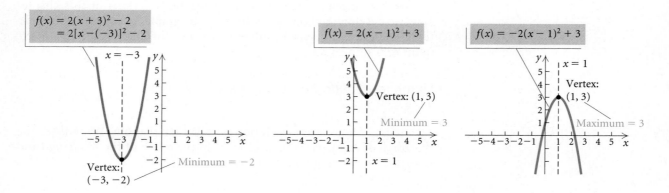

CONNECTING THE CONCEPTS

GRAPHING QUADRATIC FUNCTIONS

The graph of $f(x) = a(x - h)^2 + k$ is a parabola that:

   opens up if $a > 0$ and down if $a < 0$;

   has (h, k) as the vertex;

   has $x = h$ as the line of symmetry;

   has k as a minimum value (output) if $a > 0$ and has k as a maximum value if $a < 0$.

If the equation is in the form $f(x) = a(x - h)^2 + k$, we can learn a great deal about the graph without graphing.

FUNCTION	$f(x) = 3\left(x - \frac{1}{4}\right)^2 - 2$ $= 3\left(x - \frac{1}{4}\right)^2 + (-2)$	$g(x) = -3(x + 5)^2 + 7$ $= -3[x - (-5)]^2 + 7$
VERTEX	$\left(\frac{1}{4}, -2\right)$	$(-5, 7)$
LINE OF SYMMETRY	$x = \frac{1}{4}$	$x = -5$
MAXIMUM	No: $3 > 0$, graph opens up.	Yes, 7 $(-3 < 0$, graph opens down.)
MINIMUM	Yes, -2 $(3 > 0$, graph opens up.)	No: $-3 < 0$ graph opens down.

Note that the vertex (h, k) is used to find the maximum or the minimum value of the function. The maximum or minimum value is the number k, *not* the ordered pair (h, k).

Graphing Quadratic Functions of the Type $f(x) = ax^2 + bx + c, \ a \neq 0$

We now use a modification of the method of completing the square as an aid in graphing and analyzing quadratic functions of the form $f(x) = ax^2 + bx + c, \ a \neq 0$.

COMPLETING THE SQUARE

SEE APPENDIX A.

EXAMPLE 1 Complete the square to find the vertex, the line of symmetry, and the maximum or minimum value of $f(x) = x^2 + 10x + 23$. Then graph the function.

Solution To express $f(x) = x^2 + 10x + 23$ in the form $f(x) = a(x - h)^2 + k$, we construct a trinomial square. To do so, we take half the coefficient of x and square it, obtaining $(10/2)^2$, or 25. We now add and subtract that number on the right-hand side. We can think of this as adding $25 - 25$, which is 0.

$f(x) = x^2 + 10x + 23$ Note that 25 completes the square for $x^2 + 10x$.

$= x^2 + 10x + 25 - 25 + 23$ Adding $25 - 25$, or 0, to the right side

$= (x^2 + 10x + 25) - 25 + 23$

$= (x + 5)^2 - 2$ Factoring and simplifying

$= [x - (-5)]^2 + (-2)$ Writing in the form $f(x) = a(x - h)^2 + k$

From this form of the function, we know the following:

Vertex: $(-5, -2)$;

Line of symmetry: $x = -5$;

Minimum value of the function $= -2$.

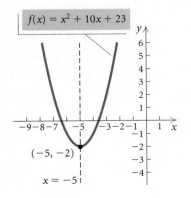

$f(x) = x^2 + 10x + 23$

$(-5, -2)$

$x = -5$

The graph of f, shown at left, is a shift of the graph of $y = x^2$ left 5 units and down 2 units.

Keep in mind that the line of symmetry is not part of the graph; it is a characteristic of the graph. If you fold the graph on its line of symmetry, the two halves of the graph will coincide.

EXAMPLE 2 Complete the square to find the vertex, the line of symmetry, and the maximum or minimum value of $g(x) = x^2/2 - 4x + 8$. Then graph the function.

Solution   To complete the square, we factor $\frac{1}{2}$ out of the first two terms. This makes the coefficient of x^2 within the parentheses 1:

$$g(x) = \frac{x^2}{2} - 4x + 8 = \frac{1}{2}(x^2 - 8x) + 8. \qquad \text{Factoring } \tfrac{1}{2} \text{ out of the first two terms}$$

Now we complete the square inside the parentheses: Half of -8 is -4, and $(-4)^2 = 16$. We add and subtract 16 inside the parentheses:

$$g(x) = \tfrac{1}{2}(x^2 - 8x + 16 - 16) + 8$$
$$= \tfrac{1}{2}(x^2 - 8x + 16) - \tfrac{1}{2} \cdot 16 + 8 \qquad \begin{array}{l}\text{Using the distributive law to} \\ \text{remove } -16 \text{ from within the} \\ \text{parentheses}\end{array}$$
$$= \tfrac{1}{2}(x - 4)^2 + 0, \text{ or } \tfrac{1}{2}(x - 4)^2. \qquad \text{Factoring and simplifying}$$

We know the following:

Vertex: $(4, 0)$;

Line of symmetry: $x = 4$;

Minimum value of the function $= 0$.

The graph of g is a vertical shrinking of the graph of $y = x^2$ followed by a shift 4 units to the right.   —

EXAMPLE 3 Complete the square to find the vertex, the line of symmetry, and the maximum or minimum value of $f(x) = -2x^2 + 10x - \frac{23}{2}$. Then graph the function.

Solution   We have

$$f(x) = -2x^2 + 10x - \tfrac{23}{2}$$
$$= -2(x^2 - 5x) - \tfrac{23}{2} \qquad \begin{array}{l}\text{Factoring } -2 \text{ out of the} \\ \text{first two terms}\end{array}$$
$$= -2\left(x^2 - 5x + \tfrac{25}{4} - \tfrac{25}{4}\right) - \tfrac{23}{2} \qquad \begin{array}{l}\text{Completing the square} \\ \text{inside the parentheses}\end{array}$$
$$= -2\left(x^2 - 5x + \tfrac{25}{4}\right) - 2\left(-\tfrac{25}{4}\right) - \tfrac{23}{2} \qquad \begin{array}{l}\text{Removing } -\tfrac{25}{4} \text{ from within} \\ \text{the parentheses}\end{array}$$
$$= -2\left(x - \tfrac{5}{2}\right)^2 + \tfrac{25}{2} - \tfrac{23}{2}$$
$$= -2\left(x - \tfrac{5}{2}\right)^2 + 1.$$

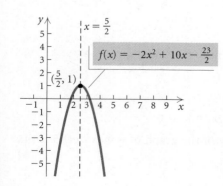

This form of the function yields the following:

Vertex: $\left(\tfrac{5}{2}, 1\right)$;

Line of symmetry: $x = \tfrac{5}{2}$;

Maximum value of the function $= 1$.

The graph is found by vertically stretching the graph of $f(x) = x^2$, reflecting the graph across the x-axis, shifting the graph to the right $\frac{5}{2}$ units, and shifting that curve up 1 unit. —

In many situations, we want to find the coordinates of the vertex directly from the equation $f(x) = ax^2 + bx + c$ using a formula. One way to develop such a formula is to consider the x-coordinate of the vertex as centered between the x-intercepts, or zeros, of the function. The zeros of a quadratic function $f(x) = ax^2 + bx + c$ are the *solutions* of the quadratic equation $ax^2 + bx + c = 0$.

> *Quadratic Equations*
>
> A **quadratic equation** is an equation equivalent to
>
> $$ax^2 + bx + c = 0, \quad a \neq 0,$$
>
> where the coefficients a, b, and c are real numbers.

SOLVING QUADRATIC EQUATIONS

SEE APPENDIX A.

The quadratic formula can be used to solve any quadratic equation.

> *The Quadratic Formula*
>
> The solutions of $ax^2 + bx + c = 0$, $a \neq 0$, are given by
>
> $$x = \frac{-b \pm \sqrt{b^2 - 4ac}}{2a}.$$

By averaging the two solutions of $ax^2 + bx + c = 0$, we find a formula for the x-coordinate of the vertex:

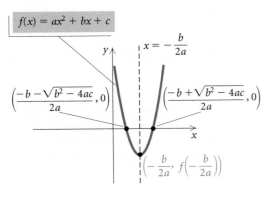

$$
\begin{aligned}
\text{x-coordinate of vertex} &= \frac{\dfrac{-b - \sqrt{b^2 - 4ac}}{2a} + \dfrac{-b + \sqrt{b^2 - 4ac}}{2a}}{2} \\[2ex]
&= \frac{\dfrac{-2b}{2a}}{2} = \frac{-\dfrac{b}{a}}{2} = -\frac{b}{2a}.
\end{aligned}
$$

We use this value of x to find the y-coordinate of the vertex,

$$f\left(-\frac{b}{2a}\right).$$

The Vertex of a Parabola

The **vertex** of the graph of $f(x) = ax^2 + bx + c$ is

$$\left(-\frac{b}{2a}, f\left(-\frac{b}{2a}\right)\right).$$

We calculate the   We substitute to
x-coordinate.   find the y-coordinate.

EXAMPLE 4   For the function $f(x) = -x^2 + 14x - 47$:

a) Find the vertex.

b) Determine whether there is a maximum or minimum value and find
that value.

c) Find the range.

d) On what intervals is the function increasing? decreasing?

Solution   There is no need to graph the function.

a) The x-coordinate of the vertex is

$$-\frac{b}{2a} = -\frac{14}{2(-1)}, \text{ or } 7.$$

Since

$$f(7) = -7^2 + 14 \cdot 7 - 47 = 2,$$

the vertex is $(7, 2)$.

b) Since a is negative ($a = -1$), the graph opens down so the second coordinate of the vertex, 2, is the maximum value of the function.

c) The range is $(-\infty, 2]$.

d) Since the graph opens down, function values increase to the left of the
vertex and decrease to the right of the vertex. Thus the function is increasing on the interval $(-\infty, 7)$ and decreasing on $(7, \infty)$.

You should check these results visually using a grapher.   —

Applications

Many real-world situations involve finding the maximum or minimum
value of a quadratic function.

EXAMPLE 5   *Maximizing Area.* A stone mason has enough stones to
enclose a rectangular patio with 60 ft of stone wall. If the house forms
one side of the rectangle, what is the maximum area that the mason can
enclose? What should the dimensions of the patio be in order to yield
this area?

Solution In this case, it is helpful to make use of a five-step problem-solving strategy.

1. **Familiarize.** We make a drawing of the situation, using w to represent the width of the patio, in feet. Then $(60 - 2w)$ feet of stone is available for the length. Suppose the patio were 10 ft wide. It would then be $60 - 2 \cdot 10 = 40$ ft long. The area would be $(10 \text{ ft})(40 \text{ ft}) = 400 \text{ ft}^2$. If the patio were 12 ft wide, it would be $60 - 2 \cdot 12 = 36$ ft long. The area would be $(12 \text{ ft})(36 \text{ ft}) = 432 \text{ ft}^2$. If it were 16 ft wide, it would be $60 - 2 \cdot 16 = 28$ ft long and the area would be $(16 \text{ ft})(28 \text{ ft}) = 448 \text{ ft}^2$. There are more combinations of length and width than we could possibly try. Instead we will find a function that represents the area and then determine the maximum value of the function.

2. **Translate.** Since the area of a rectangle is given by length times width, we have

$$A(w) = (60 - 2w)w \qquad A = lw; \; l = 60 - 2w$$
$$= -2w^2 + 60w,$$

where $A(w)$ is the area of the patio, in square feet, as a function of the width, w.

3. **Carry out.** To solve this problem, we need to determine the maximum value of $A(w)$ and find the dimensions for which that maximum occurs. Since A is a quadratic function and w^2 has a negative coefficient, we know that the function has a maximum value that occurs at the vertex of the graph of the function. The first coordinate of the vertex, $(w, A(w))$, is

$$w = -\frac{b}{2a} = -\frac{60}{2(-2)} = 15 \text{ ft.}$$

Thus, if $w = 15$ ft, then the length $l = 60 - 2 \cdot 15 = 30$ ft; and the area is $15 \cdot 30$, or 450 ft^2.

4. **Check.** As a partial check, we note that $450 \text{ ft}^2 > 448 \text{ ft}^2$, which is the largest area we found in a guess in the *Familiarize* step. A more complete check, assuming that the function $A(w)$ is correct, would examine a table of values for $A(w) = (60 - 2w)w$ and/or examine its graph.

X	Y1
14.7	449.82
14.8	449.92
14.9	449.98
15	450
15.1	449.98
15.2	449.92
15.3	449.82
X = 15	

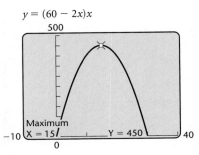

5. **State.** The maximum possible area is 450 ft^2 when the patio is 15 ft wide and 30 ft long.

Exercise Set 1.5

In Exercises 1 and 2, use the given graph to find each of the following:

a) *the vertex;*
b) *the line of symmetry; and*
c) *the maximum or minimum value of the function.*

1.

2.

In Exercises 3–10, complete the square to:

a) *find the vertex;*
b) *find the line of symmetry; and*
c) *determine whether there is a maximum or minimum value and find that value.*

Then check your answers using a grapher.

3. $f(x) = x^2 - 8x + 12$

4. $g(x) = x^2 + 7x - 8$

5. $f(x) = x^2 - 7x + 12$

6. $g(x) = x^2 - 5x + 6$

7. $g(x) = 2x^2 + 6x + 8$

8. $f(x) = 2x^2 - 10x + 14$

9. $g(x) = -2x^2 + 2x + 1$

10. $f(x) = -3x^2 - 3x + 1$

In Exercises 11–18, match the equation with one of the figures (a)–(h), which follow. Use a grapher only as a check.

a)

b)

c)

d)

e)

f)

g)

h)

11. $y = (x + 3)^2$

12. $y = -(x - 4)^2 + 3$

13. $y = 2(x - 4)^2 - 1$

14. $y = x^2 - 3$

15. $y = -\frac{1}{2}(x + 3)^2 + 4$

16. $y = (x - 3)^2$

17. $y = -(x + 3)^2 + 4$

18. $y = 2(x - 1)^2 - 4$

In Exercises 19–26:

a) *Find the vertex.*

b) *Determine whether there is a maximum or minimum value and find that value.*

c) *Find the range.*

d) *Find the intervals on which the function is increasing and the intervals on which the function is decreasing.*

Then check your answers using a grapher.

19. $f(x) = x^2 - 6x + 5$ **20.** $f(x) = x^2 + 4x - 5$

21. $f(x) = 2x^2 + 4x - 16$ **22.** $f(x) = \frac{1}{2}x^2 - 3x + \frac{5}{2}$

23. $f(x) = -\frac{1}{2}x^2 + 5x - 8$

24. $f(x) = -3x^2 + 24x - 49$

25. *Maximizing Volume.* Plasco Manufacturing plans to produce a one-compartment vertical file by bending the long side of an 10-in. by 18-in. sheet of plastic along two lines to form a U-shape. How tall should the file be in order to maximize the volume that the file can hold?

26. *Maximizing Area.* A fourth-grade class decides to enclose a rectangular garden, using the side of the school as one side of the rectangle. What is the maximum area that the class can enclose with 32 ft of fence? What should the dimensions of the garden be in order to yield this area?

27. *Maximizing Area.* The sum of the base and the height of a triangle is 20 cm. Find the dimensions for which the area is a maximum.

28. *Maximizing Area.* The sum of the base and the height of a parallelogram is 69 cm. Find the dimensions for which the area is a maximum.

29. *Minimizing Cost.* Aki's Bicycle Designs has determined that when x hundred bicycles are built, the average cost per bicycle is given by

$$C(x) = 0.1x^2 - 0.7x + 2.425,$$

where $C(x)$ is in hundreds of dollars. How many bicycles should be built in order to minimize the average cost per bicycle?

Maximizing Profit. *In business, profit is the difference between revenue and cost, that is,*

$$Total\ profit = Total\ revenue - Total\ cost,$$
$$P(x) = R(x) - C(x),$$

where x is the number of units sold. Find the maximum profit and the number of units that must be sold in order to yield the maximum profit for each of the following.

30. $R(x) = 5x,\ C(x) = 0.001x^2 + 1.2x + 60$

31. $R(x) = 50x - 0.5x^2,\ C(x) = 10x + 3$

32. $R(x) = 20x - 0.1x^2,\ C(x) = 4x + 2$

33. *Maximizing Area.* A rancher needs to enclose two adjacent rectangular corrals, one for sheep and one for cattle. If a river forms one side of the corrals and 240 yd of fencing is available, what is the largest total area that can be enclosed?

34. *Norman Window.* A Norman window is a rectangle with a semicircle on top. Sky Blue Windows is designing a Norman window that will require 24 ft of trim. What dimensions will allow the maximum amount of light to enter a house?

A Norman window

Discussion and Writing

35. Write a problem for a classmate to solve. Design it so that it is a maximum or minimum problem using a quadratic function.

36. Suppose that the graph of $f(x) = ax^2 + bx + c$ has x-intercepts $(x_1, 0)$ and $(x_2, 0)$. What are the x-intercepts of $g(x) = -ax^2 - bx - c$? Explain.

Skill Maintenance

For each function f, construct and simplify the difference quotient

$$\frac{f(x + h) - f(x)}{h}.$$

37. $f(x) = 3x - 7$

38. $f(x) = 2x^2 - x + 4$

A graph of $y = f(x)$ follows. No formula is given for f. Make a hand-drawn graph of each of the following.

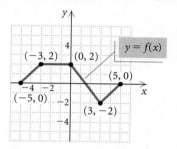

39. $g(x) = -2f(x)$

40. $g(x) = f(2x)$

Synthesis

41. Find c such that
$$f(x) = -0.2x^2 - 3x + c$$
has a maximum value of -225.

42. Find b such that
$$f(x) = -4x^2 + bx + 3$$
has a maximum value of 50.

43. Graph: $f(x) = (|x| - 5)^2 - 3$.

44. Find a quadratic function with vertex $(4, -5)$ and containing the point $(-3, 1)$.

45. *Minimizing Area.* A 24-in. piece of string is cut into two pieces. One piece is used to form a circle while the other is used to form a square. How should the string be cut so that the sum of the areas is a minimum?

Composite and Inverse Functions

1.6

- *Find the composition of two functions and the domain of the composition; decompose a function as a composition of two functions.*
- *Determine whether a function is one-to-one, and if it is, find a formula for its inverse.*
- *Simplify expressions of the type $(f \circ f^{-1})(x)$ and $(f^{-1} \circ f)(x)$.*

The Composition of Functions

In the real world, it is not uncommon for a function's output to depend on some input that is itself an output of another function. For instance, the amount a person pays as state income tax usually depends on the amount of adjusted gross income on the person's federal tax return, which, in turn, depends on his or her annual earnings. Such functions are called **composite functions.**

To illustrate how composite functions work, suppose a chemistry student needs a formula to convert Fahrenheit temperatures to Kelvin units. The formula $c(t) = \frac{5}{9}(t - 32)$ gives the Celsius temperature $c(t)$ that corresponds to the Fahrenheit temperature t. The formula $k(c) = c + 273$ gives the Kelvin temperature $k(c)$ that corresponds to the Celsius temperature c. Thus, 50° Fahrenheit corresponds to

$$c(50) = \frac{5}{9}(50 - 32)$$
$$= \frac{5}{9}(18) = 10° \text{ Celsius}$$

and 10° Celsius corresponds to

$$k(10) = 10 + 273$$
$$= 283° \text{ Kelvin.}$$

We see that 50° Fahrenheit is the same as 283° Kelvin. This two-step procedure can be used to convert any Fahrenheit temperature to Kelvin units.

$y_1 = \frac{5}{9}(x - 32), \quad y_2 = y_1 + 273$

X	Y₁	Y₂
50	10	283
59	15	288
68	20	293
77	25	298
86	30	303
95	35	308
104	40	313

X = 50

In the table shown at left, we use a grapher to convert Fahrenheit temperatures, x, to Celsius temperatures, y_1, using $y_1 = \frac{5}{9}(x - 32)$. We also convert Celsius temperatures to Kelvin units, y_2, using $y_2 = y_1 + 273$.

A student making numerous conversions might look for a formula that converts directly from Fahrenheit to Kelvin. Such a formula can be found by substitution:

$$y_2 = y_1 + 273$$
$$= \left[\frac{5}{9}(x - 32)\right] + 273 \qquad \text{Substituting}$$
$$= \frac{5}{9}x - \frac{160}{9} + 273 = \frac{5}{9}x - \frac{160}{9} + \frac{2457}{9}$$
$$= \frac{5x + 2297}{9}. \qquad \text{Simplifying}$$

We can show on a grapher that the same values that appear in the table for y_2 will appear when y_2 is entered as

$$y_2 = \frac{5x + 2297}{9}.$$

In the more commonly used function notation, we have

$$k(c(t)) = c(t) + 273$$

$$= \frac{5}{9}(t - 32) + 273 \qquad \text{Substituting}$$

$$= \frac{5t + 2297}{9}. \qquad \text{Simplifying as above}$$

Since the last equation expresses the Kelvin temperature as a new function, K, of the Fahrenheit temperature, t, we can write

$$K(t) = \frac{5t + 2297}{9},$$

where $K(t)$ is the Kelvin temperature corresponding to the Fahrenheit temperature, t. Here we have $K(t) = k(c(t))$. The new function K is called the **composition** of k and c and can be denoted $k \circ c$ (read "k composed with c," "the composition of k and c," or "k circle c").

Composition of Functions

The **composite function** $f \circ g$, the **composition** of f and g, is defined as

$$(f \circ g)(x) = f(g(x)),$$

where x is in the domain of g and $g(x)$ is in the domain of f.

EXAMPLE 1   Given that $f(x) = 2x - 5$ and $g(x) = x^2 - 3x + 8$, find each of the following.

a) $(f \circ g)(7)$ and $(g \circ f)(7)$
b) $(f \circ g)(x)$ and $(g \circ f)(x)$

Solution   Consider each function separately:

$f(x) = 2x - 5$ This function multiplies each input by 2 and subtracts 5.

and

$g(x) = x^2 - 3x + 8.$ This function squares an input, subtracts 3 times the input from the result, and then adds 8.

a) To find $(f \circ g)(7)$, we first find $g(7)$. Then we use $g(7)$ as an input for f:

$$(f \circ g)(7) = f(g(7)) = f(7^2 - 3 \cdot 7 + 8)$$
$$= f(36)$$
$$= 2 \cdot 36 - 5$$
$$= 67.$$

To find $(g \circ f)(7)$, we first find $f(7)$. Then we use $f(7)$ as an input for g:

$$(g \circ f)(7) = g(f(7)) = g(2 \cdot 7 - 5)$$
$$= g(9) = 9^2 - 3 \cdot 9 + 8$$
$$= 62.$$

We can check our work using a grapher by entering on the equation-editor screen

$$y_1 = f(x) = 2x - 5$$

and

$$y_2 = g(x) = x^2 - 3x + 8.$$

Then, on the home screen, we find $(f \circ g)(7)$ and $(g \circ f)(7)$ using the function notation $Y_1(Y_2(7))$ and $Y_2(Y_1(7))$, respectively.

b) To find $(f \circ g)(x)$, we substitute $g(x)$ for x in the equation for $f(x)$:

$$(f \circ g)(x) = f(g(x)) = f(x^2 - 3x + 8)$$

Substituting $x^2 - 3x + 8$ for $g(x)$

$$= 2(x^2 - 3x + 8) - 5$$
$$= 2x^2 - 6x + 16 - 5$$
$$= 2x^2 - 6x + 11.$$

To find $(g \circ f)(x)$, we substitute $f(x)$ for x in the equation for $g(x)$:

$$(g \circ f)(x) = g(f(x)) = g(2x - 5)$$

Substituting $2x - 5$ for $f(x)$

$$= (2x - 5)^2 - 3(2x - 5) + 8$$
$$= 4x^2 - 20x + 25 - 6x + 15 + 8$$
$$= 4x^2 - 26x + 48.$$

Note in Example 1 that, as a rule, $(f \circ g)(x) \neq (g \circ f)(x)$. We can check this graphically, as shown in the graphs at left.

EXAMPLE 2 Given that $f(x) = \sqrt{x}$ and $g(x) = x - 3$:

a) Find $h(x)$ and $k(x)$ if $h = f \circ g$ and $k = g \circ f$.

b) Find the domains of h and k.

Solution

a) $h(x) = (f \circ g)(x) = f(g(x)) = f(x - 3) = \sqrt{x - 3}$

$k(x) = (g \circ f)(x) = g(f(x)) = g(\sqrt{x}) = \sqrt{x} - 3$

b) The domain of f is $\{x \mid x \geq 0\}$, or the interval $[0, \infty)$. The domain of g is $(-\infty, \infty)$.

To find the domain of $h = f \circ g$, we must consider that the outputs for g will serve as inputs for f. Since the inputs for f cannot be negative, we must have

$$g(x) = x - 3, \quad \text{where } x - 3 \geq 0, \text{ or } x \geq 3.$$

Thus the domain of $h = \{x \mid x \geq 3\}$, or the interval $[3, \infty)$, as the graph in Fig. 1 confirms.

$y_1 = 2x - 5, \quad y_2 = x^2 - 3x + 8$

```
Y1(Y2(7))
                      67
Y2(Y1(7))
                      62
```

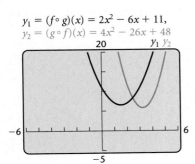

$y_1 = (f \circ g)(x) = 2x^2 - 6x + 11,$
$y_2 = (g \circ f)(x) = 4x^2 - 26x + 48$

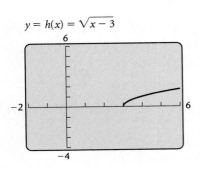

$y = h(x) = \sqrt{x - 3}$

FIGURE 1

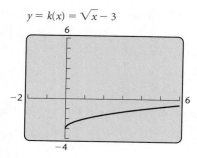

$y = k(x) = \sqrt{x} - 3$

FIGURE 2

To find the domain of $k = g \circ f$, we must consider that the outputs for f will serve as inputs for g. Since g can accept *any* real number as an input, any output from f is acceptable. Thus the domain of $k = \{x \mid x \geq 0\}$, as the graph in Fig. 2 confirms.

Decomposing a Function as a Composition

In calculus, one often needs to recognize how a function can be expressed as the composition of two functions. In this way, we are "decomposing" the function.

EXAMPLE 3 If $h(x) = (2x - 3)^5$, find $f(x)$ and $g(x)$ such that $h(x) = (f \circ g)(x)$.

Solution The function $h(x)$ raises $(2x - 3)$ to the 5th power. Two functions that can be used for the composition are

$$f(x) = x^5 \quad \text{and} \quad g(x) = 2x - 3.$$

We can check by forming the composition:

$$h(x) = (f \circ g)(x) = f(g(x)) = f(2x - 3) = (2x - 3)^5.$$

This is the most "obvious" answer to the question. There can be other less obvious answers. For example, if

$$f(x) = (x + 7)^5 \quad \text{and} \quad g(x) = 2x - 10,$$

then

$$h(x) = (f \circ g)(x) = f(g(x))$$
$$= f(2x - 10) = [(2x - 10) + 7]^5 = (2x - 3)^5.$$

Inverses

When we go from an output of a function back to its input or inputs, we get an inverse relation. When that relation is a function, we have an inverse function. We now study inverse functions and how to find their formulas when the original function has a formula.

Consider the relation h given as follows:

$$h = \{(-8, 5), (4, -2), (-7, 1), (3.8, 6.2)\}.$$

Suppose we *interchange* the first and second coordinates. The relation we obtain is called the **inverse** of the relation h and is given as follows:

$$\text{Inverse of } h = \{(5, -8), (-2, 4), (1, -7), (6.2, 3.8)\}.$$

Inverse Relation

Interchanging the first and second coordinates of each ordered pair in a relation produces the **inverse relation.**

EXAMPLE 4   Consider the relation g given by

$$g = \{(2, 4), (-1, 3), (-2, 0)\}.$$

Graph the relation in blue. Find the inverse and graph it in red.

Solution   The relation g is shown in blue in the figure below. The inverse of the relation is

$$\{(4, 2), (3, -1), (0, -2)\}$$

and is shown in red. The pairs in the inverse are reflections across the line $y = x$.

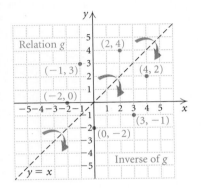

Inverse Relation

If a relation is defined by an equation, interchanging the variables produces an equation of the **inverse relation.**

EXAMPLE 5   Find an equation for the inverse of the relation:

$$y = x^2 - 5x.$$

Solution   We interchange x and y and obtain an equation of the inverse:

$$x = y^2 - 5y.$$

Interactive Discovery

Graph each of the following relations. Then find the inverse of each and graph it using the same set of axes. What pattern do you see? Use a square viewing window for a true geometric perspective.

$$y = 3x + 2,$$
$$y = x,$$
$$xy = 2,$$
$$x^2 + 3y^2 = 4,$$
$$y^2 = 4x - 5.$$

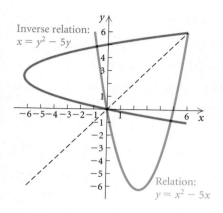

Inverse relation:
$x = y^2 - 5y$

Relation:
$y = x^2 - 5x$

If a relation is given by an equation, then the solutions of the inverse can be found from those of the original equation by interchanging the first and second coordinates of each ordered pair. Thus the graphs of a relation and its inverse are always reflections of each other across the line $y = x$. This is illustrated with the equations of Example 5 in the graph at left. We will explore inverses and their graphs later in this section.

Inverses and One-to-One Functions

Let's consider the following two functions.

YEAR (DOMAIN)	FIRST-CLASS POSTAGE COST, IN CENTS (RANGE)
1978	⟶ 15
1983	⟶ 20
1984	
1989	⟶ 25
1991	⟶ 29
1995	⟶ 32
1999	⟶ 33

NUMBER (DOMAIN)	CUBING (RANGE)
−3	⟶ −27
−2	⟶ −8
−1	⟶ −1
0	⟶ 0
1	⟶ 1
2	⟶ 8
3	⟶ 27

Source: U.S. Postal Service

Suppose we reverse the arrows. Are these inverse relations functions?

YEAR (RANGE)	FIRST-CLASS POSTAGE COST, IN CENTS (DOMAIN)
1978	⟵ 15
1983	⟵ 20
1984	⟵
1989	⟵ 25
1991	⟵ 29
1995	⟵ 32
1999	⟵ 33

NUMBER (RANGE)	CUBING (DOMAIN)
−3	⟵ −27
−2	⟵ −8
−1	⟵ −1
0	⟵ 0
1	⟵ 1
2	⟵ 8
3	⟵ 27

We see that the inverse of the postage function is not a function. Like all functions, each input in the postage function has exactly one output. However, the outputs for both 1983 and 1984 are the same, 20. Thus in the inverse of the postage function, the input 20 has *two* outputs, 1983 and 1984. When the same output of a function comes from two or more different inputs, the inverse cannot be a function. In the cubing function, each output corresponds to exactly one input, so its inverse is also a function. The cubing function is an example of a **one-to-one function.** If the inverse of a function f is also a function, it is named f^{-1} (read "f-inverse").

The -1 in f^{-1} is *not* an exponent!

One-to-One Functions and Inverses

A function f is **one-to-one** if different inputs have different outputs—that is,

if $a \neq b$, then $f(a) \neq f(b)$. Or,

A function f is **one-to-one** if when the outputs are the same, the inputs are the same—that is,

if $f(a) = f(b)$, then $a = b$.

If a function is one-to-one, then its inverse is a function.

The domain of a one-to-one function f is the range of the inverse f^{-1}.

The range of a one-to-one function f is the domain of the inverse f^{-1}.

EXAMPLE 6   Given the function f described by $f(x) = 2x - 3$, prove that f is one-to-one (that is, it has an inverse that is a function).

Solution   To show that f is one-to-one, we show that if $f(a) = f(b)$, then $a = b$. Assume that $f(a) = f(b)$ for any numbers a and b in the domain of f. Since $f(a) = 2a - 3$ and $f(b) = 2b - 3$, we have

$$2a - 3 = 2b - 3$$
$$2a = 2b \qquad \text{Adding 3}$$
$$a = b. \qquad \text{Dividing by 2}$$

Thus, if $f(a) = f(b)$, then $a = b$. This shows that f is one-to-one. ▬

EXAMPLE 7   Given the function g described by $g(x) = x^2$, prove that g is not one-to-one.

Solution   To prove that g is not one-to-one, we need to find two numbers a and b for which $a \neq b$ and $g(a) = g(b)$. Two such numbers are -3 and 3, because $-3 \neq 3$ and $g(-3) = g(3) = 9$. Thus g is not one-to-one. ▬

The graph at the top of the next page shows a function, in blue, and its inverse, in red. To determine whether the inverse is a function, we can apply the vertical-line test to its graph. By reflecting each such vertical line back across the line $y = x$, we obtain an equivalent **horizontal-line test** for the original function.

The vertical-line test
shows that the inverse
is not a function.

The horizontal-line test
shows that the function
is not one-to-one.

Horizontal-Line Test

If it is possible for a horizontal line to intersect the graph of a function more than once, then the function is *not* one-to-one and its inverse is *not* a function.

EXAMPLE 8 Graph each of the following functions. Then determine whether each is one-to-one and thus has an inverse that is a function.

a) $f(x) = 4 - x$

b) $f(x) = x^2$

c) $f(x) = \sqrt[3]{x + 2} + 3$

d) $f(x) = 3x^5 - 20x^3$

Solution   We graph each function using a grapher. Then we apply the horizontal-line test.

a)

$y = 4 - x$

b)

$y = x^2$

c)

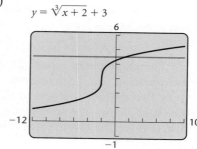

$y = \sqrt[3]{x + 2} + 3$

d)

$y = 3x^5 - 20x^3$

RESULT	REASON
a) One-to-one; inverse is a function	No horizontal line intersects the graph more than once.
b) Not one-to-one; inverse is not a function	There are many horizontal lines that intersect the graph more than once. Note that where the line $y = 4$ intersects the graph, the first coordinates are -2 and 2. Although these are different inputs, they have the same output, 4.
c) One-to-one; inverse is a function	No horizontal line intersects the graph more than once.
d) Not one-to-one; inverse is not a function	There are many horizontal lines that intersect the graph more than once.

Finding Formulas for Inverses

Suppose that a function is described by a formula. If it has an inverse that is a function, we proceed as follows to find a formula for f^{-1}.

Obtaining a Formula for an Inverse

If a function f is one-to-one, a formula for its inverse can generally be found as follows:

1. Replace $f(x)$ with y.

2. Interchange x and y.

3. Solve for y.

4. Replace y with $f^{-1}(x)$.

EXAMPLE 9 Determine whether the function $f(x) = 2x - 3$ is one-to-one, and if it is, find a formula for $f^{-1}(x)$.

Solution The graph of f is shown at left. It passes the horizontal-line test. Thus it is one-to-one and its inverse is a function.
    We also proved that f is one-to-one in Example 6.

$y = 2x - 3$

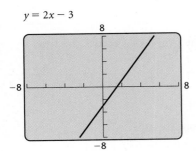

1. Replace $f(x)$ with y: $\qquad y = 2x - 3$

2. Interchange x and y: $\qquad x = 2y - 3$

3. Solve for y: $\qquad x + 3 = 2y$

$$\frac{x + 3}{2} = y$$

4. Replace y with $f^{-1}(x)$: $\quad f^{-1}(x) = \dfrac{x + 3}{2}.$

Consider

$$f(x) = 2x - 3 \quad \text{and} \quad f^{-1}(x) = \frac{x + 3}{2}$$

from Example 9. For the input 5, we have

$$f(5) = 2 \cdot 5 - 3 = 10 - 3 = 7.$$

The output is 7. Now we use 7 for the input in the inverse:

$$f^{-1}(7) = \frac{7 + 3}{2}$$

$$= \frac{10}{2} = 5.$$

The function f takes the number 5 to 7. The inverse function f^{-1} takes the number 7 back to 5.

EXAMPLE 10 Graph

$$f(x) = 2x - 3 \quad \text{and} \quad f^{-1}(x) = \frac{x + 3}{2}$$

using the same set of axes. Then compare the two graphs.

Solution   The graphs of f and f^{-1} are shown at left. They are graphed in a square window. Note that the graph of f^{-1} can be drawn by reflecting the graph of f across the line $y = x$. That is, if we were to graph $f(x) = 2x - 3$ in wet ink and fold along the line $y = x$, the graph of $f^{-1}(x) = (x + 3)/2$ would be formed by the ink transferred from f.

When we interchange x and y in finding a formula for the inverse of $f(x) = 2x - 3$, we are in effect reflecting the graph of that function across the line $y = x$. For example, when the coordinates of the y-intercept, $(0, -3)$, of the graph of f are reversed, we get the x-intercept, $(-3, 0)$, of the graph of f^{-1}.

On some graphers, we can graph the inverse of a function after graphing the function itself by accessing a drawing feature. Consult your user's manual or the *Graphing Calculator Manual* that accompanies this text for the procedure.

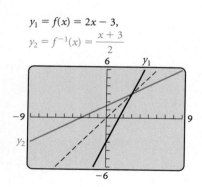

$y_1 = f(x) = 2x - 3,$
$y_2 = f^{-1}(x) = \dfrac{x + 3}{2}$

The graph of f^{-1} is a reflection of the graph of f across the line $y = x$.

EXAMPLE 11   Consider $g(x) = x^3 + 2.$

a) Determine whether the function is one-to-one.

b) If it is one-to-one, find a formula for its inverse.

c) Graph the function and its inverse.

Solution

a) The graph of $g(x) = x^3 + 2$ is shown below. It passes the horizontal-line test and thus has an inverse that is a function.

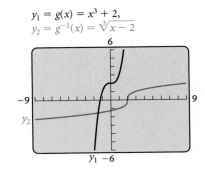

b) We follow the procedure for finding an inverse.

1. Replace $g(x)$ with y: $y = x^3 + 2$
2. Interchange x and y: $x = y^3 + 2$
3. Solve for y: $x - 2 = y^3$
   $$\sqrt[3]{x - 2} = y$$
4. Replace y with $g^{-1}(x)$: $g^{-1}(x) = \sqrt[3]{x - 2}$.

c) To find the graph, we reflect the graph of $g(x) = x^3 + 2$ across the line $y = x$. This can be done by plotting points or by using a grapher, as shown in part (a) above.

Inverse Functions and Composition

Suppose that we were to use some input a for a one-to-one function f and find its output, $f(a)$. The function f^{-1} would then take that output back to a. Similarly, if we began with an input b for the function f^{-1} and found its output, $f^{-1}(b)$, the original function f would then take that output back to b. This is summarized as follows.

If a function f is one-to-one, then f^{-1} is the unique function such that each of the following holds:

$$(f^{-1} \circ f)(x) = f^{-1}(f(x)) = x, \quad \text{for each } x \text{ in the domain of } f, \text{ and}$$

$$(f \circ f^{-1})(x) = f(f^{-1}(x)) = x, \quad \text{for each } x \text{ in the domain of } f^{-1}.$$

EXAMPLE 12   Given that $f(x) = 5x + 8$, use composition of functions to show that $f^{-1}(x) = (x - 8)/5$.

Solution   We find $(f^{-1} \circ f)(x)$ and $(f \circ f^{-1})(x)$ and check to see that each is x:

$$(f^{-1} \circ f)(x) = f^{-1}(f(x))$$

$$= f^{-1}(5x + 8) = \frac{(5x + 8) - 8}{5}$$

$$= \frac{5x}{5} = x;$$

$$(f \circ f^{-1})(x) = f(f^{-1}(x))$$

$$= f\left(\frac{x - 8}{5}\right) = 5\left(\frac{x - 8}{5}\right) + 8$$

$$= x - 8 + 8 = x.$$

Restricting a Domain

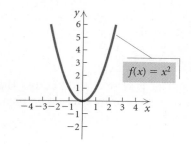

In the case in which the inverse of a function is not a function, the domain of the function can be restricted to allow the inverse to be a function. We saw in Example 8 that $f(x) = x^2$ is not one-to-one. The graph is shown at left.

Suppose that we had tried to find a formula for the inverse as follows:

$$y = x^2 \qquad \text{Replacing } f(x) \text{ with } y$$
$$x = y^2 \qquad \text{Interchanging } x \text{ and } y$$
$$\pm\sqrt{x} = y. \qquad \text{Solving for } y$$

This is not the equation of a function. An input of, say, 4 would yield two outputs, -2 and 2. In such cases, it is convenient to consider "part" of the function by restricting the domain of $f(x)$. For example, if we restrict the domain of $f(x) = x^2$ to nonnegative numbers, then its inverse is a function as shown with the graphs of $f(x) = x^2$, $x \geq 0$, and $f^{-1}(x) = \sqrt{x}$, $x \geq 0$ below.

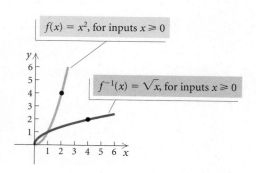

Exercise Set 1.6

Find $(f \circ g)(x)$ and $(g \circ f)(x)$.

1. $f(x) = x + 3$, $g(x) = x - 3$
2. $f(x) = \frac{4}{5}x$, $g(x) = \frac{5}{4}x$
3. $f(x) = 3x - 7$, $g(x) = \dfrac{x + 7}{3}$
4. $f(x) = \frac{2}{3}x - \frac{4}{5}$, $g(x) = 1.5x + 1.2$
5. $f(x) = 20$, $g(x) = 0.05$
6. $f(x) = x^4$, $g(x) = \sqrt[4]{x}$
7. $f(x) = \sqrt{x + 5}$, $g(x) = x^2 - 5$
8. $f(x) = x^5 - 2$, $g(x) = \sqrt[5]{x + 2}$
9. $f(x) = \dfrac{1 - x}{x}$, $g(x) = \dfrac{1}{1 + x}$
10. $f(x) = \dfrac{x^2 - 1}{x^2 + 1}$, $g(x) = \dfrac{3x - 4}{5x - 2}$
11. $f(x) = x^3 - 5x^2 + 3x + 7$, $g(x) = x + 1$
12. $f(x) = x - 1$, $g(x) = x^3 + 2x^2 - 3x - 9$

Find $f(x)$ and $g(x)$ such that $h(x) = (f \circ g)(x)$. Answers may vary, but try to select the most obvious answer.

13. $h(x) = (4 + 3x)^5$
14. $h(x) = \sqrt[3]{x^2 - 8}$
15. $h(x) = \dfrac{1}{(x - 2)^4}$
16. $h(x) = \dfrac{1}{\sqrt{3x + 7}}$
17. $h(x) = \dfrac{x^3 - 1}{x^3 + 1}$
18. $h(x) = |9x^2 - 4|$
19. $h(x) = \left(\dfrac{2 + x^3}{2 - x^3}\right)^6$
20. $h(x) = (\sqrt{x} - 3)^4$
21. $h(x) = \sqrt{\dfrac{x - 5}{x + 2}}$
22. $h(x) = \sqrt{1 + \sqrt{1 + x}}$
23. $h(x) = (x + 2)^3 - 5(x + 2)^2 + 3(x + 2) - 1$
24. $h(x) = 2(x - 1)^{5/3} + 5(x - 1)^{2/3}$

25. *Dress Sizes.* A dress that is size x in France is size $s(x)$ in the United States, where $s(x) = x - 32$. A dress that is size x in the United States is size $y(x)$ in Italy, where $y(x) = 2(x + 12)$. Find a function that will convert French dress sizes to Italian dress sizes.

26. *Ripple Spread.* A stone is thrown into a pond. A circular ripple is spreading over the pond in such a way that the radius is increasing at the rate of 3 ft/sec.

a) Find a function $r(t)$ for the radius in terms of t.
b) Find a function $A(r)$ for the area of the ripple in terms of the radius r.
c) Find $(A \circ r)(t)$. Explain the meaning of this function.

Find the inverse of the relation.

27. $\{(7, 8), (-2, 8), (3, -4), (8, -8)\}$
28. $\{(0, 1), (5, 6), (-2, -4)\}$
29. $\{(-1, -1), (-3, 4)\}$
30. $\{(-1, 3), (2, 5), (-3, 5), (2, 0)\}$

Find an equation of the inverse relation.

31. $y = 4x - 5$ 32. $2x^2 + 5y^2 = 4$

33. $x^3y = -5$

34. $y = 3x^2 - 5x + 9$

Graph the equation by substituting and plotting points. Then reflect the graph across the line $y = x$ to obtain the graph of its inverse.

35. $x = y^2 - 3$

36. $y = x^2 + 1$

37. $y = |x|$

38. $x = |y|$

Using the horizontal-line test, determine whether the function is one-to-one.

39. $f(x) = 2.7^x$

40. $f(x) = 2^{-x}$

41. $f(x) = 4 - x^2$

42. $f(x) = x^3 - 3x + 1$

43. $f(x) = \dfrac{8}{x^2 - 4}$

44. $f(x) = \sqrt{\dfrac{10}{4 + x}}$

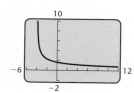

45. $f(x) = \sqrt[3]{x + 2} - 2$

46. $f(x) = \dfrac{8}{x}$

Determine whether the function is one-to-one. Use a grapher and the horizontal-line test or use an algebraic procedure.

47. $f(x) = 5x - 8$

48. $f(x) = 3 + 4x$

49. $f(x) = 1 - x^2$

50. $f(x) = |x| - 2$

51. $f(x) = |x + 2|$

52. $f(x) = -0.8$

53. $f(x) = -\dfrac{4}{x}$

54. $f(x) = \dfrac{2}{x + 3}$

Graph the function and its inverse using a grapher. Use an inverse drawing feature, if available. Find the domain and the range of f. Find the domain and the range of the inverse f^{-1}.

55. $f(x) = 0.8x + 1.7$

56. $f(x) = 2.7 - 1.08x$

57. $f(x) = \frac{1}{2}x - 4$

58. $f(x) = x^3 - 1$

59. $f(x) = \sqrt{x - 3}$

60. $f(x) = -\dfrac{2}{x}$

61. $f(x) = x^2 - 4, \ x \geq 0$

62. $f(x) = 3 - x^2, \ x \geq 0$

63. $f(x) = (3x - 9)^3$

64. $f(x) = \sqrt[3]{\dfrac{x - 3.2}{1.4}}$

In Exercises 65–80, for each function:

a) *Determine whether it is one-to-one, using a grapher if desired.*

b) *If it is one-to-one, find a formula for the inverse.*

65. $f(x) = x + 4$

66. $f(x) = 7 - x$

67. $f(x) = 2x - 1$

68. $f(x) = 5x + 8$

69. $f(x) = \dfrac{4}{x + 7}$

70. $f(x) = -\dfrac{3}{x}$

71. $f(x) = \dfrac{x + 4}{x - 3}$

72. $f(x) = \dfrac{5x - 3}{2x + 1}$

73. $f(x) = x^3 - 1$

74. $f(x) = (x + 5)^3$

75. $f(x) = x\sqrt{4 - x^2}$

76. $f(x) = 4x^5 - 20x^3 + 2x^2 - 5x + 1$

77. $f(x) = 5x^2 - 2, \ x \geq 0$

78. $f(x) = 4x^2 + 3, \ x \geq 0$

79. $f(x) = \sqrt{x + 1}$

80. $f(x) = \sqrt[3]{x - 8}$

Find the inverse by thinking about the operations of the function and then reversing, or undoing, them. Check your work algebraically.

FUNCTION	INVERSE
81. $f(x) = 3x$	$f^{-1}(x) = $ ▓▓▓
82. $f(x) = \frac{1}{4}x + 7$	$f^{-1}(x) = $ ▓▓▓
83. $f(x) = -x$	$f^{-1}(x) = $ ▓▓▓
84. $f(x) = \sqrt[3]{x} - 5$	$f^{-1}(x) = $ ▓▓▓

85. $f(x) = \sqrt[3]{x - 5}$ $f^{-1}(x) = $ �▮▮▮

86. $f(x) = x^{-1}$ $f^{-1}(x) = $ ▪▪▪

For the function f, use composition of functions to show that f^{-1} is as given.

87. $f(x) = \dfrac{7}{8}x,\ f^{-1}(x) = \dfrac{8}{7}x$

88. $f(x) = \dfrac{(x + 5)}{4},\ f^{-1}(x) = 4x - 5$

89. $f(x) = \dfrac{(1 - x)}{x},\ f^{-1}(x) = \dfrac{1}{x + 1}$

90. $f(x) = \sqrt[3]{x + 4},\ f^{-1}(x) = x^3 - 4$

91. Find $f(f^{-1}(5))$ and $f^{-1}(f(a))$:
$$f(x) = x^3 - 4.$$

92. Find $f^{-1}(f(p))$ and $f(f^{-1}(1253))$:
$$f(x) = \sqrt[5]{\dfrac{2x - 7}{3x + 4}}.$$

93. *Dress Sizes in the United States and Italy.* A function that will convert dress sizes in the United States to those in Italy is
$$g(x) = 2(x + 12).$$

  a) Find the dress sizes in Italy that correspond to sizes 6, 8, 10, 14, and 18 in the United States.
  b) Find a formula for the inverse of the function.
  c) Use the inverse function to find the dress sizes in the United States that correspond to 36, 40, 44, 52, and 60 in Italy.

94. *Bus Chartering.* An organization determines that the cost per person of chartering a bus is given by the formula
$$C(x) = \dfrac{100 + 5x}{x},$$
where *x* is the number of people in the group and *C(x)* is in dollars. Determine $C^{-1}(x)$ and explain what it represents.

95. *Reaction Distance.* You are driving a car when a deer suddenly darts across the road in front of you. Your brain registers the emergency and sends a signal to your foot to hit the brake. The car travels a distance *D*, in feet, during this time, where *D* is a function of the speed *r*, in miles per hour, that the car is traveling when you see the deer. That reaction distance *D* is a linear function given by
$$D(r) = \dfrac{11r + 5}{10}.$$

  a) Find $D(0)$, $D(10)$, $D(20)$, $D(50)$, and $D(65)$.
  b) Graph $D(r)$.
  c) Find $D^{-1}(r)$ and explain what it represents.
  d) Graph the inverse.

96. *Bread Consumption.* The number *N* of 1-lb loaves of bread consumed per person per year *t* years after 1995 is given by the function
$$N(t) = 0.6514t + 53.1599$$
(*Source:* U.S. Department of Agriculture).

  a) Find the consumption of bread per person in 1998 and 2000.
  b) Use a grapher to graph the function and its inverse.
  c) Explain what the inverse represents.

Discussion and Writing

97. Suppose that you have graphed a function using a grapher and you see that it is one-to-one. How could you then use the TRACE feature to make a hand-drawn graph of the inverse?

98. The following formulas for the conversion between Fahrenheit and Celsius temperatures have been considered several times in this text:
$$C = \tfrac{5}{9}(F - 32)$$
and
$$F = \tfrac{9}{5}C + 32.$$
Discuss these formulas from the standpoint of inverses.

Skill Maintenance

Graph.

99. $y = \dfrac{3}{2}x - 4$ **100.** $y = -x^2 + x - 4$

101. Write an equation of the line perpendicular to the graph of the line $3x + y = 7$ and containing the point $(5, -2)$.

102. Find the slope of the *y*-intercept of the line with equation $6 - 8y = 17x$.

Synthesis

Using only a grapher, determine whether the functions are inverses of each other.

103. $f(x) = \sqrt[3]{\dfrac{x - 3.2}{1.4}}$, $g(x) = 1.4x^3 + 3.2$

104. $f(x) = \dfrac{2x - 5}{4x + 7}$, $g(x) = \dfrac{7x - 4}{5x + 2}$

105. $f(x) = \dfrac{2}{3}$, $g(x) = \dfrac{3}{2}$

106. $f(x) = x^4$, $x \geq 0$; $g(x) = \sqrt[4]{x}$

107. The function $f(x) = x^2 - 3$ is not one-to-one. Restrict the domain of f so that its inverse is a function. Find the inverse and state the restriction on the domain of the inverse.

108. Consider the function f given by
$$f(x) = \begin{cases} x^3 + 2, & x \leq -1, \\ x^2, & -1 < x < 1, \\ x + 1, & x \geq 1. \end{cases}$$
Does f have an inverse that is a function? Why or why not?

109. Find three examples of functions that are their own inverses, that is, $f = f^{-1}$.

Chapter Summary and Review 1

Important Properties and Formulas

Terminology about Lines

Slope: $m = \dfrac{y_2 - y_1}{x_2 - x_1}$

The Slope–Intercept Equation:
$$y = mx + b$$
The Point–Slope Equation:
$$y - y_1 = m(x - x_1)$$
Horizontal Lines: $y = b$
Vertical Lines: $x = a$
Parallel Lines: $m_1 = m_2$, $b_1 \neq b_2$
Perpendicular Lines:
$$m_1 m_2 = -1, \text{ or}$$
$$x = a, y = b$$

The Algebra of Functions

The Sum of Two Functions:
$$(f + g)(x) = f(x) + g(x)$$
The Difference of Two Functions:
$$(f - g)(x) = f(x) - g(x)$$
The Product of Two Functions:
$$(fg)(x) = f(x) \cdot g(x)$$
The Quotient of Two Functions:
$$(f/g)(x) = f(x)/g(x), \; g(x) \neq 0$$

Tests for Symmetry

x-axis: If replacing y with $-y$ produces an equivalent equation, then the graph is symmetric with respect to the x-axis.

y-axis: If replacing x with $-x$ produces an equivalent equation, then the graph is symmetric with respect to the y-axis.

Origin: If replacing x with $-x$ and y with $-y$ produces an equivalent equation, then the graph is symmetric with respect to the origin.

Even Function: $f(-x) = f(x)$

Odd Function: $f(-x) = -f(x)$

Transformations

Vertical Translation: $y = f(x) \pm b$

Horizontal Translation: $y = f(x \pm d)$

Reflection across the x-axis: $y = -f(x)$

Reflection across the y-axis: $y = f(-x)$

Vertical Stretching or Shrinking:
$$y = af(x)$$

Horizontal Stretching or Shrinking:
$$y = f(cx)$$

Quadratic Equation:
$$ax^2 + bx + c = 0, \ a \neq 0, \ a, b, c \text{ real}$$

Quadratic Function:
$$f(x) = ax^2 + bx + c, \ a \neq 0, \ a, b, c \text{ real}$$

Quadratic Formula:
For $ax^2 + bx + c = 0, \ a \neq 0$,
$$x = \frac{-b \pm \sqrt{b^2 - 4ac}}{2a}.$$

The Vertex of a Parabola:
The **vertex** of the graph of $f(x) = ax^2 + bx + c$ is
$$\left(-\frac{b}{2a}, f\left(-\frac{b}{2a} \right) \right).$$

The Composition of Two Functions:
$$(f \circ g)(x) = f(g(x))$$

One-to-One Function:
$$f(a) = f(b) \rightarrow a = b$$

REVIEW EXERCISES

Determine whether the relation is a function. Identify the domain and the range.

1. $\{(3, 1), (5, 3), (7, 7), (3, 5)\}$

2. $\{(2, 7), (-2, -7), (7, -2), (0, 2), (1, -4)\}$

Determine whether the graph is that of a function.

3.

4.

5.

6.

Find the domain of the function. Do not use a grapher.

7. $f(x) = 4 - 5x + x^2$ **8.** $f(x) = \dfrac{3}{x} + 2$

9. $f(x) = \dfrac{1}{x^2 - 6x + 5}$ **10.** $f(x) = \dfrac{-5x}{|16 - x^2|}$

Use a grapher to graph the function. Then visually estimate the domain and the range.

11. $f(x) = \sqrt{16 - (x + 1)^2}$

12. $f(x) = |x - 3| + |x + 4| - 12$

13. $f(x) = x^3 - 5x^2 + 2x - 7$

14. $f(x) = x^4 - 8x^2 - 3$

15. Given that $f(x) = x^2 - x - 3$, find each of the following.

   a) $f(0)$ **b)** $f(-3)$

   c) $f(a - 1)$ **d)** $\dfrac{f(x + h) - f(x)}{h}$

In Exercises 16 and 17, the table of data contains input–output values for a function. Answer the following questions.

a) *Is the change in the inputs, x, the same?*
b) *Is the change in the outputs, y, the same?*
c) *Is the function linear?*

16.

x	y
−3	8
−2	11
−1	14
0	17
1	20
2	22
3	26

17.

x	y
20	11.8
30	24.2
40	36.6
50	49.0
60	61.4
70	73.8
80	86.2

18. Find the slope and the y-intercept of the graph of $-2x - y = 7$.

Write a point–slope equation for a line with the following characteristics.

19. $m = -\frac{2}{3}$, y-intercept $(0, -4)$

20. $m = 3$, passes through $(-2, -1)$

21. Passes through $(4, 1)$ and $(-2, -1)$

Determine whether the lines are parallel, perpendicular, or neither.

22. $3x - 2y = 8$,
    $6x - 4y = 2$

23. $y - 2x = 4$,
    $2y - 3x = -7$

24. $y = \frac{3}{2}x + 7$,
    $y = -\frac{2}{3}x - 4$

Given the point $(1, -1)$ and the line $2x + 3y = 4$:

25. Find an equation of the line containing the given point and parallel to the given line.

26. Find an equation of the line containing the given point and perpendicular to the given line.

27. *Total Cost.* Clear County Cable Television charges a $25 installation fee and $20 per month for "basic" service. Write and graph an equation that can be used to determine the total cost, $C(t)$, of t months of basic cable television service. Find the total cost of 6 months of service.

28. *Temperature and Depth of the Earth.* The function T given by $T(d) = 10d + 20$ can be used to determine the temperature T, in degrees Celsius, at a depth d, in kilometers, inside the earth.

   a) Find $T(5)$, $T(20)$, and $T(1000)$.
   b) Graph T.
   c) The radius of the earth is about 5600 km. Use this fact to determine the domain of the function.

Make a hand-drawn graph of each of the following. Check your results using a grapher.

29. $f(x) = \begin{cases} x^3, & \text{for } x < -2, \\ |x|, & \text{for } -2 \le x \le 2, \\ \sqrt{x - 1}, & \text{for } x > 2 \end{cases}$

30. $f(x) = \begin{cases} \dfrac{x^2 - 1}{x + 1}, & \text{for } x \ne -1, \\ 3, & \text{for } x = -1 \end{cases}$

31. $f(x) = [\![x]\!]$

32. $f(x) = [\![x - 3]\!]$

In Exercises 33–35, use a grapher to estimate the following for each function.

a) *The intervals on which the function is increasing or decreasing*
b) *Any relative maxima or minima*

33. $f(x) = 1.1x^3 - 9.04x^2 - 18x + 702$

34. $f(x) = \dfrac{20}{(x + 4)^2 + 1} - 5$

35. $f(x) = 0.8x\sqrt{16 - (x + 1)^2}$

36. *Inscribed Rectangle.* A rectangle is inscribed in a semicircle of radius 2, as shown. The variable $x =$ half the length of the rectangle. Express the area of the rectangle as a function of x.

37. *Minimizing Surface Area.* A container firm is designing an open-top rectangular box, with a square base, that will hold 108 in³. Let $x =$ the length of a side of the base.

a) Express the surface area as a function of x.
b) Find the domain of the function.
c) Graph the function.
d) What dimensions minimize the surface area of the box?

First, graph the equation and determine visually whether it is symmetric with respect to the x-axis, the y-axis, and the origin. Then verify your assertion algebraically.

38. $x^2 + y^2 = 4$ 　　　　**39.** $y^2 = x^2 + 3$

40. $x + y = 3$ 　　　　**41.** $y = x^2$

42. $y = x^3$ 　　　　**43.** $y = x^4 - x^2$

Determine visually whether the function is even, odd, or neither even nor odd.

44. 　　　　　　　　　　**45.**

46. 　　　　　　　　　　**47.**

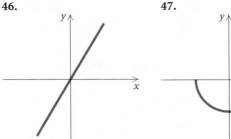

Test algebraically whether the function is even, odd, or neither even nor odd. Then check your work graphically, where possible, using a grapher.

48. $f(x) = 9 - x^2$ 　　　**49.** $f(x) = x^3 - 2x + 4$

50. $f(x) = x^7 - x^5$ 　　　**51.** $f(x) = |x|$

52. $f(x) = \sqrt{16 - x^2}$ 　　**53.** $f(x) = \dfrac{10x}{x^2 + 1}$

Write an equation for a function that has a graph with the given characteristics. Check your answer using a grapher.

54. The shape of $y = x^2$, but shifted left 3 units

55. The shape of $y = \sqrt{x}$, but upside down and shifted right 3 units and up 4 units

56. The shape of $y = |x|$, but stretched vertically by a factor of 2 and shifted right 3 units

A graph of $y = f(x)$ is shown below. No formula for f is given. Make a hand-drawn graph of each of the following.

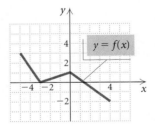

57. $y = f(x - 1)$ 　　　　**58.** $y = f(2x)$

59. $y = -2f(x)$ 　　　　**60.** $y = 3 + f(x)$

For each pair of functions in Exercises 61 and 62:

a) *Find the domain of f, g, $f + g$, $f - g$, fg, and f/g.*
b) *Find $(f + g)(x)$, $(f - g)(x)$, $fg(x)$, and $(f/g)(x)$.*

61. $f(x) = \dfrac{4}{x^2}$; $g(x) = 3 - 2x$

62. $f(x) = 3x^2 + 4x$; $g(x) = 2x - 1$

63. Given the total-revenue and total-cost functions
$$R(x) = 120x - 0.5x^2 \quad \text{and} \quad C(x) = 15x + 6,$$
find total profit $P(x)$.

In Exercises 64 and 65, complete the square to:

a) *find the vertex;*
b) *find the line of symmetry; and*
c) *determine whether there is a maximum or minimum value and find that value.*

Then check your answers using a grapher.

64. $f(x) = -4x^2 + 3x - 1$

65. $f(x) = 5x^2 - 10x + 3$

In Exercises 66–69, match the equation with one of the figures (a)–(d), which follow. Do not use a grapher except as a check.

a)

b)

c)

d)

66. $y = (x - 2)^2$

67. $y = (x + 3)^2 - 6$

68. $y = -2(x + 3)^2 + 4$

69. $y = -\frac{1}{2}(x - 2)^2 + 5$

70. *Maximizing Volume.* The Berniers have 24 ft of flexible fencing with which to build a rectangular "toy corral." If the fencing is 2 ft high, what dimensions should the corral have in order to maximize its volume?

71. *Dimensions of a Box.*   An open box is made from a 10-cm by 20-cm piece of aluminum by cutting a square from each corner and folding up the edges. The area of the resulting base is 90 cm². What is the length of the sides of the squares?

72. *Maximizing Box Dimensions.*   An open box is to be made from a 10-cm by 20-cm piece of aluminum by cutting a square from each corner and folding up the edges. What should the length of the sides of the squares be in order to maximize the volume?

In Exercises 73 and 74, for the pair of functions:

a) *Find the domain of $f \circ g$ and $g \circ f$.*
b) *Find $(f \circ g)(x)$ and $(g \circ f)(x)$.*

73. $f(x) = \dfrac{4}{x^2}; \; g(x) = 3 - 2x$

74. $f(x) = 3x^2 + 4x; \; g(x) = 2x - 1$

Find $f(x)$ and $g(x)$ such that $h(x) = (f \circ g)x$.

75. $h(x) = \sqrt{5x + 2}$

76. $h(x) = 4(5x - 1)^2 + 9$

77. Find the inverse of the relation
$$\{(1.3, -2.7), (8, -3), (-5, 3), (6, -3), (7, -5)\}.$$

78. Find an equation of the inverse relation.

a) $y = 3x^2 + 2x - 1$
b) $0.8x^3 - 5.4y^2 = 3x$

In Exercises 79–82, given the function:

a) *Determine whether it is one-to-one, using a grapher if desired.*
b) *If it is one-to-one, find a formula for the inverse.*

79. $f(x) = \sqrt{x - 6}$ **80.** $f(x) = x^3 - 8$

81. $f(x) = 3x^2 + 2x - 1$ **82.** $f(x) = e^x$

83. Find $f(f^{-1}(657))$: $f(x) = \dfrac{4x^5 - 16x^{37}}{119x}, \; x > 1.$

Discussion and Writing

84. Given the graph of $y = f(x)$, explain and contrast the effect of the constant c on the graphs of $y = f(cx)$ and $y = cf(x)$.

85. a) Graph several functions of the type $y_1 = f(x)$ and $y_2 = |f(x)|$ using a grapher. Describe a procedure, involving transformations, for creating hand-drawn graphs of y_2 from y_1.
  b) Describe a procedure, involving transformations, for creating hand-drawn graphs of $y_2 = f(|x|)$ from $y_1 = f(x)$.

86. If the graphs of
$$f(x) = a_1(x - h_1)^2 + k_1$$
and
$$g(x) = a_2(x - h_2)^2 + k_2$$
have the same shape, what, if anything, can you conclude about the a's, the h's, and the k's? Explain your answer.

87. Describe the difference between $f^{-1}(x)$ and $[f(x)]^{-1}$.

Synthesis

Find the domain.

88. $f(x) = \dfrac{\sqrt{1 - x}}{x - |x|}$

89. $f(x) = (x - 9x^{-1})^{-1}$

90. Prove that the sum of two odd functions is odd.

91. Describe how the graph of $y = -f(-x)$ is obtained from the graph of $y = f(x)$.

92. Find b such that $f(x) = -3x^2 + bx - 1$ has a maximum value of 2.

93. Use only a grapher. Determine whether the following functions are inverses of each other:
$$f(x) = \frac{4 + 3x}{x - 2}, \qquad g(x) = \frac{x + 4}{x - 3}.$$

The Trigonometric Functions 2

APPLICATION

Musical instruments can generate complex sine waves. Consider two tones whose graphs are $y = 2 \sin x$ and $y = \sin 2x$. The combination of the two tones produces a new sound whose graph is $y = 2 \sin x + \sin 2x$.

This problem appears as Example 9 in Section 2.6.

We now consider an important class of functions called *trigonometric*, or *circular*, *functions*. Historically, these functions arose from a study of triangles; hence the name trigonometric. We will begin our study with right triangles and degree measure and solve applied problems involving right triangles. Then we will consider trigonometric functions of angles or rotations of any size with both degree and radian measure. A circle of radius 1 (a unit circle) is then used to define the six basic trigonometric functions; hence the name circular functions. The domains and ranges of these functions consist of real numbers.

$y_1 = 2 \sin x, \quad y_2 = \sin 2x,$
$y_3 = 2 \sin x + \sin 2x$

111

Trigonometric Functions of Acute Angles

2.1

- • *Determine the six trigonometric ratios for a given acute angle of a right triangle.*
- • *Determine the trigonometric function values of 30°, 45°, and 60°.*
- • *Using a grapher, find function values for any acute angle, and given a function value of an acute angle, find the angle.*
- • *Given the function values of an acute angle, find the function values of its complement.*

The Trigonometric Ratios

We begin our study of trigonometry by considering right triangles and acute angles measured in degrees. An **acute angle** is an angle with measure greater than 0° and less than 90°. Greek letters such as α (alpha), β (beta), γ (gamma), θ (theta), and ϕ (phi) are used to denote an angle. Consider a right triangle with one of its acute angles labeled θ. The side opposite the right angle is called the **hypotenuse**. The other sides of the triangle are referenced by their position relative to the acute angle θ. One side is opposite θ and one is adjacent to θ.

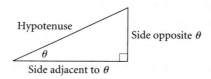

The *lengths* of the sides of the triangle are used to define the six trigonometric ratios:

sine (sin),   cosecant (csc),

cosine (cos),   secant (sec),

tangent (tan),   cotangent (cot).

The **sine of θ** is the *length* of the side opposite θ divided by the *length* of the hypotenuse:

$$\sin \theta = \frac{\text{length of side opposite } \theta}{\text{length of hypotenuse}}.$$

The ratio depends on the measure of angle θ and thus is a function of θ. The notation $\sin \theta$ actually means $\sin (\theta)$, where sin, or sine, is the name of the function.

The **cosine of θ** is the *length* of the side adjacent to θ divided by the *length* of the hypotenuse:

$$\cos \theta = \frac{\text{length of side adjacent to } \theta}{\text{length of hypotenuse}}.$$

The six trigonometric ratios, or trigonometric functions, are defined as follows.

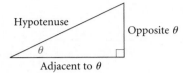

Trigonometric Function Values of an Acute Angle θ

Let θ be an acute angle of a right triangle. Then the six trigonometric functions of θ are as follows:

$$\sin \theta = \frac{\text{side opposite } \theta}{\text{hypotenuse}}, \qquad \csc \theta = \frac{\text{hypotenuse}}{\text{side opposite } \theta},$$

$$\cos \theta = \frac{\text{side adjacent to } \theta}{\text{hypotenuse}}, \qquad \sec \theta = \frac{\text{hypotenuse}}{\text{side adjacent to } \theta},$$

$$\tan \theta = \frac{\text{side opposite } \theta}{\text{side adjacent to } \theta}, \qquad \cot \theta = \frac{\text{side adjacent to } \theta}{\text{side opposite } \theta}.$$

EXAMPLE 1   In the right triangle shown at left, find the six trigonometric function values of (a) θ and (b) α.

Solution   We use the definitions.

a) $\sin \theta = \dfrac{\text{opp}}{\text{hyp}} = \dfrac{12}{13}, \qquad \csc \theta = \dfrac{\text{hyp}}{\text{opp}} = \dfrac{13}{12},$

$\cos \theta = \dfrac{\text{adj}}{\text{hyp}} = \dfrac{5}{13}, \qquad \sec \theta = \dfrac{\text{hyp}}{\text{adj}} = \dfrac{13}{5},$

$\tan \theta = \dfrac{\text{opp}}{\text{adj}} = \dfrac{12}{5}, \qquad \cot \theta = \dfrac{\text{adj}}{\text{opp}} = \dfrac{5}{12}$

> The references to opposite, adjacent, and hypotenuse are relative to θ.

b) $\sin \alpha = \dfrac{\text{opp}}{\text{hyp}} = \dfrac{5}{13}, \qquad \csc \alpha = \dfrac{\text{hyp}}{\text{opp}} = \dfrac{13}{5},$

$\cos \alpha = \dfrac{\text{adj}}{\text{hyp}} = \dfrac{12}{13}, \qquad \sec \alpha = \dfrac{\text{hyp}}{\text{adj}} = \dfrac{13}{12},$

$\tan \alpha = \dfrac{\text{opp}}{\text{adj}} = \dfrac{5}{12}, \qquad \cot \alpha = \dfrac{\text{adj}}{\text{opp}} = \dfrac{12}{5}$

> The references to opposite, adjacent, and hypotenuse are relative to α.

In Example 1(a), we note that the value of $\sin \theta$, $\frac{12}{13}$, is the reciprocal of $\frac{13}{12}$, the value of $\csc \theta$. Likewise, we see the same reciprocal relationship between the values of $\cos \theta$ and $\sec \theta$ and between the values of $\tan \theta$ and $\cot \theta$. For any angle, the cosecant, secant, and cotangent values are the reciprocals of the sine, cosine, and tangent function values, respectively.

Reciprocal Functions

$$\csc \theta = \frac{1}{\sin \theta}, \qquad \sec \theta = \frac{1}{\cos \theta}, \qquad \cot \theta = \frac{1}{\tan \theta}$$

If we know the values of the sine, cosine, and tangent functions of an angle, we can use these reciprocal relationships to find the values of the cosecant, secant, and cotangent functions of that angle.

EXAMPLE 2 Given that $\sin \phi = \frac{4}{5}$, $\cos \phi = \frac{3}{5}$, and $\tan \phi = \frac{4}{3}$, find $\csc \phi$, $\sec \phi$, and $\cot \phi$.

Solution Using the reciprocal relationships, we have

$$\csc \phi = \frac{1}{\sin \phi} = \frac{1}{\frac{4}{5}} = \frac{5}{4}, \qquad \sec \phi = \frac{1}{\cos \phi} = \frac{1}{\frac{3}{5}} = \frac{5}{3},$$

and

$$\cot \phi = \frac{1}{\tan \phi} = \frac{1}{\frac{4}{3}} = \frac{3}{4}.$$

Triangles are said to be **similar** if their corresponding angles have the *same* measure. In similar triangles, the lengths of corresponding sides are in the same ratio. The right triangles shown below are similar. Note that the corresponding angles are equal and the length of each side of the second triangle is four times the length of the corresponding side of the first triangle.

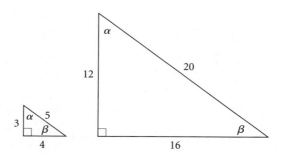

Let's observe the sine, cosine, and tangent values of β in each triangle. Can we expect corresponding function values to be the same?

FIRST TRIANGLE	SECOND TRIANGLE
$\sin \beta = \dfrac{3}{5}$	$\sin \beta = \dfrac{12}{20} = \dfrac{3}{5}$
$\cos \beta = \dfrac{4}{5}$	$\cos \beta = \dfrac{16}{20} = \dfrac{4}{5}$
$\tan \beta = \dfrac{3}{4}$	$\tan \beta = \dfrac{12}{16} = \dfrac{3}{4}$

For the two triangles, the corresponding values of $\sin \beta$, $\cos \beta$, and $\tan \beta$ are the same. The lengths of the sides are proportional—thus the *ratios* are the same. This must be the case because in order for the sine, cosine, and tangent to be functions, there must be only one output (the ratio) for each input (the angle β).

> The trigonometric function values of θ depend only on the measure of the angle, not on the size of the triangle.

The Six Functions Related

We can find the other five trigonometric function values of an acute angle when one of the function-value ratios is known.

EXAMPLE 3 If $\sin \beta = \frac{6}{7}$ and β is an acute angle, find the other five trigonometric function values of β.

Solution We know from the definition of the sine function that the ratio

$$\frac{6}{7} \quad \text{is} \quad \frac{\text{opp}}{\text{hyp}}.$$

Using this information, let's consider a right triangle in which the hypotenuse has length 7 and the side opposite β has length 6. To find the length of the side adjacent to β, we recall the *Pythagorean theorem*:

$$a^2 + b^2 = c^2$$
$$a^2 + 6^2 = 7^2$$
$$a^2 + 36 = 49$$
$$a^2 = 49 - 36 = 13$$
$$a = \sqrt{13}.$$

We now use the lengths of the three sides to find the other five ratios:

$$\sin \beta = \frac{6}{7}, \qquad\qquad \csc \beta = \frac{7}{6},$$

$$\cos \beta = \frac{\sqrt{13}}{7}, \qquad\qquad \sec \beta = \frac{7}{\sqrt{13}}, \quad \text{or} \quad \frac{7\sqrt{13}}{13},$$

$$\tan \beta = \frac{6}{\sqrt{13}}, \quad \text{or} \quad \frac{6\sqrt{13}}{13}, \qquad \cot \beta = \frac{\sqrt{13}}{6}.$$

Function Values of 30°, 45°, and 60°

In Examples 1 and 3, we found the trigonometric function values of an acute angle of a right triangle when the lengths of the three sides were known. In most situations, we are asked to find the function values when the measure of the acute angle is given. For certain special angles such as 30°, 45°, and 60°, which are frequently seen in applications, we can use geometry to determine the function values.

A right triangle with a 45° angle actually has two 45° angles. Thus the triangle is *isosceles*, and the legs are the same length. Let's consider such a triangle whose legs have length 1. Then we can find the length of its hypotenuse, c, using the Pythagorean theorem as follows:

$$1^2 + 1^2 = c^2, \quad \text{or} \quad c^2 = 2, \quad \text{or} \quad c = \sqrt{2}.$$

Such a triangle is shown below. From this diagram, we can easily determine the trigonometric function values of 45°.

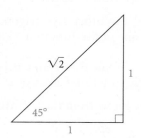

$$\sin 45° = \frac{\text{opp}}{\text{hyp}} = \frac{1}{\sqrt{2}} = \frac{\sqrt{2}}{2} \approx 0.7071,$$

$$\cos 45° = \frac{\text{adj}}{\text{hyp}} = \frac{1}{\sqrt{2}} = \frac{\sqrt{2}}{2} \approx 0.7071,$$

$$\tan 45° = \frac{\text{opp}}{\text{adj}} = \frac{1}{1} = 1$$

It is sufficient to find only the function values of the sine, cosine, and tangent, since the others are their reciprocals.

It is also possible to determine the function values of 30° and 60°. A right triangle with 30° and 60° acute angles is half of an equilateral triangle, as shown in the following figure. Thus if we choose an equilateral triangle whose sides have length 2 and take half of it, we obtain a right triangle that has a hypotenuse of length 2 and a leg of length 1. The other leg has length a, which can be found as follows:

$$a^2 + 1^2 = 2^2$$
$$a^2 + 1 = 4$$
$$a^2 = 3$$
$$a = \sqrt{3}.$$

We can now determine the function values of 30° and 60°:

$$\sin 30° = \frac{1}{2} = 0.5, \qquad\qquad \sin 60° = \frac{\sqrt{3}}{2} \approx 0.8660,$$

$$\cos 30° = \frac{\sqrt{3}}{2} \approx 0.8660, \qquad\qquad \cos 60° = \frac{1}{2} = 0.5,$$

$$\tan 30° = \frac{1}{\sqrt{3}} = \frac{\sqrt{3}}{3} \approx 0.5774, \qquad\qquad \tan 60° = \frac{\sqrt{3}}{1} = \sqrt{3} \approx 1.7321.$$

Since we will often use the function values of 30°, 45°, and 60°, either the triangles that yield them or the values themselves should be memorized.

Let's now use what we have learned about trigonometric functions of special angles to solve problems. We will consider such applications in greater detail in Section 2.2.

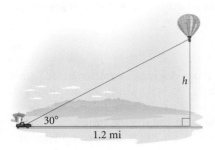

EXAMPLE 4 *Height of a Hot-air Balloon.* As a hot-air balloon began to rise, the ground crew drove 1.2 mi to an observation station. The initial observation from the station estimated the angle between the ground and the line of sight to be 30°. Approximately how high was the balloon at that point? (We are assuming that the wind velocity was low and that the balloon rose vertically for the first few minutes.)

Solution We begin with a sketch of the situation. We know the measure of an acute angle and the length of its adjacent side. Since we want to determine the length of the opposite side, we can use the tangent ratio, or the cotangent ratio. Here we use the tangent ratio:

$$\tan 30° = \frac{\text{opp}}{\text{adj}} = \frac{h}{1.2}$$

$$1.2 \tan 30° = h$$

$$1.2\left(\frac{\sqrt{3}}{3}\right) = h \qquad \text{Substituting; } \tan 30° = \frac{\sqrt{3}}{3}$$

$$0.4\sqrt{3} = h$$

$$0.7 \approx h.$$

The balloon is approximately 0.7 mi, or 3696 ft, high.

Function Values of Any Acute Angle

Historically, the measure of an angle has been expressed in degrees, minutes, and seconds. One minute, denoted $1'$, is such that $60' = 1°$, or $1' = \frac{1}{60} \cdot (1°)$. One second, denoted $1''$, is such that $60'' = 1'$, or $1'' = \frac{1}{60} \cdot (1')$. Then 61 degrees, 27 minutes, 4 seconds could be written as $61°27'4''$. This D°M'S'' form was common before the widespread use of graphers. Now the preferred notation is to express fractional parts of degrees in *decimal degree form*. Although the D°M'S'' notation is still widely used in navigation, we will most often use the decimal form in this text.

Most graphers can convert D°M'S'' notation to decimal degree notation and vice versa. This is often accomplished using a feature in the ANGLE menu. Procedures among graphers vary. (See the *Graphing Calculator Manual* that accompanies this text for the keystrokes.)

EXAMPLE 5 Convert $5°42'30''$ to decimal degree notation.

Solution On a grapher, we enter $5°42'30''$. Pressing ⎡ENTER⎤ gives us

$$5°42'30'' \approx 5.71°,$$

rounded to the nearest hundredth of a degree.

EXAMPLE 6 Convert $72.18°$ to D°M'S'' notation.

Solution   On a grapher, we enter 72.18 and use the ▶DMS feature in the ANGLE menu. The result is

$$72.18° = 72°10'48''.$$

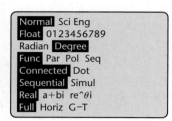

So far we have measured angles using degrees. Another useful unit for angle measure is the radian, which we will study in Section 2.4. Graphers work with either degrees or radians. Be sure to use whichever mode is appropriate. In this section, we use the degree mode.

Keep in mind the difference between an exact answer and an approximation. For example,

$$\sin 60° = \frac{\sqrt{3}}{2}.\quad \textbf{This is exact!}$$

But using a grapher, you might get an answer like

$$\sin 60° \approx 0.8660254038.\quad \textbf{This is an approximation!}$$

Graphers generally provide values only of the sine, cosine, and tangent functions. You can find values of the cosecant, secant, and cotangent by taking reciprocals of the sine, cosine, and tangent functions, respectively.

EXAMPLE 7   Find the trigonometric function value, rounded to four decimal places, of each of the following.

a) tan 29.7°   **b)** sec 48°   **c)** sin 84°10'39''

Solution

a) We check to be sure that the grapher is in *degree mode*. The function value is

$$\tan 29.7° \approx 0.5703899297$$
$$\approx 0.5704.\quad \text{Rounded to four decimal places}$$

b) The secant function value can be found by taking the reciprocal of the cosine function value:

$$\sec 48° = \frac{1}{\cos 48°} \approx 1.49447655 \approx 1.4945.$$

c) Using the ANGLE menu on a grapher, we enter sin 84°10'39''. Pressing ENTER gives us

$$\sin 84°10'39'' \approx 0.9948409474 \approx 0.9948.$$

We can use the TABLE feature on a grapher to find an angle for which we know a trigonometric function value.

EXAMPLE 8   Find the acute angle, to the nearest tenth of a degree, whose sine value is approximately 0.20113.

Solution   With a grapher set in degree mode, we first enter the equation $y = \sin x$. With a minimum value of 0 and a step-value of 0.1, we scroll through the table of values looking for the y-value closest to 0.20113.

X	Y₁	
11.1	.19252	
11.2	.19423	
11.3	.19595	
11.4	.19766	
11.5	.19937	
11.6	.20108 ←	——— sin 11.6° ≈ 0.20108
11.7	.20279	
X = 11.6		

We find that 11.6° is the angle whose sine value is about 0.20113.

The quickest way to find the angle with a grapher is to use an inverse function key. (We first studied inverse functions in Section 1.6 and will consider inverse *trigonometric* functions in Section 3.4.) Usually two keys must be pressed in sequence. For this example, if we press

$\boxed{\text{2ND}}$ $\boxed{\text{SIN}}$ .20113 $\boxed{\text{ENTER}}$,

we find that the acute angle whose sine is 0.20113 is approximately 11.60304613°, or 11.6°.

EXAMPLE 9   *Ladder Safety.* A paint crew has purchased new 30-ft extension ladders. The manufacturer states that the safest placement on a wall is to extend the ladder to 25 ft and to position the base $6\frac{1}{2}$ ft from the wall. (*Source*: R. D. Werner Co., Inc.) What angle does the ladder make with the ground in this position?

Solution   We make a drawing and then use the most convenient trigonometric function. Because we know the length of the side adjacent to θ and the length of the hypotenuse, we choose the cosine function.

From the definition of the cosine function, we have

$$\cos \theta = \frac{6.5 \text{ ft}}{25 \text{ ft}} = 0.26.$$

Using a grapher, we find that

$$\theta \approx 74.92993786°.$$

Thus when the ladder is in its safest position, it makes an angle of about 75° with the ground.

Cofunctions and Complements

We recall that two angles are **complementary** whenever the sum of their measures is 90°. Each is the complement of the other. In a right triangle, the acute angles are complementary, since the sum of all three angle measures is 180° and the right angle accounts for 90° of this total. Thus if one acute angle of a right triangle is θ, the other is $90° - \theta$.

Find the six trigonometric function values of the acute angles given in this triangle. Note that 53° and 37° are complementary angles since 53° + 37° = 90°.

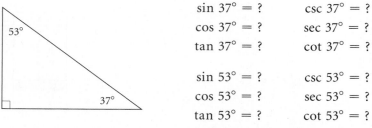

sin 37° = ?	csc 37° = ?
cos 37° = ?	sec 37° = ?
tan 37° = ?	cot 37° = ?
sin 53° = ?	csc 53° = ?
cos 53° = ?	sec 53° = ?
tan 53° = ?	cot 53° = ?

Try this with the acute, complementary angles 20.3° and 69.7° as well. What pattern do you observe? Look for this same pattern in Example 1 earlier in this section.

Note that the sine of an angle is also the cosine of the angle's complement. Similarly, the tangent of an angle is the cotangent of the angle's complement, and the secant of an angle is the cosecant of the angle's complement. These pairs of functions are called **cofunctions**. A list of cofunction identities follows.

Cofunction Identities

$$\sin \theta = \cos (90° - \theta), \qquad \cos \theta = \sin (90° - \theta),$$
$$\tan \theta = \cot (90° - \theta), \qquad \cot \theta = \tan (90° - \theta),$$
$$\sec \theta = \csc (90° - \theta), \qquad \csc \theta = \sec (90° - \theta)$$

EXAMPLE 10 Given that $\sin 18° \approx 0.3090$, $\cos 18° \approx 0.9511$, and $\tan 18° \approx 0.3249$, find the six trigonometric function values of 72°.

Solution Using reciprocal relationships, we know that

$$\csc 18° = \frac{1}{\sin 18°} \approx 3.2361,$$

$$\sec 18° = \frac{1}{\cos 18°} \approx 1.0515,$$

and $\cot 18° = \dfrac{1}{\tan 18°} \approx 3.0777.$

Since 72° and 18° are complementary, we have

$\sin 72° = \cos 18° \approx 0.9511,$	$\cos 72° = \sin 18° \approx 0.3090,$
$\tan 72° = \cot 18° \approx 3.0777,$	$\cot 72° = \tan 18° \approx 0.3249,$
$\sec 72° = \csc 18° \approx 3.2361,$	$\csc 72° = \sec 18° \approx 1.0515.$

Exercise Set 2.1

In Exercises 1–6, find the six trigonometric function values of the specified angle.

1.

2.

3.

4.

5.

6.

Given a function value of an acute angle, find the other five trigonometric function values.

7. $\sin \theta = \frac{24}{25}$

8. $\cos \sigma = 0.7$

9. $\tan \phi = 2$

10. $\cot \theta = \frac{1}{3}$

11. $\csc \theta = 1.5$

12. $\sec \beta = \sqrt{17}$

Find the exact function value.

13. $\cos 45°$

14. $\tan 30°$

15. $\sec 60°$

16. $\sin 45°$

17. $\cot 60°$

18. $\csc 45°$

19. $\sin 30°$

20. $\cos 60°$

21. *Distance Between Bases.* A baseball diamond is actually a square 90 ft on a side. If a line is drawn from third base to first base, then a right triangle *QPH* is formed, where $\angle QPH$ is 45°. Using a trigonometric function, find the distance from third base to first base.

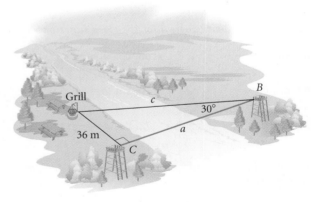

22. *Distance Across a River.* Find the distance a across the river.

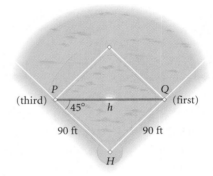

Convert to decimal degree notation. Round to two decimal places.

23. 9°43′

24. 52°15′

25. 35°50″

26. 64°53′

27. 3°2′

28. 19°47′23″

29. 49°38′46″

30. 76°11′34″

Convert to degrees, minutes, and seconds. Round to the nearest second.

31. 17.6°

32. 20.14°

33. 83.025°   **34.** 67.84°

35. 11.75°   **36.** 29.8°

37. 47.8268°   **38.** 0.253°

Find the function value. Round to four decimal places.

39. cos 51°   **40.** cot 17°

41. tan 4°13′   **42.** sin 26.1°

43. sec 38.43°   **44.** cos 74°10′40″

45. cos 40.35°   **46.** csc 45.2°

47. sin 69°   **48.** tan 63°48′

49. tan 85.4°   **50.** cos 4°

51. csc 89.5°   **52.** sec 35.28°

53. cot 30°25′6″   **54.** sin 59.2°

Find the acute angle θ, to the nearest tenth of a degree, for the given function value.

55. sin θ = 0.5125   **56.** tan θ = 2.032

57. tan θ = 0.2226   **58.** cos θ = 0.3842

59. sin θ = 0.9022   **60.** tan θ = 3.056

61. cos θ = 0.6879   **62.** sin θ = 0.4005

63. cot θ = 2.127   **64.** csc θ = 1.147

$$\left(\text{Hint: } \tan\theta = \frac{1}{\cot\theta}.\right)$$

65. sec θ = 1.279   **66.** cot θ = 1.351

Find the exact acute angle θ for the given function value.

67. $\sin\theta = \dfrac{\sqrt{2}}{2}$ **68.** $\cot\theta = \dfrac{\sqrt{3}}{3}$

69. $\cos\theta = \dfrac{1}{2}$ **70.** $\sin\theta = \dfrac{1}{2}$

71. tan θ = 1   **72.** $\cos\theta = \dfrac{\sqrt{3}}{2}$

Use the cofunction and reciprocal identities to complete each of the following.

73. $\cos 20° = \underline{\quad\quad} 70° = \dfrac{1}{\underline{\quad} 20°}$

74. $\sin 64° = \underline{\quad\quad} 26° = \dfrac{1}{\underline{\quad} 64°}$

75. $\tan 52° = \cot \underline{\quad\quad} = \dfrac{1}{\underline{\quad} 52°}$

76. $\sec 13° = \csc \underline{\quad\quad} = \dfrac{1}{\underline{\quad} 13°}$

77. Given that
$$\sin 65° \approx 0.9063, \qquad \cos 65° \approx 0.4226,$$
$$\tan 65° \approx 2.145, \qquad \cot 65° \approx 0.4663,$$
$$\sec 65° \approx 2.366, \qquad \csc 65° \approx 1.103,$$
find the six function values of 25°.

78. Given that sin 38.7° ≈ 0.6252, cos 38.7° ≈ 0.7804, and tan 38.7° ≈ 0.8012, find the six function values of 51.3°.

79. Given that sin 82° = *p*, cos 82° = *q*, and tan 82° = *r*, find the six function values of 8° in terms of *p*, *q*, and *r*.

Discussion and Writing

80. Explain the difference between two functions being reciprocal functions and two functions being cofunctions.

81. Explain why it is not necessary to memorize the function values for both 30° and 60°.

Skill Maintenance

82. Graph: $f(x) = \begin{cases} x - 2, & \text{for } x \le -1, \\ 3, & \text{for } -1 < x \le 2, \\ x, & \text{for } x > 2. \end{cases}$

Determine algebraically whether the graph is symmetric with respect to the x-axis, the y-axis, and the origin.

83. $y = 3x^4 - 3$ **84.** $y^2 = x$

85. $2x - 5y = 0$

Synthesis

86. Given that cos θ = 0.9651, find csc (90° − θ).

87. Given that sec β = 1.5304, find sin (90° − β).

88. Find the six trigonometric function values of α.

89. Show that the area of this right triangle is $\frac{1}{2}bc \sin A$.

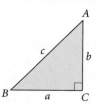

90. Show that the area of this triangle is $\frac{1}{2}ab\sin\theta$.

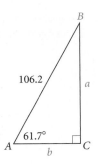

Applications of Right Triangles

2.2

- *Solve right triangles.*
- *Solve applied problems involving right triangles and trigonometric functions.*

Solving Right Triangles

Now that we can find function values for any acute angle, it is possible to *solve* right triangles. To **solve** a triangle means to find the lengths of all sides and the measures of all angles.

EXAMPLE 1 In $\triangle ABC$ (shown at left), find a, b, and B, where a and b represent lengths of sides and B represents the measure of $\angle B$. Here we use standard lettering for naming the sides and angles of a right triangle; side a is opposite angle A, side b is opposite angle B, where a and b are the legs, and side c, the hypotenuse, is opposite angle C, the right angle.

Solution In $\triangle ABC$, we know three of the measures:

$$A = 61.7°, \qquad a = ?,$$
$$B = ?, \qquad b = ?,$$
$$C = 90°, \qquad c = 106.2.$$

Since the sum of the angle measures of any triangle is 180° and $C = 90°$, the sum of A and B is 90°. Thus,

$$B = 90° - A = 90° - 61.7° = 28.3°.$$

We are given an acute angle and the hypotenuse. This suggests that we can use the sine and cosine ratios to find a and b, respectively:

$$\sin 61.7° = \frac{\text{opp}}{\text{hyp}} = \frac{a}{106.2} \quad \text{and} \quad \cos 61.7° = \frac{\text{adj}}{\text{hyp}} = \frac{b}{106.2}.$$

Solving for a and b, we get

$$a = 106.2\sin 61.7° \quad \text{and} \quad b = 106.2\cos 61.7°$$
$$a \approx 93.5 \qquad\qquad\qquad b \approx 50.3.$$

Thus,

$$A = 61.7°, \qquad a \approx 93.5,$$
$$B = 28.3°, \qquad b \approx 50.3,$$
$$C = 90°, \qquad c = 106.2.$$

EXAMPLE 2   In △*DEF* (shown at left), find *D* and *F*. Then find *d*.

Solution   In △*DEF*, we know three of the measures:

$$D = ?, \qquad d = ?,$$
$$E = 90°, \qquad e = 23,$$
$$F = ?, \qquad f = 13.$$

We know the side adjacent to *D* and the hypotenuse. This suggests the use of the cosine ratio:

$$\cos D = \frac{\text{adj}}{\text{hyp}} = \frac{13}{23}.$$

We now find the angle whose cosine is $\frac{13}{23}$. To the nearest hundredth of a degree,

$$D \approx 55.58°. \qquad \text{Pressing } \boxed{\text{2ND}}\ \boxed{\text{cos}}\ (13/23)\ \boxed{\text{ENTER}}$$

Since the sum of *D* and *F* is 90°, we can find *F* by subtracting:

$$F = 90° - D \approx 90° - 55.58° \approx 34.42°.$$

To find *d*, we can use either the Pythagorean theorem or a trigonometric function. We could use cos *F*, sin *D*, or the tangent or cotangent ratios for either *D* or *F*. Let's use tan *D*:

$$\tan D = \frac{\text{opp}}{\text{adj}} = \frac{d}{13}, \quad \text{or} \quad \tan 55.58° \approx \frac{d}{13}.$$

Then

$$d \approx 13 \tan 55.58° \approx 19.$$

The six measures are

$$D \approx 55.58°, \qquad d \approx 19,$$
$$E = 90°, \qquad e = 23,$$
$$F \approx 34.42°, \qquad f = 13.$$

Applications

Right triangles can be used to model and solve many applied problems in the real world.

EXAMPLE 3   *Hiking at the Grand Canyon.* A backpacker hiking east along the north rim of the Grand Canyon notices an unusual rock formation directly across the canyon. She decides to continue watching the landmark while hiking along the rim. In 2 hr, she has gone 10 km due east and the landmark is still visible but at approximately a 50° angle to the north rim. (See the figure at left.)

a) How many kilometers is she from the rock formation?

b) How far is it across the canyon at *her starting point*?

Solution

a) We know the side adjacent to the 50° angle and want to find the hypotenuse. We can use the cosine function:

$$\cos 50° = \frac{10 \text{ km}}{c}$$

$$c = \frac{10 \text{ km}}{\cos 50°} \approx 15.6 \text{ km}.$$

After hiking 10 km, she is approximately 15.6 km from the rock formation.

b) We know the side adjacent to the 50° angle and want to find the opposite side. We can use the tangent function:

$$\tan 50° = \frac{b}{10 \text{ km}}$$

$$b = 10 \text{ km} \cdot \tan 50° \approx 11.9 \text{ km}.$$

Thus it is approximately 11.9 km across the canyon at her starting point.

Many applications with right triangles involve an *angle of elevation* or an *angle of depression.* The angle between the horizontal and a line of sight above the horizontal is called an **angle of elevation.** The angle between the horizontal and a line of sight below the horizontal is called an **angle of depression.** For example, suppose that you are looking straight ahead and then you move your eyes up to look at an approaching airplane. The angle that your eyes pass through is an angle of elevation. If the pilot of the plane is looking forward and then looks down, the pilot's eyes pass through an angle of depression.

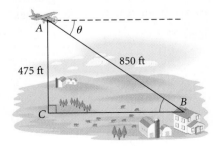

EXAMPLE 4 *Aerial Photography.* An aerial photographer who photographs farm properties for a real estate company has determined from experience that the best photo is taken at a height of approximately 475 ft and a distance of 850 ft from the farmhouse. What is the angle of depression from the plane to the house?

Solution When parallel lines are cut by a transversal, alternate interior angles are equal. Thus the angle of depression from the plane to the house, θ, is equal to the angle of elevation from the house to the plane, so we can use the right triangle shown in the figure. Since we know the side opposite $\angle B$ and the hypotenuse, we can find θ by using the sine function:

$$\sin \theta = \sin B = \frac{475 \text{ ft}}{850 \text{ ft}}.$$

Using the inverse sine key, we find that

$$\theta \approx 34°.$$

Thus the angle of depression is approximately 34°. ▬

EXAMPLE 5 *Cloud Height.* To measure cloud height at night, a vertical beam of light is directed on a spot on the cloud. From a point 135 ft away from the light source, the angle of elevation to the spot is found to be 67.35°. Find the height of the cloud.

Solution From the figure, we have

$$\tan 67.35° = \frac{h}{135 \text{ ft}}$$

$$h = 135 \text{ ft} \cdot \tan 67.35° \approx 324 \text{ ft}.$$

The height of the cloud is about 324 ft. ▬

Some applications of trigonometry involve the concept of direction, or bearing. In this text we present two ways of giving direction, the first below and the second in Exercise Set 2.3.

BEARING: FIRST-TYPE One method of giving direction, or bearing, involves reference to a north–south line using an acute angle. For example, N55°W means 55° west of north and S67°E means 67° east of south.

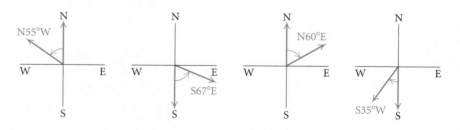

EXAMPLE 6 *Distance to a Forest Fire.* A forest ranger at point *A* sights a fire directly south. A second ranger at point *B*, 7.5 mi east, sights the same fire at a bearing of S27°23′W. How far from *A* is the fire?

Solution We first find the complement of 27°23′:

$$B = 90° - 27°23' \qquad \text{Angle } B \text{ is opposite side } d \text{ in the right triangle.}$$

$$= 62°37'$$

$$\approx 62.62°.$$

From the figure shown at the top of the following page, we see that the

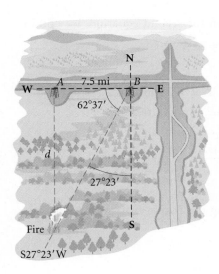

desired distance d is part of a right triangle. We have

$$\frac{d}{7.5 \text{ mi}} \approx \tan 62.62°$$

$$d \approx 7.5 \text{ mi} \tan 62.62° \approx 14.5 \text{ mi}.$$

The forest ranger at A is about 14.5 mi from the fire.

EXAMPLE 7 *Comiskey Park.* In the new Comiskey Park, the home of the Chicago White Sox baseball team, the first row of seats in the upper deck is farther away from home plate than the last row of seats in the old Comiskey Park. Although there is no obstructed view in the new park, some of the fans still complain about the present distance from home plate to the upper deck of seats. (*Source: Chicago Tribune*, September 19, 1993) From a seat in the last row of the upper deck directly behind the batter, the angle of depression to home plate is 29.9° and the angle of depression to the pitcher's mound is 24.2°. Find (a) the viewing distance to home plate and (b) the viewing distance to the pitcher's mound.

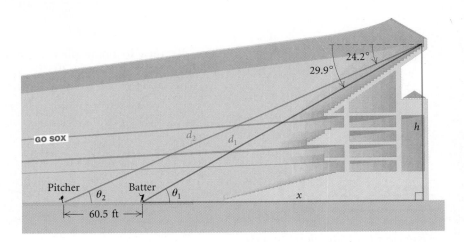

Solution From geometry we know that $\theta_1 = 29.9°$ and $\theta_2 = 24.2°$. The standard distance from home plate to the pitcher's mound is 60.5 ft. In the drawing, we let d_1 be the viewing distance to home plate, d_2 the viewing distance to the pitcher's mound, h the elevation of the last row, and x the horizontal distance from the batter to a point directly below the seat in the last row of the upper deck.

We begin by determining the distance x. We use the tangent function with $\theta_1 = 29.9°$ and $\theta_2 = 24.2°$:

$$\tan 29.9° = \frac{h}{x} \qquad \text{and} \quad \tan 24.2° = \frac{h}{x + 60.5}$$

or

$$h = x \tan 29.9° \quad \text{and} \qquad h = (x + 60.5) \tan 24.2°.$$

Then substituting $x \tan 29.9°$ for h in the second equation, we obtain

$$x \tan 29.9° = (x + 60.5) \tan 24.2°.$$

Solving for x, we get

$$x \tan 29.9° = x \tan 24.2° + 60.5 \tan 24.2°$$

$$x \tan 29.9° - x \tan 24.2° = x \tan 24.2° + 60.5 \tan 24.2° - x \tan 24.2°$$

$$x(\tan 29.9° - \tan 24.2°) = 60.5 \tan 24.2°$$

$$x = \frac{60.5 \tan 24.2°}{\tan 29.9° - \tan 24.2°}$$

$$x \approx 216.5.$$

We can then find d_1 and d_2 using the cosine function:

$$\cos 29.9° = \frac{216.5}{d_1} \quad \text{and} \quad \cos 24.2° = \frac{216.5 + 60.5}{d_2}$$

or

$$d_1 = \frac{216.5}{\cos 29.9°} \quad \text{and} \quad d_2 = \frac{277}{\cos 24.2°}$$

$$d_1 \approx 249.7 \qquad\qquad d_2 \approx 303.7.$$

The distance to home plate is about 250 ft,* and the distance to the pitcher's mound is about 304 ft.   ▬

Exercise Set **2.2**

In Exercises 1–6, solve the right triangle.

1.

2.

3.

4.

5.

6.

*In the old Comiskey Park, the distance to home plate was only 150 ft.

In Exercises 7–16, solve the right triangle. (Standard lettering has been used.)

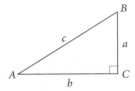

7. $A = 87°43'$, $a = 9.73$ **8.** $a = 12.5$, $b = 18.3$

9. $b = 100$, $c = 450$ **10.** $B = 56.5°$, $c = 0.0447$

11. $A = 47.58°$, $c = 48.3$ **12.** $B = 20.6°$, $a = 7.5$

13. $A = 35°$, $b = 40$ **14.** $B = 69.3°$, $b = 93.4$

15. $b = 1.86$, $c = 4.02$ **16.** $a = 10.2$, $c = 20.4$

17. *Safety Line to Raft.*   Each spring Madison uses his vacation time to ready his lake property for the summer. He wants to run a new safety line from point *B* on the shore to the corner of the anchored diving raft. The current safety line, which runs perpendicular to the shore line to point *A*, is 40 ft long. He estimates the angle from *B* to the corner of the raft to be 50°. Approximately how much rope does he need for the new safety line if he allows 5 ft of rope at each end to fasten the rope?

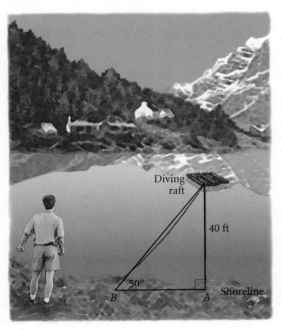

18. *Enclosing an Area.*   Glynis is enclosing a triangular area in a corner of her fenced rectangular backyard for her Labrador retriever. In order for a certain tree to be included in this pen, one side needs to be 14.5 ft and make a 53° angle with the new side. How long is the new side?

19. *Easel Display.*   A marketing group is designing an easel to display posters advertising their newest products. They want the easel to be 6 ft tall and fit flush against a wall. For optimal eye contact, the best angle between the front and back legs of the easel is 23°. How far from the wall should the front legs be placed in order to obtain this angle?

20. *Height of a Tree.*   A supervisor must train a new team of loggers to estimate the heights of trees. As an example, she walks off 40 ft from the base of a tree and estimates the angle of elevation to the tree's peak to be 70°. Approximately how tall is the tree?

21. *Sand Dunes National Park.*   While visiting the Sand Dunes National Park in Colorado, Ian approximated the angle of elevation to the top of a sand dune to be 20°. After walking 800 ft closer, he guessed that the angle of elevation had increased by 15°. Approximately how tall is the dune he was observing?

22. *Tee-Shirt Design.* A new tee-shirt design is to have a regular pentagon inscribed in a circle, as shown in the figure. Each side of the pentagon is to be 3.5 in. long. Find the radius of the circumscribed circle.

3.5 in.

23. *Inscribed Octagon.* A regular octagon is inscribed in a circle of radius 15.8 cm. Find the perimeter of the octagon.

24. *Height of a Weather Balloon.* A weather balloon is directly west of two observing stations that are 10 mi apart. The angles of elevation of the balloon from the two stations are 17.6° and 78.2°. How high is the balloon?

25. *Height of a Kite.* For a science fair project, a group of students tested different materials used to construct kites. Their instructor provided an instrument that accurately measures the angle of elevation. In one of the tests, the angle of elevation was 63.4° with 670 ft of string out. Assuming the string was taut, how high was the kite?

26. *Height of a Building.* A window washer on a ladder looks at a nearby building 100 ft away, noting that the angle of elevation of the top of the building is 18.7° and the angle of depression of the bottom of the building is 6.5°. How tall is the nearby building?

18.7° 100 ft
6.5°

27. *Distance Between Towns.* From a hot-air balloon 2 km high, the angles of depression to two towns, in line with the balloon, are 81.2° and 13.5°. How far apart are the towns?

28. *Angle of Elevation.* What is the angle of elevation of the sun when a 35-ft mast casts a 20-ft shadow?

35 ft
20 ft

29. *Distance from a Lighthouse.* From the top of a lighthouse 55 ft above sea level, the angle of depression to a small boat is 11.3°. How far from the foot of the lighthouse is the boat?

30. *Lightning Detection.* In extremely large forests, it is not cost-effective to position forest rangers in towers or to use small aircraft to continually watch for fires. Since lightning is a frequent cause of fire, lightning detectors are now commonly used instead. These devices not only give a bearing on the location but also measure the intensity of the lightning. A detector at point Q is situated 15 mi west of a central fire station at point R. The bearing from Q to where lightning hits due south of R is S37.6°E. How far is the hit from point R?

31. *Lobster Boat.* A lobster boat is situated due west of a lighthouse. A barge is 12 km south of the lobster boat. From the barge, the bearing to the lighthouse is N63°20′E. How far is the lobster boat from the lighthouse?

12 km
63°20′
N

32. *Length of an Antenna.* A vertical antenna is mounted atop a 50-ft pole. From a point on the level ground 75 ft from the base of the pole, the antenna subtends an angle of 10.5°. Find the length of the antenna.

Discussion and Writing

33. In this section, the trigonometric functions have been defined as functions of acute angles. Thus the set of angles whose measures are greater than 0° and less than 90°. is the domain for each function. What appear to be the ranges for the sine, the cosine, and the tangent functions considering this limited domain?

34. Explain in your own words five ways in which length c can be determined in this triangle. Which way seems the most efficient?

Skill Maintenance

In Exercises 35 and 36, complete the square to:

a) *find the vertex;*
b) *find the line of symmetry; and*
c) *determine whether there is a maximum or minimum value and find that value.*

35. $f(x) = x^2 - 8x + 10$ **36.** $f(x) = 3x^2 - 6x - 1$

Given that $f(x) = 2x - 1$ and $g(x) = x^2 - 5$, find each of the following.

37. $(f + g)(x)$ **38.** $(fg)(-1)$

Synthesis

39. Find h, to the nearest tenth.

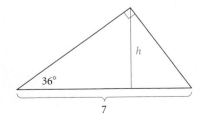

40. Find a, to the nearest tenth.

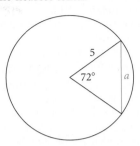

41. *Construction of Picnic Pavilions.* A construction company is mass-producing picnic pavilions for national parks, as shown in the figure. The rafter ends are to be sawed in such a way that they will be vertical when in place. The front wall is 8 ft high, the back wall is $6\frac{1}{2}$ ft high, and the distance between walls is 8 ft. At what angle should the rafters be cut?

42. *Diameter of a Pipe.* A V-gauge is used to find the diameter of a pipe. The advantage of such a device is that it is rugged, it is accurate, and it has no moving parts to break down. In the figure, the measure of angle AVB is 54°. A pipe is placed in the V-shaped slot and the distance VP is used to estimate the diameter. The line VP is calibrated by listing as its units the corresponding diameters. This, in effect, establishes a function between VP and d.

a) Suppose that the diameter of a pipe is 2 cm. What is the distance VP?

b) Suppose that the distance VP is 3.93 cm. What is the diameter of the pipe?

c) Find a formula for d in terms of VP.

d) Find a formula for VP in terms of d.

43. *Sound of an Airplane.* It is common experience to hear the sound of a low-flying airplane, and look at the wrong place in the sky to see the plane. Suppose that a plane is traveling directly at you at a speed of 200 mph and an altitude of 3000 ft, and you hear the sound at what seems to be an angle of inclination of 20°. At what angle θ should you actually look in order to see the plane? Consider the speed of sound to be 1100 ft/sec.

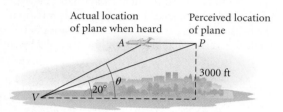

Actual location of plane when heard

Perceived location of plane

A

P

3000 ft

θ

20°

V

44. *Measuring the Radius of the Earth.*   One way to measure the radius of the earth is to climb to the top of a mountain whose height above sea level is known and measure the angle between a vertical line to the center of the earth from the top of the mountain and a line drawn from the top of the mountain to the horizon, as shown in the figure. The height of Mt. Shasta in California is 14,162 ft. From the top of Mt. Shasta, one can see the horizon on the Pacific Ocean. The angle formed between a line to the horizon and the vertical is found to be 87°53′. Use this information to estimate the radius of the earth, in miles.

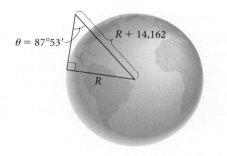

$\theta = 87°53'$

$R + 14{,}162$

R

Trigonometric Functions of Any Angle

2.3

- *Find angles that are coterminal with a given angle and find the complement and the supplement of a given angle.*
- *Determine the six trigonometric function values for any angle in standard position when the coordinates of a point on the terminal side are given.*
- *Find the function values for any angle whose terminal side lies on an axis.*
- *Find the function values for an angle whose terminal side makes an angle of 30°, 45°, or 60° with the x-axis.*
- *Use a grapher to find function values and angles.*

Angles, Rotations, and Degree Measure

An *angle* is a familiar figure in the world around us.

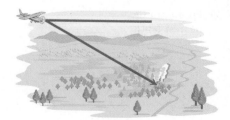

An **angle** is the union of two rays with a common endpoint called the **vertex**. In trigonometry, we often think of an angle as a **rotation**. To do so, think of locating a ray along the positive x-axis with its endpoint at the origin. This ray is called the **initial side** of the angle. Though we leave that ray fixed, think of making a copy of it and rotating it. A rotation *counterclockwise* is a **positive rotation,** and a rotation *clockwise* is a **negative rotation.** The ray at the end of the rotation is called the **terminal side** of the angle. The angle formed is said to be in **standard position.**

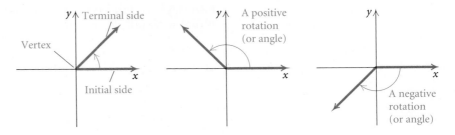

The measure of an angle or rotation may be given in degrees. The Babylonians developed the idea of dividing the circumference of a circle into 360 equal parts, or degrees. If we let the measure of one of these parts be $1°$, then one complete positive revolution or rotation has a measure of $360°$. One half of a revolution has a measure of $180°$, one fourth of a revolution has a measure of $90°$, and so on. We can also speak of an angle of measure $60°$, $135°$, $330°$, or $420°$. The terminal sides of these angles lie in quadrants I, II, IV, and I, respectively. The negative rotations $-30°$, $-110°$, and $-225°$ represent angles with terminal sides in quadrants IV, III, and II, respectively.

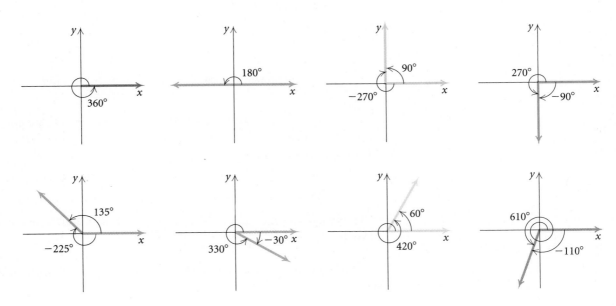

If two or more angles have the same terminal side, the angles are said to be **coterminal**. To find angles coterminal with a given angle, we add or subtract multiples of 360°. For example, 420°, shown above, has the same terminal side as 60°, since 420° = 360° + 60°. Thus we say that angles of measure 60° and 420° are coterminal. The negative rotation that measures −300° is also coterminal with 60° because 60° − 360° = −300°. Other examples of coterminal angles shown above are 90° and −270°, −90° and 270°, 135° and −225°, −30° and 330°, and −110° and 610°.

EXAMPLE 1   Find two positive and two negative angles that are coterminal with (a) 51° and (b) −7°.

Solution

a) We add and subtract multiples of 360°. Many answers are possible.

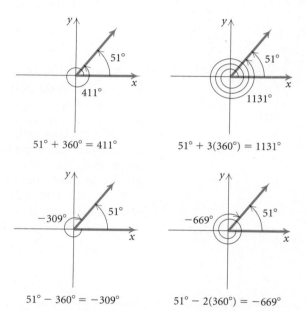

$$51° + 360° = 411°$$

$$51° + 3(360°) = 1131°$$

$$51° − 360° = −309°$$

$$51° − 2(360°) = −669°$$

Thus angles of measure 411°, 1131°, −309°, and −669° are coterminal with 51°.

b) We have the following:

$$-7° + 360° = 353°, \qquad -7° + 2(360°) = 713°,$$
$$-7° - 360° = -367°, \qquad -7° - 10(360°) = -3607°.$$

Thus angles of measure 353°, 713°, −367°, and −3607° are coterminal with −7°.

Angles can be classified by their measures, as seen in the following figure.

Right: $\theta = 90°$ Acute: $0° < \theta < 90°$ Obtuse: $90° < \theta < 180°$ Straight: $\theta = 180°$

Two acute angles are **complementary** if their sum is 90°. For example, angles that measure 10° and 80° are complementary because $10° + 80° = 90°$. Two positive angles are **supplementary** if their sum is 180°. For example, angles that measure 45° and 135° are supplementary because $45° + 135° = 180°$.

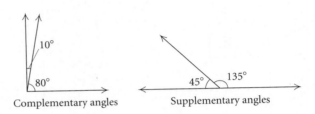

Complementary angles   Supplementary angles

EXAMPLE 2 Find the complement and the supplement of 71.46°.

Solution We have

$$90° - 71.46° = 18.54°,$$
$$180° - 71.46° = 108.54°.$$

Thus the complement of 71.46° is 18.54° and the supplement is 108.54°.

Trigonometric Functions of Angles or Rotations

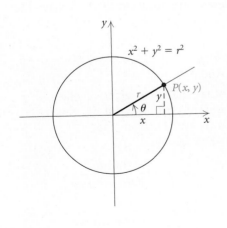

Many applied problems in trigonometry involve the use of angles that are not acute. Thus we need to extend the domains of the trigonometric functions defined in Section 2.1 to angles, or rotations, of any size. To do this, we first consider a right triangle with one vertex at the origin of a coordinate system and one vertex on the positive x-axis. (See the figure at left.) The other vertex is at P, a point on the circle whose center is at the origin and whose radius r is the length of the hypotenuse of the triangle. This triangle is a **reference triangle** for angle θ, which is in standard position. Note that y is the length of the side opposite θ and x is the length of the side adjacent to θ.

Recalling the definitions in Section 2.1, we note that three of the trigonometric functions of angle θ are defined as follows:

$$\sin \theta = \frac{\text{opp}}{\text{hyp}} = \frac{y}{r}, \qquad \cos \theta = \frac{\text{adj}}{\text{hyp}} = \frac{x}{r}, \qquad \tan \theta = \frac{\text{opp}}{\text{adj}} = \frac{y}{x}.$$

Since x and y are the coordinates of the point P and the length of the radius is the length of the hypotenuse, we can also define these functions as follows:

$$\sin\theta = \frac{y\text{-coordinate}}{\text{radius}}, \qquad \cos\theta = \frac{x\text{-coordinate}}{\text{radius}},$$

$$\tan\theta = \frac{y\text{-coordinate}}{x\text{-coordinate}}.$$

We will use these definitions for functions of angles of any measure. The following figures show angles whose terminal sides lie in quadrants II, III, and IV.

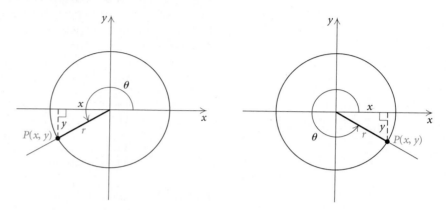

A reference triangle can be drawn for any angle, as shown. Note that the angle is in standard position; that is, it is always measured from the positive half of the x-axis. The point $P(x, y)$ is a point, other than the vertex, on the terminal side of the angle. Each of its two coordinates may be positive, negative, or zero, depending on the location of the terminal side. *The length of the radius, which is also the length of the hypotenuse of the reference triangle, is always considered positive.* (Note that $x^2 + y^2 = r^2$, or $r = \sqrt{x^2 + y^2}$.) Regardless of the location of P, we have the following definitions.

Trigonometric Functions of Any Angle θ

Suppose that $P(x, y)$ is any point other than the vertex on the terminal side of any angle θ in standard position, and r is the radius, or distance from the origin to $P(x, y)$. Then the trigonometric functions are defined as follows.

$$\sin\theta = \frac{y\text{-coordinate}}{\text{radius}} = \frac{y}{r} \qquad \csc\theta = \frac{\text{radius}}{y\text{-coordinate}} = \frac{r}{y}$$

$$\cos\theta = \frac{x\text{-coordinate}}{\text{radius}} = \frac{x}{r} \qquad \sec\theta = \frac{\text{radius}}{x\text{-coordinate}} = \frac{r}{x}$$

$$\tan\theta = \frac{y\text{-coordinate}}{x\text{-coordinate}} = \frac{y}{x} \qquad \cot\theta = \frac{x\text{-coordinate}}{y\text{-coordinate}} = \frac{x}{y}$$

Function values of the trigonometric functions can be positive, negative, or zero, depending on where the terminal side of the angle lies. The length of the radius is always positive. Thus the signs of the function values depend only on the coordinates of the point P on the terminal side of the angle. In the first quadrant, all function values are positive because both coordinates are positive. In the second quadrant, first coordinates are negative and second coordinates are positive; thus only the sine and the cosecant values are positive. Similarly, we can determine the signs of the function values in the third and fourth quadrants. Because of the reciprocal relationships, we need to learn only the signs for the sine, cosine, and tangent functions.

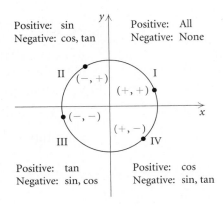

EXAMPLE 3 Find the six trigonometric function values for each angle shown.

a)

b)

c)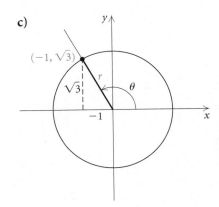

Solution

a) We first determine r, the distance from the origin $(0, 0)$ to the point $(-4, -3)$. The distance between $(0, 0)$ and any point (x, y) on the terminal side of the angle is

$$r = \sqrt{(x - 0)^2 + (y - 0)^2}$$
$$= \sqrt{x^2 + y^2}.$$

Substituting -4 for x and -3 for y, we find

$$r = \sqrt{(-4)^2 + (-3)^2}$$
$$= \sqrt{16 + 9} = \sqrt{25} = 5.$$

Using the definitions of the trigonometric functions, we can now find the function values of θ. We substitute -4 for x, -3 for y, and 5 for r:

$$\sin \theta = \frac{y}{r} = \frac{-3}{5} = -\frac{3}{5}, \qquad \csc \theta = \frac{r}{y} = \frac{5}{-3} = -\frac{5}{3},$$

$$\cos \theta = \frac{x}{r} = \frac{-4}{5} = -\frac{4}{5}, \qquad \sec \theta = \frac{r}{x} = \frac{5}{-4} = -\frac{5}{4},$$

$$\tan \theta = \frac{y}{x} = \frac{-3}{-4} = \frac{3}{4}, \qquad \cot \theta = \frac{x}{y} = \frac{-4}{-3} = \frac{4}{3}.$$

As expected, the tangent and the cotangent values are positive and the other four are negative. This is true for all angles in quadrant III.

b) We first determine r, the distance from the origin to the point $(1, -1)$:

$$r = \sqrt{1^2 + (-1)^2} = \sqrt{1 + 1} = \sqrt{2}.$$

Substituting 1 for x, -1 for y, and $\sqrt{2}$ for r, we find

$$\sin \theta = \frac{y}{r} = \frac{-1}{\sqrt{2}} = -\frac{\sqrt{2}}{2}, \qquad \csc \theta = \frac{r}{y} = \frac{\sqrt{2}}{-1} = -\sqrt{2},$$

$$\cos \theta = \frac{x}{r} = \frac{1}{\sqrt{2}} = \frac{\sqrt{2}}{2}, \qquad \sec \theta = \frac{r}{x} = \frac{\sqrt{2}}{1} = \sqrt{2},$$

$$\tan \theta = \frac{y}{x} = \frac{-1}{1} = -1, \qquad \cot \theta = \frac{x}{y} = \frac{1}{-1} = -1.$$

c) We determine r, the distance from the origin to the point $(-1, \sqrt{3})$:

$$r = \sqrt{(-1)^2 + (\sqrt{3})^2} = \sqrt{1 + 3} = \sqrt{4} = 2.$$

Substituting -1 for x, $\sqrt{3}$ for y, and 2 for r, we find the trigonometric function values of θ are

$$\sin \theta = \frac{\sqrt{3}}{2}, \qquad \csc \theta = \frac{2}{\sqrt{3}} = \frac{2\sqrt{3}}{3},$$

$$\cos \theta = \frac{-1}{2} = -\frac{1}{2}, \qquad \sec \theta = \frac{2}{-1} = -2,$$

$$\tan \theta = \frac{\sqrt{3}}{-1} = -\sqrt{3}, \qquad \cot \theta = \frac{-1}{\sqrt{3}} = -\frac{\sqrt{3}}{3}.$$

Any point other than the origin on the terminal side of an angle can be used to determine the trigonometric function values of that angle. The function values are the same regardless of which point is used. To illustrate this, let's consider an angle θ in standard position whose terminal side lies on the line $y = -\frac{1}{2}x$. We can determine two second-quadrant solutions of the equation, find the length r for each point, and then compare the sine, cosine, and tangent function values using each point.

If $x = -4$, then $y = -\frac{1}{2}(-4) = 2$.

If $x = -8$, then $y = -\frac{1}{2}(-8) = 4$.

For $(-4, 2)$, $r = \sqrt{(-4)^2 + 2^2} = \sqrt{20} = 2\sqrt{5}$.

For $(-8, 4)$, $r = \sqrt{(-8)^2 + 4^2} = \sqrt{80} = 4\sqrt{5}$.

Using $(-4, 2)$ and $r = 2\sqrt{5}$, we find that

$$\sin \theta = \frac{2}{2\sqrt{5}} = \frac{1}{\sqrt{5}} = \frac{\sqrt{5}}{5}, \qquad \cos \theta = \frac{-4}{2\sqrt{5}} = \frac{-2}{\sqrt{5}} = -\frac{2\sqrt{5}}{5},$$

and $\quad \tan \theta = \dfrac{2}{-4} = -\dfrac{1}{2}$.

Using $(-8, 4)$ and $r = 4\sqrt{5}$, we find that

$$\sin \theta = \frac{4}{4\sqrt{5}} = \frac{1}{\sqrt{5}} = \frac{\sqrt{5}}{5}, \qquad \cos \theta = \frac{-8}{4\sqrt{5}} = \frac{-2}{\sqrt{5}} = -\frac{2\sqrt{5}}{5},$$

and $\quad \tan \theta = \dfrac{4}{-8} = -\dfrac{1}{2}$.

We see that the function values are the same using either point. Any point other than the origin on the terminal side of an angle can be used to determine the trigonometric function values.

> The trigonometric function values of θ depend only on the angle, not on the choice of the point on the terminal side that is used to compute them.

The Six Functions Related

When we know one of the function values of an angle, we can find the other five if we know the quadrant in which the terminal side lies. The procedure is to sketch a reference triangle in the appropriate quadrant, use the Pythagorean theorem as needed to find the lengths of its sides, and then read off the ratios of the sides.

EXAMPLE 4 Given that $\tan \theta = -\frac{2}{3}$ and θ is in the second quadrant, find the other function values.

Solution We first sketch a second-quadrant angle. Since

$$\tan \theta = \frac{y}{x} = -\frac{2}{3} = \frac{2}{-3}, \qquad \text{Expressing } -\frac{2}{3} \text{ as } \frac{2}{-3} \text{ since } \theta \text{ is in quadrant II}$$

we make the legs of lengths 2 and 3. The hypotenuse must then have length $\sqrt{2^2 + 3^2}$, or $\sqrt{13}$. Now we can read off the appropriate ratios:

$$\sin \theta = \frac{2}{\sqrt{13}}, \quad \text{or} \quad \frac{2\sqrt{13}}{13}, \qquad\qquad \csc \theta = \frac{\sqrt{13}}{2},$$

$$\cos \theta = -\frac{3}{\sqrt{13}}, \quad \text{or} \quad -\frac{3\sqrt{13}}{13}, \qquad \sec \theta = -\frac{\sqrt{13}}{3},$$

$$\tan \theta = -\frac{2}{3}, \qquad\qquad\qquad\qquad \cot \theta = -\frac{3}{2}.$$

Terminal Side on an Axis

An angle whose terminal side falls on one of the axes is a **quadrantal angle.** Then one of the coordinates of any point on that side is 0. The definitions of the trigonometric functions still apply, but in some cases, function values will not be defined because a denominator will be 0.

EXAMPLE 5   Find the sine, cosine, and tangent values for 90°, 180°, 270°, and 360°.

Solution   We first draw a sketch of each angle in standard position and label a point on the terminal side. Since the function values are the same for all points on the terminal side, we choose (0, 1), (−1, 0), (0, −1), and (1, 0) for convenience. Note also that $r = 1$ for each choice.

Then by the definitions we get

$$\sin 90° = \frac{1}{1} = 1, \qquad \sin 180° = \frac{0}{1} = 0, \qquad \sin 270° = \frac{-1}{1} = -1, \qquad \sin 360° = \frac{0}{1} = 0,$$

$$\cos 90° = \frac{0}{1} = 0, \qquad \cos 180° = \frac{-1}{1} = -1, \qquad \cos 270° = \frac{0}{1} = 0, \qquad \cos 360° = \frac{1}{1} = 1,$$

$$\tan 90° = \frac{1}{0}, \quad \text{Undefined} \qquad \tan 180° = \frac{0}{-1} = 0, \qquad \tan 270° = \frac{-1}{0}, \quad \text{Undefined} \qquad \tan 360° = \frac{0}{1} = 0.$$

In Example 5, all the values can be found using a grapher, but you will find that it is convenient to be able to compute them mentally. It is also helpful to note that coterminal angles have the same function values.

EXAMPLE 6   Find each of the following.

a) $\sin(-90°)$ **b)** $\csc 540°$

Solution

a) We note that $-90°$ is coterminal with $270°$. Thus,

$$\sin(-90°) = \sin 270° = \frac{-1}{1} = -1.$$

b) Since $540° = 180° + 360°$, $540°$ and $180°$ are coterminal. Thus,

$$\csc 540° = \csc 180° = \frac{1}{\sin 180°} = \frac{1}{0}, \quad \text{which is undefined.} \quad \rule{12pt}{3pt}$$

Trigonometric values can always be checked using a grapher. When the value is undefined, the grapher will display a message like

```
ERR: DIVIDE BY 0
1: Quit
2: Goto
```

Reference Angles: 30°, 45°, and 60°

We can also mentally determine trigonometric function values whenever the terminal side makes a 30°, 45°, or 60° angle with the x-axis. Consider, for example, an angle of 150°. The terminal side makes a 30° angle with the x-axis, since $180° - 150° = 30°$.

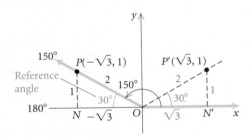

As the figure shows, $\triangle ONP$ is congruent to $\triangle ON'P'$; therefore, the ratios of the sides of the two triangles are the same. Thus the trigonometric function values are the same except perhaps for the sign. We could determine the function values directly from $\triangle ONP$, but this is not necessary. If we remember that in quadrant II, the sine is positive and the cosine and the tangent are negative, we can simply use the function values of 30° that we already know and prefix the appropriate sign. Thus,

$$\sin 150° = \sin 30° = \frac{1}{2}, \qquad \cos 150° = -\cos 30° = -\frac{\sqrt{3}}{2},$$

and $\tan 150° = -\tan 30° = -\dfrac{\sqrt{3}}{3}.$

Triangle *ONP* is the reference triangle and its acute angle, ∠*NOP*, is called a *reference angle*.

> ### Reference Angle
>
> The **reference angle** for an angle is the acute angle formed by the terminal side of the angle and the *x*-axis.

EXAMPLE 7   Find the sine, cosine, and tangent function values for each of the following.

a) 225°   **b)** −780°

Solution

a) We draw a figure showing the terminal side of a 225° angle. The reference angle is 225° − 180°, or 45°.

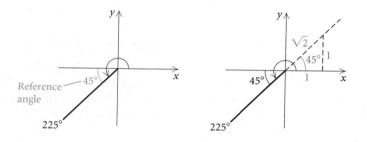

Recall from Section 2.1 that sin 45° = $\sqrt{2}/2$, cos 45° = $\sqrt{2}/2$, and tan 45° = 1. Also note that in the third quadrant, the sine and the cosine are negative and the tangent is positive. Thus we have

$$\sin 225° = -\frac{\sqrt{2}}{2}, \qquad \cos 225° = -\frac{\sqrt{2}}{2}, \quad \text{and} \quad \tan 225° = 1.$$

b) We draw a figure showing the terminal side of a −780° angle. Since −780° + 2(360°) = −60°, we know that −780° and −60° are coterminal.

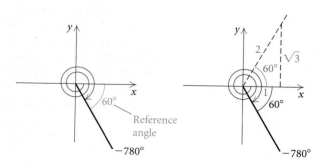

The reference angle for $-60°$ is the acute angle formed by the terminal side of the angle and the x-axis. Thus the reference angle for $-60°$ is $60°$. We know that since $-780°$ is a fourth-quadrant angle, the cosine is positive and the sine and the tangent are negative. Recalling that $\sin 60° = \sqrt{3}/2$, $\cos 60° = 1/2$, and $\tan 60° = \sqrt{3}$, we have

$$\sin(-780°) = -\frac{\sqrt{3}}{2}, \qquad \cos(-780°) = \frac{1}{2},$$

and $\tan(-780°) = -\sqrt{3}.$

Function Values for Any Angle

When the terminal side of an angle falls on one of the axes or makes a 30°, 45°, or 60° angle with the x-axis, we can find exact function values without the use of a grapher. But this group is only a small subset of *all* angles. Using a grapher, we can approximate the trigonometric function values of *any* angle. In fact, we can approximate or find exact function values of all angles without using a reference angle.

EXAMPLE 8   Find each of the following function values using a grapher and round the answer to four decimal places.

a) $\cos 112°$ b) $\sec 500°$ c) $\tan(-83.4°)$

d) $\csc 351.75°$ e) $\cos 2400°$ f) $\sin 175°40'9''$

g) $\cot(-135°)$

Solution   Using a grapher set in DEGREE mode, we find the values.

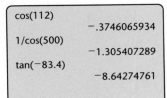

a) $\cos 112° \approx -0.3746$

b) $\sec 500° = \dfrac{1}{\cos 500°} \approx -1.3054$

c) $\tan(-83.4°) \approx -8.6427$

d) $\csc 351.75° = \dfrac{1}{\sin 351.75°} \approx -6.9690$

e) $\cos 2400° = -0.5$

f) $\sin 175°40'9'' \approx 0.0755$

g) $\cot(-135°) = \dfrac{1}{\tan(-135°)} = 1$

In many applications, we have a trigonometric function value and want to find the measure of a corresponding angle. When only acute angles are considered, there is only one angle for each trigonometric function value. This is not the case when we extend the domain of the trigonometric functions to the set of *all* angles. For a given function value, there is an infinite number of angles that have that function value. There can be two such angles for each value in the range from 0° to 360°. To determine a unique answer, the quadrant in which the terminal side lies must be specified.

The grapher gives the reference angle as an output for each function value that is entered as an input. Knowing the reference angle and the quadrant in which the terminal side lies, we can find the specified angle.

EXAMPLE 9 Given the function value and the quadrant restriction, find θ.

a) $\sin \theta = 0.2812$, $90° < \theta < 180°$

b) $\cot \theta = -0.1611$, $270° < \theta < 360°$

Solution

a) We first sketch the angle in the second quadrant. We use the grapher to find the acute angle (reference angle) whose sine is 0.2812. The reference angle is approximately 16.33°. We find the angle θ by subtracting 16.33° from 180°:

$$180° - 16.33° = 163.67°.$$

Thus, $\theta \approx 163.67°$.

b) We begin by sketching the angle in the fourth quadrant. Because the cotangent value is the reciprocal of the tangent value, we know that

$$\tan \theta \approx \frac{1}{-0.1611} \approx -6.2073.$$

We use the grapher to find the acute angle (reference angle) whose tangent is 6.2073, ignoring the fact that $\tan \theta$ is negative. The reference angle is approximately 80.85°. We find angle θ by subtracting 80.85° from 360°:

$$360° - 80.85° = 279.15°.$$

Thus, $\theta \approx 279.15°$.

Exercise Set 2.3

For angles of the following measures, state in which quadrant the terminal side lies. It helps to sketch the angle in standard position.

1. 187°   **2.** −14.3°   **3.** 245°15′

4. −120°   **5.** 800°   **6.** 1075°

7. −460.5°   **8.** 315°   **9.** −912°

10. 13°15′60″   **11.** 537°   **12.** −345.14°

Find two positive angles and two negative angles that are coterminal with the given angle. Answers may vary.

13. 74°   **14.** −81°   **15.** 115.3°

16. 275°10′   **17.** −180°   **18.** −310°

Find the complement and the supplement.

19. 17.11°   **20.** 47°38′   **21.** 12°3′14″

22. 9.038°   **23.** 45.2°   **24.** 67.31°

Find the six trigonometric function values for the angle shown.

25.

26.

27. **28.**

The terminal side of angle θ in standard position lies on the given line in the given quadrant. Find sin θ, cos θ, and tan θ.

29. $2x + 3y = 0$; quadrant IV
30. $4x + y = 0$; quadrant II
31. $5x - 4y = 0$; quadrant I
32. $y = 0.8x$; quadrant III

A function value and a quadrant are given. Find the other five function values. Give exact answers.

33. $\sin \theta = -\dfrac{1}{3}$, quadrant III

34. $\tan \beta = 5$, quadrant I
35. $\cot \theta = -2$, quadrant IV

36. $\cos \alpha = -\dfrac{4}{5}$, quadrant II

37. $\cos \phi = \dfrac{3}{5}$, quadrant IV

38. $\sin \theta = -\dfrac{5}{13}$, quadrant III

Find the reference angle and the exact function value if it exists.

39. $\cos 150°$ **40.** $\sec (-225°)$ **41.** $\tan (-135°)$
42. $\sin (-45°)$ **43.** $\sin 7560°$ **44.** $\tan 270°$
45. $\cos 495°$ **46.** $\tan 675°$ **47.** $\csc (-210°)$
48. $\sin 300°$ **49.** $\cot 570°$ **50.** $\cos (-120°)$
51. $\tan 330°$ **52.** $\cot 855°$ **53.** $\sec (-90°)$
54. $\sin 90°$ **55.** $\cos (-180°)$ **56.** $\csc 90°$
57. $\tan 240°$ **58.** $\cot (-180°)$ **59.** $\sin 495°$
60. $\sin 1050°$ **61.** $\csc 225°$ **62.** $\sin (-450°)$
63. $\cos 0°$ **64.** $\tan 480°$ **65.** $\cot (-90°)$
66. $\sec 315°$ **67.** $\cos 90°$ **68.** $\sin (-135°)$
69. $\cos 270°$ **70.** $\tan 0°$

Find the signs of the six trigonometric function values for the given angles.

71. $319°$ **72.** $-57°$ **73.** $194°$
74. $-620°$ **75.** $-215°$ **76.** $290°$
77. $-272°$ **78.** $91°$

Use a calculator in Exercises 79–82, but do not use the trigonometric function keys.

79. Given that
$$\sin 41° = 0.6561,$$
$$\cos 41° = 0.7547,$$
$$\tan 41° = 0.8693,$$
find the trigonometric function values for 319°.

80. Given that
$$\sin 27° = 0.4540,$$
$$\cos 27° = 0.8910,$$
$$\tan 27° = 0.5095,$$
find the trigonometric function values for 333°.

81. Given that
$$\sin 65° = 0.9063,$$
$$\cos 65° = 0.4226,$$
$$\tan 65° = 2.1445,$$
find the trigonometric function values for 115°.

82. Given that
$$\sin 35° = 0.5736,$$
$$\cos 35° = 0.8192,$$
$$\tan 35° = 0.7002,$$
find the trigonometric function values for 215°.

Aerial Navigation. In aerial navigation, directions are given in degrees clockwise from north. Thus, east is 90°, south is 180°, and west is 270°. Several aerial directions or bearings are given below.

83. An airplane flies 150 km from an airport in a direction of 120°. How far east of the airport is the plane then? How far south?

84. An airplane leaves an airport and travels for 100 mi in a direction of 300°. How far north of the airport is the plane then? How far west?

85. An airplane travels at 150 km/h for 2 hr in a direction of 138° from Omaha. At the end of this time, how far south of Omaha is the plane?

86. An airplane travels at 120 km/h for 2 hr in a direction of 319° from Chicago. At the end of this time, how far north of Chicago is the plane?

Find the function value. Round to four decimal places.

87. tan 310.8°   **88.** cos 205.5°

89. cot 146.15°   **90.** sin (−16.4°)

91. sin 118°42′   **92.** cos 273°45′

93. cos (−295.8°)   **94.** tan 1086.2°

95. cos 5417°   **96.** sec 240°55′

97. csc 520°   **98.** sin 3824°

Given the function value and the quadrant restriction, find θ.

FUNCTION VALUE	INTERVAL	θ
99. $\sin \theta = -0.9956$	(270°, 360°)	
100. $\tan \theta = 0.2460$	(180°, 270°)	
101. $\cos \theta = -0.9388$	(180°, 270°)	
102. $\sec \theta = -1.0485$	(90°, 180°)	
103. $\tan \theta = -3.0545$	(270°, 360°)	
104. $\sin \theta = -0.4313$	(180°, 270°)	
105. $\csc \theta = 1.0480$	(0°, 90°)	
106. $\cos \theta = -0.0990$	(90°, 180°)	

Discussion and Writing

107. Why do the function values of θ depend only on the angle and not on the choice of a point on the terminal side?

108. Why is the domain of the tangent function different from the domains of the sine and the cosine functions?

Skill Maintenance

Find the slope of the line containing the given points.

109. (2, −8) and (−5, −1)

110. (5, 7) and (5, −7)

Determine the domain of the function.

111. $f(x) = \dfrac{x - 4}{x + 2}$

112. $g(x) = \dfrac{x^2 - 9}{2x^2 - 7x - 15}$

Synthesis

113. *Valve Cap on a Bicycle.* The valve cap on a bicycle wheel is 12.5 in. from the center of the wheel. From the position shown, the wheel starts to roll. After the wheel has turned 390°, how far above the ground is the valve cap? Assume that the outer radius of the tire is 13.375 in.

114. *Seats of a Ferris Wheel.* The seats of a ferris wheel are 35 ft from the center of the wheel. When you board the wheel, you are 5 ft above the ground. After you have rotated through an angle of 765°, how far above the ground are you?

Another useful unit of angle measure is called a *radian*. To introduce radian measure, we use a circle centered at the origin with a radius of length 1. Such a circle is called a **unit circle.** Its equation is $x^2 + y^2 = 1$.

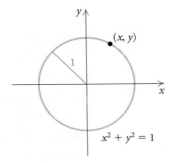

Distances on the Unit Circle

The circumference of a circle of radius r is $2\pi r$. Thus for the unit circle, where $r = 1$, the circumference is 2π. If a point starts at A and travels around the circle (Fig. 1), it will travel a distance of 2π. If it travels halfway around the circle (Fig. 2), it will travel a distance of $\frac{1}{2} \cdot 2\pi$, or π.

FIGURE 1

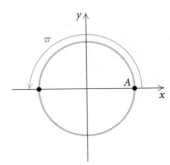

FIGURE 2

If a point C travels $\frac{1}{8}$ of the way around the circle (Fig. 3), it will travel a distance of $\frac{1}{8} \cdot 2\pi$, or $\pi/4$. Note that C is $\frac{1}{4}$ of the way from A to B. If a point D travels $\frac{1}{6}$ of the way around the circle (Fig. 4), it will travel a distance of $\frac{1}{6} \cdot 2\pi$, or $\pi/3$. Note that D is $\frac{1}{3}$ of the way from A to B.

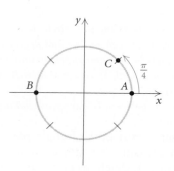

FIGURE 3   FIGURE 4

EXAMPLE 1   How far will a point travel if it goes (a) $\frac{1}{4}$, (b) $\frac{1}{12}$, (c) $\frac{3}{8}$, and (d) $\frac{5}{6}$ of the way around the unit circle?

Solution

a) $\frac{1}{4}$ of the total distance around the circle is $\frac{1}{4} \cdot 2\pi$, which is $\frac{1}{2} \cdot \pi$, or $\pi/2$.

b) The distance will be $\frac{1}{12} \cdot 2\pi$, which is $\frac{1}{6}\pi$, or $\pi/6$.

c) The distance will be $\frac{3}{8} \cdot 2\pi$, which is $\frac{3}{4}\pi$, or $3\pi/4$.

d) The distance will be $\frac{5}{6} \cdot 2\pi$, which is $\frac{5}{3}\pi$, or $5\pi/3$. Think of $5\pi/3$ as $\pi + \frac{2}{3}\pi$.

These distances are illustrated in the following figures.

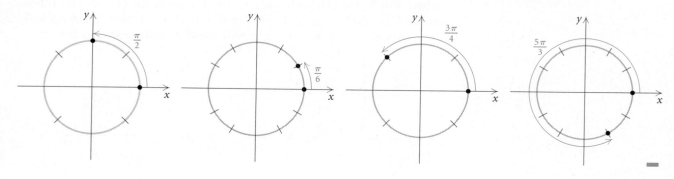

A point may travel completely around the circle and then continue. For example, if it goes around once and then continues $\frac{1}{4}$ of the way around, it will have traveled a distance of $2\pi + \frac{1}{4} \cdot 2\pi$, or $5\pi/2$ (Fig. 5). *Every* real number determines a point on the unit circle. For the positive number 10, for example, we start at A and travel counterclockwise a distance of 10. The point at which we stop is the point "determined" by the number 10 (Fig. 6).

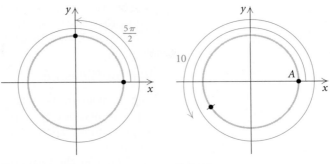

FIGURE 5   FIGURE 6

For a negative number, we move clockwise around the circle. Points for $-\pi/4$ and $-3\pi/2$ are shown in the figure below. The number 0 determines the point A.

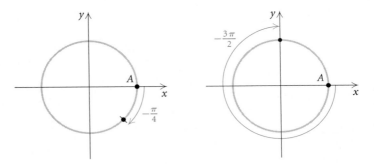

EXAMPLE 2   On the unit circle, mark the point determined by each of the following real numbers.

a) $\dfrac{9\pi}{4}$ **b)** $-\dfrac{7\pi}{6}$

Solution

a) Think of $9\pi/4$ as $2\pi + \frac{1}{4}\pi$ (see the figure on the left below). Since $9\pi/4 > 0$, the point moves counterclockwise. The point goes completely around once and then continues $\frac{1}{4}$ of the way from A to B.

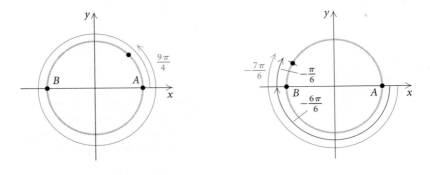

b) The number is negative, so the point moves clockwise. From A to B, the distance is π, or $\frac{6}{6}\pi$, so we need to go beyond B another distance of $\pi/6$, clockwise. (See the figure on the right above.)

Radian Measure

Degree measure is a common unit of angle measure in many everyday applications. But in many scientific fields and in mathematics (calculus, in particular), there is another commonly used unit of measure called the *radian*.

Consider the unit circle. Recall that this circle has radius 1. Suppose we measure, moving counterclockwise, an arc of length 1, and mark a point T on the circle.

If we draw a ray from the origin through T, we have formed an angle. The measure of that angle is 1 **radian**. The word radian comes from the word *radius*. Thus measuring 1 "radius" along the circumference of the circle determines an angle whose measure is 1 *radian*. One radian is about 57.3°. Angles that measure 2 radians, 3 radians, and 6 radians are shown below.

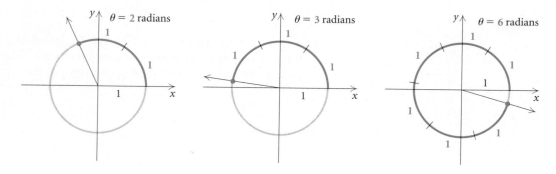

When we make a complete (counterclockwise) revolution, the terminal side coincides with the initial side on the positive x-axis. We then have an angle whose measure is 2π radians, or about 6.28 radians, which is the circumference of the circle:

$$2\pi r = 2\pi(1) = 2\pi.$$

Thus a rotation of 360° (1 revolution) has a measure of 2π radians. A half revolution is a rotation of 180°, or π radians. A quarter revolution is a rotation of 90°, or $\pi/2$ radians, and so on.

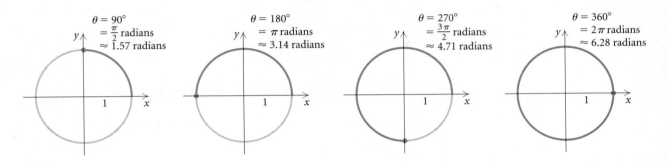

To convert between degrees and radians, we first note that

$$360° = 2\pi \text{ radians.}$$

It follows that

$$180° = \pi \text{ radians.}$$

To make conversions, we multiply by 1, noting that:

$$\frac{\pi \text{ radians}}{180°} = \frac{180°}{\pi \text{ radians}} = 1.$$

To convert from degree to radian measure, multiply by $\dfrac{\pi \text{ radians}}{180°}$.

To convert from radian to degree measure, multiply by $\dfrac{180°}{\pi \text{ radians}}$.

EXAMPLE 3   Convert each of the following to radians.

a) 120°

b) −297.25°

Solution

a) $120° = 120° \cdot \dfrac{\pi \text{ radians}}{180°}$ Multiplying by 1

$\qquad = \dfrac{120°}{180°} \pi \text{ radians}$

$\qquad = \dfrac{2\pi}{3} \text{ radians,} \quad \text{or about} \quad 2.09 \text{ radians}$

b) $-297.25° = -297.25° \cdot \dfrac{\pi \text{ radians}}{180°}$

$$= -\dfrac{297.25°}{180°} \, \pi \text{ radians}$$

$$= -\dfrac{297.25\pi}{180} \text{ radians}$$

$$\approx -5.19 \text{ radians}$$

EXAMPLE 4   Convert each of the following to degrees.

a) $\dfrac{3\pi}{4}$ radians

b) 8.5 radians

Solution

a) $\dfrac{3\pi}{4}$ radians $= \dfrac{3\pi}{4}$ radians $\cdot \dfrac{180°}{\pi \text{ radians}}$ Multiplying by 1

$$= \dfrac{3\pi}{4\pi} \cdot 180° = \dfrac{3}{4} \cdot 180° = 135°$$

b) 8.5 radians $= 8.5$ radians $\cdot \dfrac{180°}{\pi \text{ radians}}$

$$= \dfrac{8.5(180°)}{\pi} \approx 487.01°$$

The radian–degree equivalents of the most commonly used angle measures are illustrated in the following figures.

When a rotation is given in radians, the word "radians" is optional and is most often omitted. *Thus if no unit is given for a rotation, the rotation is understood to be in radians.*

We can also find coterminal, complementary, and supplementary angles in radian measure just as we did for degree measure in Section 2.3.

EXAMPLE 5   Find a positive angle and a negative angle that are coterminal with $2\pi/3$. Many answers are possible.

Solution   To find angles coterminal with a given angle, we add or subtract multiples of 2π:

$$\frac{2\pi}{3} + 2\pi = \frac{2\pi}{3} + \frac{6\pi}{3} = \frac{8\pi}{3},$$

$$\frac{2\pi}{3} - 3(2\pi) = \frac{2\pi}{3} - \frac{18\pi}{3} = -\frac{16\pi}{3}.$$

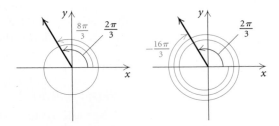

Thus, $8\pi/3$ and $-16\pi/3$ are coterminal with $2\pi/3$.

EXAMPLE 6   Find the complement and the supplement of $\pi/6$.

Solution   Since 90° equals $\pi/2$ radians, the complement of $\pi/6$ is

$$\frac{\pi}{2} - \frac{\pi}{6} = \frac{3\pi}{6} - \frac{\pi}{6} = \frac{2\pi}{6}, \quad \text{or} \quad \frac{\pi}{3}.$$

Since 180° equals π radians, the supplement of $\pi/6$ is

$$\pi - \frac{\pi}{6} = \frac{6\pi}{6} - \frac{\pi}{6} = \frac{5\pi}{6}.$$

Thus the complement of $\pi/6$ is $\pi/3$ and the supplement is $5\pi/6$.

Arc Length and Central Angles

Radian measure can be determined using a circle other than a unit circle. In the figure at left, a unit circle (with radius 1) is shown along with another circle (with radius r, $r \neq 1$). The angle shown is a **central angle** of both circles.

From geometry, we know that the arcs that the angle subtends have their lengths in the same ratio as the radii of the circles. The radii of the circles are r and 1. The corresponding arc lengths are s and s_1. Thus we have the proportion

$$\frac{s}{s_1} = \frac{r}{1},$$

which can also be written as

$$\frac{s_1}{1} = \frac{s}{r}.$$

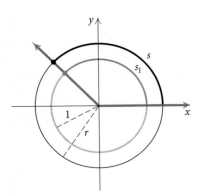

Now s_1 is the *radian measure* of the rotation in question. It is common to use a Greek letter, such as θ, for the measure of an angle or rotation and the letter s for arc length. Adopting this convention, we rewrite the proportion above as

$$\theta = \frac{s}{r}.$$

In any circle, the measure (in radians) of a central angle, the arc length the angle subtends, and the length of the radius are related in this fashion. Or, in general, the following is true.

Radian Measure

The **radian measure** θ of a rotation is the ratio of the distance s traveled by a point at a radius r from the center of rotation, to the length of the radius r:

$$\theta = \frac{s}{r}.$$

When using the formula $\theta = s/r$, you must make sure that θ is in radians and that s and r are expressed in the same unit.

EXAMPLE 7   Find the measure of a rotation in radians where a point 2 m from the center of rotation travels 4 m.

Solution

$$\theta = \frac{s}{r}$$

$$= \frac{4 \text{ m}}{2 \text{ m}} = 2 \qquad \text{The unit is understood to be radians.}$$

EXAMPLE 8   Find the length of an arc of a circle of radius 5 cm associated with an angle of $\pi/3$ radians.

Solution   We have

$$\theta = \frac{s}{r}, \quad \text{or} \quad s = r\theta.$$

Thus, $s = 5 \cdot \pi/3$ cm, or about 5.24 cm.

Linear Speed and Angular Speed

Linear speed is defined to be distance traveled per unit of time. If we use v for linear speed, s for distance, and t for time, then

$$v = \frac{s}{t}.$$

Similarly, **angular speed** is defined to be amount of rotation per unit of time. For example, we might speak of the angular speed of a bicycle wheel as 150 revolutions per minute or the angular speed of the earth as 2π radians per day. The Greek letter ω (omega) is generally used for angular speed. Thus angular speed is defined as

$$\omega = \frac{\theta}{t}.$$

As an example of how these definitions can be applied, let's consider the refurbished carousel at the Children's Museum in Indianapolis, Indiana. It consists of three circular rows of animals. All animals, regardless of the row, travel at the same angular speed. But the animals in the outer row travel at a greater linear speed than those in the inner rows. What is the relationship between the linear speed v and the angular speed ω?

To develop the relationship we seek, recall that, for rotations measured in radians, $\theta = s/r$. This is equivalent to

$$s = r\theta.$$

We divide by time, t, to obtain

$$\frac{s}{t} = r\,\frac{\theta}{t}.$$

Now s/t is linear speed v and θ/t is angular speed ω. Thus we have the relationship we seek, $v = r\omega$.

Linear Speed in Terms of Angular Speed

The **linear speed** v of a point a distance r from the center of rotation is given by

$$v = r\omega,$$

where ω is the **angular speed** in radians per unit of time.

For our new formula $v = r\omega$, the units of distance for v and r must be the same, ω must be in radians per unit of time, and the units of time for v and ω must be the same.

EXAMPLE 9 *Linear Speed of an Earth Satellite.* An earth satellite in circular orbit 1200 km high makes one complete revolution every 90 min. What is its linear speed? Use 6400 km for the length of a radius of the earth.

Solution To use the formula $v = r\omega$, we need to know r and ω:

$$r = 6400 \text{ km} + 1200 \text{ km}$$ Radius of earth plus height of satellite

$$= 7600 \text{ km},$$

$$\omega = \frac{\theta}{t} = \frac{2\pi}{90 \text{ min}} = \frac{\pi}{45 \text{ min}}.$$ We have, as usual, omitted the word radians.

Now, using $v = r\omega$, we have

$$v = 7600 \text{ km} \cdot \frac{\pi}{45 \text{ min}} = \frac{7600\pi}{45} \cdot \frac{\text{km}}{\text{min}} \approx 531 \frac{\text{km}}{\text{min}}.$$

Thus the linear speed of the satellite is approximately 531 km/min. ▬

EXAMPLE 10 *Angular Speed of a Capstan.* An anchor is hoisted at a rate of 2 ft/sec as the chain is wound around a capstan with a 1.8-yd diameter. What is the angular speed of the capstan?

Solution We will use the formula $v = r\omega$ in the form $\omega = v/r$, taking care to use the proper units. Since v is given in feet per second, we need r in feet:

$$r = \frac{d}{2} = \frac{1.8}{2} \text{ yd} \cdot \frac{3 \text{ ft}}{1 \text{ yd}} = 2.7 \text{ ft}.$$

Then ω will be in radians per second:

$$\omega = \frac{v}{r} = \frac{2 \text{ ft/sec}}{2.7 \text{ ft}} = \frac{2 \text{ ft}}{\text{sec}} \cdot \frac{1}{2.7 \text{ ft}} \approx 0.741/\text{sec}.$$

Thus the angular speed is approximately 0.741 radian/sec. ▬

The formulas $\theta = \omega t$ and $v = r\omega$ can be used in combination to find distances and angles in various situations involving rotational motion.

EXAMPLE 11 *Angle of Revolution.* A Porsche 911 is traveling at a speed of 65 mph. Its tires have an outside diameter of 25.086 in. Find the angle through which a tire turns in 10 sec.

STUDY TIP

The *Student's Solutions Manual* is an excellent resource if you need additional help with an exercise in the exercise sets. It contains worked-out solutions to the odd-numbered exercises in the exercise sets.

Solution   Recall that $\omega = \theta/t$, or $\theta = \omega t$. Thus we can find θ if we know ω and t. To find ω, we use the formula $v = r\omega$. The linear speed v of a point on the outside of the tire is the speed of the Porsche, 65 mph. For convenience, we first convert 65 mph to feet per second:

$$v = 65\frac{\text{mi}}{\text{hr}} \cdot \frac{1 \text{ hr}}{60 \text{ min}} \cdot \frac{1 \text{ min}}{60 \text{ sec}} \cdot \frac{5280 \text{ ft}}{1 \text{ mi}}$$

$$\approx 95.333 \frac{\text{ft}}{\text{sec}}.$$

The radius of the tire is half the diameter. Now $r = d/2 = 25.086$ in./2 $=$ 12.543 in. We will convert to feet, since v is in feet per second:

$$r = 12.543 \text{ in.} \cdot \frac{1 \text{ ft}}{12 \text{ in.}}$$

$$= \frac{12.543}{12} \text{ ft} \approx 1.045 \text{ ft.}$$

Using $v = r\omega$, we have

$$95.333 \frac{\text{ft}}{\text{sec}} = 1.045 \text{ ft} \cdot \omega,$$

so

$$\omega = \frac{95.333 \text{ ft/sec}}{1.045 \text{ ft}} \approx \frac{91.228}{\text{sec}}.$$

Then in 10 sec,

$$\theta = \omega t = \frac{91.228}{\text{sec}} \cdot 10 \text{ sec} \approx 912.$$

Thus the angle, in radians, through which a tire turns in 10 sec is 912.

Exercise Set 2.4

For each of Exercises 1–4, sketch a unit circle and mark the points determined by the given real numbers.

1. a) $\dfrac{\pi}{4}$ **b)** $\dfrac{3\pi}{2}$ **c)** $\dfrac{3\pi}{4}$

   d) π **e)** $\dfrac{11\pi}{4}$ **f)** $\dfrac{17\pi}{4}$

2. a) $\dfrac{\pi}{2}$ **b)** $\dfrac{5\pi}{4}$ **c)** 2π

   d) $\dfrac{9\pi}{4}$ **e)** $\dfrac{13\pi}{4}$ **f)** $\dfrac{23\pi}{4}$

3. a) $\dfrac{\pi}{6}$ **b)** $\dfrac{2\pi}{3}$ **c)** $\dfrac{7\pi}{6}$

   d) $\dfrac{10\pi}{6}$ **e)** $\dfrac{14\pi}{6}$ **f)** $\dfrac{23\pi}{4}$

4. a) $-\dfrac{\pi}{2}$ **b)** $-\dfrac{3\pi}{4}$ **c)** $-\dfrac{5\pi}{6}$

   d) $-\dfrac{5\pi}{2}$ **e)** $-\dfrac{17\pi}{6}$ **f)** $-\dfrac{9\pi}{4}$

Find two real numbers between -2π *and* 2π *that determine each of the points on the unit circle.*

5.

6.

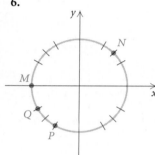

For Exercises 7 and 8, sketch a unit circle and mark the approximate location of the point determined by the given real number.

7. a) 2.4   **b)** 7.5
   c) 32   **d)** 320

8. a) 0.25   **b)** 1.8
   c) 47   **d)** 500

Find a positive angle and a negative angle that are coterminal with the given angle. Answers may vary.

9. $\dfrac{\pi}{4}$ **10.** $\dfrac{5\pi}{3}$

11. $\dfrac{7\pi}{6}$ **12.** π

Find the complement and the supplement.

13. $\dfrac{\pi}{3}$ **14.** $\dfrac{5\pi}{12}$

15. $\dfrac{3\pi}{8}$ **16.** $\dfrac{\pi}{4}$

Convert to radian measure. Leave the answer in terms of π.

17. 75°   **18.** 30°

19. 200°   **20.** -135°

21. -214.6°   **22.** 37.71°

23. -180°   **24.** 90°

Convert to radian measure. Round the answer to two decimal places.

25. 240°   **26.** 15°

27. -60°   **28.** 145°

29. 117.8°   **30.** -231.2°

31. 1.354°   **32.** 584°

Convert to degree measure. Round the answer to two decimal places.

33. $-\dfrac{3\pi}{4}$ **34.** $\dfrac{7\pi}{6}$

35. 8π **36.** $-\dfrac{\pi}{3}$

37. 1   **38.** -17.6

39. 2.347   **40.** 25

41. Certain positive angles are marked here in degrees. Find the corresponding radian measures.

42. Certain negative angles are marked here in degrees. Find the corresponding radian measures.

43. In a circle with a 120-cm radius, an arc 132 cm long subtends an angle of how many radians? how many degrees, to the nearest degree?

44. In a circle with a 10-ft diameter, an arc 20 ft long subtends an angle of how many radians? how many degrees, to the nearest degree?

45. In a circle with a 2-yd radius, how long is an arc associated with an angle of 1.6 radians?

46. In a circle with a 5-m radius, how long is an arc associated with an angle of 2.1 radians?

47. *Angle of Revolution.* Through how many radians does the minute hand of a clock rotate from 12:40 P.M. to 1:30 P.M.?

48. *Angle of Revolution.* A tire on a Dodge Neon has an outside diameter of 23.468 in. Through what angle (in radians) does the tire turn while traveling 1 mi?

23.468 in.

49. *Linear Speed.* A flywheel with a 15-cm diameter is rotating at a rate of 7 radians/sec. What is the linear speed of a point on its rim, in centimeters per minute?

50. *Linear Speed.* A wheel with a 30-cm radius is rotating at a rate of 3 radians/sec. What is the linear speed of a point on its rim, in meters per minute?

51. *Angular Speed on a Printing Press.* This text was printed on a four-color web heatset offset press. A cylinder on this press has a 13.37-in. diameter. The linear speed of a point on the cylinder's surface is 18.33 feet per second. What is the angular speed of the cylinder, in revolutions per hour? Printers often refer to the angular speed as impressions per hour

(IPH). (*Source:* Scott Coulter, Quebecor World, Taunton, MA)

52. *Linear Speeds on a Carousel.* When Alicia and Zoe ride the carousel described earlier in this section, Alicia always selects a horse on the outside row, whereas Zoe prefers the row closest to the center. These rows are 19 ft 3 in. and 13 ft 11 in. from the center, respectively. The angular speed of the carousel is 2.4 revolutions per minute. What is the difference, in miles per hour, in the linear speeds of Alicia and Zoe? (*Source:* The Children's Museum, Indianapolis, IN)

53. *Linear Speed at the Equator.* The earth has a 4000-mi radius and rotates one revolution every 24 hr. What is the linear speed of a point on the equator, in miles per hour?

54. *Linear Speed of the Earth.* The earth is 93,000,000 mi from the sun and traverses its orbit, which is nearly circular, every 365.25 days. What is the linear velocity of the earth in its orbit, in miles per hour?

55. *Determining the Speed of a River.* A water wheel has a 10-ft radius. To get a good approximation of the speed of the river, you count the revolutions of the wheel and find that it makes 14 revolutions per minute (rpm). What is the speed of the river, in miles per hour?

10 ft

56. *The Tour de France.*   Lance Armstrong won the 1999 Tour de France bicycle race. The wheel of his bicycle had a 58-cm diameter. His overall average linear speed during the race was 40.273 km/h. What was the angular speed of the wheel, in revolutions per hour? (*Source:* Wilcockson, John, and Charles Pelkey, *Lance Armstrong & the 1999 Tour de France.* Boulder, CO: VeloPress, 1999, p. 173)

57. *John Deere Tractor.*   A rear wheel on a John Deere 8300 farm tractor has a 23-in. radius. Find the angle (in radians) through which a wheel rotates in 12 sec if the tractor is traveling at a speed of 22 mph.

Discussion and Writing

58. Explain in your own words why it is preferable to omit the word, or unit, *radians* in radian measures.

59. In circular motion with a fixed angular speed, the length of the radius is directly proportional to the linear speed. Explain why with an example.

60. Two new cars are each driven at an average speed of 60 mph for an extended highway test drive of 2000 mi. The diameter of the wheels of the two cars are 15 in. and 16 in., respectively. If the cars use tires of equal durability and profile, differing only by the diameter, which car will probably need new tires first? Explain your answer.

Skill Maintenance

Convert to decimal degree notation. Round to two decimal places.

61. 37°18′   **62.** 5°26″

Convert to degrees, minutes, and seconds. Round to the nearest second.

63. 81.21°   **64.** 3.1°

Synthesis

65. A **mil** is a unit of angle measure. A right angle has a measure of 1600 mils. Convert each of the following to degrees, minutes, and seconds.

   a) 100 mils   **b)** 350 mils

66. A **grad** is a unit of angle measure similar to a degree. A right angle has a measure of 100 grads. Convert each of the following to grads.

   a) 48°   **b)** $\dfrac{5\pi}{7}$

67. *Angular Speed of a Gear Wheel.*   One gear wheel turns another, the teeth being on the rims. The wheels have 40-cm and 50-cm radii, and the smaller wheel rotates at 20 rpm. Find the angular speed of the larger wheel, in radians per second.

68. *Angular Speed of a Pulley.*   Two pulleys, 50 cm and 30 cm in diameter, respectively, are connected by a belt. The larger pulley makes 12 revolutions per minute. Find the angular speed of the smaller pulley, in radians per second.

69. *Distance between Points on the Earth.*   To find the distance between two points on the earth when their latitude and longitude are known, we can use a right triangle for an excellent approximation if the points are not too far apart. Point *A* is at latitude 38°27′30″ N, longitude 82°57′15″ W; and point *B* is at latitude 38°28′45″ N, longitude 82°56′30″ W. Find the distance from *A* to *B* in nautical miles (one minute of latitude is one nautical mile).

70. *Hands of a Clock.*   At what times between noon and 1:00 P.M. are the hands of a clock perpendicular?

Circular Functions: Graphs and Properties

2.5

- Given the coordinates of a point on the unit circle, find its reflections across the x-axis, the y-axis, and the origin.
- Determine the six trigonometric function values for a real number when the coordinates of the point on the unit circle determined by that real number are given.
- Find function values for any real number using a grapher.
- Graph the six circular functions and state their properties.

The domains of the trigonometric functions, defined in Sections 2.1 and 2.3, have been sets of angles or rotations measured in a real number of degree units. We can also consider the domains to be sets of real numbers, or radians, introduced in Section 2.4. Many applications in calculus that use the trigonometric functions refer only to radians.

Let's again consider radian measure and the unit circle. We defined radian measure for θ as

$$\theta = \frac{s}{r}.$$

When $r = 1$,

$$\theta = \frac{s}{1}, \quad \text{or} \quad \theta = s.$$

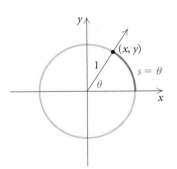

STUDY TIP

Take advantage of the numerous detailed art pieces in this text. They provide a visual image of the concept being discussed. Taking the time to study each figure is an efficient way to learn and retain the concepts.

The arc length *s* on the unit circle is the same as the radian measure of the angle θ.

In the figure above, the point (x, y) is the point where the terminal side of the angle with radian measure s intersects the unit circle. We can now extend our definitions of the trigonometric functions using domains composed of real numbers, or radians.

In the definitions, s can be considered the radian measure of an angle or the measure of an arc length on the unit circle. Either way, s is a real number. To each real number s, there corresponds an arc length s on the unit circle. Trigonometric functions with domains composed of real numbers are called **circular functions.**

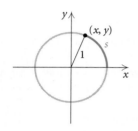

Basic Circular Functions

For a real number s that determines a point (x, y) on the unit circle:

$$\sin s = \text{second coordinate} = y,$$
$$\cos s = \text{first coordinate} = x,$$
$$\tan s = \frac{\text{second coordinate}}{\text{first coordinate}} = \frac{y}{x} \ (x \neq 0),$$
$$\csc s = \frac{1}{\text{second coordinate}} = \frac{1}{y} \ (y \neq 0),$$
$$\sec s = \frac{1}{\text{first coordinate}} = \frac{1}{x} \ (x \neq 0),$$
$$\cot s = \frac{\text{first coordinate}}{\text{second coordinate}} = \frac{x}{y} \ (y \neq 0).$$

We can consider the domains of trigonometric functions to be real numbers rather than angles. We can determine these values for a specific real number if we know the coordinates of the point on the unit circle determined by that number. As with degree measure, we can also find these function values directly using a grapher.

Reflections on the Unit Circle

Let's consider the unit circle and a few of its points. For any point (x, y) on the unit circle, $x^2 + y^2 = 1$, we know that $-1 \leq x \leq 1$ and $-1 \leq y \leq 1$. If we know the x- or y-coordinate of a point on the unit circle, we can find the other coordinate. If $x = \frac{3}{5}$, then

$$\left(\tfrac{3}{5}\right)^2 + y^2 = 1$$
$$y^2 = 1 - \tfrac{9}{25} = \tfrac{16}{25}$$
$$y = \pm\tfrac{4}{5}.$$

Thus, $\left(\frac{3}{5}, \frac{4}{5}\right)$ and $\left(\frac{3}{5}, -\frac{4}{5}\right)$ are points on the unit circle. There are two points with an x-coordinate of $\frac{3}{5}$.

Now let's consider the radian measure $\pi/3$ and determine the coordinates of the point on the unit circle determined by $\pi/3$. We construct a right triangle by dropping a perpendicular segment from the point to the x-axis.

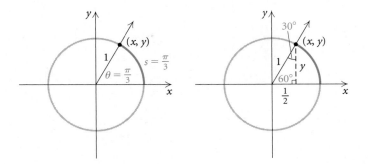

Since $\pi/3 = 60°$, we have a 30°–60° right triangle in which the side opposite the 30° angle is one half of the hypotenuse. The hypotenuse, or radius, is 1, so the side opposite the 30° angle is $\frac{1}{2}$. Using the Pythagorean theorem, we can find the other side:

$$\left(\frac{1}{2}\right)^2 + y^2 = 1$$

$$y^2 = 1 - \frac{1}{4} = \frac{3}{4}$$

$$y = \sqrt{\frac{3}{4}} = \frac{\sqrt{3}}{2}.$$

We know that y is positive since the point is in the first quadrant. Thus the coordinates of the point determined by $\pi/3$ are $x = 1/2$ and $y = \sqrt{3}/2$, or $(1/2, \sqrt{3}/2)$. We can always check to see if a point is on the unit circle by substituting into the equation $x^2 + y^2 = 1$.

Because a unit circle is symmetric with respect to the x-axis, the y-axis, and the origin, we can use the coordinates of one point on the unit circle to find coordinates of its reflections.

EXAMPLE 1   Each of the following points lies on the unit circle. Find their reflections across the x-axis, the y-axis, and the origin.

a) $\left(\dfrac{3}{5}, \dfrac{4}{5}\right)$ b) $\left(\dfrac{\sqrt{2}}{2}, \dfrac{\sqrt{2}}{2}\right)$

c) $\left(\dfrac{1}{2}, \dfrac{\sqrt{3}}{2}\right)$

Solution

a)

b)

c)

Finding Function Values

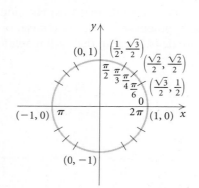

Knowing the coordinates of only a few points on the unit circle along with their reflections allows us to find trigonometric function values of the most frequently used real numbers, or radians.

EXAMPLE 2   Find each of the following function values.

a) $\tan \dfrac{\pi}{3}$

b) $\cos \dfrac{3\pi}{4}$

c) $\sin \left(-\dfrac{\pi}{6} \right)$

d) $\cos \dfrac{4\pi}{3}$

e) $\cot \pi$

f) $\csc \left(-\dfrac{7\pi}{2} \right)$

Solution   We locate the point on the unit circle determined by the rotation, and then find its coordinates using reflection if necessary.

a) The coordinates of the point determined by $\pi/3$ are $(1/2, \sqrt{3}/2)$.

b) The reflection of $(\sqrt{2}/2, \sqrt{2}/2)$ across the y-axis is $(-\sqrt{2}/2, \sqrt{2}/2)$.

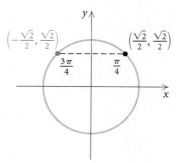

Thus, $\tan \dfrac{\pi}{3} = \dfrac{y}{x} = \dfrac{\sqrt{3}/2}{1/2} = \sqrt{3}.$

Thus, $\cos \dfrac{3\pi}{4} = x = -\dfrac{\sqrt{2}}{2}.$

c) The reflection of $(\sqrt{3}/2, 1/2)$ across the x-axis is $(\sqrt{3}/2, -1/2)$.

Thus, $\sin\left(-\dfrac{\pi}{6}\right) = y = -\dfrac{1}{2}$.

d) The reflection of $(1/2, \sqrt{3}/2)$ across the origin is $(-1/2, -\sqrt{3}/2)$.

Thus, $\cos\dfrac{4\pi}{3} = x = -\dfrac{1}{2}$.

e) The coordinates of the point determined by π are $(-1, 0)$.

We can also think of $\cot \pi$ as the reciprocal of $\tan \pi$. Since $\tan \pi = y/x = 0/-1 = 0$ and the reciprocal of 0 is not defined, we know that $\cot \pi$ is undefined.

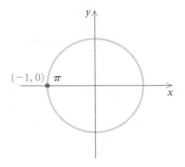

Thus, $\cot \pi = \dfrac{x}{y} = \dfrac{-1}{0}$, which is undefined.

f) The coordinates of the point determined by $-7\pi/2$ are $(0, 1)$.

Thus, $\csc\left(-\dfrac{7\pi}{2}\right) = \dfrac{1}{y} = \dfrac{1}{1} = 1$.

Using a grapher, we can find trigonometric function values of any real number without knowing the coordinates of the point that it determines on the unit circle. Most graphers have both degree and radian modes. When finding function values of radian measures, or real numbers, we *must* set the grapher in RADIAN mode. (See the window at left.)

EXAMPLE 3 Find each of the following function values of radian measures using a grapher. Round the answers to four decimal places.

a) $\cos \dfrac{2\pi}{5}$ b) $\tan(-3)$ c) $\sin 24.9$ d) $\sec \dfrac{\pi}{7}$

cos(2π/5)
                .3090169944
tan(−3)
                .1425465431
sin(24.9)
                −.2306457059

Solution Using a grapher set in RADIAN mode, we find the values.

a) $\cos \dfrac{2\pi}{5} \approx 0.3090$

b) $\tan\,(-3) \approx 0.1425$

c) $\sin 24.9 \approx -0.2306$

d) $\sec \dfrac{\pi}{7} = \dfrac{1}{\cos \dfrac{\pi}{7}} \approx 1.1099$

Note in part (d) that the secant function value can be found by taking the reciprocal of the cosine value. Thus we can enter cos $\pi/7$ and use the reciprocal key. ▬

Interactive Discovery

We can graph the unit circle using a grapher. We use PARAMETRIC mode with the following window and let X₁ᴛ = cos T and Y₁ᴛ = sin T. Here we use DEGREE mode.

WINDOW
  Tmin = 0
  Tmax = 360
  Tstep = 15
  Xmin = −1.5
  Xmax = 1.5
  Xscl = 1
  Ymin = −1
  Ymax = 1
  Yscl = 1

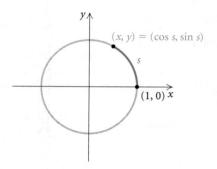

X₁ᴛ=cos(T)   Y₁ᴛ=sin(T)

T = 30
X = .8660254   Y = .5

Using the trace key and an arrow key to move the cursor around the unit circle, we see the T, X, and Y values appear on the screen. What do they represent? Repeat this exercise in RADIAN mode. What do the T, X, and Y values represent? (For more on parametric equations, see Section 5.8.)

From the definitions on p. 162 and from the preceding exploration, we can relabel any point (x, y) on the unit circle as $(\cos s, \sin s)$, where s is any real number.

$(x, y) = (\cos s, \sin s)$

s

$(1, 0)$

Graphs of the Sine and Cosine Functions

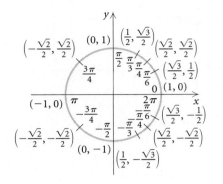

Properties of functions can be observed from their graphs. We begin by graphing the sine and cosine functions. We make a table of values, plot the points, and then connect those points with a smooth curve. It is helpful to first draw a unit circle and label a few points with coordinates. We can either use the coordinates as the function values or find approximate sine and cosine values directly with a grapher.

s	$\sin s$	$\cos s$
0	0	1
$\pi/6$	0.5	0.8660
$\pi/4$	0.7071	0.7071
$\pi/3$	0.8660	0.5
$\pi/2$	1	0
$3\pi/4$	0.7071	−0.7071
π	0	−1
$5\pi/4$	−0.7071	−0.7071
$3\pi/2$	−1	0
$7\pi/4$	−0.7071	0.7071
2π	0	1

s	$\sin s$	$\cos s$
0	0	1
$-\pi/6$	−0.5	0.8660
$-\pi/4$	−0.7071	0.7071
$-\pi/3$	−0.8660	0.5
$-\pi/2$	−1	0
$-3\pi/4$	−0.7071	−0.7071
$-\pi$	0	−1
$-5\pi/4$	0.7071	−0.7071
$-3\pi/2$	1	0
$-7\pi/4$	0.7071	0.7071
-2π	0	1

The graphs are as follows.

The sine function

The cosine function

We can check these graphs using a grapher.

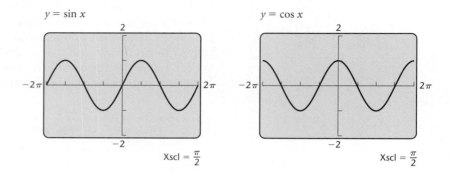

The sine and cosine functions are continuous functions. Note in the graph of the sine function that function values increase from 0 at $s = 0$ to 1 at $s = \pi/2$, then decrease to 0 at $s = \pi$, decrease further to -1 at $s = 3\pi/2$, and increase to 0 at 2π. The reverse pattern follows when s decreases from 0 to -2π. Note in the graph of the cosine function that function values start at 1 when $s = 0$, and decrease to 0 at $s = \pi/2$. They decrease further to -1 at $s = \pi$, then increase to 0 at $s = 3\pi/2$, and increase further to 1 at $s = 2\pi$. An identical pattern follows when s decreases from 0 to -2π.

From the unit circle and the graphs of the functions, we know that the domain of both the sine and cosine functions is the entire set of real numbers, $(-\infty, \infty)$. The range of each function is the set of all real numbers from -1 to 1, $[-1, 1]$.

The *domain* of the sine and cosine functions is $(-\infty, \infty)$.

The *range* of the sine and cosine functions is $[-1, 1]$.

Interactive Discovery

Another way to construct the sine and cosine graphs is by considering the unit circle and transferring vertical distances for the sine function and horizontal distances for the cosine function. Using a grapher, we can visualize the transfer of these distances. We use the grapher set in RADIAN mode and let $X_{1T} = \cos T - 1$ and $Y_{1T} = \sin T$ for the unit circle centered at $(-1, 0)$ and $X_{2T} = T$ and $Y_{2T} = \sin T$ for the sine curve.

Tmin $= 0$	Xmin $= -2$	Ymin $= -3$
Tmax $= 2\pi$	Xmax $= 2\pi$	Ymax $= 3$
Tstep $= .1$	Xscl $= \pi/2$	Yscl $= 1$

With the grapher set to SIMULTANEOUS mode, we can actually watch the sine function (in red) "unwind" from the unit circle (in blue). In the two screens below, we partially illustrate this animated procedure for the sine function.

Consult your grapher's instruction manual for specific keystrokes and graph both the sine curve and the cosine curve in this manner. (For more on parametric equations, see Section 5.8.)

A function with a repeating pattern is called **periodic**. The sine and cosine functions are examples of periodic functions. The values of each function repeat themselves every 2π units. In other words, for any s, we have

$$\sin (s + 2\pi) = \sin s \quad \text{and} \quad \cos (s + 2\pi) = \cos s.$$

To see this another way, think of the part of the graph between 0 and 2π and note that the rest of the graph consists of copies of it. If we translate the graph of $y = \sin x$ or $y = \cos x$ to the left or right 2π units, we will obtain the original graph. We say that each of these functions has a period of 2π.

Periodic Function

A function f is said to be **periodic** if there exists a positive constant p such that

$$f(s + p) = f(s)$$

for all s in the domain of f. The smallest such positive number p is called the period of the function.

The period p can be thought of as the length of the shortest recurring interval.

We can also use the unit circle to verify that the period of the sine and cosine functions is 2π. Consider any real number s and the point T that it determines on a unit circle, as shown at left. If we increase s by 2π, the point determined by $s + 2\pi$ is again the point T. Hence for any real number s,

$$\sin (s + 2\pi) = \sin s \quad \text{and} \quad \cos (s + 2\pi) = \cos s.$$

It is also true that $\sin (s + 4\pi) = \sin s$, $\sin (s + 6\pi) = \sin s$, and so on. In fact, for *any* integer k, the following equations are identities:

$$\sin [s + k(2\pi)] = \sin s \quad \text{and} \quad \cos [s + k(2\pi)] = \cos s,$$

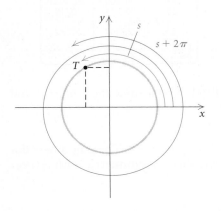

or

$$\sin s = \sin (s + 2k\pi) \quad \text{and} \quad \cos s = \cos (s + 2k\pi).$$

The **amplitude** of a periodic function is defined as one half of the distance between its maximum and minimum function values. It is always positive. Both the graphs and the unit circle verify that the maximum value of the sine and cosine functions is 1, whereas the minimum value of each is −1. Thus,

$$\text{the amplitude of the sine function} = \tfrac{1}{2}|1 - (-1)| = 1$$

and

$$\text{the amplitude of the cosine function is } \tfrac{1}{2}|1 - (-1)| = 1.$$

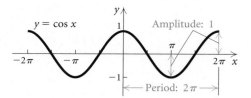

Interactive Discovery

Using the TABLE feature on a grapher, compare the y-values for $y_1 = \sin x$ and $y_2 = \sin (-x)$ and for $y_3 = \cos x$ and $y_4 = \cos (-x)$. We set TblStart $= 0$ and $\triangle$Tbl $= \pi/12$.

X	Y₁	Y₂
	0	0
.2618	.25882	−.2588
.5236	.5	−.5
.7854	.70711	−.7071
1.0472	.86603	−.866
1.309	.96593	−.9659
1.5708	1	−1
X = 0		

X	Y₃	Y₄
	1	1
.2618	.96593	.96593
.5236	.86603	.86603
.7854	.70711	.70711
1.0472	.5	.5
1.309	.25882	.25882
1.5708	0	0
X = 0		

What appears to be the relationship between $\sin x$ and $\sin (-x)$ and between $\cos x$ and $\cos (-x)$?

Consider any real number s and its opposite, $-s$. These numbers determine points T and T_1 on a unit circle that are symmetric with respect to the x-axis.

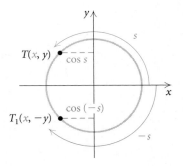

Because their second coordinates are opposites of each other, we know that for any number s,

$$\sin(-s) = -\sin s.$$

Because their first coordinates are the same, we know that for any number s,

$$\cos(-s) = \cos s.$$

Thus we have shown that the sine function is odd and the cosine function is even.

The following is a summary of the properties of the sine and cosine functions.

CONNECTING THE CONCEPTS

COMPARING THE SINE AND COSINE FUNCTIONS

SINE FUNCTION

1. Continuous
2. Period: 2π
3. Domain: All real numbers
4. Range: $[-1, 1]$
5. Amplitude: 1
6. Odd: $\sin(-s) = -\sin s$

COSINE FUNCTION

1. Continuous
2. Period: 2π
3. Domain: All real numbers
4. Range: $[-1, 1]$
5. Amplitude: 1
6. Even: $\cos(-s) = \cos s$

Graphs of the Tangent, Cotangent, Cosecant, and Secant Functions

To graph the tangent function, we could make a table of values using a grapher, but in this case it is easier to begin with the definition of tangent and the coordinates of a few points on the unit circle. We recall that

$$\tan s = \frac{y}{x} = \frac{\sin s}{\cos s}.$$

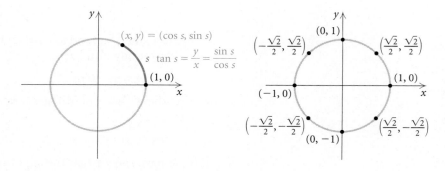

The tangent function is undefined when x, the first coordinate, is 0. That is, it is undefined for any number s whose cosine is 0:

$$s = \pm\frac{\pi}{2}, \pm\frac{3\pi}{2}, \pm\frac{5\pi}{2}, \ldots.$$

We draw vertical asymptotes at these locations (see Fig. 1 below).

FIGURE 1   FIGURE 2

We also note that

$$\tan s = 0 \text{ at } s = 0, \pm\pi, \pm 2\pi, \pm 3\pi, \ldots,$$

$$\tan s = 1 \text{ at } s = \ldots -\frac{7\pi}{4}, -\frac{3\pi}{4}, \frac{\pi}{4}, \frac{5\pi}{4}, \frac{9\pi}{4}, \ldots,$$

$$\tan s = -1 \text{ at } s = \ldots -\frac{9\pi}{4}, -\frac{5\pi}{4}, -\frac{\pi}{4}, \frac{3\pi}{4}, \frac{7\pi}{4}, \ldots.$$

We can add these ordered pairs to the graph (see Fig. 2 above) and investigate the values in $(-\pi/2, \pi/2)$ using a grapher. Note that the function value is 0 when $s = 0$, and the values increase without bound as s increases toward $\pi/2$. The graph gets closer and closer to an asymptote as s gets closer to $\pi/2$, but it never touches the line. As s decreases from 0 to $-\pi/2$, the values decrease without bound. Again the graph gets closer and closer to an asymptote, but it never touches it. We now complete the graph.

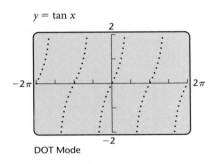

$y = \tan x$

DOT Mode

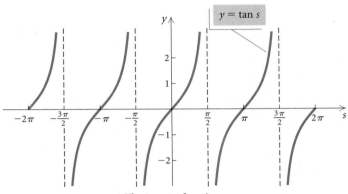

The tangent function

From the graph, we see that the tangent function is continuous except where it is not defined. The period of the tangent function is π. Note that although there is a period, there is no amplitude because there are no maximum and minimum values. Tan s is not defined where $\cos s = 0$. Thus the domain of the tangent function is the set of all real numbers except $(\pi/2) + k\pi$, where k is an integer. The range of the function is the set of all real numbers.

The cotangent function ($\cot s = \cos s/\sin s$) is undefined when y, the second coordinate, is 0—that is, it is undefined for any number s whose sine is 0. Thus the cotangent is undefined for $s = 0, \pm\pi, \pm2\pi, \pm3\pi, \ldots$. The graph of the function is shown below.

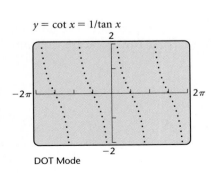

$y = \cot x = 1/\tan x$

DOT Mode

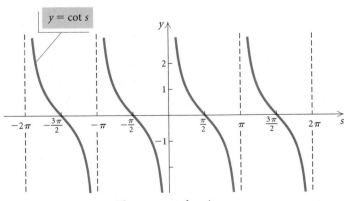

The cotangent function

The cosecant and sine functions are reciprocal functions, as are the secant and cosine functions. The graphs of the cosecant and secant functions can be constructed by finding the reciprocals of the values of the sine and cosine functions, respectively. Thus the functions will be positive together and negative together. The cosecant function is not defined for those numbers s whose sine is 0. The secant function is not defined for those numbers s whose cosine is 0. In the graphs below, the sine and cosine functions are shown by the gray curves for reference.

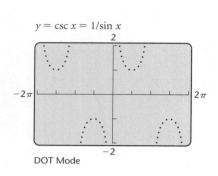

$y = \csc x = 1/\sin x$

DOT Mode

The cosecant function

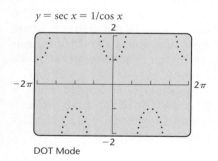

$y = \sec x = 1/\cos x$

DOT Mode

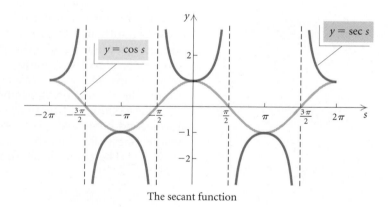

The secant function

The following is a summary of the basic properties of the tangent, cotangent, cosecant, and secant functions. These functions are continuous except where they are not defined.

CONNECTING THE CONCEPTS

COMPARING THE TANGENT, COTANGENT, COSECANT, AND SECANT FUNCTIONS

TANGENT FUNCTION

1. Period: π
2. Domain: All real numbers except $(\pi/2) + k\pi$, where k is an integer
3. Range: All real numbers

COTANGENT FUNCTION

1. Period: π
2. Domain: All real numbers except $k\pi$, where k is an integer
3. Range: All real numbers

COSECANT FUNCTION

1. Period: 2π
2. Domain: All real numbers except $k\pi$, where k is an integer
3. Range: $(-\infty, -1] \cup [1, \infty)$

SECANT FUNCTION

1. Period: 2π
2. Domain: All real numbers except $(\pi/2) + k\pi$, where k is an integer
3. Range: $(-\infty, -1] \cup [1, \infty)$

In this chapter, we have used the letter s for arc length and have avoided the letters x and y, which generally represent first and second coordinates. Nevertheless, we can represent the arc length on a unit circle by any variable, such as s, t, x, or θ. Each arc length determines a point that can be labeled with an ordered pair. The first coordinate of that ordered pair is the cosine of the arc length, and the second coordinate is the sine of the arc length. The identities we have developed hold no matter what symbols are used for variables—for example, $\cos(-s) = \cos s$, $\cos(-x) = \cos x$, $\cos(-\theta) = \cos \theta$, and $\cos(-t) = \cos t$.

Exercise Set 2.5

The following points are on the unit circle. Find the coordinates of their reflections across (a) the x-axis, (b) the y-axis, and (c) the origin.

1. $\left(-\dfrac{3}{4}, \dfrac{\sqrt{7}}{4}\right)$

2. $\left(\dfrac{2}{3}, \dfrac{\sqrt{5}}{3}\right)$

3. $\left(\dfrac{2}{5}, -\dfrac{\sqrt{21}}{5}\right)$

4. $\left(-\dfrac{\sqrt{3}}{2}, -\dfrac{1}{2}\right)$

5. The number $\pi/4$ determines a point on the unit circle with coordinates $(\sqrt{2}/2, \sqrt{2}/2)$. What are the coordinates of the point determined by $-\pi/4$?

6. A number β determines a point on the unit circle with coordinates $(-2/3, \sqrt{5}/3)$. What are the coordinates of the point determined by $-\beta$?

Find the function value using coordinates of points on the unit circle. Give exact answers.

7. $\sin \pi$

8. $\cos\left(-\dfrac{\pi}{3}\right)$

9. $\cot \dfrac{7\pi}{6}$

10. $\tan \dfrac{11\pi}{4}$

11. $\sin(-3\pi)$

12. $\csc \dfrac{3\pi}{4}$

13. $\cos \dfrac{5\pi}{6}$

14. $\tan\left(-\dfrac{\pi}{4}\right)$

15. $\cos 10\pi$

16. $\sec \dfrac{\pi}{2}$

17. $\cos \dfrac{\pi}{6}$

18. $\sin \dfrac{2\pi}{3}$

19. $\sin \dfrac{5\pi}{4}$

20. $\cos \dfrac{11\pi}{6}$

21. $\sin(-5\pi)$

22. $\tan \dfrac{3\pi}{2}$

23. $\cot \dfrac{5\pi}{2}$

24. $\tan \dfrac{5\pi}{3}$

Find the function value using a grapher. Round the answer to four decimal places.

25. $\tan \dfrac{\pi}{7}$

26. $\cos\left(-\dfrac{2\pi}{5}\right)$

27. $\sec 37$

28. $\sin 11.7$

29. $\cot 342$

30. $\tan 1.3$

31. $\cos 6\pi$

32. $\sin \dfrac{\pi}{10}$

33. $\csc 4.16$

34. $\sec \dfrac{10\pi}{7}$

35. $\tan \dfrac{7\pi}{4}$

36. $\cos 2000$

37. $\sin\left(-\dfrac{\pi}{4}\right)$

38. $\cot 7\pi$

39. $\sin 0$

40. $\cos(-29)$

41. $\tan \dfrac{2\pi}{9}$

42. $\sin \dfrac{8\pi}{3}$

In Exercises 43–48, make hand-drawn graphs.

43. **a)** Sketch a graph of $y = \sin x$.
    b) By reflecting the graph in part (a), sketch a graph of $y = \sin(-x)$.
    c) By reflecting the graph in part (a), sketch a graph of $y = -\sin x$.
    d) How do the graphs in parts (b) and (c) compare?

44. **a)** Sketch a graph of $y = \cos x$.

b) By reflecting the graph in part (a), sketch a graph of $y = \cos(-x)$.
c) By reflecting the graph in part (a), sketch a graph of $y = -\cos x$.
d) How do the graphs in parts (a) and (b) compare?

45. **a)** Sketch a graph of $y = \sin x$.
    b) By translating, sketch a graph of $y = \sin(x + \pi)$.
    c) By reflecting the graph of part (a), sketch a graph of $y = -\sin x$.
    d) How do the graphs of parts (b) and (c) compare?

46. **a)** Sketch a graph of $y = \sin x$.
    b) By translating, sketch a graph of $y = \sin(x - \pi)$.
    c) By reflecting the graph of part (a), sketch a graph of $y = -\sin x$.
    d) How do the graphs of parts (b) and (c) compare?

47. **a)** Sketch a graph of $y = \cos x$.
    b) By translating, sketch a graph of $y = \cos(x + \pi)$.
    c) By reflecting the graph of part (a), sketch a graph of $y = -\cos x$.
    d) How do the graphs of parts (b) and (c) compare?

48. **a)** Sketch a graph of $y = \cos x$.
    b) By translating, sketch a graph of $y = \cos(x - \pi)$.
    c) By reflecting the graph of part (a), sketch a graph of $y = -\cos x$.
    d) How do the graphs of parts (b) and (c) compare?

49. Of the six circular functions, which are even? Which are odd?

50. Of the six circular functions, which have period π? Which have period 2π?

Consider the coordinates on the unit circle for Exercises 51–54.

51. In which quadrants is the tangent function positive? negative?

52. In which quadrants is the sine function positive? negative?

53. In which quadrants is the cosine function positive? negative?

54. In which quadrants is the cosecant function positive? negative?

Discussion and Writing

55. Describe how the graphs of the sine and cosine functions are related.

56. Explain why both the sine and cosine functions are continuous, but the tangent function, defined as sine/cosine, is not continuous.

Skill Maintenance

Graph both functions in the same viewing window and describe how g is a transformation of f.

57. $f(x) = x^2$, $g(x) = 2x^2 - 3$

58. $f(x) = x^2$, $g(x) = (x - 2)^2$

59. $f(x) = |x|$, $g(x) = \frac{1}{2}|x - 4| + 1$

60. $f(x) = x^3$, $g(x) = -x^3$

Write an equation for a function that has a graph with the given characteristics. Check using a grapher.

61. The shape of $y = x^3$, but reflected across the x-axis, shifted right 2 units, and shifted down 1 unit

62. The shape of $y = 1/x$, but shrunk vertically by a factor of $\frac{1}{4}$ and shifted up 3 units

Synthesis

Complete. (For example, $\sin (x + 2\pi) = \sin x$.)

63. $\cos (-x) = $ _____

64. $\sin (-x) = $ _____

65. $\sin (x + 2k\pi)$, $k \in \mathbb{Z} = $ _____

66. $\cos (x + 2k\pi)$, $k \in \mathbb{Z} = $ _____

67. $\sin (\pi - x) = $ _____

68. $\cos (\pi - x) = $ _____

69. $\cos (x - \pi) = $ _____

70. $\cos (x + \pi) = $ _____

71. $\sin (x + \pi) = $ _____

72. $\sin (x - \pi) = $ _____

73. Find all numbers x that satisfy the following. Check using a grapher.

a) $\sin x = 1$
b) $\cos x = -1$
c) $\sin x = 0$
d) $\sin x < \cos x$

74. Find $f \circ g$ and $g \circ f$, where $f(x) = x^2 + 2x$ and $g(x) = \cos x$.

Use a grapher to determine the domain, the range, the period, and the amplitude of the function.

75. $y = (\sin x)^2$

76. $y = |\cos x| + 1$

Determine the domain of the function.

77. $f(x) = \sqrt{\cos x}$

78. $g(x) = \dfrac{1}{\sin x}$

79. $f(x) = \dfrac{\sin x}{\cos x}$

80. $g(x) = \log (\sin x)$

81. Consider $(\sin x)/x$, where x is between 0 and $\pi/2$. As x approaches 0, this function approaches a limiting value. What is it?

82. One of the motivations for developing trigonometry with a unit circle is that you can actually "see" $\sin \theta$ and $\cos \theta$ on the circle. Note in the figure below that $AP = \sin \theta$ and $OA = \cos \theta$. It turns out that you can also "see" the other four trigonometric functions. Prove each of the following.

a) $BD = \tan \theta$
b) $OD = \sec \theta$
c) $OE = \csc \theta$
d) $CE = \cot \theta$

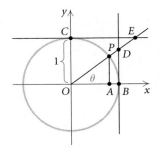

Graph.

83. $y = 3 \sin x$

84. $y = \sin |x|$

85. $y = \sin x + \cos x$

86. $y = |\cos x|$

Graphs of Transformed Sine and Cosine Functions

2.6

- *Graph transformations of $y = \sin x$ and $y = \cos x$ in the form*
  $$y = A \sin (Bx - C) + D$$
  and
  $$y = A \cos (Bx - C) + D$$
  and determine the amplitude, the period, and the phase shift.
- *Graph sums of functions.*

Variations of Basic Graphs

In Section 2.5, we graphed all six trigonometric functions. In this section, we will consider variations of the graphs of the sine and cosine functions. For example, we will graph equations like the following:

$$y = 5 \sin \tfrac{1}{2}x, \qquad y = \cos (2x - \pi), \quad \text{and} \quad y = \tfrac{1}{2} \sin x - 3.$$

In particular, we are interested in graphs of function in the form

$$y = A \sin (Bx - C) + D$$

and

$$y = A \cos (Bx - C) + D,$$

TRANSFORMATIONS OF FUNCTIONS

REVIEW SECTION 1.4.

where A, B, C, and D are constants. These constants have the effect of translating, reflecting, stretching, and shrinking the basic graphs. Let's first examine the effect of each constant individually. Then we will consider the combined effects of more than one constant.

We first consider the effect of the constant D.

Interactive Discovery

Graph the first two functions in each group and then predict what the graph of the third function looks like. Then graph that function.

$$y_1 = \sin x, \qquad y_2 = \sin x + 3.5, \qquad y_3 = \sin x - 4;$$
$$y_1 = \cos x, \qquad y_2 = \cos x - 2, \qquad y_3 = \cos x + 0.5.$$

What is the effect of the constant D on the graph of the basic function?

EXAMPLE 1   Sketch a graph of $y = \sin x + 3$.

Solution   The graph of $y = \sin x + 3$ is a *vertical* translation of the graph of $y = \sin x$ up 3 units. One way to sketch the graph is to first consider $y = \sin x$ on an interval of length 2π, say, $[0, 2\pi]$. The zeros of the function and the maximum and minimum values can be considered key points. These are

$$(0, 0), \quad \left(\frac{\pi}{2}, 1\right), \quad (\pi, 0), \quad \left(\frac{3\pi}{2}, -1\right), \quad (2\pi, 0).$$

These key points are transformed up 3 units to obtain the key points of the graph of $y = \sin x + 3$. These are

$$(0, 3), \quad \left(\frac{\pi}{2}, 4\right), \quad (\pi, 3), \quad \left(\frac{3\pi}{2}, 2\right), \quad (2\pi, 3).$$

The graph of $y = \sin x + 3$ can be sketched on the interval $[0, 2\pi]$ and extended to obtain the rest of the graph by repeating the graph on intervals of length 2π.

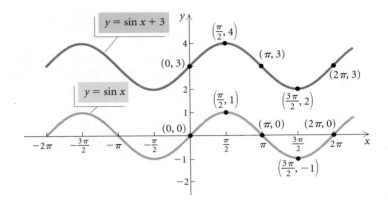

The constant D in

$$y = A \sin (Bx - C) + D \quad \text{and} \quad y = A \cos (Bx - C) + D$$

translates the graphs up D units if $D > 0$ or down $|D|$ units if $D < 0$. Next, we consider the effect of the constant A.

Interactive Discovery

Graph the first two functions in each group and then predict what the graph of the third function looks like. Then graph that function.

$$y_1 = \sin x, \quad y_2 = 2 \sin x, \quad y_3 = \tfrac{1}{2} \sin x;$$
$$y_1 = \cos x, \quad y_2 = -1.5 \cos x, \quad y_3 = -0.8 \cos x.$$

What is the effect of the constant A on the graph of the basic function (a) when $0 < A < 1$? (b) when $A > 1$? (c) when $-1 < A < 0$? (d) when $A < -1$?

EXAMPLE 2 Sketch a graph of $y = 2 \cos x$. What is the amplitude?

Solution The constant 2 in $y = 2 \cos x$ has the effect of stretching the graph of $y = \cos x$ vertically by a factor of 2 units. Since the function values of $y = \cos x$ are such that $-1 \le \cos x \le 1$, the function values of $y = 2 \cos x$ are such that $-2 \le 2 \cos x \le 2$. The maximum value of $y = 2 \cos x$ is 2, and the minimum value is -2. Thus the *amplitude*, A, is $\tfrac{1}{2}|2 - (-2)|$, or 2.

We draw the graph of $y = \cos x$ and consider its key points,

$$(0, 1), \quad \left(\frac{\pi}{2}, 0\right), \quad (\pi, -1), \quad \left(\frac{3\pi}{2}, 0\right), \quad (2\pi, 1),$$

on the interval $[0, 2\pi]$.

We then multiply the second coordinates by 2 to obtain the key points of $y = 2 \cos x$. These are

$$(0, 2), \quad \left(\frac{\pi}{2}, 0\right), \quad (\pi, -2), \quad \left(\frac{3\pi}{2}, 0\right), \quad (2\pi, 2).$$

We plot these points and sketch the graph on the interval $[0, 2\pi]$. Then we repeat this part of the graph on adjacent intervals of length 2π.

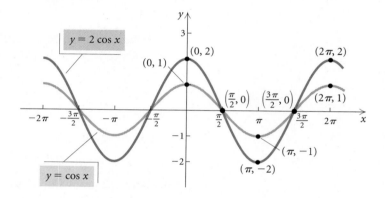

If $|A| > 1$, then there will be a vertical stretching. If $|A| < 1$, then there will be a vertical shrinking. If $A < 0$, the graph is also reflected across the x-axis.

Amplitude

The **amplitude** of the graphs of $y = A \sin (Bx - C) + D$ and $y = A \cos (Bx - C) + D$ is $|A|$.

EXAMPLE 3   Sketch a graph of $y = -\frac{1}{2} \sin x$.

Solution   The amplitude of the graph is $\left|-\frac{1}{2}\right|$, or $\frac{1}{2}$. The graph of $y = -\frac{1}{2} \sin x$ is a vertical shrinking and a reflection of the graph of $y = \sin x$ across the x-axis. In graphing, the key points of $y = \sin x$,

$$(0, 0), \quad \left(\frac{\pi}{2}, 1\right), \quad (\pi, 0), \quad \left(\frac{3\pi}{2}, -1\right), \quad (2\pi, 0),$$

are transformed to

$$(0, 0), \quad \left(\frac{\pi}{2}, -\frac{1}{2}\right), \quad (\pi, 0), \quad \left(\frac{3\pi}{2}, \frac{1}{2}\right), \quad (2\pi, 0).$$

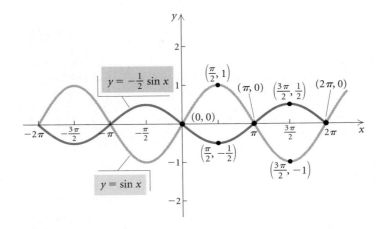

Now we consider the effect of the constant B.

Interactive Discovery

In general, changes in the constants A and D do not change the period. To see what effect a change in the constant B will have on the basic functions, graph

$$y_1 = \cos x, \qquad y_2 = \cos (-2x), \qquad y_3 = \cos \frac{1}{2} x.$$

Then visualize the graphs of $y_4 = \cos 4x$ and $y_5 = \cos \left(-\frac{1}{2} x\right)$ and check your guesses. What is the period of each of these graphs? What is the effect of the constant B on the graph of the basic function?

EXAMPLE 4   Sketch a graph of $y = \sin 4x$. What is the period?

Solution   The constant B has the effect of changing the period. The graph of $y = f(4x)$ is obtained from the graph of $y = f(x)$ by shrinking the graph horizontally. The new graph is obtained by dividing the first coordinate of each ordered-pair solution of $y = f(x)$ by 4. The key points of $y = \sin x$ are

$$(0, 0), \quad \left(\frac{\pi}{2}, 1\right), \quad (\pi, 0), \quad \left(\frac{3\pi}{2}, -1\right), \quad (2\pi, 0).$$

These are transformed to the key points of $y = \sin 4x$, which are

$$(0, 0), \quad \left(\frac{\pi}{8}, 1\right), \quad \left(\frac{\pi}{4}, 0\right), \quad \left(\frac{3\pi}{8}, -1\right), \quad \left(\frac{\pi}{2}, 0\right).$$

We plot these key points and sketch in the graph on the shortened interval $[0, \pi/2]$. Then we repeat the graph on other intervals of length $\pi/2$.

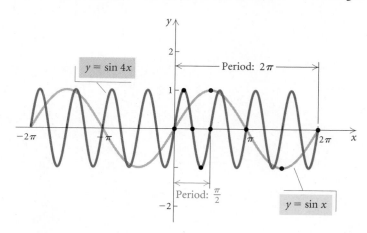

If $|B| < 1$, then there will be a horizontal stretching. If $|B| > 1$, then there will be a horizontal shrinking. If $B < 0$, the graph is also reflected across the y-axis.

Period

The **period** of the graphs of $y = A \sin (Bx - C) + D$ and $y = A \cos (Bx - C) + D$ is $\left| \dfrac{2\pi}{B} \right|$.

Now we examine the effect of the constant C.

Interactive Discovery

Graph

$$y_1 = \cos x, \qquad y_2 = \cos (x - \pi),$$
$$y_3 = \cos \left(x + \frac{\pi}{2} \right), \qquad y_4 = \cos \left(x - \frac{\pi}{4} \right).$$

Each curve has an amplitude of 1 and a period of 2π, but there are four distinct graphs. What is the effect of the constant C?

EXAMPLE 5   Sketch a graph of $y = \sin \left(x - \dfrac{\pi}{2} \right)$.

Solution   The amplitude is 1, and the period is 2π. The graph of $y = f(x - c)$ is obtained from the graph of $y = f(x)$ by translating the graph horizontally—to the right c units if $c > 0$ and to the left $|c|$ units if $c < 0$. The graph of $y = \sin (x - \pi/2)$ is a translation of the graph of $y = \sin x$ to the right $\pi/2$ units. The value $\pi/2$ is called the *phase shift*. The key points of $y = \sin x$,

$$(0, 0), \quad \left(\frac{\pi}{2}, 1 \right), \quad (\pi, 0), \quad \left(\frac{3\pi}{2}, -1 \right), \quad (2\pi, 0),$$

are transformed by adding $\pi/2$ to each of the first coordinates to obtain the following key points of $y = \sin (x - \pi/2)$:

$$\left(\frac{\pi}{2}, 0 \right), \quad (\pi, 1), \quad \left(\frac{3\pi}{2}, 0 \right), \quad (2\pi, -1), \quad \left(\frac{5\pi}{2}, 0 \right).$$

We plot these key points and sketch the curve on the interval $[\pi/2, 5\pi/2]$. Then we repeat the graph on other intervals of length 2π.

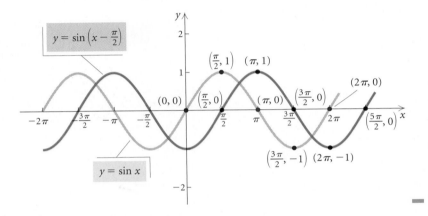

Now we consider combined transformations of graphs. It is helpful to rewrite

$$y = A \sin (Bx - C) + D \qquad \text{and} \quad y = A \cos (Bx - C) + D$$

as

$$y = A \sin \left[B \left(x - \frac{C}{B} \right) \right] + D \quad \text{and} \quad y = A \cos \left[B \left(x - \frac{C}{B} \right) \right] + D.$$

EXAMPLE 6   Sketch a graph of $y = \cos (2x - \pi)$.

Solution   The graph of

$$y = \cos (2x - \pi)$$

is the same as the graph of

$$y = 1 \cdot \cos \left[2 \left(x - \frac{\pi}{2} \right) \right] + 0.$$

The amplitude is 1. The factor 2 shrinks the period by half, making the period $|2\pi/2|$, or π. The $\pi/2$ translates the graph of $y = \cos 2x$ to the right $\pi/2$ units. Thus to form the graph, we first graph $y = \cos x$, followed by $y = \cos 2x$ and then $y = \cos [2(x - \pi/2)]$.

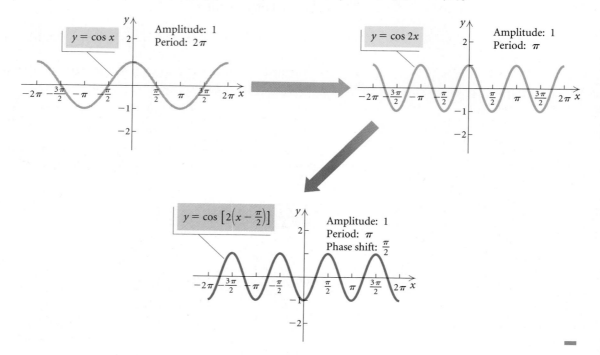

Phase Shift

The **phase shift** of the graphs

$$y = A \sin (Bx - C) + D = A \sin \left[B\left(x - \frac{C}{B} \right) \right] + D$$

and

$$y = A \cos (Bx - C) + D = A \cos \left[B\left(x - \frac{C}{B} \right) \right] + D$$

is the quantity $\dfrac{C}{B}$.

If $C/B > 0$, the graph is translated to the right C/B units. If $C/B < 0$, the graph is translated to the left $|C/B|$ units. Be sure that the horizontal stretching or shrinking based on the constant B is done before the translation based on the phase shift C/B.

Let's now summarize the effect of the constants. We carry out the procedures in the order listed.

Transformations of Sine and Cosine Functions

To graph

$$y = A \sin (Bx - C) + D = A \sin \left[B\left(x - \frac{C}{B} \right) \right] + D$$

and

$$y = A \cos (Bx - C) + D = A \cos \left[B\left(x - \frac{C}{B} \right) \right] + D,$$

follow the steps listed below in the order in which they are listed.

1. Stretch or shrink the graph horizontally according to B.

$|B| < 1$	Stretch horizontally
$|B| > 1$	Shrink horizontally
$B < 0$	Reflect across the y-axis

The *period* is $\left| \dfrac{2\pi}{B} \right|$.

2. Stretch or shrink the graph vertically according to A.

$|A| < 1$	Shrink vertically
$|A| > 1$	Stretch vertically
$A < 0$	Reflect across the x-axis

The *amplitude* is $|A|$.

3. Translate the graph horizontally according to C/B.

$\dfrac{C}{B} < 0$	$\left\| \dfrac{C}{B} \right\|$ units to the left
$\dfrac{C}{B} > 0$	$\dfrac{C}{B}$ units to the right

The *phase shift* is $\dfrac{C}{B}$.

4. Translate the graph vertically according to D.

$D < 0$	$|D|$ units down
$D > 0$	D units up

EXAMPLE 7   Sketch a graph of $y = 3 \sin (2x + \pi/2) + 1$. Find the amplitude, the period, and the phase shift.

Solution   We first note that

$$y = 3 \sin \left(2x + \frac{\pi}{2} \right) + 1 = 3 \sin \left[2\left(x - \left(-\frac{\pi}{4} \right) \right) \right] + 1.$$

Then we have the following:

$$\text{Amplitude} = |A| = |3| = 3,$$

$$\text{Period} = \left|\frac{2\pi}{B}\right| = \left|\frac{2\pi}{2}\right| = \pi,$$

$$\text{Phase shift} = \frac{C}{B} = \frac{-\pi/2}{2} = -\frac{\pi}{4}.$$

To create the final graph, we begin with the basic sine curve, $y = \sin x$. Then we sketch graphs of each of the following equations in sequence.

1. $y = \sin 2x$

2. $y = 3 \sin 2x$

3. $y = 3 \sin \left[2\left(x - \left(-\frac{\pi}{4}\right)\right)\right]$

4. $y = 3 \sin \left[2\left(x - \left(-\frac{\pi}{4}\right)\right)\right] + 1$

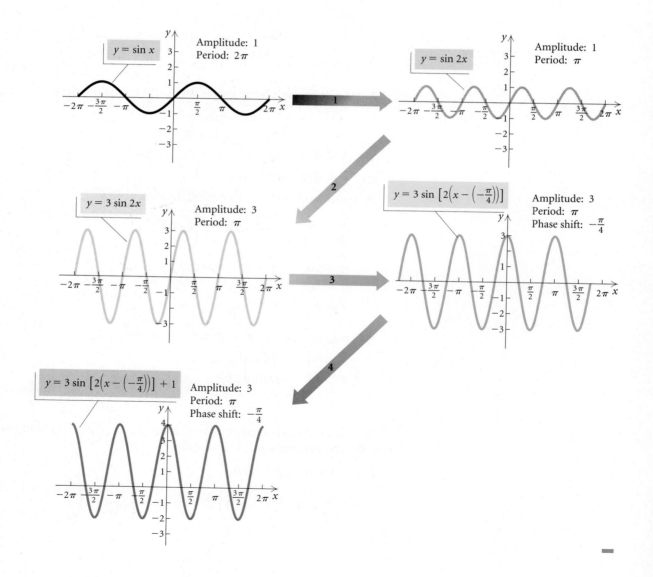

All the graphs in Examples 1–7 can be checked using a grapher. Even though it is faster and more accurate to graph using a grapher, graphing by hand gives us a greater understanding of the effect of changing the constants *A*, *B*, *C*, and *D*.

Graphers are especially convenient when a period or a phase shift is not a multiple of $\pi/4$.

EXAMPLE 8   Graph $y = 3 \cos 2\pi x - 1$. Find the amplitude, the period, and the phase shift.

Solution   First we note the following:

$$\text{Amplitude} = |A| = |3| = 3,$$

$$\text{Period} = \left|\frac{2\pi}{B}\right| = \left|\frac{2\pi}{2\pi}\right| = |1| = 1,$$

$$\text{Phase shift} = \frac{C}{B} = \frac{0}{2\pi} = 0.$$

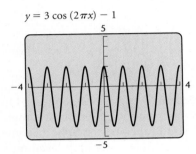

$y = 3 \cos (2\pi x) - 1$

There is no phase shift in this case because the constant $C = 0$. The graph has a vertical translation of the graph of the cosine function down 1 unit, an amplitude of 3, and a period of 1, so we can use $[-4, 4, -5, 5]$ as the viewing window.

The transformation techniques that we learned in this section for graphing the sine and cosine functions can also be applied in the same manner to the other trigonometric functions. Transformations of this type appear in the synthesis exercises in Exercise Set 2.6.

An **oscilloscope** is an electronic device that converts electrical signals into graphs like those in the preceding examples. These graphs are often called sine waves. By manipulating the controls, we can change the amplitude, the period, and the phase of sine waves. The oscilloscope has many applications, and the trigonometric functions play a major role in many of them.

Graphs of Sums: Addition of Ordinates

The output of an electronic synthesizer used in the recording and playing of music can be converted into sine waves by an oscilloscope. The following graphs illustrate simple tones of different frequencies. The frequency of a simple tone is the number of vibrations in the signal of the tone per second. The loudness or intensity of the tone is reflected in the height of the graph (its amplitude). The three tones in the diagrams below all have the same intensity but different frequencies.

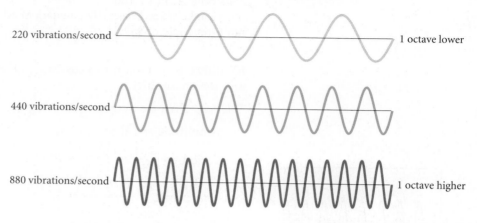

Musical instruments can generate extremely complex sine waves. On a single instrument, overtones can become superimposed on a simple tone. When multiple notes are played simultaneously, graphs become very complicated. This can happen when multiple notes are played on a single instrument or a group of instruments, or even when the same simple note is played on different instruments.

Combinations of simple tones produce interesting curves. Consider two tones whose graphs are $y_1 = 2 \sin x$ and $y_2 = \sin 2x$. The combination of the two tones produces a new sound whose graph is $y = 2 \sin x + \sin 2x$, shown in the following example.

EXAMPLE 9 Graph: $y = 2 \sin x + \sin 2x$.

Solution We graph $y = 2 \sin x$ and $y = \sin 2x$ using the same set of axes.

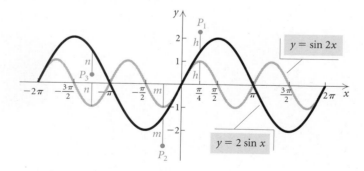

Now we graphically add some y-coordinates, or ordinates, to obtain points on the graph that we seek. At $x = \pi/4$, we transfer the distance h, which is the value of $\sin 2x$, up to add it to the value of $2 \sin x$. Point P_1 is on the graph that we seek. At $x = -\pi/4$, we use a similar procedure, but this time both ordinates are negative. Point P_2 is on the graph. At $x = -5\pi/4$, we add the negative ordinate of sin $2x$ to the positive ordinate of 2 sin x. Point P_3 is also on the graph. We continue to plot points in this fashion and then connect them to get the desired graph, shown below. This method is called *addition of ordinates*, because we add the y-values (ordinates) of $y =$ sin $2x$ to the y-values (ordinates) of $y = 2$ sin x.

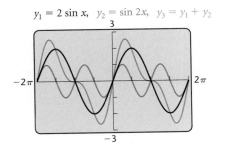

$y_1 = 2$ sin x, $y_2 =$ sin $2x$, $y_3 = y_1 + y_2$

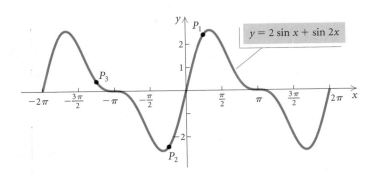

$y = 2$ sin $x +$ sin $2x$

Using a grapher, we can quickly determine the period of a trigonometric function that is a combination of sine and cosine functions.

EXAMPLE 10   Graph $y = 2$ cos $x -$ sin $3x$ and determine its period.

Solution   We graph $y = 2$ cos $x -$ sin $3x$ with appropriate dimensions. The period appears to be 2π.

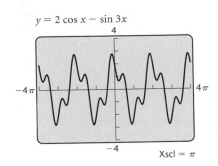

$y = 2$ cos $x -$ sin $3x$

Xscl $= \pi$

Exercise Set **2.6**

Determine the amplitude, the period, and the phase shift of the function, and, without a grapher, sketch the graph of the function. Then check the graph using a grapher.

1. $y =$ sin $x + 1$

2. $y = \frac{1}{4}$ cos x

3. $y = -3$ cos x

4. $y =$ sin $(-2x)$

5. $y = 2$ sin $\left(\frac{1}{2} x\right)$

6. $y =$ cos $\left(x - \frac{\pi}{2}\right)$

7. $y = \frac{1}{2}$ sin $\left(x + \frac{\pi}{2}\right)$

8. $y =$ cos $x - \frac{1}{2}$

9. $y = 3$ cos $(x - \pi)$

10. $y = -$sin $\left(\frac{1}{4} x\right) + 1$

11. $y = \frac{1}{3}$ sin $x - 4$

12. $y =$ cos $\left(\frac{1}{2} x + \frac{\pi}{2}\right)$

13. $y = -$cos $(-x) + 2$

14. $y = \frac{1}{2}$ sin $\left(2x - \frac{\pi}{4}\right)$

Determine the amplitude, the period, and the phase shift of the function. Then graph the function using a grapher. Try to visualize the graph before creating it.

15. $y = 2 \cos \left(\dfrac{1}{2} x - \dfrac{\pi}{2} \right)$

16. $y = 4 \sin \left(\dfrac{1}{4} x + \dfrac{\pi}{8} \right)$

17. $y = -\dfrac{1}{2} \sin \left(2x + \dfrac{\pi}{2} \right)$

18. $y = -3 \cos (4x - \pi) + 2$

19. $y = 2 + 3 \cos (\pi x - 3)$

20. $y = 5 - 2 \cos \left(\dfrac{\pi}{2} x + \dfrac{\pi}{2} \right)$

21. $y = -\dfrac{1}{2} \cos (2\pi x) + 2$

22. $y = -2 \sin (-2x + \pi) - 2$

23. $y = -\sin \left(\dfrac{1}{2} x - \dfrac{\pi}{2} \right) + \dfrac{1}{2}$

24. $y = \dfrac{1}{3} \cos (-3x) + 1$

25. $y = \cos (-2\pi x) + 2$

26. $y = \dfrac{1}{2} \sin (2\pi x + \pi)$

27. $y = -\dfrac{1}{4} \cos (\pi x - 4)$

28. $y = 2 \sin (2\pi x + 1)$

In Exercises 29–36, without a grapher, match the function with one of graphs (a)–(h), which follow. Then check your work using a grapher.

a)

b)

c)

d)

e)

f)

g)

h)

29. $y = -\cos 2x$

30. $y = \dfrac{1}{2} \sin x - 2$

31. $y = 2 \cos \left(x + \dfrac{\pi}{2} \right)$

32. $y = -3 \sin \dfrac{1}{2} x - 1$

33. $y = \sin (x - \pi) - 2$

34. $y = -\dfrac{1}{2} \cos \left(x - \dfrac{\pi}{4} \right)$

35. $y = \dfrac{1}{3} \sin 3x$

36. $y = \cos \left(x - \dfrac{\pi}{2} \right)$

In Exercises 37–40, determine the equation of the function that is graphed.

37.

38.

39.

40.

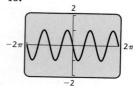

Graph using addition of ordinates. Then check your work using a grapher.

41. $y = 2 \cos x + \cos 2x$ **42.** $y = 3 \cos x + \cos 3x$

43. $y = \sin x + \cos 2x$ **44.** $y = 2 \sin x + \cos 2x$

45. $y = \sin x - \cos x$ **46.** $y = 3 \cos x - \sin x$

47. $y = 3 \cos x + \sin 2x$ **48.** $y = 3 \sin x - \cos 2x$

Use a grapher to graph the function.

49. $y = x + \sin x$ **50.** $y = -x - \sin x$

51. $y = \cos x - x$ **52.** $y = -(\cos x - x)$

53. $y = \cos 2x + 2x$ **54.** $y = \cos 3x + \sin 3x$

55. $y = 4 \cos 2x - 2 \sin x$ **56.** $y = 7.5 \cos x + \sin 2x$

Discussion and Writing

57. In the equations $y = A \sin (Bx - C) + D$ and $y = A \cos (Bx - C) + D$, which constants translate the graphs and which constants stretch and shrink the graphs? Describe in your own words the effect of each constant.

58. In the transformation steps listed in this section, why must step (1) precede step (3)? Give an example that illustrates this.

Skill Maintenance

Find the complement and the supplement.

59. $73°5'10''$ **60.** $23.1°$

Convert to radian measure. Leave the answer in terms of π.

61. $-60°$ **62.** $118°$

Synthesis

The transformation techniques that we learned in this section for graphing the sine and cosine functions can also be applied to the other trigonometric functions. Sketch a graph of each of the following. Then check your work using a grapher.

63. $y = -\tan x$ **64.** $y = \tan (-x)$

65. $y = -2 + \cot x$ **66.** $y = -\dfrac{3}{2} \csc x$

67. $y = 2 \tan \dfrac{1}{2} x$ **68.** $y = \cot 2x$

69. $y = 2 \sec (x - \pi)$ **70.** $y = 4 \tan \left(\dfrac{1}{4} x + \dfrac{\pi}{8} \right)$

71. $y = 2 \csc \left(\dfrac{1}{2} x - \dfrac{3\pi}{4} \right)$ **72.** $y = 4 \sec (2x - \pi)$

Use a grapher to graph each of the following on the given interval and approximate the zeros.

73. $f(x) = \dfrac{\sin x}{x}$; $[-12, 12]$

74. $f(x) = \dfrac{\cos x - 1}{x}$; $[-12, 12]$

75. $f(x) = x^3 \sin x$; $[-5, 5]$

76. $f(x) = \dfrac{(\sin x)^2}{x}$; $[-4, 4]$

77. *Temperature During an Illness.* The temperature T of a patient during a 12-day illness is given by

$$T(t) = 101.6° + 3° \sin \left(\dfrac{\pi}{8} t \right).$$

a) Graph the function on the interval $[0, 12]$.

b) What are the maximum and the minimum temperatures during the illness?

78. *Periodic Sales.* A company in a northern climate has sales of skis as given by

$$S(t) = 10 \left(1 - \cos \dfrac{\pi}{6} t \right),$$

where t is the time, in months ($t = 0$ corresponds to July 1), and $S(t)$ is in thousands of dollars.

a) Graph the function on a 12-month interval $[0, 12]$.

b) What is the period of the function?

c) What is the minimum amount of sales and when does it occur?

d) What is the maximum amount of sales and when does it occur?

79. *Satellite Location.* A satellite circles the earth in such a way that it is y miles from the equator (north or south, height not considered) t minutes after its launch, where

$$y(t) = 3000 \left[\cos \dfrac{\pi}{45} (t - 10) \right].$$

a) Graph the function.

b) What are the amplitude, the period, and the phase shift?

80. *Water Wave.*   The cross-section of a water wave is given by

$$y = 3 \sin\left(\frac{\pi}{4} x + \frac{\pi}{4}\right),$$

where y is the vertical height of the water wave and x is the distance from the origin to the wave.

a) Graph the function.
b) What are the amplitude, the period, and the phase shift?

81. *Damped Oscillations.*   Suppose that the motion of a spring is given by

$$d(t) = 6e^{-0.8t} \cos(6\pi t) + 4,$$

where d is the distance, in inches, of a weight from the point at which the spring is attached to a ceiling, after t seconds.

a) Graph the function on the interval $[0, 10]$.
b) How far do you think the spring is from the ceiling when the spring stops bobbing?

82. *Rotating Beacon.*   A police car is parked 10 ft from a wall. On top of the car is a beacon rotating in such a way that the light is at a distance $d(t)$ from point Q after t seconds, where

$$d(t) = 10 \tan(2\pi t).$$

When d is positive, the light is pointing north of Q, and when d is negative, the light is pointing south of Q.

a) Graph the function on the interval $[0, 2]$.
b) Explain the meaning of the values of t for which the function is undefined.

Chapter Summary and Review   2

Important Properties and Formulas

Trigonometric Function Values of an Acute Angle θ

Let θ be an acute angle of a right triangle. The six trigonometric functions of θ are as follows:

$$\sin\theta = \frac{\text{opp}}{\text{hyp}}, \quad \cos\theta = \frac{\text{adj}}{\text{hyp}}, \quad \tan\theta = \frac{\text{opp}}{\text{adj}},$$

$$\csc\theta = \frac{\text{hyp}}{\text{opp}}, \quad \sec\theta = \frac{\text{hyp}}{\text{adj}}, \quad \cot\theta = \frac{\text{adj}}{\text{opp}}.$$

Reciprocal Functions

$$\csc \theta = \frac{1}{\sin \theta}, \qquad \sec \theta = \frac{1}{\cos \theta}, \qquad \cot \theta = \frac{1}{\tan \theta}$$

Function Values of Special Angles

	0°	30°	45°	60°	90°
sin	0	1/2	$\sqrt{2}/2$	$\sqrt{3}/2$	1
cos	1	$\sqrt{3}/2$	$\sqrt{2}/2$	1/2	0
tan	0	$\sqrt{3}/3$	1	$\sqrt{3}$	Undefined

Cofunction Identities

$$\sin \theta = \cos (90° - \theta), \qquad \cos \theta = \sin (90° - \theta),$$
$$\tan \theta = \cot (90° - \theta), \qquad \cot \theta = \tan (90° - \theta),$$
$$\sec \theta = \csc (90° - \theta), \qquad \csc \theta = \sec (90° - \theta)$$

Trigonometric Functions of Any Angle θ

$P(x, y)$ is any point on the terminal side of any angle θ in standard position, and r is the distance from the origin to $P(x, y)$, where $r = \sqrt{x^2 + y^2}$.

$$\sin \theta = \frac{y}{r}, \qquad \cos \theta = \frac{x}{r}, \qquad \tan \theta = \frac{y}{x}$$

Signs of Function Values

The signs of the function values depend only on the coordinates of the point P on the terminal side.

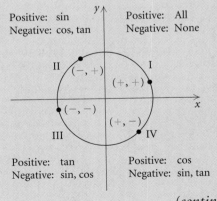

Positive: sin
Negative: cos, tan

Positive: All
Negative: None

Positive: tan
Negative: sin, cos

Positive: cos
Negative: sin, tan

(continued)

Radian–Degree Equivalents

Linear Speed in Terms of Angular Speed

$$v = r\omega$$

Basic Circular Functions

For a real number s that determines a point (x, y) on the unit circle:

$$\sin s = y, \qquad \cos s = x, \qquad \tan s = \frac{y}{x}.$$

Sine (Odd Function): $\sin(-s) = -\sin s$
Cosine (Even Function): $\cos(-s) = \cos s$

Transformations of Sine and Cosine Functions

To graph $y = A \sin(Bx - C) + D$ and $y = A \cos(Bx - C) + D$:

1. Stretch or shrink the graph horizontally according to B.
2. Stretch or shrink the graph vertically according to A.
3. Translate the graph horizontally according to C/B.
4. Translate the graph vertically according to D.

REVIEW EXERCISES

1. Find the six trigonometric function values of the specified angle.

Find the exact function value, if it exists.

2. $\cos 45°$

3. $\cot 60°$

4. $\cos 495°$

5. $\sin 150°$

6. $\sec(-270°)$

7. $\tan(-600°)$

8. Convert $22.27°$ to degrees, minutes, and seconds. Round to the nearest second.

9. Convert $47°33'27''$ to decimal degree notation. Round to two decimal places.

Find the function value. Round to four decimal places.

10. $\tan 2184°$ **11.** $\sec 27.9°$ **12.** $\cos 18°13'42''$

13. $\sin 245°24'$ **14.** $\cot(-33.2°)$ **15.** $\sin 556.13°$

Find θ in the interval indicated. Round the answer to the nearest tenth of a degree.

16. $\cos \theta = -0.9041$, $(180°, 270°)$

17. $\tan \theta = 1.0799$, $(0°, 90°)$

Find the exact acute angle θ, in degrees, given the function value.

18. $\sin \theta = \dfrac{\sqrt{3}}{2}$

19. $\tan \theta = \sqrt{3}$

20. Given that $\sin 59.1° \approx 0.8581$, $\cos 59.1° \approx 0.5135$, and $\tan 59.1° \approx 1.6709$, find the six function values for $30.9°$.

Solve each of the following right triangles. Standard lettering has been used.

21. $a = 7.3$, $c = 8.6$

22. $a = 30.5$, $B = 51.17°$

23. One leg of a right triangle bears east. The hypotenuse is 734 m long and bears N57°23′E. Find the perimeter of the triangle.

24. An observer's eye is 6 ft above the floor. A mural is being viewed. The bottom of the mural is at floor level. The observer looks down 13° to see the bottom and up 17° to see the top. How tall is the mural?

Find a positive angle and a negative angle that are coterminal with the given angle. Answers may vary.

25. $65°$

26. $\dfrac{7\pi}{3}$

Find the complement and the supplement.

27. $13.4°$

28. $\dfrac{\pi}{6}$

29. Find the six trigonometric function values for the angle θ shown.

30. Given that $\tan \theta = 2/\sqrt{5}$ and that the terminal side is in quadrant III, find the other five function values.

31. An airplane travels at 530 mph for $3\frac{1}{2}$ hr in a direction of 160° from Minneapolis, Minnesota. At the end of that time, how far south of Minneapolis is the airplane?

32. On a unit circle, mark and label the points determined by $7\pi/6$, $-3\pi/4$, $-\pi/3$, and $9\pi/4$.

For angles of the following measures, state in which quadrant the terminal side lies, convert to radian measure in terms of π, and convert to radian measure not in terms of π.

33. $145.2°$

34. $-30°$

Convert to degree measure. Round the answer to two decimal places.

35. $\dfrac{3\pi}{2}$

36. 3

37. Find the length of an arc of a circle, given a central angle of $\pi/4$ and a radius of 7 cm.

38. An arc 18 m long on a circle of radius 8 m subtends an angle of how many radians? how many degrees, to the nearest degree?

39. Inside La Madeleine French Bakery and Cafe in Houston, Texas, there is one of the few remaining working watermills in the world. The 300-yr-old French-built waterwheel has a radius of 7 ft and makes one complete revolution in 70 sec. What is the linear speed in feet per minute of a point on the rim? (*Source:* La Madeleine French Bakery and Cafe, Houston, TX)

40. An automobile wheel has a diameter of 14 in. If the car travels at a speed of 55 mph, what is the angular velocity, in radians per hour, of a point on the edge of the wheel?

41. The point $\left(\frac{3}{5}, -\frac{4}{5}\right)$ is on a unit circle. Find the coordinates of its reflections across the x-axis, the y-axis, and the origin.

Find the exact function value, if it exists.

42. $\cos \pi$

43. $\tan \dfrac{5\pi}{4}$

44. $\sin \dfrac{5\pi}{3}$

45. $\sin \left(-\dfrac{7\pi}{6}\right)$

46. $\tan \dfrac{\pi}{6}$

47. $\cos \left(-13\pi\right)$

Find the function value. Round to four decimal places.

48. $\sin 24$

49. $\cos \left(-75\right)$

50. $\cot 16\pi$

51. $\tan \dfrac{3\pi}{7}$

52. $\sec 14.3$

53. $\cos \left(-\dfrac{\pi}{5} \right)$

54. Make a hand-drawn graph from -2π to 2π for each of the six trigonometric functions.

55. What is the period of each of the six trigonometric functions?

56. Complete the following table.

FUNCTION	DOMAIN	RANGE
sine		
cosine		
tangent		

57. Complete the following table with the sign of the specified trigonometric function value in each of the four quadrants.

FUNCTION	I	II	III	IV
sine				
cosine				
tangent				

Determine the amplitude, the period, and the phase shift of the function; and without a grapher, sketch the graph of the function. Then check the graph using a grapher.

58. $y = \sin \left(x + \dfrac{\pi}{2} \right)$

59. $y = 3 + \dfrac{1}{2} \cos \left(2x - \dfrac{\pi}{2} \right)$

In Exercises 60–63, without a grapher, match the function with one of graphs (a)–(d), which follow. Then check your work using a grapher.

a)

b)

c)

d)

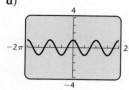

60. $y = \cos 2x$

61. $y = \dfrac{1}{2} \sin x + 1$

62. $y = -2 \sin \dfrac{1}{2} x - 3$

63. $y = -\cos \left(x - \dfrac{\pi}{2} \right)$

64. Sketch a graph of $y = 3 \cos x + \sin x$ for values of x between 0 and 2π.

Discussion and Writing

65. Compare the terms radian and degree.

66. Describe the shape of the graph of the cosine function. How many maximum values are there of the cosine function? Where do they occur?

67. Explain the disadvantage of a grapher when graphing a function like

$$f(x) = \frac{\sin x}{x}.$$

68. Does $5 \sin x = 7$ have a solution for x? Why or why not?

Synthesis

69. For what values of x in $(0, \pi/2]$ is $\sin x < x$ true?

70. Graph $y = 3 \sin (x/2)$ and determine the domain, the range, and the period.

71. In the window below, $y_1 = \sin x$ is shown and y_2 is shown in red. Express y_2 as a transformation of the graph of y_1.

$y_1 = \sin x, \quad y_2 = ?$

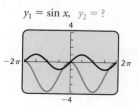

72. Find the domain of $y = \log (\cos x)$.

73. Given that $\sin x = 0.6144$ and that the terminal side is in quadrant II, find the other basic circular function values.

Trigonometric Identities, Inverse Functions, and Equations

3

APPLICATION

The number of daylight hours in Kajaani, Finland, varies from about 4.3 hr to 20.7 hr (*Source: The Astronomical Almanac*). The function

$$H(d) = 7.8787 \sin (0.0166d - 1.2723) + 12.1840$$

can be used to approximate the number of daylight hours H on a certain day d in Kajaani. We can use this function to determine on which day of the year there will be about 10.5 hr of daylight.

This problem appears as Exercise 62 in Exercise Set 3.5.

There are a number of relationships among the trigonometric functions, expressed as identities, that are important in algebraic and trigonometric manipulations. A large part of this chapter is devoted to those identities and their use in solving trigonometric equations. We also provide a detailed examination of inverses of the trigonometric functions. This chapter provides not only a basis for applications but also a foundation for more advanced work in mathematics.

$y = 7.8787 \sin (0.0166x - 1.2723) + 12.1840$

Identities: Pythagorean and Sum and Difference

3.1

- State the Pythagorean identities.
- Simplify and manipulate expressions containing trigonometric expressions.
- Use the sum and difference identities to find function values.

An **identity** is an equation that is true for all *possible* replacements of the variables. The following is a list of the identities studied in Chapter 2.

Basic Identities

$$\sin x = \frac{1}{\csc x},$$ $$\csc x = \frac{1}{\sin x},$$ $$\sin(-x) = -\sin x,$$
$$\cos(-x) = \cos x,$$

$$\cos x = \frac{1}{\sec x},$$ $$\sec x = \frac{1}{\cos x},$$ $$\tan(-x) = -\tan x,$$

$$\tan x = \frac{1}{\cot x},$$ $$\cot x = \frac{1}{\tan x},$$ $$\tan x = \frac{\sin x}{\cos x},$$

$$\cot x = \frac{\cos x}{\sin x}$$

In this section, we will develop some other important identities.

Pythagorean Identities

We now consider three other identities that are fundamental to a study of trigonometry. They are called the *Pythagorean identities.* Recall that the equation of a unit circle in the xy-plane is

$$x^2 + y^2 = 1. \qquad \text{Unit circle}$$

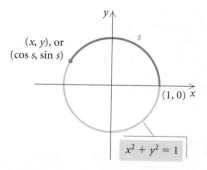

For any point on the unit circle, the coordinates x and y satisfy this equation. Suppose that a real number s determines a point on the unit circle with coordinates (x, y), or $(\cos s, \sin s)$. Then $x = \cos s$ and $y = \sin s$. Substituting $\cos s$ for x and $\sin s$ for y in the equation of the unit circle gives us the identity

$$(\cos s)^2 + (\sin s)^2 = 1,$$

which can be expressed as

$$\sin^2 s + \cos^2 s = 1.$$

It is conventional in trigonometry to use the notation $\sin^2 s$ rather than $(\sin s)^2$, which uses parentheses. Note that $\sin^2 s \neq \sin s^2$.

$y = (\sin x)^2 = (\sin x)(\sin x)$

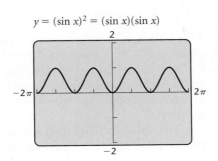

$y = \sin x^2 = \sin(x \cdot x)$

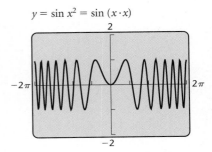

The identity $\sin^2 s + \cos^2 s = 1$ gives a relationship between the sine and the cosine of any real number s. It is an important **Pythagorean identity.**

Interactive Discovery

Addition of y-values provides a unique way of developing the identity $\sin^2 x + \cos^2 x = 1$. First, graph $y_1 = \sin^2 x$ and $y_2 = \cos^2 x$. By visually adding the y-values, sketch the graph of the sum, $y_3 = \sin^2 x + \cos^2 x$. Then graph y_3 using a grapher and check your sketch. The resulting graph appears to be the line $y_4 = 1$, which is the graph of $\sin^2 x + \cos^2 x$. These graphs *do not* prove the identity, but they do provide a check in the interval shown.

$y_1 = \sin^2 x$

$y_2 = \cos^2 x$

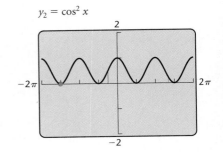

$y_3 = \sin^2 x + \cos^2 x, \quad y_4 = 1$

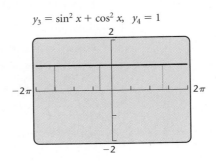

We can divide by $\sin^2 s$ on both sides of the preceding identity:

$$\frac{\sin^2 s}{\sin^2 s} + \frac{\cos^2 s}{\sin^2 s} = \frac{1}{\sin^2 s}. \qquad \text{Dividing by } \sin^2 s$$

Simplifying gives us a second identity:

$$1 + \cot^2 s = \csc^2 s.$$

This equation is true for any replacement of s with a real number for which $\sin^2 s \neq 0$, since we divided by $\sin^2 s$. But the numbers for which $\sin^2 s = 0$ (or $\sin s = 0$) are exactly the ones for which the cotangent and cosecant functions are undefined. Hence our new equation holds for all real numbers s for which $\cot s$ and $\csc s$ are defined and is thus an identity.

The third Pythagorean identity, obtained by dividing by $\cos^2 s$ on both sides of the first Pythagorean identity, is

$$\tan^2 s + 1 = \sec^2 s.$$

The identities we have developed hold no matter what symbols are used for the variables. For example, we could write $\sin^2 s + \cos^2 s = 1$, $\sin^2 \theta + \cos^2 \theta = 1$, or $\sin^2 x + \cos^2 x = 1$.

Pythagorean Identities

$\sin^2 x + \cos^2 x = 1$,

$1 + \cot^2 x = \csc^2 x$,

$1 + \tan^2 x = \sec^2 x$

It is often helpful to express the Pythagorean identities in equivalent forms. For example, $\sin^2 x + \cos^2 x = 1$ is frequently seen as $\sin^2 x = 1 - \cos^2 x$ or $\cos^2 x = 1 - \sin^2 x$.

Simplifying Trigonometric Expressions

We can factor, simplify, and manipulate trigonometric expressions in the same way that we manipulate strictly algebraic expressions.

EXAMPLE 1 Multiply and simplify: $\cos x (\tan x - \sec x)$.

Solution

$\cos x (\tan x - \sec x)$

$= \cos x \tan x - \cos x \sec x \qquad \text{Multiplying}$

$= \cos x \dfrac{\sin x}{\cos x} - \cos x \dfrac{1}{\cos x} \qquad \text{Recalling the identities } \tan x = \dfrac{\sin x}{\cos x}$

$\qquad\qquad\qquad\qquad\qquad\qquad \text{and } \sec x = \dfrac{1}{\cos x} \text{ and substituting}$

$= \sin x - 1 \qquad\qquad\qquad \text{Simplifying}$

There is no general procedure for manipulating trigonometric expressions, but it is often helpful to write everything in terms of sines and cosines, as we did in Example 1. We also look for the Pythagorean identity, $\sin^2 x + \cos^2 x = 1$, within a trigonometric expression.

EXAMPLE 2 Factor and simplify: $\sin^2 x \cos^2 x + \cos^4 x$.

Solution

$$\sin^2 x \cos^2 x + \cos^4 x$$
$$= \cos^2 x \,(\sin^2 x + \cos^2 x) \qquad \text{Removing a common factor}$$
$$= \cos^2 x \qquad\qquad\qquad\quad \text{Using } \sin^2 x + \cos^2 x = 1 \quad \blacksquare$$

A grapher can be used to perform a partial check of an identity. First, we graph the expression on the left side of the equals sign. Then we graph the expression on the right side using the same screen. If the two graphs are indistinguishable, then we have a partial verification that the equation is an identity. Of course, we can never see the entire graph, so there can always be some doubt. Also, the graphs may not overlap precisely, but you may not be able to tell because the difference between the graphs may be less than the width of a pixel. However, if the graphs are obviously different, we know that a mistake has been made.

For example, consider the identity in Example 1:

$$\cos x \,(\tan x - \sec x) = \sin x - 1.$$

Recalling that $\sec x = 1/\cos x$, we enter

$$y_1 = \cos x \,[\tan x - (1/\cos x)] \quad \text{and} \quad y_2 = \sin x - 1.$$

To graph, we first select SEQUENTIAL mode. Then we select the "line"-graph style for y_1 and the "path"-graph style, denoted by -0, for y_2. The grapher will graph y_1 first. Then it will graph y_2 as the circular cursor traces the leading edge of the graph, allowing us to determine whether the graphs coincide. As you can see in the screen on the left below, the graphs appear to be identical. Thus, $\cos x \,(\tan x - \sec x) = \sin x - 1$ is most likely an identity.

$y_1 = \cos x \,[\tan x - (1/\cos x)]$,
$y_2 = \sin x - 1$

X	Y₁	Y₂
−6.283	−1	−1
−5.498	−.2929	−.2929
−4.712	ERROR	0
−3.927	−.2929	−.2929
−3.142	−1	−1
−2.356	−1.707	−1.707
−1.571	ERROR	−2

X = −6.28318530718

TblStart = -2π
ΔTbl = $\pi/4$

The TABLE feature can also be used to check identities. Note in the table on the right above that the function values are the same except for those values of x for which $\cos x = 0$. The domain of y_1 excludes these values. The domain of y_2 is the set of all real numbers. Thus all real numbers except $\pm\pi/2, \pm3\pi/2, \pm5\pi/2, \ldots$ are possible replacements for x in the identity. Recall that an identity is an equation that is true for all *possible* replacements.

Suppose that we had simplified incorrectly in Example 1 and had gotten $\cos x - 1$ on the right instead of $\sin x - 1$. Then two different graphs would have appeared in the window. Thus we would have known that we did not have an identity and that $\cos x (\tan x - \sec x) \neq \cos x - 1$.

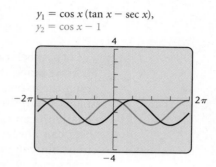

$y_1 = \cos x (\tan x - \sec x),$
$y_2 = \cos x - 1$

EXAMPLE 3   Simplify each of the following trigonometric expressions.

a) $\dfrac{\cot (-\theta)}{\csc (-\theta)}$ **b)** $\dfrac{2 \sin^2 t + \sin t - 3}{1 - \cos^2 t - \sin t}$

Solution

a) $\dfrac{\cot (-\theta)}{\csc (-\theta)} = \dfrac{\dfrac{\cos (-\theta)}{\sin (-\theta)}}{\dfrac{1}{\sin (-\theta)}}$ Rewriting in terms of sines and cosines

$= \dfrac{\cos (-\theta)}{\sin (-\theta)} \cdot \sin (-\theta)$ Multiplying by the reciprocal

$= \cos (-\theta) = \cos \theta$ The cosine function is even.

b) $\dfrac{2 \sin^2 t + \sin t - 3}{1 - \cos^2 t - \sin t}$

$= \dfrac{2 \sin^2 t + \sin t - 3}{\sin^2 t - \sin t}$ Using $\sin^2 x = 1 - \cos^2 x$, and substituting $\sin^2 t$ for $1 - \cos^2 t$

$= \dfrac{(2 \sin t + 3)(\sin t - 1)}{\sin t (\sin t - 1)}$ Factoring in both numerator and denominator

$= \dfrac{2 \sin t + 3}{\sin t}$ Simplifying

We can add and subtract trigonometric rational expressions in the same way that we did algebraic expressions.

EXAMPLE 4 Add and simplify: $\dfrac{\cos x}{1 + \sin x} + \tan x$.

Solution

$$\frac{\cos x}{1 + \sin x} + \tan x = \frac{\cos x}{1 + \sin x} + \frac{\sin x}{\cos x} \qquad \text{Using } \tan x = \frac{\sin x}{\cos x}$$

$$= \frac{\cos x}{1 + \sin x} \cdot \frac{\cos x}{\cos x} + \frac{\sin x}{\cos x} \cdot \frac{1 + \sin x}{1 + \sin x}$$

$$\text{Multiplying by forms of 1}$$

$$= \frac{\cos^2 x + \sin x + \sin^2 x}{\cos x \,(1 + \sin x)} \qquad \text{Adding}$$

$$= \frac{1 + \sin x}{\cos x \,(1 + \sin x)} \qquad \text{Using } \sin^2 x + \cos^2 x = 1$$

$$= \frac{1}{\cos x}, \quad \text{or} \quad \sec x \qquad \text{Simplifying}$$

When radicals occur, the use of absolute value is sometimes necessary, but it can be difficult to determine when to use it. In Examples 5 and 6, we will assume that all radicands are nonnegative. This means that the identities are meant to be confined to certain quadrants.

EXAMPLE 5 Multiply and simplify: $\sqrt{\sin^3 x \cos x} \cdot \sqrt{\cos x}$.

Solution

$$\sqrt{\sin^3 x \cos x} \cdot \sqrt{\cos x} = \sqrt{\sin^3 x \cos^2 x}$$

$$= \sqrt{\sin^2 x \cos^2 x \sin x}$$

$$= \sin x \cos x \sqrt{\sin x}$$

EXAMPLE 6 Rationalize the denominator: $\sqrt{\dfrac{2}{\tan x}}$.

Solution

$$\sqrt{\frac{2}{\tan x}} = \sqrt{\frac{2}{\tan x} \cdot \frac{\tan x}{\tan x}}$$

$$= \sqrt{\frac{2 \tan x}{\tan^2 x}}$$

$$= \frac{\sqrt{2 \tan x}}{\tan x}$$

Often in calculus, a substitution is a useful manipulation, as we show in the following example.

EXAMPLE 7 Express $\sqrt{9 + x^2}$ as a trigonometric function of θ without using radicals by letting $x = 3 \tan \theta$. Assume that $0 < \theta < \pi/2$. Then find $\sin \theta$ and $\cos \theta$.

Solution   We have

$$\sqrt{9 + x^2} = \sqrt{9 + (3 \tan \theta)^2} \qquad \text{Substituting } 3 \tan \theta \text{ for } x$$
$$= \sqrt{9 + 9 \tan^2 \theta}$$
$$= \sqrt{9(1 + \tan^2 \theta)} \qquad \text{Factoring}$$
$$= \sqrt{9 \sec^2 \theta} \qquad \text{Using } 1 + \tan^2 x = \sec^2 x$$
$$= 3|\sec \theta| = 3 \sec \theta. \qquad \text{For } 0 < \theta < \pi/2, \sec \theta > 0.$$

We can express $\sqrt{9 + x^2} = 3 \sec \theta$ as

$$\sec \theta = \frac{\sqrt{9 + x^2}}{3}.$$

In a right triangle, we know that $\sec \theta$ is hypotenuse/adjacent, when θ is one of the acute angles. Using the Pythagorean theorem, we can determine that the side opposite θ is x. Then from the right triangle, we see that

$$\sin \theta = \frac{x}{\sqrt{9 + x^2}} \quad \text{and} \quad \cos \theta = \frac{3}{\sqrt{9 + x^2}}.$$

Sum and Difference Identities

We now develop some important identities involving sums or differences of two numbers (or angles), beginning with an identity for the cosine of the difference of two numbers. We use the letters u and v for these numbers.

Let's consider a real number u in the interval $[\pi/2, \pi]$ and a real number v in the interval $[0, \pi/2]$. These determine points A and B on the unit circle, as shown below. The arc length s is $u - v$, and we know that $0 \le s \le \pi$. Recall that the coordinates of A are $(\cos u, \sin u)$, and the coordinates of B are $(\cos v, \sin v)$.

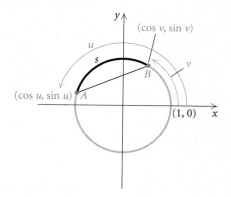

Using the distance formula, we can write an expression for the distance AB:

$$AB = \sqrt{(\cos u - \cos v)^2 + (\sin u - \sin v)^2}.$$

This can be simplified as follows:

$$AB = \sqrt{\cos^2 u - 2\cos u \cos v + \cos^2 v + \sin^2 u - 2\sin u \sin v + \sin^2 v}$$
$$= \sqrt{(\sin^2 u + \cos^2 u) + (\sin^2 v + \cos^2 v) - 2(\cos u \cos v + \sin u \sin v)}$$
$$= \sqrt{2 - 2(\cos u \cos v + \sin u \sin v)}.$$

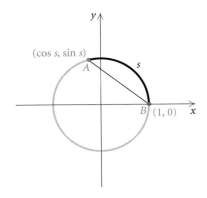

Now let's imagine rotating the circle above so that point B is at $(1, 0)$. Although the coordinates of point A are now $(\cos s, \sin s)$, the distance AB has not changed.

Again we use the distance formula to write an expression for the distance AB:

$$AB = \sqrt{(\cos s - 1)^2 + (\sin s - 0)^2}.$$

This can be simplified as follows:

$$AB = \sqrt{\cos^2 s - 2\cos s + 1 + \sin^2 s}$$
$$= \sqrt{(\sin^2 s + \cos^2 s) + 1 - 2\cos s}$$
$$= \sqrt{2 - 2\cos s}.$$

Equating our two expressions for AB, we obtain

$$\sqrt{2 - 2(\cos u \cos v + \sin u \sin v)} = \sqrt{2 - 2\cos s}.$$

Solving this equation for $\cos s$ gives

$$\cos s = \cos u \cos v + \sin u \sin v. \tag{1}$$

But $s = u - v$, so we have the equation

$$\cos(u - v) = \cos u \cos v + \sin u \sin v. \tag{2}$$

Formula (1) above holds when s is the length of the shortest arc from A to B. Given any real numbers u and v, the length of the shortest arc from A to B is not always $u - v$. In fact, it could be $v - u$. However, since $\cos(-x) = \cos x$, we know that $\cos(v - u) = \cos(u - v)$. Thus, $\cos s$ is always equal to $\cos(u - v)$. Formula (2) holds for all real numbers u and v. That formula is thus the identity we sought:

$$\mathbf{\cos(\textit{u} - \textit{v}) = \cos \textit{u} \cos \textit{v} + \sin \textit{u} \sin \textit{v}.}$$

Using a grapher, graph

$$y_1 = \cos(x - 3)$$

and

$$y_2 = \cos x \cos 3 + \sin x \sin 3$$

to illustrate this result.

The cosine sum formula follows easily from the one we have just derived. Let's consider $\cos(u + v)$. This is equal to $\cos[u - (-v)]$, and by the identity above, we have

$$\cos(u + v) = \cos[u - (-v)]$$
$$= \cos u \cos(-v) + \sin u \sin(-v).$$

But $\cos(-v) = \cos v$ and $\sin(-v) = -\sin v$, so the identity we seek is

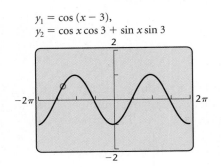

$y_1 = \cos(x - 3),$
$y_2 = \cos x \cos 3 + \sin x \sin 3$

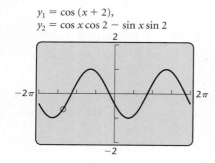

$y_1 = \cos{(x + 2)},$
$y_2 = \cos{x} \cos{2} - \sin{x} \sin{2}$

the following:

$$\cos{(u + v)} = \cos{u} \cos{v} - \sin{u} \sin{v}.$$

Using a grapher, graph

$$y_1 = \cos{(x + 2)}$$

and $y_2 = \cos{x} \cos{2} - \sin{x} \sin{2}$

to illustrate this result.

EXAMPLE 8   Find $\cos{(5\pi/12)}$ exactly.

Solution   We can express $5\pi/12$ as a difference of two numbers whose sine and cosine values are known:

$$\frac{5\pi}{12} = \frac{9\pi}{12} - \frac{4\pi}{12}, \quad \text{or} \quad \frac{3\pi}{4} - \frac{\pi}{3}.$$

Then, using $\cos{(u - v)} = \cos{u} \cos{v} + \sin{u} \sin{v}$, we have

$$\cos{\frac{5\pi}{12}} = \cos{\left(\frac{3\pi}{4} - \frac{\pi}{3}\right)} = \cos{\frac{3\pi}{4}} \cos{\frac{\pi}{3}} + \sin{\frac{3\pi}{4}} \sin{\frac{\pi}{3}}$$

$$= -\frac{\sqrt{2}}{2} \cdot \frac{1}{2} + \frac{\sqrt{2}}{2} \cdot \frac{\sqrt{3}}{2}$$

$$= -\frac{\sqrt{2}}{4} + \frac{\sqrt{6}}{4} = \frac{\sqrt{6} - \sqrt{2}}{4}.$$

```
cos(5π/12)
              .2588190451
(√(6)−√(2))/4
              .2588190451
```

We can check this result using a grapher set in RADIAN mode. ▬

Consider $\cos{(\pi/2 - \theta)}$. We can use the identity for the cosine of a difference to simplify as follows:

$$\cos{\left(\frac{\pi}{2} - \theta\right)} = \cos{\frac{\pi}{2}} \cos{\theta} + \sin{\frac{\pi}{2}} \sin{\theta}$$

$$= 0 \cdot \cos{\theta} + 1 \cdot \sin{\theta} = \sin{\theta}.$$

Thus we have developed the identity

$$\sin{\theta} = \cos{\left(\frac{\pi}{2} - \theta\right)}. \qquad \text{This cofunction identity first appeared in Section 2.1.} \qquad (3)$$

This identity holds for any real number θ. From it we can obtain an identity for the sine function. We first let α be any real number. Then we replace θ in $\sin{\theta} = \cos{(\pi/2 - \theta)}$ with $\pi/2 - \alpha$. This gives us

$$\sin{\left(\frac{\pi}{2} - \alpha\right)} = \cos{\left[\frac{\pi}{2} - \left(\frac{\pi}{2} - \alpha\right)\right]} = \cos{\alpha},$$

which yields the identity

$$\cos{\alpha} = \sin{\left(\frac{\pi}{2} - \alpha\right)}. \qquad (4)$$

Using identities (3) and (4) and the identity for the cosine of a difference, we can obtain an identity for the sine of a sum. We start with identity (3) and substitute $u + v$ for θ:

$$\sin \theta = \cos \left(\frac{\pi}{2} - \theta \right) \qquad \text{Identity (3)}$$

$$\sin (u + v) = \cos \left[\frac{\pi}{2} - (u + v) \right] \qquad \text{Substituting } u + v \text{ for } \theta$$

$$= \cos \left[\left(\frac{\pi}{2} - u \right) - v \right]$$

$$= \cos \left(\frac{\pi}{2} - u \right) \cos v + \sin \left(\frac{\pi}{2} - u \right) \sin v$$

Using the identity for the cosine of a difference

$$= \sin u \cos v + \cos u \sin v. \qquad \text{Using identities (3) and (4)}$$

Thus the identity we seek is

$$\sin (u + v) = \sin u \cos v + \cos u \sin v.$$

To find a formula for the sine of a difference, we can use the identity just derived, substituting $-v$ for v:

$$\sin (u + (-v)) = \sin u \cos (-v) + \cos u \sin (-v).$$

Simplifying gives us

$$\sin (u - v) = \sin u \cos v - \cos u \sin v.$$

EXAMPLE 9   Find $\sin 105°$ exactly.

Solution   We express $105°$ as the sum of two measures:

$$105° = 45° + 60°.$$

Then

$$\sin 105° = \sin (45° + 60°)$$

$$= \sin 45° \cos 60° + \cos 45° \sin 60°$$

Using $\sin (u + v) = \sin u \cos v + \cos u \sin v$

$$= \frac{\sqrt{2}}{2} \cdot \frac{1}{2} + \frac{\sqrt{2}}{2} \cdot \frac{\sqrt{3}}{2}$$

$$= \frac{\sqrt{2} + \sqrt{6}}{4}.$$

```
sin(105)
              .9659258263
(√(2)+√(6))/4
              .9659258263
```

We can check this result using a grapher set in DEGREE mode.   ▬

Formulas for the tangent of a sum or a difference can be derived using identities already established. A summary of the sum and difference identities follows.*

Sum and Difference Identities

$$\sin (u \pm v) = \sin u \cos v \pm \cos u \sin v,$$

$$\cos (u \pm v) = \cos u \cos v \mp \sin u \sin v,$$

$$\tan (u \pm v) = \frac{\tan u \pm \tan v}{1 \mp \tan u \tan v}$$

EXAMPLE 10   Find $\tan 15°$ exactly.

Solution   We rewrite $15°$ as $45° - 30°$ and use the identity for the tangent of a difference:

$$\tan 15° = \tan (45° - 30°) = \frac{\tan 45° - \tan 30°}{1 + \tan 45° \tan 30°}$$

$$= \frac{1 - \sqrt{3}/3}{1 + 1 \cdot \sqrt{3}/3} = \frac{3 - \sqrt{3}}{3 + \sqrt{3}}.$$

EXAMPLE 11   Assume that $\sin \alpha = \frac{2}{3}$ and $\sin \beta = \frac{1}{3}$ and that α and β are between 0 and $\pi/2$. Then evaluate $\sin (\alpha + \beta)$.

Solution   Using the identity for the sine of a sum, we have

$$\sin (\alpha + \beta) = \sin \alpha \cos \beta + \cos \alpha \sin \beta$$

$$= \tfrac{2}{3} \cos \beta + \tfrac{1}{3} \cos \alpha.$$

To finish, we need to know the values of $\cos \beta$ and $\cos \alpha$. Using reference triangles and the Pythagorean theorem, we can determine these values from the diagrams:

$$\cos \alpha = \frac{\sqrt{5}}{3} \quad \text{and} \quad \cos \beta = \frac{2\sqrt{2}}{3}.$$ Cosine values are positive in the first quadrant.

Substituting these values gives us

$$\sin (\alpha + \beta) = \frac{2}{3} \cdot \frac{2\sqrt{2}}{3} + \frac{1}{3} \cdot \frac{\sqrt{5}}{3}$$

$$= \frac{4}{9}\sqrt{2} + \frac{1}{9}\sqrt{5}, \quad \text{or} \quad \frac{4\sqrt{2} + \sqrt{5}}{9}.$$

---

*There are six identities here, half of them obtained by using the signs shown in color.

Exercise Set 3.1

Multiply and simplify. Check your result using a grapher.

1. $(\sin x - \cos x)(\sin x + \cos x)$
2. $\tan x \,(\cos x - \csc x)$
3. $\cos y \sin y \,(\sec y + \csc y)$
4. $(\sin x + \cos x)(\sec x + \csc x)$
5. $(\sin \phi - \cos \phi)^2$
6. $(1 + \tan x)^2$
7. $(\sin x + \csc x)(\sin^2 x + \csc^2 x - 1)$
8. $(1 - \sin t)(1 + \sin t)$

Factor and simplify. Check your result using a grapher.

9. $\sin x \cos x + \cos^2 x$
10. $\tan^2 \theta - \cot^2 \theta$
11. $\sin^4 x - \cos^4 x$
12. $4 \sin^2 y + 8 \sin y + 4$
13. $2 \cos^2 x + \cos x - 3$
14. $3 \cot^2 \beta + 6 \cot \beta + 3$
15. $\sin^3 x + 27$
16. $1 - 125 \tan^3 s$

Simplify and check using a grapher.

17. $\dfrac{\sin^2 x \cos x}{\cos^2 x \sin x}$

18. $\dfrac{30 \sin^3 x \cos x}{6 \cos^2 x \sin x}$

19. $\dfrac{\sin^2 x + 2 \sin x + 1}{\sin x + 1}$

20. $\dfrac{\cos^2 \alpha - 1}{\cos \alpha + 1}$

21. $\dfrac{4 \tan t \sec t + 2 \sec t}{6 \tan t \sec t + 2 \sec t}$

22. $\dfrac{\csc (-x)}{\cot (-x)}$

23. $\dfrac{\sin^4 x - \cos^4 x}{\sin^2 x - \cos^2 x}$

24. $\dfrac{4 \cos^3 x}{\sin^2 x} \cdot \left(\dfrac{\sin x}{4 \cos x}\right)^2$

25. $\dfrac{5 \cos \phi}{\sin^2 \phi} \cdot \dfrac{\sin^2 \phi - \sin \phi \cos \phi}{\sin^2 \phi - \cos^2 \phi}$

26. $\dfrac{\tan^2 y}{\sec y} \div \dfrac{3 \tan^3 y}{\sec y}$

27. $\dfrac{1}{\sin^2 s - \cos^2 s} - \dfrac{2}{\cos s - \sin s}$

28. $\left(\dfrac{\sin x}{\cos x}\right)^2 - \dfrac{1}{\cos^2 x}$

29. $\dfrac{\sin^2 \theta - 9}{2 \cos \theta + 1} \cdot \dfrac{10 \cos \theta + 5}{3 \sin \theta + 9}$

30. $\dfrac{9 \cos^2 \alpha - 25}{2 \cos \alpha - 2} \cdot \dfrac{\cos^2 \alpha - 1}{6 \cos \alpha - 10}$

Simplify and check using a grapher. Assume that all radicands are nonnegative.

31. $\sqrt{\sin^2 x \cos x} \cdot \sqrt{\cos x}$
32. $\sqrt{\cos^2 x \sin x} \cdot \sqrt{\sin x}$
33. $\sqrt{\cos \alpha \sin^2 \alpha} - \sqrt{\cos^3 \alpha}$
34. $\sqrt{\tan^2 x - 2 \tan x \sin x + \sin^2 x}$
35. $(1 - \sqrt{\sin y})(\sqrt{\sin y} + 1)$
36. $\sqrt{\cos \theta}\,(\sqrt{2 \cos \theta} + \sqrt{\sin \theta \cos \theta})$

Rationalize the denominator.

37. $\sqrt{\dfrac{\sin x}{\cos x}}$

38. $\sqrt{\dfrac{\cos x}{\tan x}}$

39. $\sqrt{\dfrac{\cos^2 y}{2 \sin^2 y}}$

40. $\sqrt{\dfrac{1 - \cos \beta}{1 + \cos \beta}}$

Rationalize the numerator.

41. $\sqrt{\dfrac{\cos x}{\sin x}}$

42. $\sqrt{\dfrac{\sin x}{\cot x}}$

43. $\sqrt{\dfrac{1 + \sin y}{1 - \sin y}}$

44. $\sqrt{\dfrac{\cos^2 x}{2 \sin^2 x}}$

Use the given substitution to express the given radical expression as a trigonometric function without radicals. Assume that $a > 0$ and $0 < \theta < \pi/2$. Then find expressions for the indicated trigonometric functions.

45. Let $x = a \sin \theta$ in $\sqrt{a^2 - x^2}$. Then find $\cos \theta$ and $\tan \theta$.

46. Let $x = 2 \tan \theta$ in $\sqrt{4 + x^2}$. Then find $\sin \theta$ and $\cos \theta$.

47. Let $x = 3 \sec \theta$ in $\sqrt{x^2 - 9}$. Then find $\sin \theta$ and $\cos \theta$.

48. Let $x = a \sec \theta$ in $\sqrt{x^2 - a^2}$. Then find $\sin \theta$ and $\cos \theta$.

Use the given substitution to express the given radical expression as a trigonometric function without radicals. Assume that $0 < \theta < \pi/2$.

49. Let $x = \sin \theta$ in $\dfrac{x^2}{\sqrt{1 - x^2}}$.

50. Let $x = 4 \sec \theta$ in $\dfrac{\sqrt{x^2 - 16}}{x^2}$.

Use the sum and difference identities to evaluate exactly. Then check using a grapher.

51. $\sin \dfrac{\pi}{12}$ **52.** $\cos 75°$ **53.** $\tan 105°$

54. $\tan \dfrac{5\pi}{12}$ **55.** $\cos 15°$ **56.** $\sin \dfrac{7\pi}{12}$

First write each of the following as a trigonometric function of a single angle; then evaluate.

57. $\sin 37° \cos 22° + \cos 37° \sin 22°$

58. $\cos 83° \cos 53° + \sin 83° \sin 53°$

59. $\dfrac{\tan 20° + \tan 32°}{1 - \tan 20° \tan 32°}$

60. $\dfrac{\tan 35° - \tan 12°}{1 + \tan 35° \tan 12°}$

Assuming that $\sin u = \frac{3}{5}$ and $\sin v = \frac{4}{5}$ and that u and v are between 0 and $\pi/2$, evaluate each of the following exactly.

61. $\cos (u + v)$ **62.** $\tan (u - v)$ **63.** $\sin (u - v)$

Assuming that $\sin \theta = 0.6249$ and $\cos \phi = 0.1102$ and that both θ and ϕ are first-quadrant angles, evaluate each of the following.

64. $\sin (\theta - \phi)$ **65.** $\tan (\theta + \phi)$ **66.** $\cos (\theta + \phi)$

Simplify.

67. $\sin (\alpha + \beta) + \sin (\alpha - \beta)$

68. $\cos (\alpha + \beta) - \cos (\alpha - \beta)$

69. $\cos (u + v) \cos v + \sin (u + v) \sin v$

70. $\sin (u - v) \cos v + \cos (u - v) \sin v$

Discussion and Writing

71. What is the difference between a trigonometric equation that is an identity and a trigonometric equation that is not an identity? Give an example of each.

72. Why is it possible with a grapher to *disprove* that an equation is an identity but not to *prove* that one is?

Skill Maintenance

Find the function value using coordinates of points on the unit circle. Give exact answers.

73. $\sin \dfrac{3\pi}{4}$ **74.** $\sec \left(-\dfrac{5\pi}{3}\right)$

Given that $\sin 31° = 0.5150$ and $\cos 31° = 0.8572$, find the specified function value.

75. $\sec 59°$ **76.** $\tan 59°$

Synthesis

Angles Between Lines. One of the identities gives an easy way to find an angle formed by two lines. Consider two lines with equations $l_1: y = m_1 x + b_1$ and $l_2: y = m_2 x + b_2$.

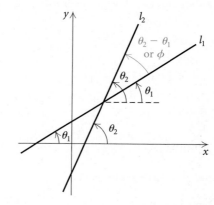

The slopes m_1 and m_2 are the tangents of the angles θ_1 and θ_2 that the lines form with the positive direction of the x-axis. Thus we have $m_1 = \tan \theta_1$ and $m_2 = \tan \theta_2$. To find the measure of $\theta_2 - \theta_1$, or ϕ, we proceed as follows:

$$\tan \phi = \tan (\theta_2 - \theta_1)$$
$$= \frac{\tan \theta_2 - \tan \theta_1}{1 + \tan \theta_2 \tan \theta_1}$$
$$= \frac{m_2 - m_1}{1 + m_2 m_1}.$$

This formula also holds when the lines are taken in the reverse order. When ϕ is acute, $\tan \phi$ will be positive. When ϕ is obtuse, $\tan \phi$ will be negative.

Find the measure of the angle from l_1 to l_2.

77. $l_1: 2x = 3 - 2y,$
    $l_2: x + y = 5$

78. $l_1: 3y = \sqrt{3}x + 3,$
    $l_2: y = \sqrt{3}x + 2$

79. $l_1: y = 3,$
    $l_2: x + y = 5$

80. $l_1: 2x + y - 4 = 0,$
    $l_2: y - 2x + 5 = 0$

81. *Circus Guy Wire.* In a circus, a guy wire A is attached to the top of a 30-ft pole. Wire B is used for performers to walk up to the tight wire, 10 ft above the ground. Find the angle ϕ between the wires if they are attached to the ground 40 ft from the pole.

82. Given that $f(x) = \sin x$, show that

$$\frac{f(x + h) - f(x)}{h} = \sin x \left(\frac{\cos h - 1}{h} \right) + \cos x \left(\frac{\sin h}{h} \right).$$

83. Given that $f(x) = \cos x$, show that

$$\frac{f(x + h) - f(x)}{h} = \cos x \left(\frac{\cos h - 1}{h} \right) - \sin x \left(\frac{\sin h}{h} \right).$$

Show that each of the following is not an identity by first finding a replacement or replacements for which the sides of the equation do not name the same number. Then use a grapher to show that the equation is not an identity.

84. $\sqrt{\sin^2 \theta} = \sin \theta$

85. $\dfrac{\sin 5x}{x} = \sin 5$

86. $\sin (-x) = \sin x$

87. $\cos (2\alpha) = 2 \cos \alpha$

88. $\tan^2 \theta + \cot^2 \theta = 1$

89. $\dfrac{\cos 6x}{\cos x} = 6$

Find the slope of line l_1, where m_2 is the slope of line l_2 and ϕ is the smallest positive angle from l_1 to l_2.

90. $m_2 = \frac{4}{3}$, $\phi = 45°$

91. $m_2 = \frac{2}{3}$, $\phi = 30°$

92. Line l_1 contains the points $(-2, 4)$ and $(5, -1)$. Find the slope of line l_2 such that the angle from l_1 to l_2 is 45°.

93. Line l_1 contains the points $(-3, 7)$ and $(-3, -2)$. Line l_2 contains $(0, -4)$ and $(2, 6)$. Find the smallest positive angle from l_1 to l_2.

94. Find an identity for $\sin 2\theta$. (*Hint:* $2\theta = \theta + \theta$.)

95. Find an identity for $\cos 2\theta$. (*Hint:* $2\theta = \theta + \theta$.)

Derive the identity. Check using a grapher.

96. $\sin \left(x - \dfrac{3\pi}{2} \right) = \cos x$

97. $\tan \left(x + \dfrac{\pi}{4} \right) = \dfrac{1 + \tan x}{1 - \tan x}$

98. $\dfrac{\sin (\alpha + \beta)}{\cos (\alpha - \beta)} = \dfrac{\tan \alpha + \tan \beta}{1 + \tan \alpha \tan \beta}$

99. $\sin (\alpha + \beta) + \sin (\alpha - \beta) = 2 \sin \alpha \cos \beta$

Identities: Cofunction, Double-Angle, and Half-Angle

3.2

- Use cofunction identities to derive other identities.
- Use the double-angle identities to find function values of twice an angle when one function value is known for that angle.
- Use the half-angle identities to find function values of half an angle when one function value is known for that angle.
- Simplify trigonometric expressions using the double-angle and half-angle identities.

Cofunction Identities

Each of the identities listed below yields a conversion to a *cofunction*. For this reason, we call them cofunction identities.

Cofunction Identities

$$\sin\left(\frac{\pi}{2} - x\right) = \cos x, \qquad \cos\left(\frac{\pi}{2} - x\right) = \sin x,$$

$$\tan\left(\frac{\pi}{2} - x\right) = \cot x, \qquad \cot\left(\frac{\pi}{2} - x\right) = \tan x,$$

$$\sec\left(\frac{\pi}{2} - x\right) = \csc x, \qquad \csc\left(\frac{\pi}{2} - x\right) = \sec x$$

We verified the first two of these identities in Section 3.1. The other four can be proved using the first two and the definitions of the trigonometric functions. These identities hold for all real numbers, and thus, for all angle measures, but if we restrict θ to values such that $0° < \theta < 90°$, or $0 < \theta < \pi/2$, then we have a special application to the acute angles of a right triangle.

Interactive Discovery

Graph $y_1 = \sin x$ and $y_2 = \sin(x + \pi/2)$ on the same screen. Note that the graph of y_2 is a translation of the graph of y_1 to the left $\pi/2$ units. Now graph $y_3 = \cos x$.

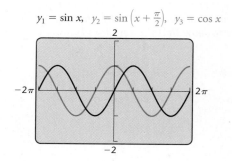

$$y_1 = \sin x, \quad y_2 = \sin\left(x + \frac{\pi}{2}\right), \quad y_3 = \cos x$$

We note that the graph of y_2 is the same as the graph of y_3. This leads to a possible identity:

$$\sin\left(x + \frac{\pi}{2}\right) = \cos x.$$

Repeat this exploration for

$$y = \sin\left(x - \frac{\pi}{2}\right), \qquad y = \cos\left(x + \frac{\pi}{2}\right), \quad \text{and} \quad y = \cos\left(x - \frac{\pi}{2}\right).$$

What possible identities do you observe?

The identity $\sin(x + \pi/2) = \cos x$ can be proved using the identity for the sine of a sum developed in Section 3.1.

EXAMPLE 1   Prove the identity $\sin(\theta + \pi/2) = \cos\theta$.

Solution

$$\sin\left(\theta + \frac{\pi}{2}\right) = \sin\theta\cos\frac{\pi}{2} + \cos\theta\sin\frac{\pi}{2} \qquad \text{Using } \sin(u+v) =$$
$$\text{sin } u \cos v +$$
$$\cos u \sin v$$

$$= \sin\theta \cdot 0 + \cos\theta \cdot 1$$

$$= \cos\theta$$

We now state four more cofunction identities. These new identities that involve the sine and cosine functions can be verified using previously established identities as seen in Example 1.

Cofunction Identities for the Sine and Cosine

$$\sin\left(x \pm \frac{\pi}{2}\right) = \pm\cos x, \qquad \cos\left(x \pm \frac{\pi}{2}\right) = \mp\sin x$$

EXAMPLE 2   Find an identity for each of the following.

a) $\tan\left(x + \dfrac{\pi}{2}\right)$ 　　　　　　　　　　　　 **b)** $\sec(x - 90°)$

Solution

a) We have

$$\tan\left(x + \frac{\pi}{2}\right) = \frac{\sin\left(x + \dfrac{\pi}{2}\right)}{\cos\left(x + \dfrac{\pi}{2}\right)} \qquad \text{Using } \tan x = \frac{\sin x}{\cos x}$$

$$= \frac{\cos x}{-\sin x} \qquad \text{Using cofunction identities}$$

$$= -\cot x.$$

Thus the identity we seek is

$$\tan\left(x + \frac{\pi}{2}\right) = -\cot x.$$

b) We have

$$\sec(x - 90°) = \frac{1}{\cos(x - 90°)} = \frac{1}{\sin x} = \csc x.$$

Thus, $\sec(x - 90°) = \csc x$.

Double-Angle Identities

If we double an angle of measure x, the new angle will have measure $2x$. **Double-angle identities** give trigonometric function values of $2x$ in terms of function values of x. To develop these identities, we will use the sum formulas from the preceding section. We first develop a formula for $\sin 2x$. Recall that

$$\sin (u + v) = \sin u \cos v + \cos u \sin v.$$

We will consider a number x and substitute it for both u and v in this identity. Doing so gives us

$$\sin (x + x) = \sin 2x$$
$$= \sin x \cos x + \cos x \sin x$$
$$= 2 \sin x \cos x.$$

Our first double-angle identity is thus

$$\textbf{sin } 2x = 2 \textbf{ sin } x \textbf{ cos } x.$$

Graphers provide visual partial checks of identities. We can graph $y_1 = \sin 2x$ and $y_2 = 2 \sin x \cos x$ and see that they appear to have the same graph. We can also use the TABLE feature.

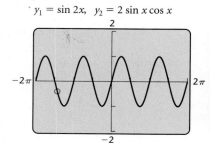

$y_1 = \sin 2x, \quad y_2 = 2 \sin x \cos x$

X	Y1	Y2
−6.283	2E−13	0
−5.498	1	1
−4.712	0	0
−3.927	−1	−1
−3.142	0	0
−2.356	1	1
−1.571	0	0

X = −1.57079632679

Double-angle identities for the cosine and tangent functions can be derived and checked using a grapher in much the same way as the identity above:

$$\textbf{cos } 2x = \textbf{cos}^2 \, x - \textbf{sin}^2 \, x, \qquad \textbf{tan } 2x = \frac{2 \textbf{ tan } x}{1 - \textbf{tan}^2 \, x}.$$

EXAMPLE 3   Given that $\tan \theta = -\frac{3}{4}$ and θ is in quadrant II, find each of the following.

a) $\sin 2\theta$ \hspace{3cm} **b)** $\cos 2\theta$

c) $\tan 2\theta$ \hspace{3cm} **d)** The quadrant in which 2θ lies

Solution   By drawing a diagram as shown, we find that

$$\sin \theta = \frac{3}{5}$$

and

$$\cos \theta = -\frac{4}{5}.$$

Thus we have the following.

a) $\sin 2\theta = 2 \sin \theta \cos \theta = 2 \cdot \frac{3}{5} \cdot \left(-\frac{4}{5}\right) = -\frac{24}{25}$

b) $\cos 2\theta = \cos^2 \theta - \sin^2 \theta = \left(-\frac{4}{5}\right)^2 - \left(\frac{3}{5}\right)^2 = \frac{16}{25} - \frac{9}{25} = \frac{7}{25}$

c) $\tan 2\theta = \frac{2 \tan \theta}{1 - \tan^2 \theta} = \frac{2 \cdot \left(-\frac{3}{4}\right)}{1 - \left(-\frac{3}{4}\right)^2} = \frac{-\frac{3}{2}}{1 - \frac{9}{16}} = -\frac{3}{2} \cdot \frac{16}{7} = -\frac{24}{7}$

Note that $\tan 2\theta$ could have been found more easily in this case by simply dividing:

$$\tan 2\theta = \frac{\sin 2\theta}{\cos 2\theta} = \frac{-\frac{24}{25}}{\frac{7}{25}} = -\frac{24}{7}.$$

d) Since $\sin 2\theta$ is negative and $\cos 2\theta$ is positive, we know that 2θ is in quadrant IV. ▬

Two other useful identities for $\cos 2x$ can be derived easily, as follows.

$$\begin{aligned}\cos 2x &= \cos^2 x - \sin^2 x \\ &= (1 - \sin^2 x) - \sin^2 x \\ &= 1 - 2 \sin^2 x\end{aligned} \qquad \begin{aligned}\cos 2x &= \cos^2 x - \sin^2 x \\ &= \cos^2 x - (1 - \cos^2 x) \\ &= 2 \cos^2 x - 1\end{aligned}$$

> **Double-Angle Identities**
>
> $\sin 2x = 2 \sin x \cos x,$ $\qquad\qquad \begin{aligned}\cos 2x &= \cos^2 x - \sin^2 x \\ &= 1 - 2 \sin^2 x \\ &= 2 \cos^2 x - 1,\end{aligned}$
>
> $\tan 2x = \dfrac{2 \tan x}{1 - \tan^2 x}$

Solving the last two cosine double-angle identities for $\sin^2 x$ and $\cos^2 x$, respectively, we obtain two more identities:

$$\sin^2 x = \frac{1 - \cos 2x}{2} \quad \text{and} \quad \cos^2 x = \frac{1 + \cos 2x}{2}.$$

Using division and these two identities gives us the following useful identity:

$$\tan^2 x = \frac{1 - \cos 2x}{1 + \cos 2x}.$$

EXAMPLE 4 Find an equivalent expression for each of the following.

a) $\sin 3\theta$ in terms of function values of θ

b) $\cos^3 x$ in terms of function values of x or $2x$, raised only to the first power

Solution

a) $\sin 3\theta = \sin (2\theta + \theta)$

$\qquad = \sin 2\theta \cos \theta + \cos 2\theta \sin \theta$

$\qquad = (2 \sin \theta \cos \theta) \cos \theta + (2 \cos^2 \theta - 1) \sin \theta$

$\qquad\qquad$ Using $\sin 2\theta = 2 \sin \theta \cos \theta$ and $\cos 2\theta = 2 \cos^2 \theta - 1$

$\qquad = 2 \sin \theta \cos^2 \theta + 2 \sin \theta \cos^2 \theta - \sin \theta$

$\qquad = 4 \sin \theta \cos^2 \theta - \sin \theta$

We could also substitute $\cos^2 \theta - \sin^2 \theta$ or $1 - 2 \sin^2 \theta$ for $\cos 2\theta$. Each substitution leads to a different result, but all results are equivalent.

b) $\cos^3 x = \cos^2 x \cos x$

$\qquad = \dfrac{1 + \cos 2x}{2} \cos x$

Half-Angle Identities

If we take half of an angle of measure x, the new angle will have measure $x/2$. **Half-angle identities** give trigonometric function values of $x/2$ in terms of function values of x. To develop these identities, we take square roots and replace x with $x/2$. For example,

$$\sin^2 x = \frac{1 - \cos 2x}{2} \rightarrow \left| \sin \frac{x}{2} \right| = \sqrt{\frac{1 - \cos x}{2}}.$$

The formula on the right above is called a *half-angle formula*. We can eliminate the absolute-value sign by using $\pm$ signs, with the understanding that our use of $+$ and $-$ depends on the quadrant in which the angle $x/2$ lies. Half-angle identities for the cosine and tangent functions can be derived in a similar manner. Two additional formulas for the half-angle tangent identity are listed below.

Half-Angle Identities

$$\sin \frac{x}{2} = \pm \sqrt{\frac{1 - \cos x}{2}},$$

$$\cos \frac{x}{2} = \pm \sqrt{\frac{1 + \cos x}{2}},$$

$$\tan \frac{x}{2} = \pm \sqrt{\frac{1 - \cos x}{1 + \cos x}}$$

$$\qquad = \frac{\sin x}{1 + \cos x} = \frac{1 - \cos x}{\sin x}$$

EXAMPLE 5 Find $\tan (\pi/8)$ exactly. Then check the answer using a grapher.

Solution

$$\tan \frac{\pi}{8} = \tan \frac{\frac{\pi}{4}}{2} = \frac{\sin \frac{\pi}{4}}{1 + \cos \frac{\pi}{4}} = \frac{\frac{\sqrt{2}}{2}}{1 + \frac{\sqrt{2}}{2}} = \frac{\frac{\sqrt{2}}{2}}{\frac{2 + \sqrt{2}}{2}}$$

$$= \frac{\sqrt{2}}{2 + \sqrt{2}} = \frac{\sqrt{2}}{2 + \sqrt{2}} \cdot \frac{2 - \sqrt{2}}{2 - \sqrt{2}} = \sqrt{2} - 1$$

```
tan(π/8)
              .4142135624
√(2)−1
              .4142135624
```

The identities that we have developed are also useful for simplifying trigonometric expressions.

EXAMPLE 6   Simplify each of the following.

a) $\dfrac{\sin x \cos x}{\frac{1}{2} \cos 2x}$ **b)** $2 \sin^2 \dfrac{x}{2} + \cos x$

Solution

a) We can obtain $2 \sin x \cos x$ in the numerator by multiplying the expression by $\frac{2}{2}$:

$$\frac{\sin x \cos x}{\frac{1}{2} \cos 2x} = \frac{2}{2} \cdot \frac{\sin x \cos x}{\frac{1}{2} \cos 2x} = \frac{2 \sin x \cos x}{\cos 2x}$$

$$= \frac{\sin 2x}{\cos 2x} \qquad \text{Using } \sin 2x = 2 \sin x \cos x$$

$$= \tan 2x.$$

b) We have

$$2 \sin^2 \frac{x}{2} + \cos x = 2\left(\frac{1 - \cos x}{2}\right) + \cos x$$

$$\text{Using } \sin \frac{x}{2} = \pm \sqrt{\frac{1 - \cos x}{2}}, \text{ or } \sin^2 \frac{x}{2} = \frac{1 - \cos x}{2}$$

$$= 1 - \cos x + \cos x = 1.$$

We can check this result using a graph or a table.

S T U D Y   T I P

It is never too late to begin reviewing for the final examination. The Skill Maintenance exercises found in each exercise set review and reinforce skill taught in earlier sections. Include all of these exercises in your weekly preparation. Answers to both even-numbered and odd-numbered exercises along with section references appear at the back of the text.

$y_1 = 2 \sin^2 \dfrac{x}{2} + \cos x, \quad y_2 = 1$

X	Y₁	Y₂
−6.283	1	1
−5.498	1	1
−4.712	1	1
−3.927	1	1
−3.142	1	1
−2.356	1	1
−1.571	1	1

X = −6.28318530718

ΔTbl = π/4

Exercise Set 3.2

1. Given that $\sin (3\pi/10) \approx 0.8090$ and $\cos (3\pi/10) \approx 0.5878$, find each of the following.
   a) The other four function values for $3\pi/10$
   b) The six function values for $\pi/5$

2. Given that
$$\sin \frac{\pi}{12} = \frac{\sqrt{2 - \sqrt{3}}}{2} \quad \text{and} \quad \cos \frac{\pi}{12} = \frac{\sqrt{2 + \sqrt{3}}}{2},$$
find exact answers for each of the following.
   a) The other four function values for $\pi/12$
   b) The six function values for $5\pi/12$

3. Given that $\sin \theta = \frac{1}{3}$ and that the terminal side is in quadrant II, find exact answers for each of the following.
   a) The other function values for θ
   b) The six function values for $\pi/2 - \theta$
   c) The six function values for $\theta - \pi/2$

4. Given that $\cos \phi = \frac{4}{5}$ and that the terminal side is in quadrant IV, find exact answers for each of the following.
   a) The other function values for ϕ
   b) The six function values for $\pi/2 - \phi$
   c) The six function values for $\phi + \pi/2$

Find an equivalent expression for each of the following.

5. $\sec \left(x + \dfrac{\pi}{2} \right)$ **6.** $\cot \left(x - \dfrac{\pi}{2} \right)$

7. $\tan \left(x - \dfrac{\pi}{2} \right)$ **8.** $\csc \left(x + \dfrac{\pi}{2} \right)$

Find $\sin 2\theta$, $\cos 2\theta$, $\tan 2\theta$, and the quadrant in which 2θ lies.

9. $\sin \theta = \dfrac{4}{5}$, θ in quadrant I

10. $\cos \theta = \dfrac{5}{13}$, θ in quadrant I

11. $\cos \theta = -\dfrac{3}{5}$, θ in quadrant III

12. $\tan \theta = -\dfrac{5}{8}$, θ in quadrant II

13. $\tan \theta = -\dfrac{5}{12}$, θ in quadrant II

14. $\sin \theta = -\dfrac{\sqrt{10}}{10}$, θ in quadrant IV

15. Find an equivalent expression for $\cos 4x$ in terms of function values of x.

16. Find an equivalent expression for $\sin^4 \theta$ in terms of function values of θ, 2θ, or 4θ, raised only to the first power.

Use the half-angle identities to evaluate exactly.

17. $\cos 15°$ **18.** $\tan 67.5°$

19. $\sin 112.5°$ **20.** $\cos \dfrac{\pi}{8}$

21. $\tan 75°$ **22.** $\sin \dfrac{5\pi}{12}$

Given that $\sin \theta = 0.3416$ and θ is in quadrant I, find each of the following using identities.

23. $\sin 2\theta$ **24.** $\cos \dfrac{\theta}{2}$

25. $\sin \dfrac{\theta}{2}$ **26.** $\sin 4\theta$

In Exercises 27–30, use a grapher to determine which of the following expressions asserts an identity. Then prove the identity algebraically.

27. $\dfrac{\cos 2x}{\cos x - \sin x} = \cdots$

   a) $1 + \cos x$ **b)** $\cos x - \sin x$

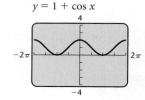

$y = 1 + \cos x$ $y = \cos x - \sin x$

   c) $-\cot x$ **d)** $\sin x \, (\cot x + 1)$

$y = -\cot x$ $y = \sin x \, (\cot x + 1)$

28. $2 \cos^2 \dfrac{x}{2} = \cdots$

   a) $\sin x \, (\csc x + \tan x)$ **b)** $\sin x - 2 \cos x$
   c) $2(\cos^2 x - \sin^2 x)$ **d)** $1 + \cos x$

29. $\dfrac{\sin 2x}{2 \cos x} = \cdots$

   a) $\cos x$ **b)** $\tan x$

   c) $\cos x + \sin x$ **d)** $\sin x$

30. $2 \sin \dfrac{\theta}{2} \cos \dfrac{\theta}{2} = \cdots$

   a) $\cos^2 \theta$ **b)** $\sin \dfrac{\theta}{2}$

   c) $\sin \theta$ **d)** $\sin \theta - \cos \theta$

Simplify. Check your results using a grapher.

31. $2 \cos^2 \dfrac{x}{2} - 1$

32. $\cos^4 x - \sin^4 x$

33. $(\sin x - \cos x)^2 + \sin 2x$

34. $(\sin x + \cos x)^2$

35. $\dfrac{2 - \sec^2 x}{\sec^2 x}$

36. $\dfrac{1 + \sin 2x + \cos 2x}{1 + \sin 2x - \cos 2x}$

37. $(-4 \cos x \sin x + 2 \cos 2x)^2 +$
$(2 \cos 2x + 4 \sin x \cos x)^2$

38. $2 \sin x \cos^3 x - 2 \sin^3 x \cos x$

Discussion and Writing

39. Discuss and compare the graphs of $y_1 = \sin x$, $y_2 = \sin 2x$, and $y_3 = \sin (x/2)$.

40. Find all errors in the following:

$$2 \sin^2 2x + \cos 4x$$
$$= 2(2 \sin x \cos x)^2 + 2 \cos 2x$$
$$= 8 \sin^2 x \cos^2 x + 2(\cos^2 x + \sin^2 x)$$
$$= 8 \sin^2 x \cos^2 x + 2.$$

Then verify with a grapher that
$$2 \sin^2 2x + \cos 4x = 8 \sin^2 x \cos^2 x + 2$$
is not an identity.

Skill Maintenance

Complete the identity.

41. $1 - \cos^2 x =$

42. $\sec^2 x - \tan^2 x =$

43. $\sin^2 x - 1 =$

44. $1 + \cot^2 x =$

Synthesis

45. Given that $\cos 51° \approx 0.6293$, find the six function values for $141°$.

Simplify. Check your results using a grapher.

46. $\sin \left(\dfrac{\pi}{2} - x \right) [\sec x - \cos x]$

47. $\cos (\pi - x) + \cot x \sin \left(x - \dfrac{\pi}{2} \right)$

48. $\dfrac{\cos x - \sin \left(\dfrac{\pi}{2} - x \right) \sin x}{\cos x - \cos (\pi - x) \tan x}$

49. $\dfrac{\cos^2 y \sin \left(y + \dfrac{\pi}{2} \right)}{\sin^2 y \sin \left(\dfrac{\pi}{2} - y \right)}$

Find $\sin \theta$, $\cos \theta$, and $\tan \theta$ under the given conditions.

50. $\cos 2\theta = \dfrac{7}{12}, \ \dfrac{3\pi}{2} \le 2\theta \le 2\pi$

51. $\tan \dfrac{\theta}{2} = -\dfrac{5}{3}, \ \pi < \theta \le \dfrac{3\pi}{2}$

52. *Nautical Mile.* *Latitude* is used to measure north–south location on the earth between the equator and the poles. For example, Chicago has latitude 42°N. (See the figure.) In Great Britain, the *nautical mile* is defined as the length of a minute of arc of the earth's radius. Since the earth is flattened slightly at the poles, a British nautical mile varies with latitude. In fact, it is given, in feet, by the function

$$N(\phi) = 6066 - 31 \cos 2\phi,$$

where ϕ is the latitude in degrees.

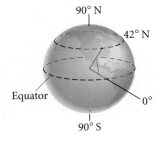

  a) What is the length of a British nautical mile at Chicago?

  b) What is the length of a British nautical mile at the North Pole?

  c) Express $N(\phi)$ in terms of $\cos \phi$ only. That is, do not use the double angle.

53. *Acceleration Due to Gravity.* The acceleration due to gravity is often denoted by g in a formula such as $S = \frac{1}{2}gt^2$, where S is the distance that an object falls in time t. The number g relates to motion near the earth's surface and is usually considered constant. In fact, however, g is not constant, but varies slightly with latitude. If ϕ stands for latitude, in degrees, g is given with good approximation by the formula

$$g = 9.78049(1 + 0.005288 \sin^2 \phi - 0.000006 \sin^2 2\phi),$$

where g is measured in meters per second per second at sea level.

a) Chicago has latitude 42°N. Find g.
b) Philadelphia has latitude 40°N. Find g.
c) Express g in terms of $\sin \phi$ only. That is, eliminate the double angle.

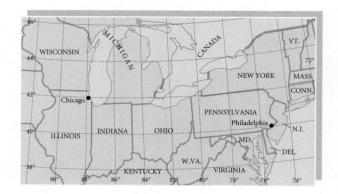

d) Where on the earth is g greatest? least? Use the MAX–MIN feature on a grapher.

Proving Trigonometric Identities

3.3

• *Prove identities using other identities.*

The Logic of Proving Identities

We outline two algebraic methods for proving identities.

Method 1. Start with either the left or the right side of the equation and obtain the other side. For example, suppose you are trying to prove that the equation $P = Q$ is an identity. You might try to produce a string of statements $(R_1, R_2, \ldots$ or $T_1, T_2, \ldots)$ like the following, which start with P and end with Q or start with Q and end with P:

$$P = R_1 \quad \text{or} \quad Q = T_1$$
$$= R_2 \qquad\qquad = T_2$$
$$\vdots \qquad\qquad\qquad \vdots$$
$$= Q \qquad\qquad = P.$$

Method 2. Work with each side separately until you obtain the same expression. For example, suppose you are trying to prove that $P = Q$ is an identity. You might be able to produce two strings of statements like the following, each ending with the same statement S.

$$P = R_1 \qquad Q = T_1$$
$$= R_2 \qquad\qquad = T_2$$
$$\vdots \qquad\qquad\qquad \vdots$$
$$= S \qquad\qquad = S.$$

The number of steps in each string might be different, but in each case the result is *S*.

A first step in learning to prove identities is to have at hand a list of the identities that you have already learned. Such a list is on the inside front cover of this text. Ask your instructor which ones you are expected to memorize. The more identities you prove, the easier it will be to prove new ones. A list of helpful hints follows.

Hints for Proving Identities

1. Use method 1 or 2 above.
2. Work with the more complex side first.
3. Carry out any algebraic manipulations, such as adding, subtracting, multiplying, or factoring.
4. Multiplying by 1 can be helpful when rational expressions are involved.
5. Converting all expressions to sines and cosines is often helpful.
6. Try something! Put your pencil to work and get involved. You will be amazed at how often this leads to success.

Proving Identities

In what follows, method 1 is used in Examples 1–3 and method 2 is used in Examples 4 and 5.

EXAMPLE 1   Prove the identity $1 + \sin 2\theta = (\sin \theta + \cos \theta)^2$.

Solution   Let's use method 1. We begin with the right side and obtain the left side:

$$(\sin \theta + \cos \theta)^2 = \sin^2 \theta + 2 \sin \theta \cos \theta + \cos^2 \theta \qquad \text{Squaring}$$
$$= 1 + 2 \sin \theta \cos \theta \qquad \begin{array}{l}\textbf{Recalling the identity}\\ \sin^2 x + \cos^2 x = 1 \textbf{ and}\\ \textbf{substituting}\end{array}$$
$$= 1 + \sin 2\theta. \qquad \begin{array}{l}\textbf{Using } \sin 2x =\\ 2 \sin x \cos x\end{array}$$

We could also begin with the left side and obtain the right side:

$$1 + \sin 2\theta = 1 + 2 \sin \theta \cos \theta \qquad \textbf{Using } \sin 2x = 2 \sin x \cos x$$
$$= \sin^2 \theta + 2 \sin \theta \cos \theta + \cos^2 \theta \qquad \begin{array}{l}\textbf{Replacing 1 with}\\ \sin^2 \theta + \cos^2 \theta\end{array}$$
$$= (\sin \theta + \cos \theta)^2. \qquad \textbf{Factoring} \qquad \blacksquare$$

EXAMPLE 2   Prove the identity

$$\frac{\sec t - 1}{t \sec t} = \frac{1 - \cos t}{t}.$$

Solution   We use method 1, starting with the left side. Note that the left side involves sec t, whereas the right side involves cos t, so it might be wise to make use of a basic identity that involves these two expressions: sec $t = 1/\cos t$.

$$\frac{\sec t - 1}{t \sec t} = \frac{\dfrac{1}{\cos t} - 1}{t \, \dfrac{1}{\cos t}} \qquad \text{Substituting } 1/\cos t \text{ for sec } t$$

$$= \left(\frac{1}{\cos t} - 1\right) \cdot \frac{\cos t}{t}$$

$$= \frac{1}{t} - \frac{\cos t}{t} \qquad \text{Multiplying}$$

$$= \frac{1 - \cos t}{t}$$

We started with the left side and obtained the right side, so the proof is complete.   ▬

EXAMPLE 3   Prove the identity

$$\frac{\sin 2x}{\sin x} - \frac{\cos 2x}{\cos x} = \sec x.$$

Solution

$$\frac{\sin 2x}{\sin x} - \frac{\cos 2x}{\cos x} = \frac{2 \sin x \cos x}{\sin x} - \frac{\cos^2 x - \sin^2 x}{\cos x} \qquad \begin{array}{l}\text{Using double-}\\\text{angle identities}\end{array}$$

$$= 2 \cos x - \frac{\cos^2 x - \sin^2 x}{\cos x} \qquad \text{Simplifying}$$

$$= \frac{2 \cos^2 x}{\cos x} - \frac{\cos^2 x - \sin^2 x}{\cos x} \qquad \begin{array}{l}\text{Multiplying } 2 \cos x\\\text{by 1, or } \cos x/\cos x\end{array}$$

$$= \frac{2 \cos^2 x - \cos^2 x + \sin^2 x}{\cos x} \qquad \text{Subtracting}$$

$$= \frac{\cos^2 x + \sin^2 x}{\cos x}$$

$$= \frac{1}{\cos x} \qquad \begin{array}{l}\text{Using a Pythagorean}\\\text{identity}\end{array}$$

$$= \sec x \qquad \begin{array}{l}\text{Recalling a basic}\\\text{identity}\end{array} \qquad ▬$$

EXAMPLE 4   Prove the identity

$$\sin^2 x \tan^2 x = \tan^2 x - \sin^2 x.$$

Solution   For this proof, we are going to work with each side separately using method 2. We try to obtain the same expression on each side. In actual practice, you might work on one side for awhile, then work on the other side, and then go back to the first side. In other words, you work back and forth until you arrive at the same expression. Let's start with the right side:

$$\tan^2 x - \sin^2 x = \frac{\sin^2 x}{\cos^2 x} - \sin^2 x$$
Recalling the identity $\tan x = \dfrac{\sin x}{\cos x}$ and substituting

$$= \frac{\sin^2 x}{\cos^2 x} - \sin^2 x \cdot \frac{\cos^2 x}{\cos^2 x}$$
Multiplying by 1 in order to subtract

$$= \frac{\sin^2 x - \sin^2 x \cos^2 x}{\cos^2 x}$$
Carrying out the subtraction

$$= \frac{\sin^2 x \, (1 - \cos^2 x)}{\cos^2 x}$$
Factoring

$$= \frac{\sin^2 x \sin^2 x}{\cos^2 x}$$
Recalling the identity $1 - \cos^2 x = \sin^2 x$ and substituting

$$= \frac{\sin^4 x}{\cos^2 x}.$$

At this point, we stop and work with the left side, $\sin^2 x \tan^2 x$, of the original identity and try to end with the same expression that we ended with on the right side:

$$\sin^2 x \tan^2 x = \sin^2 x \frac{\sin^2 x}{\cos^2 x}$$
Recalling the identity $\tan x = \dfrac{\sin x}{\cos x}$ and substituting

$$= \frac{\sin^4 x}{\cos^2 x}.$$

We have obtained the same expression from each side, so the proof is complete.   ▬

EXAMPLE 5   Prove the identity

$$\cot \phi + \csc \phi = \frac{\sin \phi}{1 - \cos \phi}.$$

Solution   We are again using method 2, beginning with the left side:

$$\cot \phi + \csc \phi = \frac{\cos \phi}{\sin \phi} + \frac{1}{\sin \phi}$$
Using basic identities

$$= \frac{1 + \cos \phi}{\sin \phi}.$$
Adding

At this point, we stop and work with the right side of the original identity:

$$\frac{\sin \phi}{1 - \cos \phi} = \frac{\sin \phi}{1 - \cos \phi} \cdot \frac{1 + \cos \phi}{1 + \cos \phi} \qquad \text{Multiplying by 1}$$

$$= \frac{\sin \phi \, (1 + \cos \phi)}{1 - \cos^2 \phi}$$

$$= \frac{\sin \phi \, (1 + \cos \phi)}{\sin^2 \phi} \qquad \text{Using } \sin^2 x = 1 - \cos^2 x$$

$$= \frac{1 + \cos \phi}{\sin \phi}. \qquad \text{Simplifying}$$

The proof is complete since we obtained the same expression from each side.

Exercise Set 3.3

Prove each of the following identities.

1. $\sec x - \sin x \tan x = \cos x$

$$y_1 = \sec x - \sin x \tan x,$$
$$y_2 = \cos x$$

2. $\dfrac{1 + \cos \theta}{\sin \theta} + \dfrac{\sin \theta}{\cos \theta} = \dfrac{\cos \theta + 1}{\sin \theta \cos \theta}$

$$y_1 = \frac{1 + \cos x}{\sin x} + \frac{\sin x}{\cos x},$$
$$y_2 = \frac{\cos x + 1}{\sin x \cos x}$$

3. $\dfrac{1 - \cos x}{\sin x} = \dfrac{\sin x}{1 + \cos x}$

4. $\dfrac{1 + \tan y}{1 + \cot y} = \dfrac{\sec y}{\csc y}$

5. $\dfrac{1 + \tan \theta}{1 - \tan \theta} + \dfrac{1 + \cot \theta}{1 - \cot \theta} = 0$

6. $\dfrac{\sin x + \cos x}{\sec x + \csc x} = \dfrac{\sin x}{\sec x}$

7. $\dfrac{\cos^2 \alpha + \cot \alpha}{\cos^2 \alpha - \cot \alpha} = \dfrac{\cos^2 \alpha \tan \alpha + 1}{\cos^2 \alpha \tan \alpha - 1}$

8. $\sec 2\theta = \dfrac{\sec^2 \theta}{2 - \sec^2 \theta}$

9. $\dfrac{2 \tan \theta}{1 + \tan^2 \theta} = \sin 2\theta$

10. $\dfrac{\cos (u - v)}{\cos u \sin v} = \tan u + \cot v$

11. $1 - \cos 5\theta \cos 3\theta - \sin 5\theta \sin 3\theta = 2 \sin^2 \theta$

12. $\cos^4 x - \sin^4 x = \cos 2x$

13. $2 \sin \theta \cos^3 \theta + 2 \sin^3 \theta \cos \theta = \sin 2\theta$

14. $\dfrac{\tan 3t - \tan t}{1 + \tan 3t \tan t} = \dfrac{2 \tan t}{1 - \tan^2 t}$

15. $\dfrac{\tan x - \sin x}{2 \tan x} = \sin^2 \dfrac{x}{2}$

16. $\dfrac{\cos^3 \beta - \sin^3 \beta}{\cos \beta - \sin \beta} = \dfrac{2 + \sin 2\beta}{2}$

17. $\sin (\alpha + \beta) \sin (\alpha - \beta) = \sin^2 \alpha - \sin^2 \beta$

18. $\cos^2 x \, (1 - \sec^2 x) = -\sin^2 x$

19. $\tan \theta \, (\tan \theta + \cot \theta) = \sec^2 \theta$

20. $\dfrac{\cos \theta + \sin \theta}{\cos \theta} = 1 + \tan \theta$

21. $\dfrac{1 + \cos^2 x}{\sin^2 x} = 2 \csc^2 x - 1$

22. $\dfrac{\tan y + \cot y}{\csc y} = \sec y$

23. $\dfrac{1 + \sin x}{1 - \sin x} + \dfrac{\sin x - 1}{1 + \sin x} = 4 \sec x \tan x$

24. $\tan \theta - \cot \theta = (\sec \theta - \csc \theta)(\sin \theta + \cos \theta)$

25. $\cos^2 \alpha \cot^2 \alpha = \cot^2 \alpha - \cos^2 \alpha$

26. $\dfrac{\tan x + \cot x}{\sec x + \csc x} = \dfrac{1}{\cos x + \sin x}$

27. $2 \sin^2 \theta \cos^2 \theta + \cos^4 \theta = 1 - \sin^4 \theta$

28. $\dfrac{\cot \theta}{\csc \theta - 1} = \dfrac{\csc \theta + 1}{\cot \theta}$

29. $\dfrac{1 + \sin x}{1 - \sin x} = (\sec x + \tan x)^2$

30. $\sec^4 s - \tan^2 s = \tan^4 s + \sec^2 s$

In Exercises 31–36, use a grapher to determine which expression (A)–(F) on the right can be used to complete the identity. Then try to prove that identity algebraically.

31. $\dfrac{\cos x + \cot x}{1 + \csc x}$ **A.** $\dfrac{\sin^3 x - \cos^3 x}{\sin x - \cos x}$

32. $\cot x + \csc x$ **B.** $\cos x$

33. $\sin x \cos x + 1$ **C.** $\tan x + \cot x$

34. $2 \cos^2 x - 1$ **D.** $\cos^3 x + \sin^3 x$

35. $\dfrac{1}{\cot x \sin^2 x}$ **E.** $\dfrac{\sin x}{1 - \cos x}$

36. $(\cos x + \sin x)(1 - \sin x \cos x)$ **F.** $\cos^4 x - \sin^4 x$

Discussion and Writing

37. What restrictions must be placed on the variable in each of the following identities? Why?

  a) $\sin 2x = \dfrac{2 \tan x}{1 + \tan^2 x}$

  b) $\dfrac{1 - \cos x}{\sin x} = \dfrac{\sin x}{1 + \cos x}$

  c) $2 \sin x \cos^3 x + 2 \sin^3 x \cos x = \sin 2x$

38. Explain why $\tan (x + 450°)$ cannot be simplified using the tangent sum formula, but can be simplified using the sine and cosine sum formulas.

Skill Maintenance

For each function:

a) *Graph the function.*
b) *Determine whether the function is one-to-one.*
c) *If the function is one-to-one, find an equation for its inverse.*
d) *Graph the inverse of the function.*

39. $f(x) = 3x - 2$ **40.** $f(x) = x^3 + 1$

41. $f(x) = x^2 - 4, \; x \geq 0$ **42.** $f(x) = \sqrt{x + 2}$

Synthesis

Prove the identity.

43. $\ln |\tan x| = -\ln |\cot x|$

44. $\ln |\sec \theta + \tan \theta| = -\ln |\sec \theta - \tan \theta|$

45. Prove the identity

$\log (\cos x - \sin x) + \log (\cos x + \sin x) = \log \cos 2x.$

46. *Mechanics.* The following equation occurs in the study of mechanics:

$$\sin \theta = \dfrac{I_1 \cos \phi}{\sqrt{(I_1 \cos \phi)^2 + (I_2 \sin \phi)^2}}.$$

It can happen that $I_1 = I_2$. Assuming that this happens, simplify the equation.

47. *Alternating Current.* In the theory of alternating current, the following equation occurs:

$$R = \dfrac{1}{\omega C(\tan \theta + \tan \phi)}.$$

Show that this equation is equivalent to

$$R = \dfrac{\cos \theta \cos \phi}{\omega C \sin (\theta + \phi)}.$$

48. *Electrical Theory.* In electrical theory, the following equations occur:

$$E_1 = \sqrt{2} E_t \cos \left(\theta + \dfrac{\pi}{P} \right)$$

and

$$E_2 = \sqrt{2} E_t \cos \left(\theta - \dfrac{\pi}{P} \right).$$

Assuming that these equations hold, show that

$$\dfrac{E_1 + E_2}{2} = \sqrt{2} E_t \cos \theta \cos \dfrac{\pi}{P}$$

and

$$\dfrac{E_1 - E_2}{2} = -\sqrt{2} E_t \sin \theta \sin \dfrac{\pi}{P}.$$

Inverses of the Trigonometric Functions

3.4

INVERSE FUNCTIONS

REVIEW SECTION 1.6.

- *Find values of the inverse trigonometric functions.*
- *Simplify expressions such as sin (sin⁻¹ x) and sin⁻¹ (sin x).*
- *Simplify expressions involving compositions such as* $\sin\left(\cos^{-1}\frac{1}{2}\right)$ *without using a calculator.*
- *Simplify expressions such as sin arctan (a/b) by making a drawing and reading off appropriate ratios.*

In this section, we develop inverse trigonometric functions. The graphs of the sine, cosine, and tangent functions follow. Do these functions have inverses that are functions? They do if they are one-to-one, which means that they pass the horizontal-line test.

Note that for each function, a horizontal line (shown in red) crosses the graph more than once. Therefore, none of them has an inverse that is a function.

The graphs of an equation and its inverse are reflections of each other across the line *y* = *x*. Let's examine the inverses of each of the three functions graphed above.

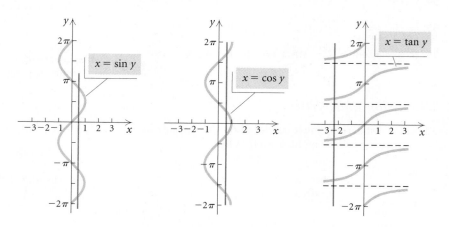

We can check again to see whether these are graphs of functions by using the vertical-line test. In each case, there is a vertical line (shown in red) that crosses the graph more than once, so each *fails* to be a function.

Restricting Ranges to Define Inverse Functions

Recall that a function like $f(x) = x^2$ does not have an inverse that is a function, but by restricting the domain of f to nonnegative numbers, we have a new squaring function, $f(x) = x^2$, $x \geq 0$, that has an inverse, $f^{-1}(x) = \sqrt{x}$. This is equivalent to restricting the range of the inverse relation to exclude ordered pairs that contain negative numbers.

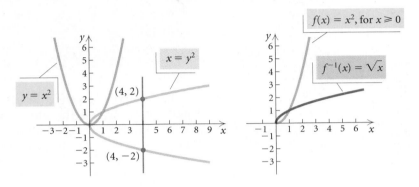

In a similar manner, we can define new trigonometric functions whose inverses are functions. We can do this by restricting either the domains of the basic trigonometric functions or the ranges of their inverse relations. This can be done in many ways, but the restrictions illustrated below with solid red curves are fairly standard in mathematics.

FIGURE 1

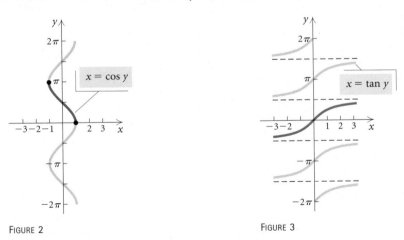

FIGURE 2

FIGURE 3

For the inverse sine function, we choose a range close to the origin that allows all inputs on the interval $[-1, 1]$ to have function values. Thus we choose the interval $[-\pi/2, \pi/2]$ for the range (Fig. 1). For the inverse cosine function, we choose a range close to the origin that allows all inputs on the interval $[-1, 1]$ to have function values. This is the interval $[0, \pi]$ (Fig. 2). For the inverse tangent function, we choose a range close to the origin that allows all real numbers to have function values. The interval $(-\pi/2, \pi/2)$ satisfies this requirement (Fig. 3).

Inverse Trigonometric Functions

FUNCTION	DOMAIN	RANGE
$y = \sin^{-1} x$ $= \arcsin x$, where $x = \sin y$	$[-1, 1]$	$[-\pi/2, \pi/2]$
$y = \cos^{-1} x$ $= \arccos x$, where $x = \cos y$	$[-1, 1]$	$[0, \pi]$
$y = \tan^{-1} x$ $= \arctan x$, where $x = \tan y$	$(-\infty, \infty)$	$(-\pi/2, \pi/2)$

The notation arcsin x arises because the function value, y, is the length of an arc on the unit circle for which the sine is x. Either of the two kinds of notation above can be read "the inverse sine of x" or "the arc sine of x" or "the number (or angle) whose sine is x."

The notation $\sin^{-1} x$ is not exponential notation.

It does *not* mean $\dfrac{1}{\sin x}$!

The graphs of the inverse trigonometric functions are as follows.

Domain: $[-1, 1]$
Range: $\left[-\frac{\pi}{2}, \frac{\pi}{2}\right]$

Domain: $[-1, 1]$
Range: $[0, \pi]$

Domain: $(-\infty, \infty)$
Range: $\left(-\frac{\pi}{2}, \frac{\pi}{2}\right)$

Interactive Discovery

Inverse trigonometric functions can be graphed using a grapher. Graph $y = \sin^{-1} x$ using the viewing window $[-3, 3, -\pi, \pi]$, with Xscl = 1 and Yscl = $\pi/2$. Now try graphing $y = \cos^{-1} x$ and $y = \tan^{-1} x$. Then use the graphs to confirm the domain and the range of each inverse.

The following diagrams show the restricted ranges for the inverse trigonometric functions on a unit circle. Compare these graphs with the graphs above. The ranges of these functions should be memorized. The missing endpoints in the graph of the arctangent function indicate inputs that are not in the domain of the original function.

arcsine
Range $\left[-\frac{\pi}{2}, \frac{\pi}{2}\right]$

arccosine
Range $[0, \pi]$

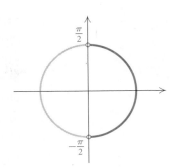

arctangent
Range $\left(-\frac{\pi}{2}, \frac{\pi}{2}\right)$

EXAMPLE 1 Find each of the following function values.

a) arcsin $\dfrac{\sqrt{2}}{2}$ 　　　　**b)** $\cos^{-1}\left(-\dfrac{1}{2}\right)$ 　　　　**c)** $\tan^{-1}\left(-\dfrac{\sqrt{3}}{3}\right)$

Solution

a) Another way to state "find arcsin $\sqrt{2}/2$" is to say "find β such that $\sin \beta = \sqrt{2}/2$." In the restricted range $[-\pi/2, \pi/2]$, the only number with a sine of $\sqrt{2}/2$ is $\pi/4$. Thus, arcsin $(\sqrt{2}/2) = \pi/4$, or 45°. (See Fig. 4 below.)

FIGURE 4

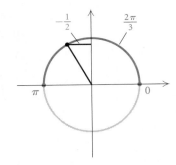

FIGURE 5

b) The only number with a cosine of $-\frac{1}{2}$ in the restricted range $[0, \pi]$ is $2\pi/3$. Thus, $\cos^{-1}\left(-\frac{1}{2}\right) = 2\pi/3$, or 120°. (See Fig. 5 above.)

c) The only number in the restricted range $(-\pi/2, \pi/2)$ with a tangent of $-\sqrt{3}/3$ is $-\pi/6$. Thus, $\tan^{-1}(-\sqrt{3}/3)$ is $-\pi/6$, or −30°. (See Fig. 6 at left.)

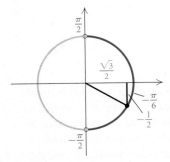

FIGURE 6

　　We can also use a grapher to find inverse trigonometric function values (see Section 2.1, p. 119). On most graphers, we can find inverse function values in either radians or degrees simply by selecting the appropriate mode. The key strokes involved in finding inverse function values vary with the grapher. Be sure to read the instructions for the particular grapher that you are using.

EXAMPLE 2 Approximate each of the following function values in both radians and degrees.

a) $\cos^{-1}(-0.2689)$

b) $\arctan(-0.2623)$

c) $\sin^{-1} 0.20345$

d) $\arccos 1.318$

e) $\csc^{-1} 8.205$

Solution

FUNCTION VALUE	MODE	READOUT	ROUNDED
a) $\cos^{-1}(-0.2689)$	Radian	1.843047111	1.8430
	Degree	105.5988209	105.6°
b) $\arctan(-0.2623)$	Radian	−.256521214	−0.2565
	Degree	−14.69758292	−14.7°
c) $\sin^{-1} 0.20345$	Radian	.2048803359	0.2049
	Degree	11.73877855	11.7°
d) $\arccos 1.318$	Radian	ERR:DOMAIN	
	Degree	ERR:DOMAIN	

The value 1.318 is not in $[-1, 1]$, the domain of the arccosine function.

e) The cosecant function is the reciprocal of the sine function:

$\csc^{-1} 8.205 = \sin^{-1}(1/8.205)$

	MODE	READOUT	ROUNDED
	Radian	.1221806653	0.1222
	Degree	7.000436462	7.0°

CONNECTING THE CONCEPTS

DOMAINS AND RANGES

The following is a summary of the domains and ranges of the trigonometric functions together with a summary of the domains and ranges of the inverse trigonometric functions. For completeness, we have included the arccosecant, the arcsecant, and the arccotangent, though there is a lack of uniformity in their definitions in mathematical literature.

FUNCTION	DOMAIN	RANGE
sin	All reals, $(-\infty, \infty)$	$[-1, 1]$
cos	All reals, $(-\infty, \infty)$	$[-1, 1]$
tan	All reals except $k\pi/2$, k odd	All reals, $(-\infty, \infty)$
csc	All reals except $k\pi$	$(-\infty, -1] \cup [1, \infty)$
sec	All reals except $k\pi/2$, k odd	$(-\infty, -1] \cup [1, \infty)$
cot	All reals except $k\pi$	All reals, $(-\infty, \infty)$

INVERSE FUNCTION	DOMAIN	RANGE
$\sin^{-1}$	$[-1, 1]$	$\left[-\dfrac{\pi}{2}, \dfrac{\pi}{2} \right]$
$\cos^{-1}$	$[-1, 1]$	$[0, \pi]$
$\tan^{-1}$	All reals, $(-\infty, \infty)$	$\left(-\dfrac{\pi}{2}, \dfrac{\pi}{2} \right)$
$\csc^{-1}$	$(-\infty, -1] \cup [1, \infty)$	$\left[-\dfrac{\pi}{2}, 0 \right) \cup \left(0, \dfrac{\pi}{2} \right]$
$\sec^{-1}$	$(-\infty, -1] \cup [1, \infty)$	$\left[0, \dfrac{\pi}{2} \right) \cup \left(\dfrac{\pi}{2}, \pi \right]$
$\cot^{-1}$	All reals, $(-\infty, \infty)$	$\left[-\dfrac{\pi}{2}, 0 \right) \cup \left(0, \dfrac{\pi}{2} \right]$

Composition of Trigonometric Functions and Their Inverses

Various compositions of trigonometric functions and their inverses often occur in practice. For example, we might want to try to simplify an expression such as

$$\sin (\sin^{-1} x) \quad \text{or} \quad \sin \left(\text{arccot } \frac{x}{2} \right).$$

In the expression on the left, we are finding "the sine of a number whose sine is x." Recall from Section 1.6 that if a function f has an inverse that is also a function, then

$$f(f^{-1}(x)) = x, \quad \text{for all } x \text{ in the domain of } f^{-1},$$

COMPOSITION OF FUNCTIONS
REVIEW SECTION 1.6.

and

$$f^{-1}(f(x)) = x, \quad \text{for all } x \text{ in the domain of } f.$$

Thus, if $f(x) = \sin x$ and $f^{-1}(x) = \sin^{-1} x$, then

$$\boldsymbol{\sin (\sin^{-1} x) = x, \quad \text{for all } x \text{ in the domain of } \sin^{-1},}$$

which is any number on the interval $[-1, 1]$. Similar results hold for the other trigonometric functions.

Composition of Trigonometric Functions

$\sin(\sin^{-1} x) = x$,   for all x in the domain of $\sin^{-1}$.

$\cos(\cos^{-1} x) = x$,   for all x in the domain of $\cos^{-1}$.

$\tan(\tan^{-1} x) = x$,   for all x in the domain of $\tan^{-1}$.

EXAMPLE 3   Simplify each of the following.

a) $\cos\left(\cos^{-1} \dfrac{\sqrt{3}}{2}\right)$ **b)** $\sin(\sin^{-1} 1.8)$

Solution

a) Since $\sqrt{3}/2$ is in $[-1, 1]$, the domain of $\cos^{-1}$, it follows that

$$\cos\left(\cos^{-1} \frac{\sqrt{3}}{2}\right) = \frac{\sqrt{3}}{2}.$$

b) Since 1.8 is not in $[-1, 1]$, the domain of $\sin^{-1}$, we cannot evaluate this expression. We know that there is no number with a sine of 1.8. Since we cannot find $\sin^{-1} 1.8$, we state that $\sin(\sin^{-1} 1.8)$ does not exist.　　　　　　　　　　　　　　　　　　　　　　　　 ▬

Now let's consider an expression like $\sin^{-1}(\sin x)$. We might also suspect that this is equal to x for any x in the domain of $\sin x$, but this is not true unless x is in the range of the $\sin^{-1}$ function. Note that in order to define $\sin^{-1}$, we had to restrict the domain of the sine function. In doing so, we restricted the range of the inverse sine function. Thus,

$$\sin^{-1}(\sin x) = x, \quad \text{for all } x \text{ in the range of } \sin^{-1}.$$

Similar results hold for the other trigonometric functions.

Special Cases

$\sin^{-1}(\sin x) = x$,   for all x in the range of $\sin^{-1}$.

$\cos^{-1}(\cos x) = x$,   for all x in the range of $\cos^{-1}$.

$\tan^{-1}(\tan x) = x$,   for all x in the range of $\tan^{-1}$.

EXAMPLE 4   Simplify each of the following.

a) $\tan^{-1}\left(\tan \dfrac{\pi}{6}\right)$ **b)** $\sin^{-1}\left(\sin \dfrac{3\pi}{4}\right)$

Solution

a) Since $\pi/6$ is in $(-\pi/2, \pi/2)$, the range of the $\tan^{-1}$ function, we can

use $\tan^{-1}(\tan x) = x$. Thus,

$$\tan^{-1}\left(\tan\frac{\pi}{6}\right) = \frac{\pi}{6}.$$

b) Note that $3\pi/4$ is not in $[-\pi/2, \pi/2]$, the range of the $\sin^{-1}$ function. Thus we *cannot* apply $\sin^{-1}(\sin x) = x$. Instead we first find $\sin(3\pi/4)$, which is $\sqrt{2}/2$, and substitute:

$$\sin^{-1}\left(\sin\frac{3\pi}{4}\right) = \sin^{-1}\left(\frac{\sqrt{2}}{2}\right) = \frac{\pi}{4}.$$

Now we find some other function compositions.

EXAMPLE 5 Simplify each of the following.

a) $\sin[\arctan(-1)]$ **b)** $\cos^{-1}\left(\sin\frac{\pi}{2}\right)$

Solution

a) Arctan (-1) is the number (or angle) θ in $(-\pi/2, \pi/2)$ whose tangent is -1. That is, $\tan\theta = -1$. Thus, $\theta = -\pi/4$ and

$$\sin[\arctan(-1)] = \sin\left[-\frac{\pi}{4}\right] = -\frac{\sqrt{2}}{2}.$$

b) $\cos^{-1}\left(\sin\frac{\pi}{2}\right) = \cos^{-1}(1) = 0$ $\sin\frac{\pi}{2} = 1$

Next, let's consider

$$\cos\left(\arcsin\frac{3}{5}\right).$$

Without using a grapher, we cannot find $\arcsin\frac{3}{5}$. However, we can still evaluate the entire expression by sketching a reference triangle. We are looking for angle θ such that $\arcsin\frac{3}{5} = \theta$, or $\sin\theta = \frac{3}{5}$. Since arcsin is defined in $[-\pi/2, \pi/2]$ and $\frac{3}{5} > 0$, we know that θ is in quadrant I. We sketch a reference right triangle, as shown at left. The angle θ in this triangle is an angle whose sine is $\frac{3}{5}$. We wish to find the cosine of this angle. Since the triangle is a right triangle, we can find the length of the base, b. It is 4. Thus we know that $\cos\theta = b/5$, or $\frac{4}{5}$. Therefore,

$$\cos\left(\arcsin\frac{3}{5}\right) = \frac{4}{5}.$$

EXAMPLE 6 Find $\sin\left(\operatorname{arccot}\frac{x}{2}\right)$.

Solution We draw a right triangle whose legs have lengths x and 2, so that $\cot\theta = x/2$. Since the range of arccot is $[-\pi/2, 0) \cup (0, \pi/2]$, we consider only positive values of x.

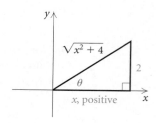

We find the length of the hypotenuse and then read off the sine ratio. We get

$$\sin\left(\operatorname{arccot}\frac{x}{2}\right) = \frac{2}{\sqrt{x^2 + 4}}.$$

In the following example, we use a sum identity to evaluate an expression.

EXAMPLE 7   Evaluate:

$$\sin\left(\sin^{-1}\frac{1}{2} + \cos^{-1}\frac{5}{13}\right).$$

Solution   Since $\sin^{-1}\frac{1}{2}$ and $\cos^{-1}\frac{5}{13}$ are both angles, the expression is the sine of a sum of two angles, so we use the identity

$$\sin(u + v) = \sin u \cos v + \cos u \sin v.$$

Thus,

$$\sin\left(\sin^{-1}\frac{1}{2} + \cos^{-1}\frac{5}{13}\right)$$

$$= \sin\left(\sin^{-1}\frac{1}{2}\right)\cdot\cos\left(\cos^{-1}\frac{5}{13}\right) + \cos\left(\sin^{-1}\frac{1}{2}\right)\cdot\sin\left(\cos^{-1}\frac{5}{13}\right)$$

$$= \frac{1}{2}\cdot\frac{5}{13} + \cos\left(\sin^{-1}\frac{1}{2}\right)\cdot\sin\left(\cos^{-1}\frac{5}{13}\right) \qquad \text{Using composition identities}$$

Now since $\sin^{-1}\frac{1}{2} = \pi/6$, $\cos\left(\sin^{-1}\frac{1}{2}\right)$ simplifies to $\cos\pi/6$, or $\sqrt{3}/2$. To find $\sin\left(\cos^{-1}\frac{5}{13}\right)$, we use a reference triangle in quadrant I and determine that the sine of the angle whose cosine is $\frac{5}{13}$ is $\frac{12}{13}$. Our expression now simplifies to

$$\frac{1}{2}\cdot\frac{5}{13} + \frac{\sqrt{3}}{2}\cdot\frac{12}{13}, \quad \text{or} \quad \frac{5 + 12\sqrt{3}}{26}.$$

Thus,

$$\sin\left(\sin^{-1}\frac{1}{2} + \cos^{-1}\frac{5}{13}\right) = \frac{5 + 12\sqrt{3}}{26}.$$

Exercise Set 3.4

Find each of the following exactly in radians and degrees.

1. $\sin^{-1}\left(-\dfrac{\sqrt{3}}{2}\right)$ **2.** $\cos^{-1}\dfrac{1}{2}$

3. $\arctan 1$ **4.** $\arcsin 0$

5. $\arccos\dfrac{\sqrt{2}}{2}$ **6.** $\sec^{-1}\sqrt{2}$

7. $\tan^{-1} 0$ **8.** $\arctan\dfrac{\sqrt{3}}{3}$

9. $\cos^{-1}\dfrac{\sqrt{3}}{2}$ **10.** $\cot^{-1}\left(-\dfrac{\sqrt{3}}{3}\right)$

11. $\csc^{-1} 2$ **12.** $\sin^{-1}\dfrac{1}{2}$

13. $\operatorname{arccot}(-\sqrt{3})$ **14.** $\tan^{-1}(-1)$

15. $\arcsin\left(-\dfrac{1}{2}\right)$ **16.** $\arccos\left(-\dfrac{\sqrt{2}}{2}\right)$

17. $\cos^{-1} 0$ **18.** $\sin^{-1}\dfrac{\sqrt{3}}{2}$

19. $\operatorname{arcsec} 2$ **20.** $\operatorname{arccsc}(-1)$

Use a grapher to find each of the following in radians, rounded to three decimal places, and in degrees, rounded to the nearest tenth of a degree.

21. $\arctan 0.3673$

22. $\cos^{-1}(-0.2935)$

23. $\sin^{-1} 0.9613$

24. $\arcsin(-0.6199)$

25. $\cos^{-1}(-0.9810)$

26. $\tan^{-1} 158$

27. $\csc^{-1}(-6.2774)$

28. $\sec^{-1} 1.1677$

29. $\tan^{-1}(1.091)$

30. $\cot^{-1} 1.265$

31. $\arcsin(-0.8192)$

32. $\arccos(-0.2716)$

33. State the domains of the inverse sine, inverse cosine, and inverse tangent functions.

34. State the ranges of the inverse sine, inverse cosine, and inverse tangent functions.

35. *Angle of Depression.* An airplane is flying at an altitude of 2000 ft toward an island. The straight-line distance from the airplane to the island is d feet. Express θ, the angle of depression, as an inverse sine function of d.

36. *Angle of Inclination.* A guy wire is attached to the top of a 50-ft pole and stretched to a point that is d feet from the bottom of the pole. Express β, the angle of inclination, as a function of d.

Evaluate.

37. $\sin(\arcsin 0.3)$ **38.** $\tan[\tan^{-1}(-4.2)]$

39. $\cos^{-1}\left[\cos\left(-\dfrac{\pi}{4}\right)\right]$ **40.** $\arcsin\left(\sin\dfrac{2\pi}{3}\right)$

41. $\sin^{-1}\left(\sin\dfrac{\pi}{5}\right)$ **42.** $\cot^{-1}\left(\cot\dfrac{2\pi}{3}\right)$

43. $\tan^{-1}\left(\tan\dfrac{2\pi}{3}\right)$ **44.** $\cos^{-1}\left(\cos\dfrac{\pi}{7}\right)$

45. $\sin\left(\arctan\dfrac{\sqrt{3}}{3}\right)$ **46.** $\cos\left(\arcsin\dfrac{\sqrt{3}}{2}\right)$

47. $\tan\left(\cos^{-1}\dfrac{\sqrt{2}}{2}\right)$ **48.** $\cos^{-1}(\sin\pi)$

49. $\arcsin\left(\cos\dfrac{\pi}{6}\right)$ **50.** $\sin^{-1}\left[\tan\left(-\dfrac{\pi}{4}\right)\right]$

51. $\tan(\arcsin 0.1)$ **52.** $\cos\left(\tan^{-1}\dfrac{\sqrt{3}}{4}\right)$

Find.

53. $\sin\left(\arctan\dfrac{a}{3}\right)$ **54.** $\tan\left(\cos^{-1}\dfrac{3}{x}\right)$

55. $\cot\left(\sin^{-1}\dfrac{p}{q}\right)$ **56.** $\sin(\cos^{-1} x)$

57. $\tan\left(\arcsin\dfrac{p}{\sqrt{p^2+9}}\right)$ **58.** $\tan\left(\dfrac{1}{2}\arcsin\dfrac{1}{2}\right)$

59. $\cos\left(\dfrac{1}{2}\arcsin\dfrac{\sqrt{3}}{2}\right)$ **60.** $\sin\left(2\cos^{-1}\dfrac{3}{5}\right)$

Evaluate.

61. $\cos\left(\sin^{-1}\dfrac{\sqrt{2}}{2}+\cos^{-1}\dfrac{3}{5}\right)$

62. $\sin\left(\sin^{-1}\dfrac{1}{2}+\cos^{-1}\dfrac{3}{5}\right)$

63. $\sin\,(\sin^{-1}x+\cos^{-1}y)$

64. $\cos\,(\sin^{-1}x-\cos^{-1}y)$

65. $\sin\,(\sin^{-1}0.6032+\cos^{-1}0.4621)$

66. $\cos\,(\sin^{-1}0.7325-\cos^{-1}0.4838)$

Discussion and Writing

67. Explain in your own words why the ranges of the inverse trigonometric functions are restricted.

68. How does the graph of $y=\sin^{-1}x$ differ from the graph of $y=\sin x$?

69. Why is it that

$$\sin\dfrac{5\pi}{6}=\dfrac{1}{2},$$

but

$$\sin^{-1}\left(\dfrac{1}{2}\right)\neq\dfrac{5\pi}{6}?$$

Skill Maintenance

Test algebraically whether the graph is symmetric with respect to the x-axis, the y-axis, and the origin. Check your work using a grapher.

70. $y^3=6x^2$

71. $3x^2+4y^2=5$

Test algebraically whether the function is even, odd, or neither even nor odd. Check your work using a grapher.

72. $f(x)=4x^2+2x-3$

73. $f(x)=2x^3-x$

Synthesis

74. Use a calculator to approximate the following expression:

$$16\tan^{-1}\dfrac{1}{5}-4\tan^{-1}\dfrac{1}{239}.$$

What number does this expression seem to approximate?

Prove the identity.

75. $\sin^{-1}x+\cos^{-1}x=\dfrac{\pi}{2}$

76. $\tan^{-1}x+\cot^{-1}x=\dfrac{\pi}{2}$

77. $\sin^{-1}x=\tan^{-1}\dfrac{x}{\sqrt{1-x^2}}$

78. $\tan^{-1}x=\sin^{-1}\dfrac{x}{\sqrt{x^2+1}}$

79. $\arcsin x=\arccos\sqrt{1-x^2},\quad$ for $x\geq0$

80. $\arccos x=\arctan\dfrac{\sqrt{1-x^2}}{x},\quad$ for $x>0$

81. *Height of a Mural.* An art student's eye is at a point A, looking at a mural of height h, with the bottom of the mural y feet above the eye. The eye is x feet from the wall. Write an expression for θ in terms of x, y, and h. Then evaluate the expression when $x=20$ ft, $y=7$ ft, and $h=25$ ft.

Solving Trigonometric Equations

3.5

• *Solve trigonometric equations.*

When an equation contains a trigonometric expression with a variable, such as cos x, it is called a trigonometric equation. Some trigonometric equations are identities, such as $\sin^2 x + \cos^2 x = 1$. Now we consider equations, such as $2 \cos x = -1$, that are usually not identities. As we have done for other types of equations, we will solve such equations by finding all values for x that make the equation true.

EXAMPLE 1   Solve: $2 \cos x = -1$.

Solution   We first solve for cos x:

$$2 \cos x = -1$$
$$\cos x = -\frac{1}{2}.$$

The solutions are numbers that have a cosine of $-\frac{1}{2}$. To find them, we use the unit circle (see Section 2.5).

   There are just two points on the unit circle for which the cosine is $-\frac{1}{2}$, as shown in the figure at left. They are the points for $2\pi/3$ and $4\pi/3$. These numbers, plus any multiple of 2π, are the solutions:

$$\frac{2\pi}{3} + 2k\pi \quad \text{or} \quad \frac{4\pi}{3} + 2k\pi,$$

where k is any integer. In degrees, the solutions are

$$120° + k \cdot 360° \quad \text{or} \quad 240° + k \cdot 360°,$$

where k is any integer.

   To check the solution to $2 \cos x = -1$, we can graph $y_1 = 2 \cos x$ and $y_2 = -1$ on the same set of axes and find the first coordinates of the points of intersection. Using $\pi/3$ as the Xscl facilitates our reading of the solutions. First, let's graph these equations on the interval from 0 to 2π, as shown in the figure on the left below. The only solutions in $[0, 2\pi)$ are $2\pi/3$ and $4\pi/3$.

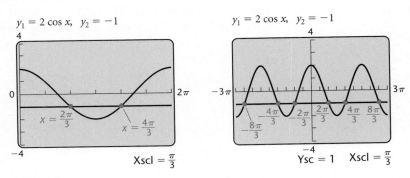

$y_1 = 2 \cos x, \ y_2 = -1$

Next, let's change the viewing window to $[-3\pi, 3\pi, -4, 4]$ and graph again. Since the cosine function is periodic, there is an infinite number of solutions. A few of these appear in the graph on the right above. From the graph, we see that the solutions are $2\pi/3 + 2k\pi$ and $4\pi/3 + 2k\pi$, where k is any integer.

EXAMPLE 2   Solve: $4 \sin^2 x = 1$.

Solution   We begin by solving for $\sin x$:

$$4 \sin^2 x = 1$$

$$\sin^2 x = \frac{1}{4}$$

$$\sin x = \pm \frac{1}{2}.$$

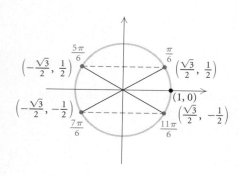

Again, we use the unit circle to find those numbers having a sine of $\frac{1}{2}$ or $-\frac{1}{2}$. The solutions are

$$\frac{\pi}{6} + 2k\pi, \quad \frac{5\pi}{6} + 2k\pi, \quad \frac{7\pi}{6} + 2k\pi, \quad \text{and} \quad \frac{11\pi}{6} + 2k\pi,$$

where k is any integer. In degrees, the solutions are

$$30° + k \cdot 360°, \quad 150° + k \cdot 360°, \quad 210° + k \cdot 360°, \quad \text{and}$$
$$330° + k \cdot 360°,$$

where k is any integer.

The general solutions listed above could be condensed using odd as well as even multiples of π:

$$\frac{\pi}{6} + k\pi \quad \text{and} \quad \frac{5\pi}{6} + k\pi,$$

or, in degrees,

$$30° + k \cdot 180° \quad \text{and} \quad 150° + k \cdot 180°,$$

where k is any integer.

Let's do a partial check using a grapher, checking only the solutions in $[0, 2\pi)$. We graph $y_1 = 4 \sin^2 x$ and $y_2 = 1$ and note that the solutions in $[0, 2\pi)$ are $\pi/6$, $5\pi/6$, $7\pi/6$, and $11\pi/6$.

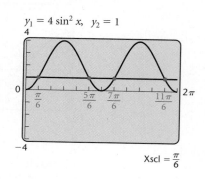

$$y_1 = 4 \sin^2 x, \quad y_2 = 1$$

$$\text{Xscl} = \frac{\pi}{6}$$

In most applications, it is sufficient to find just the solutions from 0 to 2π or from 0° to 360°. We then remember that any multiple of 2π, or 360°, can be added to obtain the rest of the solutions.

We must be careful to find all solutions in $[0, 2\pi)$ when solving trigonometric equations involving double angles.

EXAMPLE 3   Solve $3 \tan 2x = -3$ in the interval $[0, 2\pi)$.

Solution   We first solve for $\tan 2x$:

$$3 \tan 2x = -3$$
$$\tan 2x = -1.$$

We are looking for solutions x to the equation for which

$$0 \le x < 2\pi.$$

Multiplying by 2, we get

$$0 \le 2x < 4\pi,$$

which is the interval we use when solving $\tan 2x = -1$.

Using the unit circle, we find points $2x$ in $[0, 4\pi)$ for which $\tan 2x = -1$. These values of $2x$ are as follows:

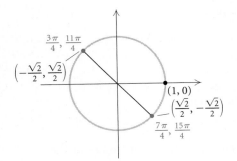

$$2x = \frac{3\pi}{4}, \quad \frac{7\pi}{4}, \quad \frac{11\pi}{4}, \quad \text{and} \quad \frac{15\pi}{4}.$$

Thus the desired values of x in $[0, 2\pi)$ are each of these values divided by 2. Therefore,

$$x = \frac{3\pi}{8}, \quad \frac{7\pi}{8}, \quad \frac{11\pi}{8}, \quad \text{and} \quad \frac{15\pi}{8}.$$

Graphers are needed to solve some trigonometric equations. Answers can be found in radians or degrees, depending on the mode setting.

EXAMPLE 4   Solve $\frac{1}{2} \cos \phi + 1 = 1.2108$ in $[0, 360°)$.

Solution   We have

$$\frac{1}{2} \cos \phi + 1 = 1.2108$$

$$\frac{1}{2} \cos \phi = 0.2108$$

$$\cos \phi = 0.4216.$$

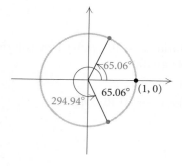

Using a grapher set in DEGREE mode, we find that the reference angle, arccos 0.4216, is

$$\phi \approx 65.06°.$$

Since $\cos \phi$ is positive, the solutions are in quadrants I and IV. The solutions in $[0, 360°)$ are

$$65.06° \quad \text{and} \quad 360° - 65.06° = 294.94°.$$

EXAMPLE 5 Solve $2 \cos^2 u = 1 - \cos u$ in $[0°, 360°)$.

Algebraic Solution

We use the principle of zero products:

$$2 \cos^2 u = 1 - \cos u$$
$$2 \cos^2 u + \cos u - 1 = 0$$
$$(2 \cos u - 1)(\cos u + 1) = 0$$
$$2 \cos u - 1 = 0 \quad \text{or} \quad \cos u + 1 = 0$$
$$2 \cos u = 1 \quad \text{or} \quad \cos u = -1$$
$$\cos u = \frac{1}{2} \quad \text{or} \quad \cos u = -1.$$

Thus,

$$u = 60°, 300° \quad \text{or} \quad u = 180°.$$

The solutions in $[0°, 360°)$ are 60°, 180°, and 300°.

Graphical Solution

INTERSECT METHOD: We graph the equations

$$y_1 = 2 \cos^2 x \quad \text{and} \quad y_2 = 1 - \cos x$$

and use the INTERSECT feature to find the first coordinates of the points of intersection.

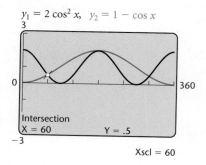

The leftmost solution is 60°. Using the INTERSECT feature two more times, we find the other solutions, 180° and 300°.

The Intersect and Zero Methods

An equation $f(x) = g(x)$ can be solved using the Intersect method by graphing $y_1 = f(x)$ and $y_2 = g(x)$ and using the INTERSECT feature to find the first coordinate of the point of intersection of the graphs. The equation can also be solved using the Zero method by rewriting it with 0 on one side of the equals sign, if necessary, and then using the ZERO feature.

ZERO METHOD: We write the equation in the form

$$2 \cos^2 u + \cos u - 1 = 0.$$

Then we graph $y = 2 \cos^2 x + \cos x - 1$ and use the ZERO feature to determine the zeros of the function.

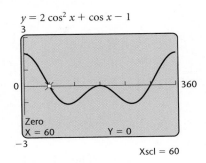

$y = 2 \cos^2 x + \cos x - 1$

Zero
X = 60   Y = 0

Xscl = 60

The leftmost zero is 60°. Using the ZERO feature two more times, we find the other zeros, 180° and 300°.

The solutions in [0°, 360°) are 60°, 180°, and 300°.

EXAMPLE 6   Solve $\sin^2 \beta - \sin \beta = 0$ in $[0, 2\pi)$.

Solution   We have

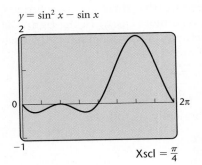

$y = \sin^2 x - \sin x$

Xscl = $\frac{\pi}{4}$

$$\sin^2 \beta - \sin \beta = 0$$
$$\sin \beta \, (\sin \beta - 1) = 0 \qquad \text{Factoring}$$
$$\sin \beta = 0 \qquad or \quad \sin \beta - 1 = 0$$
$$\sin \beta = 0 \qquad or \qquad \sin \beta = 1$$
$$\beta = 0, \pi \quad or \qquad \beta = \frac{\pi}{2}.$$

The solutions in $[0, 2\pi)$ are 0, $\pi/2$, and π.

If a trigonometric equation is quadratic but difficult or impossible to factor, we use the quadratic formula.

EXAMPLE 7   Solve $10 \sin^2 x - 12 \sin x - 7 = 0$ in $[0°, 360°)$.

Solution   This equation is quadratic in $\sin x$ with $a = 10$, $b = -12$, and $c = -7$. Substituting into the quadratic formula, we get

$$\sin x = \frac{12 \pm \sqrt{144 + 280}}{20} \qquad \textbf{Using the quadratic formula}$$

$$= \frac{12 \pm \sqrt{424}}{20}$$

$$\approx \frac{12 \pm 20.5913}{20}$$

$$\sin x \approx 1.6296 \quad or \quad \sin x \approx -0.4296.$$

Since sine values are never greater than 1, the first of the equations has no solution. Using the other equation, we find the reference angle to be 25.44°. Since sin x is negative, the solutions are in quadrants III and IV.

Thus the solutions in [0°, 360°) are

$$180° + 25.44° = 205.44° \quad \text{and} \quad 360° - 25.44° = 334.56°.$$

Trigonometric equations can involve more than one function.

EXAMPLE 8   Solve $2 \cos^2 x \tan x = \tan x$ in [0, 2π).

Solution   Using a grapher, we can determine that there are six solutions. If we let Xscl = $\pi/4$, the solutions are read more easily. In the figures at left, we show the Intersect and Zero methods of solving graphically. Each illustrates that the solutions in [0, 2π) are

$$0, \quad \frac{\pi}{4}, \quad \frac{3\pi}{4}, \quad \pi, \quad \frac{5\pi}{4}, \quad \text{and} \quad \frac{7\pi}{4}.$$

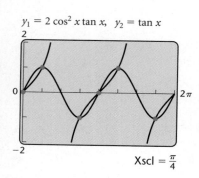

$y_1 = 2 \cos^2 x \tan x, \quad y_2 = \tan x$

Xscl = $\frac{\pi}{4}$

$y_1 = 2 \cos^2 x \tan x - \tan x$

Xscl = $\frac{\pi}{4}$

We can verify these solutions algebraically, as follows:

$$2 \cos^2 x \tan x = \tan x$$
$$2 \cos^2 x \tan x - \tan x = 0$$
$$\tan x (2 \cos^2 x - 1) = 0$$
$$\tan x = 0 \qquad \textit{or} \quad 2 \cos^2 x - 1 = 0$$
$$\cos^2 x = \frac{1}{2}$$
$$\cos x = \pm \frac{\sqrt{2}}{2}$$
$$x = 0, \pi \quad \textit{or} \qquad x = \frac{\pi}{4}, \frac{3\pi}{4}, \frac{5\pi}{4}, \frac{7\pi}{4}.$$

These solutions check with those found graphically. Thus, $x = 0$, $\pi/4$, $3\pi/4$, π, $5\pi/4$, and $7\pi/4$.

When a trigonometric equation involves more than one function, it is sometimes helpful to use identities to rewrite the equation in terms of a single function.

EXAMPLE 9 Solve $\sin x + \cos x = 1$ in $[0, 2\pi)$.

Algebraic Solution

We have

$$\sin x + \cos x = 1$$
$$\sin^2 x + 2 \sin x \cos x + \cos^2 x = 1$$

Squaring both sides

$$2 \sin x \cos x + 1 = 1$$

Using $\sin^2 x + \cos^2 x = 1$

$$2 \sin x \cos x = 0$$
$$\sin 2x = 0.$$

Using $2 \sin x \cos x = \sin 2x$

We are looking for solutions x to the equation for which $0 \le x < 2\pi$. Multiplying by 2, we get $0 \le 2x < 4\pi$, which is the interval we consider to solve $\sin 2x = 0$. These values of $2x$ are 0, π, 2π, and 3π. Thus the desired values of x in $[0, 2\pi)$ satisfying this equation are 0, $\pi/2$, π, and $3\pi/2$. Now we check these in the original equation:

$$\sin 0 + \cos 0 = 0 + 1 = 1,$$

$$\sin \frac{\pi}{2} + \cos \frac{\pi}{2} = 1 + 0 = 1,$$

$$\sin \pi + \cos \pi = 0 + (-1) = -1,$$

$$\sin \frac{3\pi}{2} + \cos \frac{3\pi}{2} = (-1) + 0 = -1.$$

We find that π and $3\pi/2$ do not check, but the other values do. Thus the solutions in $[0, 2\pi)$ are

$$0 \quad \text{and} \quad \frac{\pi}{2}.$$

When the solution process involves squaring both sides, values are sometimes obtained that are not solutions of the original equation. As we saw in this example, it is important to check the possible solutions.

Graphical Solution

We can graph the left side and then the right side of the equation as seen in the first window below. Then we look for points of intersection. We could also rewrite the equation as $\sin x + \cos x - 1 = 0$, graph the left side, and look for the zeros of the function, as illustrated in the second window below. In each window, we see the solutions in $[0, 2\pi)$ as 0 and $\pi/2$.

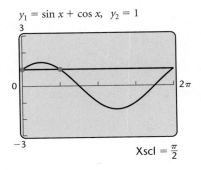

$y_1 = \sin x + \cos x, \quad y_2 = 1$

$Xscl = \frac{\pi}{2}$

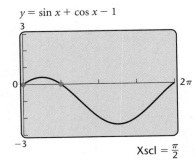

$y = \sin x + \cos x - 1$

$Xscl = \frac{\pi}{2}$

This example illustrates a valuable advantage of the grapher—that is, with a grapher, extraneous solutions do not appear.

EXAMPLE 10   Solve $\cos 2x + \sin x = 1$ in $[0, 2\pi)$.

Algebraic Solution

We have

$$\cos 2x + \sin x = 1$$
$$1 - 2\sin^2 x + \sin x = 1 \qquad \text{Using the identity } \cos 2x = 1 - 2\sin^2 x$$
$$-2\sin^2 x + \sin x = 0$$
$$\sin x\,(1 - 2\sin x) = 0 \qquad \text{Factoring}$$
$$\sin x = 0 \quad or \quad 1 - 2\sin x = 0 \qquad \text{Principle of zero products}$$
$$\sin x = 0 \quad or \quad \sin x = \frac{1}{2}$$
$$x = 0, \pi \quad or \quad x = \frac{\pi}{6}, \frac{5\pi}{6}.$$

All values check. The solutions in $[0, 2\pi)$ are 0, $\pi/6$, $5\pi/6$, and π.

Graphical Solution

We graph $y_1 = \cos 2x + \sin x - 1$ and look for the zeros of the function.

$y = \cos 2x + \sin x - 1$

Xscl $= \frac{\pi}{6}$

The solutions in $[0, 2\pi)$ are 0, $\pi/6$, $5\pi/6$, and π, or, approximately, 0, 0.524, 2.618, and 3.142.

EXAMPLE 11   Solve $\tan^2 x + \sec x - 1 = 0$ in $[0, 2\pi)$.

Algebraic Solution

We have

$$\tan^2 x + \sec x - 1 = 0$$
$$\sec^2 x - 1 + \sec x - 1 = 0 \qquad \text{Using the identity } 1 + \tan^2 x = \sec^2 x$$
$$\sec^2 x + \sec x - 2 = 0$$
$$(\sec x + 2)(\sec x - 1) = 0 \qquad \text{Factoring}$$
$$\sec x = -2 \qquad or \quad \sec x = 1 \qquad \text{Principle of zero products}$$
$$x = \frac{2\pi}{3}, \frac{4\pi}{3} \quad or \qquad x = 0$$

All these values check. The solutions in $[0, 2\pi)$ are 0, $2\pi/3$, and $4\pi/3$.

Graphical Solution

We graph $y = \tan^2 x + \sec x - 1$, but we enter this equation in the form

$$y_1 = \tan^2 x + \frac{1}{\cos x} - 1.$$

We use the ZERO feature to find zeros of the function.

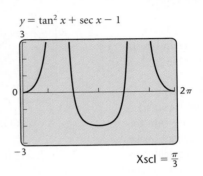

$$y = \tan^2 x + \sec x - 1$$

Xscl $= \frac{\pi}{3}$

The solutions in $[0, 2\pi)$ are 0, $2\pi/3$, and $4\pi/3$.   ▬

Sometimes we cannot find solutions algebraically, but we can approximate them with a grapher.

EXAMPLE 12   Solve each of the following in $[0, 2\pi)$.

a) $x^2 - 1.5 = \cos x$

b) $\sin x - \cos x = \cot x$

Solution

a) In the screen on the left, we graph $y_1 = x^2 - 1.5$ and $y_2 = \cos x$ and look for points of intersection. In the screen on the right, we graph $y_1 = x^2 - 1.5 - \cos x$ and look for the zeros of the function.

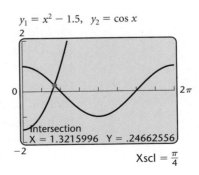

$$y_1 = x^2 - 1.5, \quad y_2 = \cos x$$

Intersection
X = 1.3215996   Y = .24662556

Xscl $= \frac{\pi}{4}$

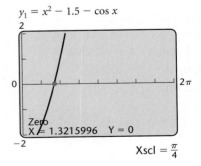

$$y_1 = x^2 - 1.5 - \cos x$$

Zero
X = 1.3215996   Y = 0

Xscl $= \frac{\pi}{4}$

We determine the solution in $[0, 2\pi)$ to be approximately 1.32.

b) In the screen on the left, we graph $y_1 = \sin x - \cos x$ and $y_2 = \cot x$ and determine the points of intersection. In the screen on the right, we graph the function $y_1 = \sin x - \cos x - \cot x$ and determine the zeros.

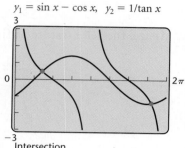

$y_1 = \sin x - \cos x, \quad y_2 = 1/\tan x$

Intersection
X = 1.1276527   Y = .47462662
X = 5.661357   Y = −1.395337

$y = \sin x - \cos x - 1/\tan x$

Zeros
X = 1.1276527   Y = 0
X = 5.661357   Y = 0

Each method leads to the approximate solutions 1.13 and 5.66 in $[0, 2\pi)$.

Exercise Set 3.5

Solve, finding all solutions. Express the solutions in both radians and degrees.

1. $\cos x = \dfrac{\sqrt{3}}{2}$ **2.** $\sin x = -\dfrac{\sqrt{2}}{2}$

3. $\tan x = -\sqrt{3}$ **4.** $\cos x = \dfrac{1}{2}$

Solve, finding all solutions in $[0, 2\pi)$ or $[0°, 360°)$. Verify your answer using a grapher.

5. $2 \cos x - 1 = -1.2814$

6. $\sin x + 3 = 2.0816$

7. $2 \sin x + \sqrt{3} = 0$

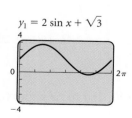

$y_1 = 2 \sin x + \sqrt{3}$

8. $2 \tan x - 4 = 1$

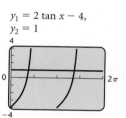

$y_1 = 2 \tan x - 4,$
$y_2 = 1$

9. $2 \cos^2 x = 1$ **10.** $\csc^2 x - 4 = 0$

11. $2 \sin^2 x + \sin x = 1$ **12.** $\cos^2 x + 2 \cos x = 3$

13. $2 \cos^2 x - \sqrt{3} \cos x = 0$

14. $2 \sin^2 \theta + 7 \sin \theta = 4$

15. $6 \cos^2 \phi + 5 \cos \phi + 1 = 0$

16. $2 \sin t \cos t + 2 \sin t - \cos t - 1 = 0$

17. $\sin 2x \cos x - \sin x = 0$

18. $5 \sin^2 x - 8 \sin x = 3$

19. $\cos^2 x + 6 \cos x + 4 = 0$

20. $2 \tan^2 x = 3 \tan x + 7$

21. $7 = \cot^2 x + 4 \cot x$

22. $3 \sin^2 x = 3 \sin x + 2$

Solve, finding all solutions in $[0, 2\pi)$. *Then check each solution using a grapher.*

23. $\cos 2x - \sin x = 1$

$y_1 = \cos 2x - \sin x,$
$y_2 = 1$

24. $2 \sin x \cos x + \sin x = 0$

$y = 2 \sin x \cos x + \sin x$

25. $\sin 4x - 2 \sin 2x = 0$

26. $\tan x \sin x - \tan x = 0$

27. $\sin 2x \cos x + \sin x = 0$

28. $\cos 2x \sin x + \sin x = 0$

29. $2 \sec x \tan x + 2 \sec x + \tan x + 1 = 0$

30. $\sin 2x \sin x - \cos 2x \cos x = -\cos x$

31. $\sin 2x + \sin x + 2 \cos x + 1 = 0$

32. $\tan^2 x + 4 = 2 \sec^2 x + \tan x$

33. $\sec^2 x - 2 \tan^2 x = 0$

34. $\cot x = \tan (2x - 3\pi)$

35. $2 \cos x + 2 \sin x = \sqrt{6}$

36. $\sqrt{3} \cos x - \sin x = 1$

37. $\sec^2 x + 2 \tan x = 6$

38. $5 \cos 2x + \sin x = 4$

39. $\cos (\pi - x) + \sin \left(x - \dfrac{\pi}{2} \right) = 1$

40. $\dfrac{\sin^2 x - 1}{\cos \left(\dfrac{\pi}{2} - x \right) + 1} = \dfrac{\sqrt{2}}{2} - 1$

Solve using a grapher, finding all solutions in $[0, 2\pi)$.

41. $x \sin x = 1$ **42.** $x^2 + 2 = \sin x$

43. $2 \cos^2 x = x + 1$ **44.** $x \cos x - 2 = 0$

45. $\cos x - 2 = x^2 - 3x$ **46.** $\sin x = \tan \dfrac{x}{2}$

Discussion and Writing

47. Jan lists her answer to a problem as $\pi/6 + k\pi$, for any integer k, while Jacob lists his answer as $\pi/6 + 2k\pi$ and $7\pi/6 + 2\pi k$, for any integer k. Are their answers equivalent? Why or why not?

48. Under what circumstances will a grapher give exact solutions of a trigonometric equation?

Skill Maintenance

Solve the right triangle.

49.

50.

Write a slope–intercept equation for a line with the given characteristics.

51. $m = \frac{3}{4}$; passes through $(0, -5)$

52. Passes through $(2, -10)$ and $(-2, -10)$

Synthesis

Solve in $[0, 2\pi)$. *Check your results using a grapher.*

53. $|\sin x| = \dfrac{\sqrt{3}}{2}$

54. $|\cos x| = \dfrac{1}{2}$

55. $\sqrt{\tan x} = \sqrt[4]{3}$

56. $12 \sin x - 7\sqrt{\sin x} + 1 = 0$

57. $\ln (\cos x) = 0$

58. $e^{\sin x} = 1$

59. $\sin (\ln x) = -1$

60. $e^{\ln (\sin x)} = 1$

Some graphers can use regression to fit a trigonometric function to a set of data. (For more on regression, see Appendix C.)

61. *Sales.*   Sales of certain products fluctuate in cycles. The data in the following table show the total sales of skis per month for a business in a northern climate.

MONTH, x		TOTAL SALES, y (IN THOUSANDS)
August,	8	0
November,	11	7
February,	2	$14
May,	5	7
August,	8	0

a) Using the SINE REGRESSION feature on a grapher, fit a sine function of the form $y = A \sin (Bx - C) + D$ to this set of data.

b) Approximate the total sales for December and for July.

62. *Daylight Hours.*   The data in the following table give the number of daylight hours for certain days in Kajaani, Finland.

DAY, x		NUMBER OF DAYLIGHT HOURS, y
January 10,	10	5.0
February 19,	50	9.1
March 3,	62	10.4
April 28,	118	16.4
May 14,	134	18.2
June 11,	162	20.7
July 17,	198	19.5
August 22,	234	15.7
September 19,	262	12.7
October 1,	274	11.4
November 14,	318	6.7
December 28,	362	4.3

Source: *The Astronomical Almanac*, 1995, Washington: U.S. Government Printing Office.

a) Using the SINE REGRESSION feature on a grapher, model these data with an equation of the form $y = A \sin (Bx - C) + D$.

b) Approximate the number of daylight hours in Kajaani for April 22, July 4, and December 15.

63. *Temperature During an Illness.*   The temperature T, in degrees Fahrenheit, of a patient t days into a 12-day illness is given by

$$T(t) = 101.6° + 3° \sin \left(\frac{\pi}{8} t \right).$$

Find the times t during the illness at which the patient's temperature was 103°.

64. *Satellite Location.*   A satellite circles the earth in such a manner that it is y miles from the equator (north or south, height from the surface not considered) t minutes after its launch, where

$$y = 5000 \left[\cos \frac{\pi}{45} (t - 10) \right].$$

At what times t in the interval [0, 240], the first 4 hr, is the satellite 3000 mi north of the equator?

65. *Nautical Mile.*   (See Exercise 52 in Exercise Set 3.2.) In Great Britain, the *nautical mile* is defined as the length of a minute of arc of the earth's radius. Since the earth is flattened at the poles, a British nautical mile varies with latitude. In fact, it is given, in feet, by the function

$$N(\phi) = 6066 - 31 \cos 2\phi,$$

where ϕ is the latitude in degrees. At what latitude north is the length of a British nautical mile found to be 6040 ft?

66. *Acceleration Due to Gravity.*   (See Exercise 53 in Exercise Set 3.2.) The acceleration due to gravity is often denoted by g in a formula such as $S = \frac{1}{2}gt^2$, where S is the distance that an object falls in t seconds. The number g is generally considered constant, but in fact it varies slightly with latitude. If ϕ stands for latitude, in degrees, an excellent approximation of g is given by the formula

$$g = 9.78049(1 + 0.005288 \sin^2 \phi - 0.000006 \sin^2 2\phi),$$

where g is measured in meters per second per second at sea level. At what latitude north does $g = 9.8$?

Solve.

67. $\arccos x = \arccos \frac{3}{5} - \arcsin \frac{4}{5}$

68. $\sin^{-1} x = \tan^{-1} \frac{1}{3} + \tan^{-1} \frac{1}{2}$

69. Suppose that $\sin x = 5 \cos x$. Find $\sin x \cos x$.

Chapter Summary and Review 3

Important Properties and Formulas

Basic Identities

$$\sin x = \frac{1}{\csc x}, \qquad \sin(-x) = -\sin x,$$

$$\cos(-x) = \cos x,$$

$$\cos x = \frac{1}{\sec x}, \qquad \tan(-x) = -\tan x$$

$$\tan x = \frac{1}{\cot x},$$

$$\tan x = \frac{\sin x}{\cos x},$$

$$\cot x = \frac{\cos x}{\sin x},$$

Pythagorean Identities

$$\sin^2 x + \cos^2 x = 1,$$
$$1 + \cot^2 x = \csc^2 x,$$
$$1 + \tan^2 x = \sec^2 x$$

Sum and Difference Identities

$$\sin(u \pm v) = \sin u \cos v \pm \cos u \sin v,$$
$$\cos(u \pm v) = \cos u \cos v \mp \sin u \sin v,$$
$$\tan(u \pm v) = \frac{\tan u \pm \tan v}{1 \mp \tan u \tan v}$$

Cofunction Identities

$$\sin\left(\frac{\pi}{2} - x\right) = \cos x,$$

$$\tan\left(\frac{\pi}{2} - x\right) = \cot x,$$

$$\sec\left(\frac{\pi}{2} - x\right) = \csc x,$$

$$\sin\left(x \pm \frac{\pi}{2}\right) = \pm\cos x,$$

$$\cos\left(x \pm \frac{\pi}{2}\right) = \mp\sin x$$

Double-Angle Identities

$$\sin 2x = 2\sin x \cos x,$$
$$\cos 2x = \cos^2 x - \sin^2 x$$
$$= 1 - 2\sin^2 x$$
$$= 2\cos^2 x - 1,$$
$$\tan 2x = \frac{2\tan x}{1 - \tan^2 x}$$

Half-Angle Identities

$$\sin \frac{x}{2} = \pm\sqrt{\frac{1 - \cos x}{2}},$$

$$\cos \frac{x}{2} = \pm\sqrt{\frac{1 + \cos x}{2}},$$

$$\tan \frac{x}{2} = \pm\sqrt{\frac{1 - \cos x}{1 + \cos x}}$$

$$= \frac{\sin x}{1 + \cos x}$$

$$= \frac{1 - \cos x}{\sin x}$$

Inverse Trigonometric Functions

FUNCTION	DOMAIN	RANGE
$y = \sin^{-1} x$	$[-1, 1]$	$\left[-\frac{\pi}{2}, \frac{\pi}{2}\right]$
$y = \cos^{-1} x$	$[-1, 1]$	$[0, \pi]$
$y = \tan^{-1} x$	$(-\infty, \infty)$	$\left(-\frac{\pi}{2}, \frac{\pi}{2}\right)$

(continued)

Composition of Trigonometric Functions
The following are true for any x in the domain of the inverse function:

$$\sin(\sin^{-1} x) = x,$$
$$\cos(\cos^{-1} x) = x,$$
$$\tan(\tan^{-1} x) = x.$$

The following are true for any x in the range of the inverse function:

$$\sin^{-1}(\sin x) = x,$$
$$\cos^{-1}(\cos x) = x,$$
$$\tan^{-1}(\tan x) = x.$$

REVIEW EXERCISES

Complete the Pythagorean identity.

1. $1 + \cot^2 x =$

2. $\sin^2 x + \cos^2 x =$

Multiply and simplify. Check using a grapher.

3. $(\tan y - \cot y)(\tan y + \cot y)$

4. $(\cos x + \sec x)^2$

Factor and simplify. Check using a grapher.

5. $\sec x \csc x - \csc^2 x$

6. $3 \sin^2 y - 7 \sin y - 20$

7. $1000 - \cos^3 u$

Simplify and check using a grapher.

8. $\dfrac{\sec^4 x - \tan^4 x}{\sec^2 x + \tan^2 x}$

9. $\dfrac{2 \sin^2 x}{\cos^3 x} \cdot \left(\dfrac{\cos x}{2 \sin x}\right)^2$

10. $\dfrac{3 \sin x}{\cos^2 x} \cdot \dfrac{\cos^2 x + \cos x \sin x}{\sin^2 x - \cos^2 x}$

11. $\dfrac{3}{\cos y - \sin y} - \dfrac{2}{\sin^2 y - \cos^2 y}$

12. $\left(\dfrac{\cot x}{\csc x}\right)^2 + \dfrac{1}{\csc^2 x}$

13. $\dfrac{4 \sin x \cos^2 x}{16 \sin^2 x \cos x}$

14. Simplify. Assume the radicand is nonnegative.

$$\sqrt{\sin^2 x + 2 \cos x \sin x + \cos^2 x}$$

15. Rationalize the denominator: $\sqrt{\dfrac{1 + \sin x}{1 - \sin x}}$.

16. Rationalize the numerator: $\sqrt{\dfrac{\cos x}{\tan x}}$.

17. Given that $x = 3 \tan \theta$, express $\sqrt{9 + x^2}$ as a trigonometric function without radicals. Assume that $0 < \theta < \pi/2$.

Use the sum and difference formulas to write equivalent expressions. You need not simplify.

18. $\cos\left(x + \dfrac{3\pi}{2}\right)$

19. $\tan(45° - 30°)$

20. Simplify: $\cos 27° \cos 16° + \sin 27° \sin 16°$.

21. Find $\cos 165°$ exactly.

22. Given that $\tan \alpha = \sqrt{3}$ and $\sin \beta = \sqrt{2}/2$ and that α and β are between 0 and $\pi/2$, evaluate $\tan(\alpha - \beta)$ exactly.

23. Assume that $\sin \theta = 0.5812$ and $\cos \phi = 0.2341$ and that both θ and ϕ are first-quadrant angles. Evaluate $\cos(\theta + \phi)$.

Complete the cofunction identity.

24. $\cos\left(x + \dfrac{\pi}{2}\right) =$

25. $\cos\left(\dfrac{\pi}{2} - x\right) =$

26. $\sin\left(x - \dfrac{\pi}{2}\right) =$

27. Given that $\cos \alpha = -\frac{3}{5}$ and that the terminal side is in quadrant III:

a) Find the other function values for α.
b) Find the six function values for $\pi/2 - \alpha$.
c) Find the six function values for $\alpha + \pi/2$.

28. Find an equivalent expression for $\csc\left(x - \dfrac{\pi}{2}\right)$.

29. Find $\tan 2\theta$, $\cos 2\theta$, and $\sin 2\theta$ and the quadrant in which 2θ lies, where $\cos \theta = -\frac{4}{5}$ and θ is in quadrant III.

30. Find $\sin \dfrac{\pi}{8}$ exactly.

31. Given that $\sin \beta = 0.2183$ and β is in quadrant I, find $\sin 2\beta$, $\cos \dfrac{\beta}{2}$, and $\cos 4\beta$.

Simplify and check using a grapher.

32. $1 - 2 \sin^2 \dfrac{x}{2}$

33. $(\sin x + \cos x)^2 - \sin 2x$

34. $2 \sin x \cos^3 x + 2 \sin^3 x \cos x$

35. $\dfrac{2 \cot x}{\cot^2 x - 1}$

Prove the identity.

36. $\dfrac{1 - \sin x}{\cos x} = \dfrac{\cos x}{1 + \sin x}$

37. $\dfrac{1 + \cos 2\theta}{\sin 2\theta} = \cot \theta$

38. $\dfrac{\tan y + \sin y}{2 \tan \theta} = \cos^2 \dfrac{y}{2}$

39. $\dfrac{\sin x - \cos x}{\cos^2 x} = \dfrac{\tan^2 x - 1}{\sin x + \cos x}$

In Exercises 40–43, use a grapher to determine which expression (A)–(D) on the right can be used to complete the identity. Then prove the identity algebraically.

40. $\csc x - \cos x \cot x$ \qquad **A.** $\dfrac{\csc x}{\sec x}$

41. $\dfrac{1}{\sin x \cos x} - \dfrac{\cos x}{\sin x}$ \qquad **B.** $\sin x$

42. $\dfrac{\cot x - 1}{1 - \tan x}$ \qquad **C.** $\dfrac{2}{\sin x}$

43. $\dfrac{\cos x + 1}{\sin x} + \dfrac{\sin x}{\cos x + 1}$ \qquad **D.** $\dfrac{\sin x \cos x}{1 - \sin^2 x}$

Find each of the following exactly in both radians and degrees.

44. $\sin^{-1}\left(-\dfrac{1}{2}\right)$ \qquad **45.** $\cos^{-1} \dfrac{\sqrt{3}}{2}$

46. $\arctan 1$ \qquad **47.** $\arcsin 0$

Use a grapher to find each of the following in radians, rounded to three decimal places, and in degrees, rounded to the nearest tenth of a degree.

48. $\arccos(-0.2194)$

49. $\cot^{-1} 2.381$

Evaluate.

50. $\cos\left(\cos^{-1} \dfrac{1}{2}\right)$ \qquad **51.** $\tan^{-1}\left(\tan \dfrac{\sqrt{3}}{3}\right)$

52. $\sin^{-1}\left(\sin \dfrac{\pi}{7}\right)$ \qquad **53.** $\cos\left(\arcsin \dfrac{\sqrt{2}}{2}\right)$

Find.

54. $\cos\left(\arctan \dfrac{b}{3}\right)$

55. $\cos\left(2 \sin^{-1} \dfrac{4}{5}\right)$

Solve, finding all solutions. Express the solutions in both radians and degrees.

56. $\cos x = -\dfrac{\sqrt{2}}{2}$

57. $\tan x = \sqrt{3}$

Solve, finding all solutions in $[0, 2\pi)$.

58. $4 \sin^2 x = 1$

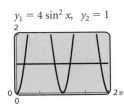

$y_1 = 4 \sin^2 x, \quad y_2 = 1$

59. $\sin 2x \sin x - \cos x = 0$

$y = \sin 2x \sin x - \cos x$

60. $2 \cos^2 x + 3 \cos x = -1$

61. $\sin^2 x - 7 \sin x = 0$

62. $\csc^2 x - 2 \cot^2 x = 0$

63. $\sin 4x + 2 \sin 2x = 0$

64. $2 \cos x + 2 \sin x = \sqrt{2}$

65. $6 \tan^2 x = 5 \tan x + \sec^2 x$

Solve using a grapher, finding all solutions in $[0, 2\pi)$.

66. $x \cos x = 1$

67. $2 \sin^2 x = x + 1$

Discussion and Writing

68. Prove the identity $2 \cos^2 x - 1 = \cos^4 x - \sin^4 x$ in three ways:

a) Start with the left side and deduce the right (method 1).

b) Start with the right side and deduce the left (method 1).

c) Work with each side separately until you deduce the same expression (method 2).

Then determine the most efficient method and explain why.

69. Why are the ranges of the inverse trigonometric functions restricted?

Synthesis

70. Find the measure of the angle from l_1 to l_2:

$$l_1: x + y = 3 \qquad l_2: 2x - y = 5.$$

71. Find an identity for $\cos(u + v)$ involving only cosines.

72. Simplify: $\cos\left(\dfrac{\pi}{2} - x\right)[\csc x - \sin x]$.

73. Find $\sin \theta$, $\cos \theta$, and $\tan \theta$ under the given conditions:

$$\sin 2\theta = \frac{1}{5}, \ \frac{\pi}{2} \le 2\theta < \pi.$$

74. Prove the following equation to be an identity and verify the results using a grapher:

$$\ln e^{\sin t} = \sin t.$$

75. Graph: $y = \sec^{-1} x$.

76. Show that

$$\tan^{-1} x = \frac{\sin^{-1} x}{\cos^{-1} x}$$

is *not* an identity.

77. Solve $e^{\cos x} = 1$ in $[0, 2\pi)$. Check using a grapher.

Applications of Trigonometry 4

APPLICATION

Fire detection lookout towers are used to observe and report forest fires. An Osborne fire finder is used to determine the location of a fire. (*Source*: National Interagency Fire Center, Boise, Idaho) If a ranger in fire tower A spots a fire at a direction of 295° and a ranger in fire tower B, located 45 mi at a direction of 045° from tower A, spots the same fire at a direction of 255°, how far from tower A is the fire? from tower B? We can use the law of sines to determine these distances.

This problem appears as Exercise 29 in Exercise Set 4.1.

Triangle trigonometry is important in applications such as large-scale construction, navigation, and surveying. In this chapter, we continue the study of triangle trigonometry that we began in Chapter 2. We will find that the trigonometric functions can also be used to solve triangles that are not right triangles.

We introduce complex numbers in this chapter. These numbers have applications in both electronics and engineering.

The study of triangles also leads to the study of vectors. A vector is a quantity that has a direction. Vectors have many practical applications in the physical sciences.

```
45 sin (30)/sin (40)
                35.0037861
45 sin (110)/sin (40)
                65.785599
```

253

The Law of Sines

4.1

- Use the law of sines to solve triangles.
- Find the area of any triangle given the lengths of two sides and the measure of the included angle.

To **solve a triangle** means to find the lengths of all its sides and the measures of all its angles. We solved right triangles in Section 2.2. For review, let's solve the right triangle shown below. We begin by listing the known measures.

$$Q = 37.1° \qquad q = ?$$
$$W = 90° \qquad w = ?$$
$$Z = ? \qquad z = 6.3$$

Since the sum of the three angle measures of any triangle is 180°, we can immediately find the third angle:

$$Z = 180° - (90° + 37.1°) = 52.9°.$$

Then using the tangent and cosine ratios, respectively, we can find q and w:

$$\tan 37.1° = \frac{q}{6.3}, \quad \text{or}$$

$$q = 6.3 \tan 37.1° \approx 4.8,$$

and $$\cos 37.1° = \frac{6.3}{w}, \quad \text{or}$$

$$w = \frac{6.3}{\cos 37.1°} \approx 7.9.$$

Now all six measures are known and we have solved triangle QWZ.

$$Q = 37.1° \qquad q \approx 4.8$$
$$W = 90° \qquad w \approx 7.9$$
$$Z = 52.9° \qquad z = 6.3$$

Solving Oblique Triangles

The trigonometric functions can also be used to solve triangles that are not right triangles. Such triangles are called **oblique**. Any triangle, right or oblique, can be solved *if at least one side and any other two measures are known*. The five possible situations are illustrated below.

1. AAS: Two angles of a triangle and a side opposite one of them are known.

2. ASA: Two angles of a triangle and the included side are known.

ASA

3. SSA: Two sides of a triangle and an angle opposite one of them are known. (In this case, there may be no solution, one solution, or two solutions. The latter is known as the ambiguous case.)

SSA

4. SAS: Two sides of a triangle and the included angle are known.

SAS

5. SSS: All three sides of the triangle are known.

SSS

The list above does not include the situation in which only the three angle measures are given. The reason for this lies in the fact that the angle measures determine *only the shape* of the triangle and *not the size*, as shown with the following triangles.

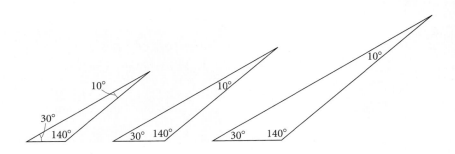

Thus we cannot solve a triangle when only the three angle measures are given.

In order to solve oblique triangles, we need to derive the *law of sines* and the *law of cosines*. The law of sines applies to the first three situations listed above. The law of cosines, which we develop in Section 4.2, applies to the last two situations.

The Law of Sines

We consider any oblique triangle. It may or may not have an obtuse angle. Although we look at only the acute-triangle case, the derivation of the obtuse-triangle case is essentially the same.

In acute $\triangle ABC$ at left, we have drawn an altitude from vertex C. It has length h. From $\triangle ADC$, we have

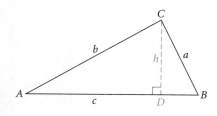

$$\sin A = \frac{h}{b}, \quad \text{or} \quad h = b \sin A.$$

From $\triangle BDC$, we have

$$\sin B = \frac{h}{a}, \quad \text{or} \quad h = a \sin B.$$

With $h = b \sin A$ and $h = a \sin B$, we now have

$$a \sin B = b \sin A$$

$$\frac{a \sin B}{\sin A \sin B} = \frac{b \sin A}{\sin A \sin B} \qquad \text{Dividing by } \sin A \sin B$$

$$\frac{a}{\sin A} = \frac{b}{\sin B}. \qquad \text{Simplifying}$$

There is no danger of dividing by 0 here because we are dealing with triangles whose angles are never 0° or 180°. Thus the sine value will never be 0.

If we were to consider altitudes from vertex A and vertex B in the triangle shown above, the same argument would give us

$$\frac{b}{\sin B} = \frac{c}{\sin C} \quad \text{and} \quad \frac{a}{\sin A} = \frac{c}{\sin C}.$$

We combine these results to obtain the law of sines.

The Law of Sines
In any triangle ABC,

$$\frac{a}{\sin A} = \frac{b}{\sin B} = \frac{c}{\sin C}.$$

In any triangle, the sides are proportional to the sines of the opposite angles.

Solving Triangles (AAS and ASA)

When two angles and a side of any triangle are known, the law of sines
can be used to solve the triangle.

EXAMPLE 1   In $\triangle EFG$, $e = 4.56$, $E = 43°$, and $G = 57°$. Solve the
triangle.

Solution   We first make a drawing. We know three of the six measures.

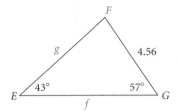

$$E = 43° \qquad e = 4.56$$
$$F = ? \qquad f = ?$$
$$G = 57° \qquad g = ?$$

From the figure, we see that we have the AAS situation. We begin by
finding F:

$$F = 180° - (43° + 57°) = 80°.$$

We can now find the other two sides, using the law of sines:

$$\frac{f}{\sin F} = \frac{e}{\sin E}$$

$$\frac{f}{\sin 80°} = \frac{4.56}{\sin 43°} \qquad \text{Substituting}$$

$$f = \frac{4.56 \sin 80°}{\sin 43°} \qquad \text{Solving for } f$$

$$f \approx 6.58;$$

$$\frac{g}{\sin G} = \frac{e}{\sin E}$$

$$\frac{g}{\sin 57°} = \frac{4.56}{\sin 43°} \qquad \text{Substituting}$$

$$g = \frac{4.56 \sin 57°}{\sin 43°} \qquad \text{Solving for } g$$

$$g \approx 5.61.$$

Thus, we have solved the triangle:

$$E = 43°, \qquad e = 4.56,$$
$$F = 80°, \qquad f \approx 6.58,$$
$$G = 57°, \qquad g \approx 5.61.$$

The law of sines is frequently used in determining distances.

EXAMPLE 2   *Rescue Mission.* During a rescue mission, a Marine fighter pilot receives data on an unidentified aircraft from an AWACS plane and is instructed to intercept the aircraft. The diagram shown below appears on the screen, but before the distance to the point of interception appears on the screen, communications are jammed. Fortunately, the pilot remembers the law of sines. How far must the pilot fly?

Solution   We let x represent the distance that the pilot must fly in order to intercept the aircraft and Z represent the point of interception. We first find angle Z:

$$Z = 180° - (115° + 27°)$$
$$= 38°.$$

Because this application involves the ASA situation, we use the law of sines to determine x:

$$\frac{x}{\sin X} = \frac{z}{\sin Z}$$

$$\frac{x}{\sin 115°} = \frac{500}{\sin 38°} \qquad \textbf{Substituting}$$

$$x = \frac{500 \sin 115°}{\sin 38°} \qquad \textbf{Solving for } x$$

$$x \approx 736.$$

Thus the pilot must fly approximately 736 km in order to intercept the unidentified aircraft.   ▬

Solving Triangles (SSA)

When two sides of a triangle and an angle opposite one of them are known, the law of sines can be used to solve the triangle.

Suppose for $\triangle ABC$ that b, c, and B are given. The various possibilities are as shown in the eight cases below: 5 cases when B is acute and 3 cases when B is obtuse. Note that $b < c$ in cases 1, 2, 3, and 6; $b = c$ in cases 4 and 7; and $b > c$ in cases 5 and 8.

Angle B Is Acute

Case 1: No solution
$b < c$; side b is too short to reach the base. No triangle is formed.

Case 2: One solution
$b < c$; side b just reaches the base and is perpendicular to it.

Case 3: Two solutions
$b < c$; an arc of radius b meets the base at two points. (This case is called the **ambiguous case.**)

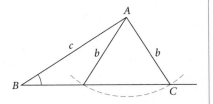

Case 4: One solution
$b = c$; an arc of radius b meets the base at just one point, other than B.

Case 5: One solution
$b > c$; an arc of radius b meets the base at just one point.

Angle B Is Obtuse

Case 6: No solution
$b < c$; side b is too short to reach the base. No triangle is formed.

Case 7: No solution
$b = c$; an arc of radius b meets the base only at point B. No triangle is formed.

Case 8: One solution
$b > c$; an arc of radius b meets the base at just one point.

The eight cases above lead us to three possibilities in the SSA situation: *no* solution, *one* solution, or *two* solutions. Let's investigate these possibilities further, looking for ways to recognize the number of solutions.

EXAMPLE 3   *No solution.* In $\triangle QRS$, $q = 15$, $r = 28$, and $Q = 43.6°$. Solve the triangle.

Solution   We make a drawing and list the known measures.

$$Q = 43.6° \qquad q = 15$$
$$R = ? \qquad r = 28$$
$$S = ? \qquad s = ?$$

We observe the SSA situation and use the law of sines to find R:

$$\frac{q}{\sin Q} = \frac{r}{\sin R}$$

$$\frac{15}{\sin 43.6°} = \frac{28}{\sin R} \qquad \text{Substituting}$$

$$\sin R = \frac{28 \sin 43.6°}{15} \qquad \text{Solving for } \sin R$$

$$\sin R \approx 1.2873.$$

Since there is no angle with a sine greater than 1, there is *no solution.* ▬

EXAMPLE 4   *One solution.* In $\triangle XYZ$, $x = 23.5$, $y = 9.8$, and $X = 39.7°$. Solve the triangle.

Solution   We make a drawing and organize the given information.

$$X = 39.7° \qquad x = 23.5$$
$$Y = ? \qquad y = 9.8$$
$$Z = ? \qquad z = ?$$

We see the SSA situation and begin by finding Y with the law of sines:

$$\frac{x}{\sin X} = \frac{y}{\sin Y}$$

$$\frac{23.5}{\sin 39.7°} = \frac{9.8}{\sin Y} \qquad \text{Substituting}$$

$$\sin Y = \frac{9.8 \sin 39.7°}{23.5} \qquad \text{Solving for } \sin Y$$

$$\sin Y \approx 0.2664.$$

Then $Y = 15.4°$ or $Y = 164.6°$, to the nearest tenth of a degree. An angle of $164.6°$ cannot be an angle of this triangle because it already has an angle of $39.7°$ and these two angles would total more than $180°$. Thus,

15.4° is the only possibility for Y. Therefore,

$$Z \approx 180° - (39.7° + 15.4°) \approx 124.9°.$$

We now find z:

$$\frac{z}{\sin Z} = \frac{x}{\sin X}$$

$$\frac{z}{\sin 124.9°} = \frac{23.5}{\sin 39.7°} \qquad \text{Substituting}$$

$$z = \frac{23.5 \sin 124.9°}{\sin 39.7°} \qquad \text{Solving for } z$$

$$z \approx 30.2.$$

We now have solved the triangle:

$$X = 39.7°, \qquad x = 23.5,$$
$$Y \approx 15.4°, \qquad y = 9.8,$$
$$Z \approx 124.9°, \qquad z \approx 30.2.$$

The next example illustrates the ambiguous case in which there are two possible solutions.

EXAMPLE 5 *Two solutions.* In $\triangle ABC$, $b = 15$, $c = 20$, and $B = 29°$. Solve the triangle.

Solution We make a drawing, list the known measures, and see that we again have the SSA situation.

$$A = ? \qquad a = ?$$
$$B = 29° \qquad b = 15$$
$$C = ? \qquad c = 20$$

We first find C:

$$\frac{b}{\sin B} = \frac{c}{\sin C}$$

$$\frac{15}{\sin 29°} = \frac{20}{\sin C} \qquad \text{Substituting}$$

$$\sin C = \frac{20 \sin 29°}{15} \approx 0.6464. \qquad \text{Solving for } \sin C$$

There are two angles less than 180° with a sine of 0.6464. They are 40° and 140°, to the nearest degree. This gives us two possible solutions.

Possible Solution I.

If $C = 40°$, then

$$A = 180° - (29° + 40°) = 111°.$$

Then we find a:

$$\frac{a}{\sin A} = \frac{b}{\sin B}$$

$$\frac{a}{\sin 111°} = \frac{15}{\sin 29°}$$

$$a = \frac{15 \sin 111°}{\sin 29°} \approx 29.$$

These measures make a triangle as shown below; thus we have a solution.

Possible Solution II.

If $C = 140°$, then

$$A = 180° - (29° + 140°) = 11°.$$

Then we find a:

$$\frac{a}{\sin A} = \frac{b}{\sin B}$$

$$\frac{a}{\sin 11°} = \frac{15}{\sin 29°}$$

$$a = \frac{15 \sin 11°}{\sin 29°} \approx 6.$$

These measures make a triangle as shown below; thus we have a second solution.

Examples 3–5 illustrate the SSA situation. Note that we need not memorize the eight cases or the procedures in finding no solution, one solution, or two solutions. When we are using the law of sines, the sine value leads us directly to the correct solution or solutions.

The Area of a Triangle

The familiar formula for the area of a triangle, $A = \frac{1}{2}bh$, can be used only when h is known. However, we can use the method used to derive the law of sines to derive an area formula that does not involve the height.

Consider a general triangle $\triangle ABC$, with area K, as shown below.

A is acute.

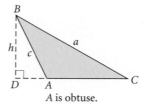

A is obtuse.

In each $\triangle ADB$,

$$\sin A = \frac{h}{c}, \quad \text{or} \quad h = c \sin A.$$

In the second triangle, $\sin A = h/c$ because $\sin A = \sin (180° - A)$. Substituting into the formula $K = \frac{1}{2}bh$, we get

$$K = \frac{1}{2} bc \sin A.$$

Any pair of sides and the included angle could have been used. Thus we also have

$$K = \frac{1}{2} ab \sin C \quad \text{and} \quad K = \frac{1}{2} ac \sin B.$$

The Area of a Triangle

The area K of any $\triangle ABC$ is one half the product of the lengths of two sides and the sine of the included angle:

$$K = \frac{1}{2} bc \sin A = \frac{1}{2} ab \sin C = \frac{1}{2} ac \sin B.$$

EXAMPLE 6 *Area of the Peace Monument.* **Through the Mentoring in the City Program sponsored by Marian College, in Indianapolis, Indiana, children have turned a vacant downtown lot into a monument for peace.[*] This community project brought together neighborhood volunteers, businesses, and government in hopes of showing children how to develop positive nonviolent ways of dealing with conflict. A landscape architect[†] used the children's drawings and ideas to design a triangular-shaped peace garden. Two sides of the property, formed by Indiana Avenue and Senate Avenue, measure 182 ft and 230 ft, respectively, and together form a 44.7° angle. The third side of the garden, formed by an apartment building, measures 163 ft. What is the area of this property?**

Solution Since we do not know a height of the triangle, we use the area formula:

$$K = \frac{1}{2} bc \sin A$$
$$K = \frac{1}{2} \cdot 182 \text{ ft} \cdot 230 \text{ ft} \cdot \sin 44.7°$$
$$K \approx 14{,}722 \text{ ft}^2.$$

The area of the property is approximately 14,722 ft².

---

[*]*The Indianapolis Star,* August 6, 1995, p. J8.
[†]Alan Day, a landscape architect with Browning Day Mullins Dierdorf, Inc., donated his time to this project.

Exercise Set 4.1

Solve the triangle, if possible.

1. $B = 38°$, $C = 21°$, $b = 24$
2. $A = 131°$, $C = 23°$, $b = 10$
3. $A = 36.5°$, $a = 24$, $b = 34$
4. $B = 118.3°$, $C = 45.6°$, $b = 42.1$
5. $C = 61°10'$, $c = 30.3$, $b = 24.2$
6. $A = 126.5°$, $a = 17.2$, $c = 13.5$
7. $c = 3$ mi, $B = 37.48°$, $C = 32.16°$
8. $a = 2345$ mi, $b = 2345$ mi, $A = 124.67°$
9. $b = 56.78$ yd, $c = 56.78$ yd, $C = 83.78°$
10. $A = 129°32'$, $C = 18°28'$, $b = 1204$ in.
11. $a = 20.01$ cm, $b = 10.07$ cm, $A = 30.3°$
12. $b = 4.157$ km, $c = 3.446$ km, $C = 51°48'$
13. $A = 89°$, $a = 15.6$ in., $b = 18.4$ in.
14. $C = 46°32'$, $a = 56.2$ m, $c = 22.1$ m
15. $a = 200$ m, $A = 32.76°$, $C = 21.97°$
16. $B = 115°$, $c = 45.6$ yd, $b = 23.8$ yd

Find the area of the triangle.

17. $B = 42°$, $a = 7.2$ ft, $c = 3.4$ ft
18. $A = 17°12'$, $b = 10$ in., $c = 13$ in.
19. $C = 82°54'$, $a = 4$ yd, $b = 6$ yd
20. $C = 75.16°$, $a = 1.5$ m, $b = 2.1$ m
21. $B = 135.2°$, $a = 46.12$ ft, $c = 36.74$ ft
22. $A = 113°$, $b = 18.2$ cm, $c = 23.7$ cm

Solve. Keep in mind the two types of bearing considered in Sections 2.2 and 2.3.

23. *Area of Back Yard.* A new homeowner has a triangular-shaped back yard. Two of the three sides measure 53 ft and 42 ft and form an included angle of 135°. To determine the amount of fertilizer and grass seed to be purchased, the owner has to know, or at least approximate, the area of the yard. Find the area of the yard to the nearest square foot.

24. *Boarding Stable.* A rancher operates a boarding stable and temporarily needs to make an extra pen. He has a piece of rope 38 ft long and plans to tie the rope to one end of the barn (S) and run the rope around a tree (T) and back to the barn (Q). The tree is 21 ft from where the rope is first tied, and the rope from the barn to the tree makes an angle of 35° with the barn. Does the rancher have enough rope if he allows $4\frac{1}{2}$ ft at each end to fasten the rope?

25. *Rock Concert.* In preparation for an outdoor rock concert, a stage crew must determine how far apart to place the two large speaker columns on stage. What generally works best is to place them at 50° angles to the center of the front row. The distance from the center of the front row to each of the speakers is 10 ft. How far apart does the crew need to place the speakers on stage?

26. *Lunar Crater.* Points A and B are on opposite sides of a lunar crater. Point C is 50 m from A. The measure of $\angle BAC$ is determined to be 112° and the measure of $\angle ACB$ is determined to be 42°. What is the width of the crater?

27. *Length of Pole.* A pole leans away from the sun at an angle of 7° to the vertical. When the angle of elevation of the sun is 51°, the pole casts a shadow 47 ft long on level ground. How long is the pole?

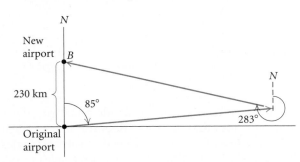

28. *Reconnaissance Airplane.* A reconnaissance airplane leaves its airport on the east coast of the United States and flies in a direction of 085°. Because of bad weather, it returns to another airport 230 km to the north of its home base. For the return trip, it flies in a direction of 283°. What is the total distance that the airplane flew?

29. *Fire Tower.* A ranger in fire tower *A* spots a fire at a direction of 295°. A ranger in fire tower *B*, located 45 mi at a direction of 045° from tower *A*, spots the same fire at a direction of 255°. How far from tower *A* is the fire? from tower *B*?

30. *Lighthouse.* A boat leaves a lighthouse *A* and sails 5.1 km. At this time it is sighted from lighthouse *B*, 7.2 km west of *A*. The bearing of the boat from *B* is N65°10′E. How far is the boat from *B*?

31. *Mackinac Island.* Mackinac Island is located 35 mi N65°20′W of Cheboygan, Michigan, where the Coast Guard cutter Mackinaw is stationed. A freighter in distress radios the Coast Guard cutter for help. It radios its position as N25°40′E of Mackinac Island and N10°10′W of Cheboygan. How far is the freighter from Cheboygan?

32. *Gears.* Three gears are arranged as shown in the figure below. Find the angle *ϕ*.

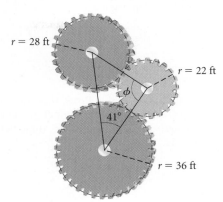

Discussion and Writing

33. Explain why the law of sines cannot be used to find the first angle when solving a triangle given three sides.

34. We considered eight cases of solving triangles given two sides and an angle opposite one of them. Describe the relationship between side *b* and the height *h* in each.

Skill Maintenance

Find the acute angle, A, in both radians and degrees, for the given function value.

35. $\cos A = 0.2213$

36. $\cos A = 1.5612$

Convert to decimal degree notation.

37. 18°14′20″

38. 125°3′42″

Synthesis

39. Prove the following area formulas for a general triangle *ABC* with area represented by *K*.

$$K = \frac{a^2 \sin B \sin C}{2 \sin A}$$

$$K = \frac{c^2 \sin A \sin B}{2 \sin C}$$

$$K = \frac{b^2 \sin C \sin A}{2 \sin B}$$

40. *Area of a Parallelogram.*   Prove that the area of a parallelogram is the product of two sides and the sine of the included angle.

41. *Area of a Quadrilateral.*   Prove that the area of a quadrilateral is one half the product of the lengths of its diagonals and the sine of the angle between the diagonals.

42. Find *d*.

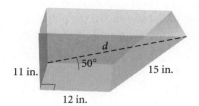

The Law of Cosines

4.2

• *Use the law of cosines to solve triangles.*
• *Determine whether the law of sines or the law of cosines should be applied to solve a triangle.*

The law of sines is used to solve triangles given a side and two angles (AAS and ASA) or given two sides and an angle opposite one of them (SSA). A second law, called the *law of cosines,* is needed to solve triangles given two sides and the included angle (SAS) or given three sides (SSS).

The Law of Cosines

To derive this property, we consider any $\triangle ABC$ placed on a coordinate system. We position the origin at one of the vertices—say, C—and the positive half of the x-axis along one of the sides—say, CB. Let (x, y) be the coordinates of vertex A. Point B has coordinates $(a, 0)$ and point C has coordinates $(0, 0)$.

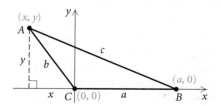

Then

$$\cos C = \frac{x}{b}, \quad \text{so} \quad x = b \cos C$$

and

$$\sin C = \frac{y}{b}, \quad \text{so} \quad y = b \sin C.$$

Thus point A has coordinates

$$(b \cos C, \, b \sin C).$$

Next, we use the distance formula to determine c^2:

$$c^2 = (x - a)^2 + (y - 0)^2,$$

or $\qquad c^2 = (b \cos C - a)^2 + (b \sin C - 0)^2.$

Now we multiply and simplify:

$$\begin{aligned} c^2 &= b^2 \cos^2 C - 2ab \cos C + a^2 + b^2 \sin^2 C \\ &= a^2 + b^2(\sin^2 C + \cos^2 C) - 2ab \cos C \\ &= a^2 + b^2 - 2ab \cos C. \end{aligned}$$

Had we placed the origin at one of the other vertices, we would have obtained

$$a^2 = b^2 + c^2 - 2bc \cos A$$

or $\qquad b^2 = a^2 + c^2 - 2ac \cos B.$

The Law of Cosines
In any triangle ABC,

$$a^2 = b^2 + c^2 - 2bc \cos A,$$
$$b^2 = a^2 + c^2 - 2ac \cos B,$$

or $\quad c^2 = a^2 + b^2 - 2ab \cos C.$

Thus, in any triangle, the square of a side is the sum of the squares of the other two sides, minus twice the product of those sides and the cosine of the included angle.

Solving Triangles (SAS)

When two sides of a triangle and the included angle are known, we can use the law of cosines to find the third side. The law of cosines or the law of sines can then be used to finish solving the triangle.

EXAMPLE 1 Solve $\triangle ABC$ if $a = 32$, $c = 48$, and $B = 125.2°$.

Solution We first label a triangle with the known and unknown measures.

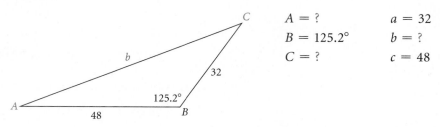

$$A = ? \qquad a = 32$$
$$B = 125.2° \qquad b = ?$$
$$C = ? \qquad c = 48$$

We can find the third side using the law of cosines, as follows:

$$b^2 = a^2 + c^2 - 2ac \cos B$$
$$b^2 = 32^2 + 48^2 - 2 \cdot 32 \cdot 48 \cos 125.2° \qquad \text{Substituting}$$
$$b^2 \approx 5098.8$$
$$b \approx 71.$$

We now have $a = 32$, $b \approx 71$, and $c = 48$, and we need to find the other two angle measures. At this point, we can find them in two ways. One way uses the law of sines. The ambiguous case may arise, however, and we would have to be alert to this possibility. The advantage of using the law of cosines again is that if we solve for the cosine and find that its value is *negative*, then we know that the angle is obtuse. If the value of the cosine is *positive*, then the angle is acute. Thus we use the law of cosines to find a second angle.

Let's find angle A. We select the formula from the law of cosines that contains $\cos A$ and substitute:

$$a^2 = b^2 + c^2 - 2bc \cos A$$
$$32^2 = 71^2 + 48^2 - 2 \cdot 71 \cdot 48 \cos A \qquad \text{Substituting}$$
$$1024 = 5041 + 2304 - 6816 \cos A$$
$$-6321 = -6816 \cos A$$
$$\cos A \approx 0.9273768$$
$$A \approx 22.0°.$$

The third angle is now easy to find:

$$C \approx 180° - (125.2° + 22.0°)$$
$$\approx 32.8°.$$

Thus,

$$A \approx 22.0°, \qquad a = 32,$$
$$B = 125.2°, \qquad b \approx 71,$$
$$C \approx 32.8°, \qquad c = 48.$$

Due to errors created by rounding, answers may vary depending on the order in which they are found. Had we found the measure of angle C first in Example 1, the angle measures would have been $C \approx 34.1°$ and $A \approx 20.7°$. Variances in rounding also change the answers. Had we used 71.4 for b in Example 1, the angle measures would have been $A \approx 21.5°$ and $C \approx 33.3°$.

Suppose we used the law of sines at the outset in Example 1 to find b. We were given only three measures: $a = 32$, $c = 48$, and $B = 125.2°$. When substituting these measures into the proportions, we see that there is not enough information to use the law of sines:

$$\frac{a}{\sin A} = \frac{b}{\sin B} \rightarrow \frac{32}{\sin A} = \frac{b}{\sin 125.2°},$$

$$\frac{b}{\sin B} = \frac{c}{\sin C} \rightarrow \frac{b}{\sin 125.2°} = \frac{48}{\sin C},$$

$$\frac{a}{\sin A} = \frac{c}{\sin C} \rightarrow \frac{32}{\sin A} = \frac{48}{\sin C}.$$

In all three situations, the resulting equation, after the substitutions, still has two unknowns. Thus we cannot use the law of sines to find b.

Solving Triangles (SSS)

When all three sides of a triangle are known, the law of cosines can be used to solve the triangle.

EXAMPLE 2   Solve $\triangle RST$ if $r = 3.5$, $s = 4.7$, and $t = 2.8$.

Solution   We sketch a triangle and label it with the given measures.

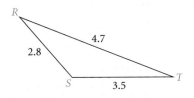

$$R = ? \qquad r = 3.5$$
$$S = ? \qquad s = 4.7$$
$$T = ? \qquad t = 2.8$$

Since we do not know any of the angle measures, we cannot use the law of sines. We begin instead by finding an angle with the law of cosines. We choose to find S first and select the formula that contains $\cos S$:

$$s^2 = r^2 + t^2 - 2rt \cos S$$
$$(4.7)^2 = (3.5)^2 + (2.8)^2 - 2(3.5)(2.8) \cos S \qquad \text{Substituting}$$
$$\cos S = \frac{(3.5)^2 + (2.8)^2 - (4.7)^2}{2(3.5)(2.8)}$$
$$\cos S \approx -0.1020408$$
$$S \approx 95.86°.$$

Similarly, we find angle R:

$$r^2 = s^2 + t^2 - 2st \cos R$$
$$(3.5)^2 = (4.7)^2 + (2.8)^2 - 2(4.7)(2.8) \cos R$$
$$\cos R = \frac{(4.7)^2 + (2.8)^2 - (3.5)^2}{2(4.7)(2.8)}$$
$$\cos R \approx 0.6717325$$
$$R \approx 47.80°.$$

Then

$$T \approx 180° - (95.86° + 47.80°) \approx 36.34°.$$

Thus,

$$R \approx 47.80°, \qquad r = 3.5,$$
$$S \approx 95.86°, \qquad s = 4.7,$$
$$T \approx 36.34°, \qquad t = 2.8.$$

0.5 cm

2 cm   2 cm

EXAMPLE 3 *Knife Bevel.* Knifemakers know that the *bevel* of the blade (the angle formed at the cutting edge of the blade) determines the cutting characteristics of the knife. A small bevel like that of a straight razor makes for a keen edge, but is impractical for heavy-duty cutting because the edge dulls quickly and is prone to chipping. A large bevel is suitable for heavy-duty work like chopping wood. Survival knives, being universal in application, are a compromise between small and large bevels. The diagram at left illustrates the blade of a hand-made Randall Model 18 survival knife. What is its bevel? (*Source:* Randall Made Knives, P.O. Box 1988, Orlando, FL 32802)

Solution   We know three sides of a triangle. We can use the law of cosines to find the bevel, angle A.

$$a^2 = b^2 + c^2 - 2bc \cos A$$
$$(0.5)^2 = 2^2 + 2^2 - 2 \cdot 2 \cdot 2 \cdot \cos A$$
$$0.25 = 4 + 4 - 8 \cos A$$
$$\cos A = \frac{4 + 4 - 0.25}{8}$$
$$\cos A = 0.96875$$
$$A \approx 14.36°.$$

Thus the bevel is approximately 14.36°.

CONNECTING THE CONCEPTS

CHOOSING THE APPROPRIATE LAW

The following summarizes the situations in which to use the law of sines and the law of cosines.

To solve an oblique triangle:

Use the *law of sines* for: Use the *law of cosines* for:

      AAS   SAS

      ASA   SSS

      SSA

The law of cosines can also be used for the SSA situation, but since the process involves solving a quadratic equation, we do not include that option in the list above.

EXAMPLE 4   In $\triangle ABC$, three measures are given. Determine which law to use when solving the triangle. You need not solve the triangle.

a) $a = 14$, $b = 23$, $c = 10$

b) $a = 207$, $B = 43.8°$, $C = 57.6°$

c) $A = 112°$, $C = 37°$, $a = 84.7$

d) $B = 101°$, $a = 960$, $c = 1042$

e) $b = 17.26$, $a = 27.29$, $A = 39°$

f) $A = 61°$, $B = 39°$, $C = 80°$

Solution   It is helpful to make a drawing of a triangle with the given information. The triangle need not be drawn to scale. The given parts are shown in color.

FIGURE	SITUATION	LAW TO USE
a)	SSS	Law of Cosines
b)	ASA	Law of Sines

FIGURE	SITUATION	LAW TO USE
c) 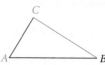	AAS	Law of Sines
d) 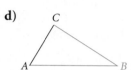	SAS	Law of Cosines
e) 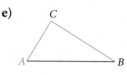	SSA	Law of Sines
f)	AAA	Cannot be solved

Exercise Set 4.2

Solve the triangle, if possible.

1. $A = 30°$, $b = 12$, $c = 24$
2. $B = 133°$, $a = 12$, $c = 15$
3. $a = 12$, $b = 14$, $c = 20$
4. $a = 22.3$, $b = 22.3$, $c = 36.1$
5. $B = 72°40'$, $c = 16$ m, $a = 78$ m
6. $C = 22.28°$, $a = 25.4$ cm, $b = 73.8$ cm
7. $a = 16$ m, $b = 20$ m, $c = 32$ m
8. $B = 72.66°$, $a = 23.78$ km, $c = 25.74$ km
9. $a = 2$ ft, $b = 3$ ft, $c = 8$ ft
10. $A = 96°13'$, $b = 15.8$ yd, $c = 18.4$ yd
11. $a = 26.12$ km, $b = 21.34$ km, $c = 19.25$ km
12. $C = 28°43'$, $a = 6$ mm, $b = 9$ mm
13. $a = 60.12$ mi, $b = 40.23$ mi, $C = 48.7°$

14. $a = 11.2$ cm, $b = 5.4$ cm, $c = 7$ cm
15. $b = 10.2$ in., $c = 17.3$ in., $A = 53.456°$
16. $a = 17$ yd, $b = 15.4$ yd, $c = 1.5$ yd

Determine which law applies. Then solve the triangle.

17. $A = 70°$, $B = 12°$, $b = 21.4$
18. $a = 15$, $c = 7$, $B = 62°$
19. $a = 3.3$, $b = 2.7$, $c = 2.8$
20. $a = 1.5$, $b = 2.5$, $A = 58°$
21. $A = 40.2°$, $B = 39.8°$, $C = 100°$
22. $a = 60$, $b = 40$, $C = 47°$
23. $a = 3.6$, $b = 6.2$, $c = 4.1$
24. $B = 110°30'$, $C = 8°10'$, $c = 0.912$

Solve.

25. *Poachers.*   A park ranger establishes an observation post from which to watch for poachers. Despite losing her map, the ranger does have a compass and a rangefinder. She observes some poachers, and the rangefinder indicates that they are 500 ft from her

position. They are headed toward big game that she knows to be 375 ft from her position. Using her compass, she finds that the poachers' azimuth (the direction measured as an angle from north) is 355° and that of the big game is 42°. What is the distance between the poachers and the game?

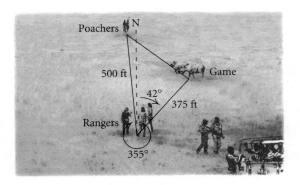

26. *Circus Highwire Act.* A circus highwire act walks up an approach wire to reach a highwire. The approach wire is 122 ft long and is currently anchored so that the angle it forms with the ground is at the maximum of 35°. A greater approach angle causes the aerialists to slip. However, the aerialists find that there is enough room to anchor the approach wire 30 ft back in order to make the approach angle less severe. When this is done, how much farther will they have to walk up the approach wire, and what will the new approach angle be?

27. *In-line Skater.* An in-line skater skates on a fitness trail along the Pacific Ocean from point *A* to point *B*. As shown below, two streets intersecting at point *C* also intersect the trail at *A* and *B*. In his car, the skater found the lengths of *AC* and *BC* to be approximately 0.5 mi and 1.3 mi, respectively. From

a map, he estimates the included angle at *C* to be 110°. How far did he skate from *A* to *B*?

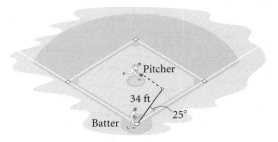

28. *Baseball Bunt.* A batter in a baseball game drops a bunt down the first-base line. It rolls 34 ft at an angle of 25° with the base path. The pitcher's mound is 60.5 ft from home plate. How far must the pitcher travel to pick up the ball? (*Hint:* A baseball diamond is a square.)

29. *Ships.* Two ships leave harbor at the same time. The first sails N15°W at 25 knots (a knot is one nautical mile per hour). The second sails N32°E at 20 knots. After 2 hr, how far apart are the ships?

30. *Survival Trip.* A group of college students is learning to navigate for an upcoming survival trip. On a map, they have been given three points at which they are to check in. The map also shows the distances between the points. However, to navigate they need to know the angle measurements. Calculate the angles for them.

31. *Airplanes.* Two airplanes leave an airport at the same time. The first flies 150 km/h in a direction of 320°. The second flies 200 km/h in a direction of 200°. After 3 hr, how far apart are the planes?

32. *Slow-pitch Softball.* A slow-pitch softball diamond is a square 65 ft on a side. The pitcher's mound is 46 ft from home. How far is it from the pitcher's mound to first base?

33. *Isosceles Trapezoid.* The longer base of an isosceles trapezoid measures 14 ft. The nonparallel sides measure 10 ft, and the base angles measure 80°.

a) Find the length of a diagonal.
b) Find the area.

34. *Area of Sail.* A sail that is in the shape of an isosceles triangle has a vertex angle of 38°. The angle is included by two sides, each measuring 20 ft. Find the area of the sail.

35. Three circles are arranged as shown in the figure below. Find the length *PQ*.

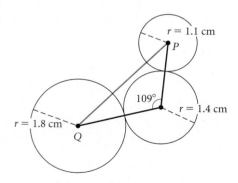

36. *Swimming Pool.* A triangular swimming pool measures 44 ft on one side and 32.8 ft on another side. These sides form an angle that measures 40.8°. How long is the other side?

Discussion and Writing

37. Try to solve this triangle using the law of cosines. Then explain why it is easier to solve it using the law of sines.

38. Explain why we cannot solve a triangle given SAS with the law of sines.

Skill Maintenance

Find the values.

39. $\sin 45°$

40. $\cos \dfrac{\pi}{6}$

41. $\cos \left(-\dfrac{2\pi}{3} \right)$

42. $\sin 300°$

Synthesis

43. *Canyon Depth.* A bridge is being built across a canyon. The length of the bridge is 5045 ft. From the deepest point in the canyon, the angles of elevation of the ends of the bridge are 78° and 72°. How deep is the canyon?

44. *Heron's Formula.* If *a*, *b*, and *c* are the lengths of the sides of a triangle, then the area *K* of the triangle is given by

$$K = \sqrt{s(s-a)(s-b)(s-c)},$$

where $s = \frac{1}{2}(a + b + c)$. The number *s* is called the *semiperimeter*. Prove Heron's formula. (*Hint:* Use the area formula $K = \frac{1}{2}bc \sin A$ developed in Section 4.1. Then use Heron's formula to find the area of the triangular swimming pool described in Exercise 36.)

45. *Area of Isosceles Triangle.* Find a formula for the area of an isosceles triangle in terms of the congruent sides and their included angle. Under what conditions will the area of a triangle with fixed congruent sides be maximum?

46. *Reconnaissance Plane.* A reconnaissance plane patrolling at 5000 ft sights a submarine at bearing 35° and at an angle of depression of 25°. A carrier is at bearing 105° and at an angle of depression of 60°. How far is the submarine from the carrier?

Introduction to Complex Numbers

4.3

• *Perform computations involving complex numbers.*

Some functions have zeros that are not real numbers. In order to find the zeros of such functions, we must consider the **complex-number system.**

The Complex-Number System

Let's begin by considering the function $f(x) = x^2 - 5$. To find the zeros of this function, we solve the associated equation $x^2 - 5 = 0$.

EXAMPLE 1 Solve: $x^2 - 5 = 0$.

Algebraic Solution

We use the principle of square roots:

$$x^2 - 5 = 0$$
$$x^2 = 5 \qquad \text{Adding 5}$$
$$x = \pm\sqrt{5}. \qquad \begin{array}{l}\text{Using the principle} \\ \text{of square roots}\end{array}$$

The solutions of the equation are the real numbers $-\sqrt{5}$ and $\sqrt{5}$, or approximately -2.236 and 2.236.

Graphical Solution

Using the Zero method, we graph the function $f(x) = x^2 - 5$ and use the ZERO feature to find its zeros.

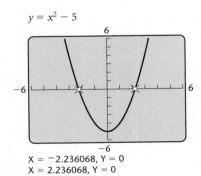

$y = x^2 - 5$

X = −2.236068, Y = 0
X = 2.236068, Y = 0

The zeros of the function $f(x) = x^2 - 5$ are the solutions of the equation $x^2 - 5 = 0$. They are approximately -2.236 and 2.236.

Note that the real solutions found with many graphers are approximations. Now let's consider an equation that has no real-number solutions.

EXAMPLE 2   Solve: $x^2 + 1 = 0$.

Algebraic Solution

We have

$x^2 + 1 = 0$

$\quad x^2 = -1$ Subtracting 1

$\quad x = \pm\sqrt{-1}.$ Using the principle of square roots

There are no real-number square roots of negative numbers. Thus the equation $x^2 + 1 = 0$ has no real-number solutions.

Graphical Solution

We graph the function $f(x) = x^2 + 1$.

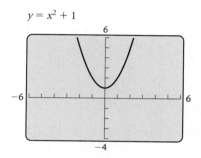

$y = x^2 + 1$

We see that since the graph does not cross the x-axis, it has no x-intercepts. This means that the function $f(x) = x^2 + 1$ has no *real-number* zeros. Thus there are no *real-number* solutions of the equation $x^2 + 1 = 0$.

We can define a number that is a solution of the equation in Example 2.

The Number i

The number i is defined such that

$$i = \sqrt{-1} \quad \text{and} \quad i^2 = -1.$$

To express roots of negative numbers in terms of i, we can use the fact that

$$\sqrt{-p} = \sqrt{-1 \cdot p} = \sqrt{-1}\,\sqrt{p} = i\sqrt{p}$$

when p is a positive real number.

EXAMPLE 3   Express each number in terms of i.

a) $\sqrt{-7}$ b) $\sqrt{-16}$

c) $-\sqrt{-13}$ d) $-\sqrt{-64}$

e) $\sqrt{-48}$

Solution

> **i is *not* under the radical.**

a) $\sqrt{-7} = \sqrt{-1 \cdot 7} = \sqrt{-1} \cdot \sqrt{7} = i\sqrt{7}$, or $\sqrt{7}i$

b) $\sqrt{-16} = \sqrt{-1 \cdot 16} = \sqrt{-1} \cdot \sqrt{16} = i \cdot 4 = 4i$

c) $-\sqrt{-13} = -\sqrt{-1 \cdot 13} = -\sqrt{-1} \cdot \sqrt{13} = -i\sqrt{13}$, or $-\sqrt{13}i$

d) $-\sqrt{-64} = -\sqrt{-1 \cdot 64} = -\sqrt{-1} \cdot \sqrt{64} = -i \cdot 8 = -8i$

e) $\sqrt{-48} = \sqrt{-1 \cdot 48} = \sqrt{-1} \cdot \sqrt{48}$
$$= i\sqrt{16 \cdot 3} = i \cdot 4\sqrt{3} = 4\sqrt{3}i, \text{ or } 4i\sqrt{3}$$

The complex numbers are formed by adding real numbers and multiples of i.

Complex Numbers

A **complex number** is a number of the form $a + bi$, where a and b are real numbers. The number a is said to be the **real part** of $a + bi$ and the number b is said to be the **imaginary part** of $a + bi$.*

Note that either a or b or both can be 0. When $b = 0$, $a + bi = a + 0i = a$, so every real number is a complex number. A complex number like $3 + 4i$ or $17i$, for which $b \neq 0$, is called an **imaginary number.** A complex number like $17i$ or $-4i$, in which $a = 0$ and $b \neq 0$, is sometimes called a **pure imaginary number.** The relationships among various types of complex numbers are shown in the figure below.

---

*Sometimes bi is considered to be the imaginary part.

Addition and Subtraction

The complex numbers obey the commutative, associative, and distributive laws. Thus we can add and subtract them as we do binomials.

EXAMPLE 4   Add or subtract and simplify each of the following.

a) $(8 + 6i) + (3 + 2i)$

b) $(4 + 5i) - (6 - 3i)$

Solution

a) $(8 + 6i) + (3 + 2i) = (8 + 3) + (6i + 2i)$

Collecting the real parts and the imaginary parts

$$= 11 + (6 + 2)i$$
$$= 11 + 8i$$

b) $(4 + 5i) - (6 - 3i) = (4 - 6) + [5i - (-3i)]$

Note that 6 and $-3i$ are both being subtracted.

$$= -2 + 8i$$

Most graphers perform operations on complex numbers as seen in the window below. Some graphers will express a complex number in the form (a, b) rather than $a + bi$.

```
(8+6i)+(3+2i)
                          11+8i
(4+5i)-(6-3i)
                          -2+8i
```

Multiplication

When $\sqrt{a}$ and $\sqrt{b}$ are real numbers, $\sqrt{a} \cdot \sqrt{b} = \sqrt{ab}$, but this is not true when $\sqrt{a}$ and $\sqrt{b}$ are not real numbers. Thus,

$$\sqrt{-2} \cdot \sqrt{-5} = \sqrt{-1} \cdot \sqrt{2} \cdot \sqrt{-1} \cdot \sqrt{5}$$
$$= i\sqrt{2} \cdot i\sqrt{5}$$
$$= i^2\sqrt{10} = -1\sqrt{10} = -\sqrt{10} \quad \text{is correct!}$$

But

$$\sqrt{-2} \cdot \sqrt{-5} = \sqrt{(-2)(-5)} = \sqrt{10} \quad \text{is wrong!}$$

Keeping this and the fact that $i^2 = -1$ in mind, we multiply in much the same way that we do with real numbers.

EXAMPLE 5   Multiply and simplify each of the following.

a) $\sqrt{-16} \cdot \sqrt{-25}$

b) $(1 + 2i)(1 + 3i)$

c) $(3 - 7i)^2$

Solution

$\sqrt{(-16)}\sqrt{(-25)}$

-20

$(1+2i)(1+3i)$

$-5+5i$

$(3-7i)^2$

$-40-42i$

a) $\sqrt{-16} \cdot \sqrt{-25} = \sqrt{-1} \cdot \sqrt{16} \cdot \sqrt{-1} \cdot \sqrt{25}$

$\qquad\qquad\qquad\quad = i \cdot 4 \cdot i \cdot 5$

$\qquad\qquad\qquad\quad = i^2 \cdot 20$

$\qquad\qquad\qquad\quad = -1 \cdot 20 \qquad i^2 = -1$

$\qquad\qquad\qquad\quad = -20$

b) $(1 + 2i)(1 + 3i) = 1 + 3i + 2i + 6i^2$ Multiplying each term of one number by every term of the other (FOIL)

$\qquad\qquad\qquad\quad = 1 + 3i + 2i - 6 \qquad i^2 = -1$

$\qquad\qquad\qquad\quad = -5 + 5i$ Collecting like terms

c) $(3 - 7i)^2 = 3^2 - 2 \cdot 3 \cdot 7i + (7i)^2$ Recall that $(A - B)^2 = A^2 - 2AB + B^2$

$\qquad\qquad\quad = 9 - 42i + 49i^2$

$\qquad\qquad\quad = 9 - 42i - 49 \qquad i^2 = -1$

$\qquad\qquad\quad = -40 - 42i$

Recall that -1 raised to an *even* power is 1, and -1 raised to an *odd* power is -1. Simplifying powers of i can then be done by using the fact that $i^2 = -1$ and expressing the given power of i in terms of i^2. Consider the following:

$i = \sqrt{-1},$ $i^5 = i^4 \cdot i = (i^2)^2 \cdot i = (-1)^2 \cdot i = i,$

$i^2 = -1,$ $i^6 = (i^2)^3 = (-1)^3 = -1,$

$i^3 = i^2 \cdot i = (-1)i = -i,$ $i^7 = (i^2)^3 \cdot i = (-1)^3 \cdot i = -i,$

$i^4 = (i^2)^2 = (-1)^2 = 1,$ $i^8 = (i^2)^4 = (-1)^4 = 1.$

Note that the powers of i cycle through the values i, -1, $-i$, and 1.

EXAMPLE 6   Simplify each of the following.

a) i^{37}

b) i^{58}

c) i^{75}

d) i^{80}

Solution

a) $i^{37} = i^{36} \cdot i = (i^2)^{18} \cdot i = (-1)^{18} \cdot i = 1 \cdot i = i$

b) $i^{58} = (i^2)^{29} = (-1)^{29} = -1$

c) $i^{75} = i^{74} \cdot i = (i^2)^{37} \cdot i = (-1)^{37} \cdot i = -1 \cdot i = -i$

d) $i^{80} = (i^2)^{40} = (-1)^{40} = 1$

Conjugates and Divison

Conjugates of complex numbers are defined as follows.

> ### Conjugate of a Complex Number
> The **conjugate** of a complex number $a + bi$ is $a - bi$. The numbers $a + bi$ and $a - bi$ are **complex conjugates.**

Each of the following pairs of numbers are complex conjugates: $-3 + 7i$ and $-3 - 7i$; $14 - 5i$ and $14 + 5i$; and $8i$ and $-8i$.

The product of a complex number and its conjugate is a real number.

EXAMPLE 7   Multiply each of the following.

a) $(5 + 7i)(5 - 7i)$ **b)** $(8i)(-8i)$

Solution

a) $(5 + 7i)(5 - 7i) = 5^2 - (7i)^2$ Using $(A + B)(A - B) = A^2 - B^2$
$$= 25 - 49i^2$$
$$= 25 - 49(-1)$$
$$= 74$$

b) $(8i)(-8i) = -64i^2$
$$= -64(-1)$$
$$= 64$$

When we are dividing complex numbers, conjugates are used.

EXAMPLE 8   Divide $2 - 5i$ by $1 - 6i$.

Solution   We write fractional notation and then multiply by 1, using the conjugate of the denominator to form the symbol for 1.

$$\frac{2 - 5i}{1 - 6i} = \frac{2 - 5i}{1 - 6i} \cdot \frac{1 + 6i}{1 + 6i}$$ Note that $1 + 6i$ is the conjugate of the divisor.

$$= \frac{(2 - 5i)(1 + 6i)}{(1 - 6i)(1 + 6i)}$$

$$= \frac{2 + 7i - 30i^2}{1 - 36i^2}$$

$$= \frac{2 + 7i + 30}{1 + 36}$$ $i^2 = -1$

$$= \frac{32 + 7i}{37}$$

$$= \frac{32}{37} + \frac{7}{37}i.$$

(5+7i)(5−7i)
         74
(8i)(−8i)
         64

(2−5i)/(1−6i) ▶ Frac
         32/37+7/37i

STUDY TIP

Prepare yourself for your homework assignment by reading the explanations of concepts and following the step-by-step solutions of examples in the text. The time you spend preparing yourself will save you valuable time when you do your assignment.

Exercise Set 4.3

Simplify. Write answers in the form a + bi, where a and b are real numbers.

1. $(-5 + 3i) + (7 + 8i)$

2. $(-6 - 5i) + (9 + 2i)$

3. $(4 - 9i) + (1 - 3i)$

4. $(7 - 2i) + (4 - 5i)$

5. $(3 + \sqrt{-16}) + (2 + \sqrt{-25})$

6. $(7 - \sqrt{-36}) + (2 + \sqrt{-9})$

7. $(10 + 7i) - (5 + 3i)$

8. $(-3 - 4i) - (8 - i)$

9. $(13 + 9i) - (8 + 2i)$

10. $(-7 + 12i) - (3 - 6i)$

11. $(1 + 3i)(1 - 4i)$

12. $(1 - 2i)(1 + 3i)$

13. $(2 + 3i)(2 + 5i)$

14. $(3 - 5i)(8 - 2i)$

15. $7i(2 - 5i)$

16. $3i(6 + 4i)$

17. $(3 + \sqrt{-16})(2 + \sqrt{-25})$

18. $(7 - \sqrt{-16})(2 + \sqrt{-9})$

19. $(5 - 4i)(5 + 4i)$

20. $(5 + 9i)(5 - 9i)$

21. $(4 + 2i)^2$

22. $(5 - 4i)^2$

23. $(-2 + 7i)^2$

24. $(-3 + 2i)^2$

25. $\dfrac{2 + \sqrt{3}i}{5 - 4i}$

26. $\dfrac{\sqrt{5} + 3i}{1 - i}$

27. $\dfrac{4 + i}{-3 - 2i}$

28. $\dfrac{5 - i}{-7 + 2i}$

29. $\dfrac{3}{5 - 11i}$

30. $\dfrac{i}{2 + i}$

31. $\dfrac{1 + i}{(1 - i)^2}$

32. $\dfrac{1 - i}{(1 + i)^2}$

33. $\dfrac{4 - 2i}{1 + i} + \dfrac{2 - 5i}{1 + i}$

34. $\dfrac{3 + 2i}{1 - i} + \dfrac{6 + 2i}{1 - i}$

Simplify.

35. i^{11}

36. i^7

37. i^{35}

38. i^{24}

39. i^{64}

40. i^{42}

41. $(-i)^{71}$

42. $(-i)^6$

43. $(5i)^4$

44. $(2i)^5$

Discussion and Writing

45. Is the sum of two imaginary numbers always an imaginary number? Explain your answer.

46. Is the product of two imaginary numbers always an imaginary number? Explain your answer.

Skill Maintenance

47. Write a slope–intercept equation for the line containing the point $(3, -5)$ and parallel to the line $3x - 6y = 7$.

Given that $f(x) = x^2 + 4$ and $g(x) = 3x + 5$, find each of the following.

48. $(f - g)(x)$

49. $(f/g)(2)$

50. For the function $f(x) = x^2 - 3x + 4$, construct and simplify the difference quotient

$$\frac{f(x + h) - f(x)}{h}.$$

Synthesis

Determine whether each of the following is true or false.

51. The sum of two numbers that are conjugates of each other is always a real number.

52. The conjugate of a sum is the sum of the conjugates of the individual complex numbers.

53. The conjugate of a product is the product of the conjugates of the individual complex numbers.

Let $z = a + bi$ and $\bar{z} = a - bi$.

54. Find a general expression for $1/z$.

55. Find a general expression for $z\bar{z}$.

56. Solve $z + 6\bar{z} = 7$ for z.

Complex Numbers: Trigonometric Form

4.4

- *Graph complex numbers.*
- *Given a complex number in standard form, find trigonometric, or polar, notation; and given a complex number in trigonometric form, find standard notation.*
- *Use trigonometric notation to multiply and divide complex numbers.*
- *Use DeMoivre's theorem to raise complex numbers to powers.*
- *Find the nth roots of a complex number.*

Graphical Representation

Just as real numbers can be graphed on a line, complex numbers can be graphed on a plane. We graph a complex number $a + bi$ in the same way that we graph an ordered pair of real numbers (a, b). However, in place of an x-axis, we have a real axis, and in place of a y-axis, we have an imaginary axis. Horizontal distances correspond to the real part of a number. Vertical distances correspond to the imaginary part.

COMPLEX NUMBERS
REVIEW SECTION 4.3.

EXAMPLE 1   Graph each of the following complex numbers.

a) $3 + 2i$ **b)** $-4 - 5i$ **c)** $-3i$

d) $-1 + 3i$ **e)** 2

Solution

We recall that the absolute value of a real number is its distance from 0 on the number line. The absolute value of a complex number is its distance from the origin in the complex plane. For example, if $z = a + bi$, then using the distance formula, we have

$$|z| = |a + bi| = \sqrt{(a - 0)^2 + (b - 0)^2} = \sqrt{a^2 + b^2}.$$

Absolute Value of a Complex Number

The **absolute value of a complex number** $a + bi$ is

$$|a + bi| = \sqrt{a^2 + b^2}.$$

EXAMPLE 2   Find the absolute value of each of the following.

a) $3 + 4i$ **b)** $-2 - i$ **c)** $\dfrac{4}{5}i$

Solution

a) $|3 + 4i| = \sqrt{3^2 + 4^2} = \sqrt{9 + 16} = \sqrt{25} = 5$

b) $|-2 - i| = \sqrt{(-2)^2 + (-1)^2} = \sqrt{5}$

c) $\left|\dfrac{4}{5}i\right| = \left|0 + \dfrac{4}{5}i\right| = \sqrt{0^2 + \left(\dfrac{4}{5}\right)^2} = \dfrac{4}{5}$

```
abs(3+4i)
                        5
abs(−2−i)
                2.236067977
abs(4/5i)
                        .8
```

We can check these results using a grapher as shown at left. Note that $\sqrt{5} \approx 2.236067977$ and $4/5 = 0.8$.

Trigonometric Notation for Complex Numbers

Now let's consider a nonzero complex number $a + bi$. Suppose that its absolute value is r. If we let θ be an angle in standard position whose terminal side passes through the point (a, b), as shown in the figure, then

$$\cos \theta = \frac{a}{r}, \quad \text{or} \quad a = r \cos \theta$$

and

$$\sin \theta = \frac{b}{r}, \quad \text{or} \quad b = r \sin \theta.$$

Substituting these values for a and b into the $(a + bi)$ notation, we get

$$a + bi = r \cos \theta + (r \sin \theta)i$$
$$= r(\cos \theta + i \sin \theta).$$

This is **trigonometric notation** for a complex number $a + bi$. The number r is called the **absolute value** of $a + bi$, and θ is called the **argument** of $a + bi$. Trigonometric notation for a complex number is also called **polar notation**.

Trigonometric Notation for Complex Numbers
$$a + bi = r(\cos \theta + i \sin \theta)$$

To find trigonometric notation for a complex number given in **standard notation,** $a + bi$, we must find r and determine the angle θ for which $\sin \theta = b/r$ and $\cos \theta = a/r$.

EXAMPLE 3 Find trigonometric notation for each of the following complex numbers.

a) $1 + i$ **b)** $\sqrt{3} - i$

Solution

a) We note that $a = 1$ and $b = 1$. Then

$$r = \sqrt{a^2 + b^2} = \sqrt{1^2 + 1^2} = \sqrt{2},$$

$$\sin \theta = \frac{b}{r} = \frac{1}{\sqrt{2}}, \quad \text{or} \quad \frac{\sqrt{2}}{2},$$

and

$$\cos \theta = \frac{1}{\sqrt{2}}, \quad \text{or} \quad \frac{\sqrt{2}}{2}.$$

Thus, $\theta = \pi/4$, or 45°, and we have

$$1 + i = \sqrt{2}\left(\cos \frac{\pi}{4} + i \sin \frac{\pi}{4}\right),$$

or

$$1 + i = \sqrt{2}(\cos 45° + i \sin 45°).$$

b) We see that $a = \sqrt{3}$ and $b = -1$. Then

$$r = \sqrt{(\sqrt{3})^2 + (-1)^2} = 2,$$

$$\sin \theta = \frac{-1}{2} = -\frac{1}{2},$$

and

$$\cos \theta = \frac{\sqrt{3}}{2}.$$

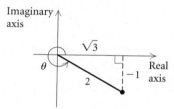

Thus, $\theta = 11\pi/6$, or 330°, and we have

$$\sqrt{3} - i = 2\left(\cos \frac{11\pi}{6} + i \sin \frac{11\pi}{6}\right),$$

or

$$\sqrt{3} - i = 2(\cos 330° + i \sin 330°).$$

As shown at left, a grapher (in DEGREE mode) can be used to determine angle values in degrees.

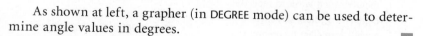

In changing to trigonometric notation, note that there are many angles satisfying the given conditions. We ordinarily choose the *smallest positive* angle.

To change from trigonometric notation to standard notation, $a + bi$, we recall that $a = r \cos \theta$ and $b = r \sin \theta$.

EXAMPLE 4   Find standard notation, $a + bi$, for each of the following complex numbers.

a) $2(\cos 120° + i \sin 120°)$

b) $\sqrt{8}\left(\cos \dfrac{7\pi}{4} + i \sin \dfrac{7\pi}{4}\right)$

Solution

a) Rewriting, we have

$$2(\cos 120° + i \sin 120°) = 2 \cos 120° + (2 \sin 120°)i.$$

Thus,

$$a = 2 \cos 120° = 2 \cdot \left(-\frac{1}{2}\right) = -1$$

and

$$b = 2 \sin 120° = 2 \cdot \frac{\sqrt{3}}{2} = \sqrt{3},$$

so

$$2(\cos 120° + i \sin 120°) = -1 + \sqrt{3}i.$$

Degree Mode

```
2(cos(120)+isin(120))
              -1+1.732050808i
```

b) Rewriting, we have

$$\sqrt{8}\left(\cos \frac{7\pi}{4} + i \sin \frac{7\pi}{4}\right) = \sqrt{8} \cos \frac{7\pi}{4} + \left(\sqrt{8} \sin \frac{7\pi}{4}\right)i.$$

Thus,

$$a = \sqrt{8} \cos \frac{7\pi}{4} = \sqrt{8} \cdot \frac{\sqrt{2}}{2} = 2$$

and

$$b = \sqrt{8} \sin \frac{7\pi}{4} = \sqrt{8} \cdot \left(-\frac{\sqrt{2}}{2}\right) = -2,$$

so

$$\sqrt{8}\left(\cos \frac{7\pi}{4} + i \sin \frac{7\pi}{4}\right) = 2 - 2i.$$

Radian Mode

```
√(8)(cos(7π/4)+isin(7π/4))
                    2−2i
```

Multiplication and Division with Trigonometric Notation

Multiplication of complex numbers is easier to manage with trigonometric notation than with standard notation. We simply multiply the absolute values and add the arguments. Let's state this in a more formal manner.

Complex Numbers: Multiplication

For any complex numbers $r_1(\cos \theta_1 + i \sin \theta_1)$ and $r_2(\cos \theta_2 + i \sin \theta_2)$,

$$r_1(\cos \theta_1 + i \sin \theta_1) \cdot r_2(\cos \theta_2 + i \sin \theta_2)$$
$$= r_1 r_2 [\cos (\theta_1 + \theta_2) + i \sin (\theta_1 + \theta_2)].$$

PROOF

$$r_1(\cos \theta_1 + i \sin \theta_1) \cdot r_2(\cos \theta_2 + i \sin \theta_2) =$$
$$r_1 r_2(\cos \theta_1 \cos \theta_2 - \sin \theta_1 \sin \theta_2) + r_1 r_2(\sin \theta_1 \cos \theta_2 + \cos \theta_1 \sin \theta_2)i$$

Now, using identities for sums of angles, we simplify, obtaining

$$r_1 r_2 \cos (\theta_1 + \theta_2) + r_1 r_2 \sin (\theta_1 + \theta_2)i,$$

or

$$r_1 r_2 [\cos (\theta_1 + \theta_2) + i \sin (\theta_1 + \theta_2)],$$

which was to be shown.

EXAMPLE 5 Multiply and express the answer to each of the following in standard notation.

a) $3(\cos 40° + i \sin 40°)$ and $4(\cos 20° + i \sin 20°)$

b) $2(\cos \pi + i \sin \pi)$ and $3\left[\cos \left(-\dfrac{\pi}{2}\right) + i \sin \left(-\dfrac{\pi}{2}\right)\right]$

Solution

a) $3(\cos 40° + i \sin 40°) \cdot 4(\cos 20° + i \sin 20°)$

$$= 3 \cdot 4 \cdot [\cos (40° + 20°) + i \sin (40° + 20°)]$$
$$= 12(\cos 60° + i \sin 60°)$$
$$= 12\left(\frac{1}{2} + \frac{\sqrt{3}}{2}i\right)$$
$$= 6 + 6\sqrt{3}i$$

Degree Mode

3(cos(40)+isin(40))*4(cos(20)+isin(20))

6+10.39230485i

b) $2(\cos \pi + i \sin \pi) \cdot 3\left[\cos \left(-\dfrac{\pi}{2}\right) + i \sin \left(-\dfrac{\pi}{2}\right)\right]$

$$= 2 \cdot 3 \cdot \left[\cos \left(\pi + \left(-\frac{\pi}{2}\right)\right) + i \sin \left(\pi + \left(-\frac{\pi}{2}\right)\right)\right]$$
$$= 6\left(\cos \frac{\pi}{2} + i \sin \frac{\pi}{2}\right)$$
$$= 6(0 + i \cdot 1)$$
$$= 6i$$

Radian Mode

2(cos(π)+isin(π))*3(cos(−π/2)+isin(−π/2))

6i

EXAMPLE 6   Convert to trigonometric notation and multiply:

$$(1 + i)(\sqrt{3} - i).$$

Solution   We first find trigonometric notation:

$$1 + i = \sqrt{2}(\cos 45° + i \sin 45°),$$ See Example 3(a).
$$\sqrt{3} - i = 2(\cos 330° + i \sin 330°).$$ See Example 3(b).

Then we multiply:

$$\sqrt{2}(\cos 45° + i \sin 45°) \cdot 2(\cos 330° + i \sin 330°)$$
$$= 2\sqrt{2}[\cos (45° + 330°) + i \sin (45° + 330°)]$$
$$= 2\sqrt{2}(\cos 375° + i \sin 375°)$$
$$= 2\sqrt{2}(\cos 15° + i \sin 15°). \qquad \text{375° has the same terminal side as 15°.}$$

To divide complex numbers, we divide the absolute values and subtract the arguments. We state this fact below, but omit the proof.

Complex Numbers: Division

For any complex numbers $r_1(\cos \theta_1 + i \sin \theta_1)$ and $r_2(\cos \theta_2 + i \sin \theta_2)$, $r_2 \neq 0$,

$$\frac{r_1(\cos \theta_1 + i \sin \theta_1)}{r_2(\cos \theta_2 + i \sin \theta_2)} = \frac{r_1}{r_2}[\cos (\theta_1 - \theta_2) + i \sin (\theta_1 - \theta_2)].$$

EXAMPLE 7   Divide

$$2\left(\cos \frac{3\pi}{2} + i \sin \frac{3\pi}{2}\right) \quad \text{by} \quad 4\left(\cos \frac{\pi}{2} + i \sin \frac{\pi}{2}\right)$$

and express the solution in standard notation.

Solution   We have

$$\frac{2\left(\cos \dfrac{3\pi}{2} + i \sin \dfrac{3\pi}{2}\right)}{4\left(\cos \dfrac{\pi}{2} + i \sin \dfrac{\pi}{2}\right)}$$

$$= \frac{2}{4}\left[\cos \left(\frac{3\pi}{2} - \frac{\pi}{2}\right) + i \sin \left(\frac{3\pi}{2} - \frac{\pi}{2}\right)\right]$$

$$= \frac{1}{2}(\cos \pi + i \sin \pi)$$

$$= \frac{1}{2}(-1 + i \cdot 0) = -\frac{1}{2}.$$

EXAMPLE 8   Convert to trigonometric notation and divide:

$$\frac{1 + i}{1 - i}.$$

Solution   We first convert to trigonometric notation:

$$1 + i = \sqrt{2}(\cos 45° + i \sin 45°), \qquad \text{See Example 3(a).}$$
$$1 - i = \sqrt{2}(\cos 315° + i \sin 315°).$$

We now divide:

$$\frac{\sqrt{2}(\cos 45° + i \sin 45°)}{\sqrt{2}(\cos 315° + i \sin 315°)}$$

$$= 1[\cos (45° - 315°) + i \sin (45° - 315°)]$$
$$= \cos (-270°) + i \sin (-270°)$$
$$= 0 + i \cdot 1 = i.$$

(1+i)/(1−i)
i

Powers of Complex Numbers

An important theorem about powers and roots of complex numbers is named for the French mathematician Abraham DeMoivre (1667–1754). Let's consider the square of a complex number $r(\cos \theta + i \sin \theta)$:

$$[r(\cos \theta + i \sin \theta)]^2 = [r(\cos \theta + i \sin \theta)] \cdot [r(\cos \theta + i \sin \theta)]$$
$$= r \cdot r \cdot [\cos (\theta + \theta) + i \sin (\theta + \theta)]$$
$$= r^2(\cos 2\theta + i \sin 2\theta).$$

Similarly, we see that

$$[r(\cos \theta + i \sin \theta)]^3$$
$$= r \cdot r \cdot r \cdot [\cos (\theta + \theta + \theta) + i \sin (\theta + \theta + \theta)]$$
$$= r^3(\cos 3\theta + i \sin 3\theta).$$

DeMoivre's theorem is the generalization of these results.

DeMoivre's Theorem

For any complex number $r(\cos \theta + i \sin \theta)$ and any natural number n,

$$[r(\cos \theta + i \sin \theta)]^n = r^n(\cos n\theta + i \sin n\theta).$$

EXAMPLE 9   Find each of the following.

a) $(1 + i)^9$

b) $(\sqrt{3} - i)^{10}$

Solution

a) We first find trigonometric notation:

$$1 + i = \sqrt{2}(\cos 45° + i \sin 45°).$$

Then

$$(1 + i)^9 = [\sqrt{2}(\cos 45° + i \sin 45°)]^9$$

$$= (\sqrt{2})^9[\cos (9 \cdot 45°) + i \sin (9 \cdot 45°)] \qquad \text{DeMoivre's theorem}$$

$$= 2^{9/2}(\cos 405° + i \sin 405°)$$

$$= 16\sqrt{2}(\cos 45° + i \sin 45°) \qquad \begin{array}{l}\text{405° has the same} \\ \text{terminal side as 45°.}\end{array}$$

$$= 16\sqrt{2}\left(\frac{\sqrt{2}}{2} + i\frac{\sqrt{2}}{2}\right)$$

$$= 16 + 16i.$$

b) We first convert to trigonometric notation:

$$\sqrt{3} - i = 2(\cos 330° + i \sin 330°).$$

Then

$$(\sqrt{3} - i)^{10} = [2(\cos 330° + i \sin 330°)]^{10}$$

$$= 2^{10}(\cos 3300° + i \sin 3300°)$$

$$= 1024(\cos 60° + i \sin 60°) \qquad \begin{array}{l}\text{3300° has the same} \\ \text{terminal side as 60°.}\end{array}$$

$$= 1024\left(\frac{1}{2} + i\frac{\sqrt{3}}{2}\right)$$

$$= 512 + 512\sqrt{3}i.$$

```
(1+i)^9
                    16+16i
(√(3)−i)^10
          512+886.8100135i
```

Roots of Complex Numbers

As we will see, every nonzero complex number has two square roots. A nonzero complex number has three cube roots, four fourth roots, and so on. In general, a nonzero complex number has n different nth roots. They can be found using the formula that we now state but do not prove.

Roots of Complex Numbers

The nth roots of a complex number $r(\cos \theta + i \sin \theta)$, $r \neq 0$, are given by

$$r^{1/n}\left[\cos\left(\frac{\theta}{n} + k \cdot \frac{360°}{n}\right) + i \sin\left(\frac{\theta}{n} + k \cdot \frac{360°}{n}\right)\right],$$

where $k = 0, 1, 2, \ldots, n - 1$.

EXAMPLE 10   Find the square roots of $2 + 2\sqrt{3}i$.

Solution   We first find trigonometric notation:

$$2 + 2\sqrt{3}i = 4(\cos 60° + i \sin 60°).$$

Then

$$[4(\cos 60° + i \sin 60°)]^{1/2}$$

$$= 4^{1/2}\left[\cos\left(\frac{60°}{2} + k \cdot \frac{360°}{2}\right) + i \sin\left(\frac{60°}{2} + k \cdot \frac{360°}{2}\right)\right], \quad k = 0, 1$$

$$= 2[\cos(30° + k \cdot 180°) + i \sin(30° + k \cdot 180°), \quad k = 0, 1.$$

Thus the roots are

$$2(\cos 30° + i \sin 30°) \text{ for } k = 0$$

and

$$2(\cos 210° + i \sin 210°) \text{ for } k = 1,$$

or

$$\sqrt{3} + i \quad \text{and} \quad -\sqrt{3} - i.$$

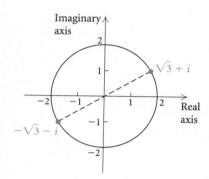

In Example 10, we see that the two square roots of the number are opposites of each other. We can illustrate this graphically. We also note that the roots are equally spaced about a circle of radius *r*—in this case, $r = 2$. The roots are $360°/2$, or $180°$ apart.

EXAMPLE 11   Find the cube roots of 1. Then locate them on a graph.

Solution   We begin by finding trigonometric notation:

$$1 = 1(\cos 0° + i \sin 0°).$$

Then

$$[1(\cos 0° + i \sin 0°)]^{1/3}$$

$$= 1^{1/3}\left[\cos\left(\frac{0°}{3} + k \cdot \frac{360°}{3}\right) + i \sin\left(\frac{0°}{3} + k \cdot \frac{360°}{3}\right)\right], \quad k = 0, 1, 2.$$

The roots are

$$1(\cos 0° + i \sin 0°), \quad 1(\cos 120° + i \sin 120°),$$

and

$$1(\cos 240° + i \sin 240°),$$

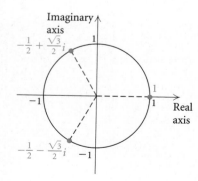

or

$$1, \quad -\frac{1}{2} + \frac{\sqrt{3}}{2}i, \quad \text{and} \quad -\frac{1}{2} - \frac{\sqrt{3}}{2}i.$$

The graphs of the cube roots lie equally spaced about a circle of radius $r = 1$. The roots are $360°/3$, or $120°$ apart.

The nth roots of 1 are often referred to as the **nth roots of unity.** In Example 11, we found the cube roots of unity.

Interactive Discovery

Using a grapher in PARAMETRIC mode, we can approximate the nth roots of a number p. We use the following window and let

$$X_{1T} = (p^\wedge(1/n)) \cos T \quad \text{and} \quad Y_{1T} = (p^\wedge(1/n)) \sin T.$$

WINDOW

  Tmin = 0

  Tmax = 360, if in degree mode, or
      $= 2\pi$, if in radian mode

  Tstep = $360/n$, or $2\pi/n$

  Xmin = -3, Xmax = 3, Xscl = 1

  Ymin = -2, Ymax = 2, Yscl = 1

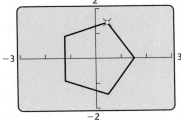

$X_{1T} = 8^{1/5} \cos T$
$Y_{1T} = 8^{1/5} \sin T$

$T = 72$
$x = .46838218, \; y = 1.4415321$

To find the fifth roots of 8, enter $X_{1T} = (8^\wedge(1/5)) \cos T$ and $Y_{1T} = (8^\wedge(1/5)) \sin T$. In this case, use DEGREE mode. After the graph has been generated, use the TRACE feature to locate the fifth roots. The T, X, and Y values appear on the screen. What do they represent?

Three of the fifth roots of 8 are approximately

$$1.5157, \quad 0.46838 + 1.44153i, \quad \text{and} \quad -1.22624 + 0.89092i.$$

Find the other two. Then use a grapher to approximate the cube roots of unity that were found in Example 11. Also approximate the fourth roots of 5 and the tenth roots of unity.

Exercise Set **4.4**

Graph the complex number and find its absolute value.

1. $4 + 3i$

2. $-2 - 3i$

3. i

4. $-5 - 2i$

5. $4 - i$

6. $6 + 3i$

7. 3

8. $-2i$

Express the indicated number in both standard notation and trigonometric notation.

9.

10.

11.

12.

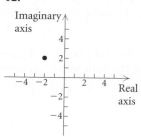

Find trigonometric notation.

13. $1 - i$

14. $-10\sqrt{3} + 10i$

15. $-3i$

16. $-5 + 5i$

17. $\sqrt{3} + i$

18. 4

19. $\dfrac{2}{5}$

20. $7.5i$

Find standard notation, $a + bi$.

21. $3(\cos 30° + i \sin 30°)$

22. $6(\cos 120° + i \sin 120°)$

23. $10(\cos 270° + i \sin 270°)$

24. $3(\cos 0° + i \sin 0°)$

25. $\sqrt{8}\left(\cos \dfrac{\pi}{4} + i \sin \dfrac{\pi}{4}\right)$

26. $5\left(\cos \dfrac{\pi}{3} + i \sin \dfrac{\pi}{3}\right)$

27. $2\left(\cos \dfrac{\pi}{2} + i \sin \dfrac{\pi}{2}\right)$

28. $3\left[\cos \left(-\dfrac{3\pi}{4}\right) + i \sin \left(-\dfrac{3\pi}{4}\right)\right]$

Multiply or divide and leave the answer in trigonometric notation.

29. $\dfrac{12(\cos 48° + i \sin 48°)}{3(\cos 6° + i \sin 6°)}$

30. $5\left(\cos \dfrac{\pi}{3} + i \sin \dfrac{\pi}{3}\right) \cdot 2\left(\cos \dfrac{\pi}{4} + i \sin \dfrac{\pi}{4}\right)$

31. $2.5(\cos 35° + i \sin 35°) \cdot 4.5(\cos 21° + i \sin 21°)$

32. $\dfrac{\dfrac{1}{2}\left(\cos \dfrac{2\pi}{3} + i \sin \dfrac{2\pi}{3}\right)}{\dfrac{3}{8}\left(\cos \dfrac{\pi}{6} + i \sin \dfrac{\pi}{6}\right)}$

Convert to trigonometric notation and then multiply or divide.

33. $(1 - i)(2 + 2i)$

34. $(1 + i\sqrt{3})(1 + i)$

35. $\dfrac{1 - i}{1 + i}$

36. $\dfrac{1 - i}{\sqrt{3} - i}$

37. $(3\sqrt{3} - 3i)(2i)$

38. $(2\sqrt{3} + 2i)(2i)$

39. $\dfrac{2\sqrt{3} - 2i}{1 + \sqrt{3}i}$

40. $\dfrac{3 - 3\sqrt{3}i}{\sqrt{3} - i}$

Raise the number to the given power and write trigonometric notation for the answer.

41. $\left[2\left(\cos \dfrac{\pi}{3} + i \sin \dfrac{\pi}{3}\right)\right]^3$

42. $[2(\cos 120° + i \sin 120°)]^4$

43. $(1 + i)^6$

44. $(-\sqrt{3} + i)^5$

Raise the number to the given power and write standard notation for the answer.

45. $[3(\cos 20° + i \sin 20°)]^3$

46. $[2(\cos 10° + i \sin 10°)]^9$

47. $(1 - i)^5$

48. $(2 + 2i)^4$

49. $\left(\dfrac{1}{\sqrt{2}} - \dfrac{1}{\sqrt{2}}i\right)^{12}$

50. $\left(\dfrac{\sqrt{3}}{2} + \dfrac{1}{2}i\right)^{10}$

Find the square roots of the number.

51. $-i$

52. $1 + i$

53. $2\sqrt{2} - 2\sqrt{2}i$

54. $-\sqrt{3} - i$

Find the cube roots of the number.

55. i

56. $-64i$

57. $2\sqrt{3} - 2i$

58. $1 - \sqrt{3}i$

59. Find and graph the fourth roots of 16.

60. Find and graph the fourth roots of i.

61. Find and graph the fifth roots of -1.

62. Find and graph the sixth roots of 1.

63. Find the tenth roots of 8.

64. Find the ninth roots of -4.

65. Find the sixth roots of -1.

66. Find the fourth roots of 12.

Find all the complex solutions of the equation.

67. $x^3 = 1$

68. $x^5 - 1 = 0$

69. $x^4 + i = 0$

70. $x^4 + 81 = 0$

71. $x^6 + 64 = 0$

72. $x^5 + \sqrt{3} + i = 0$

Discussion and Writing

73. Find and graph the square roots of $1 - i$. Explain geometrically why they are the opposites of each other.

74. Explain why trigonometric notation for a complex number is not unique, but rectangular, or standard, notation is unique.

Skill Maintenance

Convert to degree measure.

75. $\dfrac{\pi}{12}$ **76.** 3π

Convert to radian measure.

77. $330°$ **78.** $-225°$

Find the slope and the y-intercept of the line with the given equation.

79. $-\frac{1}{5}x - y = 15$ **80.** $y = 7$

Synthesis

Solve.

81. $x^2 + (1 - i)x + i = 0$

82. $3x^2 + (1 + 2i)x + 1 - i = 0$

83. Find polar notation for $(\cos \theta + i \sin \theta)^{-1}$.

84. Show that for any complex number z,
$$|z| = |-z|.$$
(*Hint*: Let $z = a + bi$ and $\bar{z} = a - bi$.)

85. Show that for any complex number z,
$$|z| = |\bar{z}|.$$
(*Hint*: Let $z = a + bi$ and $\bar{z} = a - bi$.)

86. Show that for any complex number z,
$$|z\bar{z}| = |z^2|.$$

87. Show that for any complex number z,
$$|z^2| = |z|^2.$$

88. Show that for any complex numbers z and w,
$$|z \cdot w| = |z| \cdot |w|.$$
(*Hint*: Let $z = r_1(\cos \theta_1 + i \sin \theta_1)$ and $w = r_2(\cos \theta_2 + i \sin \theta_2)$.)

89. Show that for any complex number z and any nonzero, complex number w,
$$\left|\frac{z}{w}\right| = \frac{|z|}{|w|}. \quad \text{(Use the hint for Exercise 88.)}$$

90. On a complex plane, graph $|z| = 1$.

91. On a complex plane, graph $z + \bar{z} = 3$.

Vectors and Applications

4.5

- *Determine whether two vectors are equivalent.*
- *Find the sum, or resultant, of two vectors.*
- *Resolve a vector into its horizontal and vertical components.*
- *Solve applied problems involving vectors.*

We measure some quantities using only their magnitudes. For example, we describe time, length, and mass using units like seconds, feet, and kilograms, respectively. However, to measure quantities like **displacement**, **velocity**, or **force**, we need to describe a *magnitude* and a *direction*. Together magnitude and direction describe a **vector**. The following are some examples.

DISPLACEMENT   An object moves a certain distance in a certain direction.

A surveyor steps 20 yd to the northeast.

A hiker follows a trail 5 mi to the west.

A batter hits a ball 100 m along the left-field line.

VELOCITY   An object travels at a certain speed in a certain direction.

A breeze is blowing 15 mph from the northwest.

An airplane is traveling 450 km/h in a direction of 243°.

FORCE   A push or pull is exerted on an object in a certain direction.

A force of 200 lb is required to pull a cart up a 30° incline.

A 25-lb force is required to lift a box upward.

A force of 15 newtons is exerted downward on the handle of a jack. (A newton, abbreviated N, is a unit of force used in physics, and 1 N ≈ 0.22 lb.)

Vectors

Vectors can be graphically represented by directed line segments. The length is chosen, according to some scale, to represent the **magnitude of the vector,** and the direction of the directed line segment represents the **direction of the vector.** For example, if we let 1 cm represent 5 km/h, then a 15-km/h wind from the northwest would be represented by a directed line segment 3 cm long, as shown in the figure at left.

Vector

A **vector** in the plane is a directed line segment. Two vectors are **equivalent** if they have the same *magnitude* and *direction*.

Consider a vector drawn from point *A* to point *B*. Point *A* is called the **initial point** of the vector, and point *B* is called the **terminal point.** Symbolic notation for this vector is $\overrightarrow{AB}$ (read "vector *AB*"). Vectors are also denoted by boldface letters such as **u**, **v**, and **w**. The four vectors in the figure at left have the *same* length and direction. Thus they represent **equivalent** vectors; that is,

$$\overrightarrow{AB} = \overrightarrow{CD} = \overrightarrow{OP} = \mathbf{v}.$$

In the context of vectors, we use = to mean equivalent.

The length, or **magnitude**, of $\overrightarrow{AB}$ is expressed as $|\overrightarrow{AB}|$. In order to determine whether vectors are equivalent, we find their magnitudes and directions.

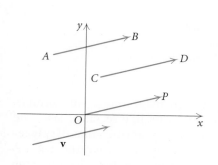

EXAMPLE 1   The vectors **u**, $\overrightarrow{OR}$, and **w** are shown in the figure below. Show that $\mathbf{u} = \overrightarrow{OR} = \mathbf{w}$.

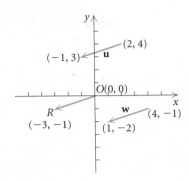

Solution   We first find the length of each vector using the distance formula.

$$|\mathbf{u}| = \sqrt{[2 - (-1)]^2 + (4 - 3)^2} = \sqrt{9 + 1} = \sqrt{10},$$
$$|\overrightarrow{OR}| = \sqrt{[0 - (-3)]^2 + [0 - (-1)]^2} = \sqrt{9 + 1} = \sqrt{10},$$
$$|\mathbf{w}| = \sqrt{(4 - 1)^2 + [-1 - (-2)]^2} = \sqrt{9 + 1} = \sqrt{10}.$$

Thus

$$|\mathbf{u}| = |\overrightarrow{OR}| = |\mathbf{w}|.$$

The vectors **u**, $\overrightarrow{OR}$, and **w** appear to go in the same direction so we check their slopes. If the lines that they are on all have the same slope, the vectors have the same direction. We calculate the slopes:

$$\begin{array}{ccc} \mathbf{u} & \overrightarrow{OR} & \mathbf{w} \\ \end{array}$$

$$\text{Slope} = \frac{4 - 3}{2 - (-1)} = \frac{0 - (-1)}{0 - (-3)} = \frac{-1 - (-2)}{4 - 1} = \frac{1}{3}.$$

Since **u**, $\overrightarrow{OR}$, and **w** have the *same* magnitude and the *same* direction,

$$\mathbf{u} = \overrightarrow{OR} = \mathbf{w}.$$

    Keep in mind that the equivalence of vectors requires only the same magnitude and the same direction—not the same location. In the illustrations at left, the first three pairs of vectors are each not equivalent. The fourth set of vectors is an example of equivalence.

Vector Addition

Suppose a person takes 4 steps east and then 3 steps north. He or she will then be 5 steps from the starting point in the direction shown at left. A vector 4 units long and pointing to the right represents 4 steps east and a vector 3 units long and pointing up represents 3 steps north. The **sum** of the two vectors is the vector 5 steps in magnitude and in the direction shown. The sum is also called the **resultant** of the two vectors.

$\mathbf{u} \ne \mathbf{v}$ (not equivalent)
*Different magnitudes;
different directions*

$\mathbf{u} \ne \mathbf{v}$
*Same magnitude;
different directions*

$\mathbf{u} \ne \mathbf{v}$
*Different magnitudes;
same direction*

$\mathbf{u} = \mathbf{v}$
*Same magnitude;
same direction*

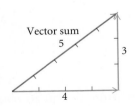

Vector sum
5
3
4

In general, two nonzero vectors **u** and **v** can be added geometrically by placing the initial point of **v** at the terminal point of **u** and then finding the vector that has the same initial point as **u** and the same terminal point as **v**, as shown in the following figure.

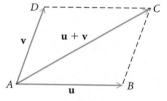

The sum **u** + **v** is the vector represented by the directed line segment from the initial point A of **u** to the terminal point C of **v**. That is, if

$$\mathbf{u} = \overrightarrow{AB} \quad \text{and} \quad \mathbf{v} = \overrightarrow{BC},$$

then

$$\mathbf{u} + \mathbf{v} = \overrightarrow{AB} + \overrightarrow{BC} = \overrightarrow{AC}.$$

We can also describe vector addition by placing the initial points of the vectors together, completing a parallelogram, and finding the diagonal of the parallelogram. (See the figure on the left below.) This description of addition is sometimes called the **parallelogram law** of vector addition. Vector addition is **commutative**. As shown in the figure on the right below, both **u** + **v** and **v** + **u** are represented by the same directed line segment.

Applications

If two forces F_1 and F_2 act on an object, the *combined* effect is the sum, or resultant, $F_1 + F_2$ of the separate forces.

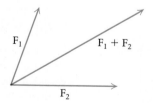

EXAMPLE 2   Forces of 15 newtons and 25 newtons act on an object at

right angles to each other. Find their sum, or resultant, giving the magnitude of the resultant and the angle that it makes with the larger force.

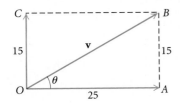

Solution We make a drawing—this time, a rectangle—using **v** or $\overrightarrow{OB}$ to represent the resultant. To find the magnitude, we use the Pythagorean theorem:

$$|\mathbf{v}|^2 = 15^2 + 25^2 \qquad \text{Here } |\mathbf{v}| \text{ denotes the length, or magnitude, of v.}$$
$$|\mathbf{v}| = \sqrt{15^2 + 25^2}$$
$$|\mathbf{v}| \approx 29.2.$$

To find the direction, we note that since OAB is a right triangle,

$$\tan \theta = \tfrac{15}{25} = 0.6.$$

Using a grapher, we find θ, the angle that the resultant makes with the larger force:

$$\theta = \tan^{-1}(0.6) \approx 31°.$$

The resultant $\overrightarrow{OB}$ has a magnitude of 29.2 and makes an angle of 31° with the larger force.

AERIAL BEARINGS

REVIEW SECTION 2.3.

Pilots must adjust the direction of their flight when there is a crosswind. Both the wind and the aircraft velocities can be described by vectors.

EXAMPLE 3 *Airplane Speed and Direction.* An airplane travels on a bearing of 100° at an airspeed of 190 km/h while a wind is blowing 48 km/h from 220°. Find the ground speed of the airplane and the direction of its track, or course, over the ground.

Airplane airspeed

Windspeed

Solution We first make a drawing. The wind is represented by $\overrightarrow{OC}$ and the velocity vector of the airplane by $\overrightarrow{OA}$. The resultant velocity vector is **v**, the sum of the two vectors. The angle θ between **v** and $\overrightarrow{OA}$ is called a **drift angle.**

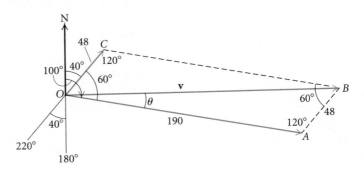

Note that the measure of $\angle COA = 100° - 40° = 60°$. Thus the measure of $\angle CBA$ is also 60° (opposite angles of a parallelogram are equal). Since the sum of all the angles of the parallelogram is 360° and $\angle OCB$ and $\angle OAB$ have the same measure, each must be 120°. By the *law*

of cosines in $\triangle OAB$, we have

$$|\mathbf{v}|^2 = 48^2 + 190^2 - 2 \cdot 48 \cdot 190 \cos 120°$$
$$|\mathbf{v}|^2 = 47{,}524$$
$$|\mathbf{v}| = 218.$$

Thus, $|\mathbf{v}|$ is 218 km/h. By the *law of sines* in the same triangle,

$$\frac{48}{\sin \theta} = \frac{218}{\sin 120°},$$

or

$$\sin \theta = \frac{48 \sin 120°}{218} \approx 0.1907$$
$$\theta \approx 11°.$$

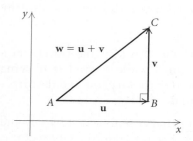

Thus, $\theta = 11°$, to the nearest degree. The ground speed of the airplane is 218 km/h, and its track is in the direction of $100° - 11°$, or $89°$.   ▬

Components

Given a vector $\mathbf{w}$, we may want to find two other vectors $\mathbf{u}$ and $\mathbf{v}$ whose sum is $\mathbf{w}$. The vectors $\mathbf{u}$ and $\mathbf{v}$ are called **components** of $\mathbf{w}$ and the process of finding them is called **resolving**, or **representing**, a vector into its vector components.

When we resolve a vector, we generally look for perpendicular components. Most often, one component will be parallel to the x-axis and the other will be parallel to the y-axis. For this reason, they are often called the **horizontal** and **vertical** components of a vector. In the figure at left, the vector $\mathbf{w} = \overrightarrow{AC}$ is resolved as the sum of $\mathbf{u} = \overrightarrow{AB}$ and $\mathbf{v} = \overrightarrow{BC}$. The horizontal component of $\mathbf{w}$ is $\mathbf{u}$ and the vertical component is $\mathbf{v}$.

EXAMPLE 4   A vector $\mathbf{w}$ has a magnitude of 130 and is inclined $40°$ with the horizontal. Resolve the vector into horizontal and vertical components.

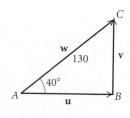

Solution   We first make a drawing showing horizontal and vertical vectors $\mathbf{u}$ and $\mathbf{v}$ whose sum is $\mathbf{w}$.

From $\triangle ABC$, we find $|\mathbf{u}|$ and $|\mathbf{v}|$ using the definitions of the cosine and sine functions:

$$\cos 40° = \frac{|\mathbf{u}|}{130}, \quad \text{or} \quad |\mathbf{u}| = 130 \cos 40° \approx 100,$$

$$\sin 40° = \frac{|\mathbf{v}|}{130}, \quad \text{or} \quad |\mathbf{v}| = 130 \sin 40° \approx 84.$$

Thus the horizontal component of $\mathbf{w}$ is 100 right, and the vertical component of $\mathbf{w}$ is 84 up.   ▬

EXAMPLE 5 *Shipping Crate.* A wooden shipping crate that weighs 816 lb is placed on a loading ramp that makes an angle of 25° with the horizontal. To keep the crate from sliding, a chain is hooked to the crate and to a pole at the top of the ramp. Find the magnitude of the components of the crate's weight (disregarding friction) perpendicular and parallel to the incline.

Solution We first make a drawing illustrating the forces with a rectangle. We let

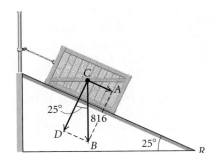

$|\overrightarrow{CB}|$ = the weight of the crate = 816 lb (force of gravity),

$|\overrightarrow{CD}|$ = the magnitude of the component of the crate's weight perpendicular to the incline (force against the ramp), and

$|\overrightarrow{CA}|$ = the magnitude of the component of the crate's weight parallel to the incline (force that pulls the crate down the ramp).

The angle at R is given to be 25° and $\angle BCD = \angle R = 25°$ because the sides of these angles are, respectively, perpendicular. Using the cosine and sine functions, we find that

$$\cos 25° = \frac{|\overrightarrow{CD}|}{816}, \quad \text{or} \quad |\overrightarrow{CD}| = 816 \cos 25° \approx 740 \text{ lb}, \quad \text{and}$$

$$\sin 25° = \frac{|\overrightarrow{CA}|}{816}, \quad \text{or} \quad |\overrightarrow{CA}| = 816 \sin 25° \approx 345 \text{ lb}.$$

Exercise Set 4.5

Sketch the pair of vectors and determine whether they are equivalent. Use the following ordered pairs for the initial and terminal points.

$A(-2, 2)$ $E(-4, 1)$ $I(-6, -3)$
$B(3, 4)$ $F(2, 1)$ $J(3, 1)$
$C(-2, 5)$ $G(-4, 4)$ $K(-3, -3)$
$D(-1, -1)$ $H(1, 2)$ $O(0, 0)$

1. $\overrightarrow{GE}, \overrightarrow{BJ}$ **2.** $\overrightarrow{DJ}, \overrightarrow{OF}$
3. $\overrightarrow{DJ}, \overrightarrow{AB}$ **4.** $\overrightarrow{CG}, \overrightarrow{FO}$
5. $\overrightarrow{DK}, \overrightarrow{BH}$ **6.** $\overrightarrow{BA}, \overrightarrow{DI}$
7. $\overrightarrow{EG}, \overrightarrow{BJ}$ **8.** $\overrightarrow{GC}, \overrightarrow{FO}$
9. $\overrightarrow{GA}, \overrightarrow{BH}$ **10.** $\overrightarrow{JD}, \overrightarrow{CG}$
11. $\overrightarrow{AB}, \overrightarrow{ID}$ **12.** $\overrightarrow{OF}, \overrightarrow{HB}$

13. Two forces of 32 N (newtons) and 45 N act on an object at right angles. Find the magnitude of the resultant and the angle that it makes with the smaller force.

14. Two forces of 50 N and 60 N act on an object at right angles. Find the magnitude of the resultant and the angle that it makes with the larger force.

15. Two forces of 410 N and 600 N act on an object. The angle between the forces is 47°. Find the magnitude of the resultant and the angle that it makes with the larger force.

16. Two forces of 255 N and 325 N act on an object. The angle between the forces is 64°. Find the magnitude of the resultant and the angle that it makes with the smaller force.

In Exercises 17–24, magnitudes of vectors **u** *and* **v** *and the angle* θ *between the vectors are given. Find the sum of* **u** + **v**. *Give the magnitude to the nearest tenth and* *give the direction by specifying to the nearest degree the angle that the resultant makes with* **u**.

17. $|\mathbf{u}| = 45$, $|\mathbf{v}| = 35$, $\theta = 90°$

18. $|\mathbf{u}| = 54$, $|\mathbf{v}| = 43$, $\theta = 150°$

19. $|\mathbf{u}| = 10$, $|\mathbf{v}| = 12$, $\theta = 67°$

20. $|\mathbf{u}| = 25$, $|\mathbf{v}| = 30$, $\theta = 75°$

21. $|\mathbf{u}| = 20$, $|\mathbf{v}| = 20$, $\theta = 117°$

22. $|\mathbf{u}| = 30$, $|\mathbf{v}| = 30$, $\theta = 123°$

23. $|\mathbf{u}| = 23$, $|\mathbf{v}| = 47$, $\theta = 27°$

24. $|\mathbf{u}| = 32$, $|\mathbf{v}| = 74$, $\theta = 72°$

25. *Hot-air Balloon.* A hot-air balloon is rising vertically 10 ft/sec while the wind is blowing horizontally 5 ft/sec. Find the speed of the balloon and the angle that it makes with the horizontal.

26. *Boat.* A boat heads 35°, propelled by a force of 750 lb. A wind from 320° exerts a force of 150 lb on the boat. How large is the resultant force, and in what direction is the boat moving?

27. *Ship.* A ship sails first N80°E for 120 nautical mi, and then S20°W for 200 nautical mi. How far is the ship, then, from the starting point, and in what direction?

28. *Airplane.* An airplane flies 032° for 210 km, and then 280° for 170 km. How far is the airplane, then, from the starting point, and in what direction?

29. *Airplane.* An airplane has an airspeed of 150 km/h. It is to make a flight in a direction of 070° while there is a 25-km/h wind from 340°. What will the airplane's actual heading be?

30. *Wind.* A wind has an easterly component (*from* the east) of 10 km/h and a southerly component (*from* the south) of 16 km/h. Find the magnitude and the direction of the wind.

31. A vector **w** of magnitude 100 points southeast. Resolve the vector into easterly and southerly components.

32. A vector **u** with a magnitude of 150 lb is inclined to the right and upward 52° from the horizontal. Resolve the vector into components.

33. *Airplane.* An airplane takes off at a speed **S** of 225 mph at an angle of 17° with the horizontal. Resolve the vector **S** into components.

34. *Wheelbarrow.* A wheelbarrow is pushed by applying a 97-lb force **F** that makes a 38° angle with the horizontal. Resolve **F** into its horizontal and vertical components. (The horizontal component is the effec-

tive force in the direction of motion and the vertical component adds weight to the wheelbarrow.)

35. *Luggage Wagon.* A luggage wagon is being pulled with vector force **V**, which has a magnitude of 780 lb at an angle of elevation of 60°. Resolve the vector **V** into components.

36. *Hot-air Balloon.* A hot-air balloon exerts a 1200-lb pull on a tether line at a 45° angle with the horizontal. Resolve the vector **B** into components.

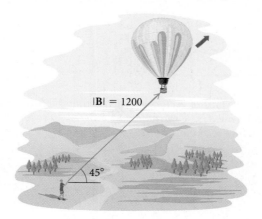

37. *Airplane.* An airplane is flying at 200 km/h in a direction of 305°. Find the westerly and northerly components of its velocity.

38. *Baseball.* A baseball player throws a baseball with a speed **S** of 72 mph at an angle of 45° with the horizontal. Resolve the vector **S** into components.

39. A block weighing 100 lb rests on a 25° incline. Find the magnitude of the components of the block's weight perpendicular and parallel to the incline.

40. A shipping crate that weighs 450 kg is placed on a loading ramp that makes an angle of 30° with the horizontal. Find the magnitude of the components of the crate's weight perpendicular and parallel to the incline.

41. An 80-lb block of ice rests on a 37° incline. What force parallel to the incline is necessary in order to keep the ice from sliding down?

42. What force is necessary to pull a 3500-lb truck up a 9° incline?

Discussion and Writing

43. Describe the concept of a vector as though you were explaining it to a classmate. Use an arrow shot from a bow in the explanation.

44. Explain why vectors $\overrightarrow{QR}$ and $\overrightarrow{RQ}$ are not equivalent.

Skill Maintenance

Find the function value using coordinates of points on the unit circle.

45. $\sin \dfrac{2\pi}{3}$

46. $\cos \dfrac{\pi}{6}$

47. $\cos \dfrac{\pi}{4}$

48. $\sin \dfrac{5\pi}{6}$

Synthesis

49. *Eagle's Flight.* An eagle flies from its nest 7 mi in the direction northeast, where it stops to rest on a cliff. It then flies 8 mi in the direction S30°W to land on top of a tree. Place an *xy*-coordinate system so that the origin is the bird's nest, the *x*-axis points east, and the *y*-axis points north.

a) At what point is the cliff located?
b) At what point is the tree?

Vector Operations

4.6

• *Perform calculations with vectors in component form.*
• *Express a vector as a linear combination of unit vectors.*
• *Express a vector in terms of its magnitude and its direction.*
• *Find the angle between two vectors using the dot product.*
• *Solve applied problems involving forces in equilibrium.*

Position Vectors

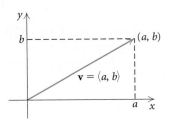

Let's consider a vector **v** whose initial point is the *origin* in an *xy*-coordinate system and whose terminal point is (a, b). We say that the vector is in **standard position** and refer to it as a position vector. Note that the ordered pair (a, b) defines the vector uniquely. Thus we can use (a, b) to denote the vector. To emphasize that we are thinking of a vector and to avoid the confusion of notation with ordered-pair and interval notation, we generally write

$$\mathbf{v} = \langle a, b \rangle.$$

The coordinate a is the *scalar* **horizontal component** of the vector, and the coordinate b is the *scalar* **vertical component** of the vector. By **scalar**, we mean a *numerical* quantity rather than a *vector* quantity. Thus, $\langle a, b \rangle$ is considered to be the *component form* of **v**. Note that a and b are *not* vectors and should not be confused with the vector component definition given in Section 4.5.

Now consider $\overrightarrow{AC}$ with $A = (x_1, y_1)$ and $C = (x_2, y_2)$. Let's see how to find the position vector equivalent to $\overrightarrow{AC}$. As you can see in the figure below, the initial point A is relocated to the origin $(0, 0)$. The coordinates of P are found by subtracting the coordinates of A from the coordinates of C. Thus, $P = (x_2 - x_1, y_2 - y_1)$ and the position vector is $\overrightarrow{OP}$.

It can be shown that $\overrightarrow{OP}$ and $\overrightarrow{AC}$ have the same magnitude and direction and are therefore equivalent. Thus, $\overrightarrow{AC} = \overrightarrow{OP} = \langle x_2 - x_1, y_2 - y_1 \rangle$.

Component Form of a Vector

The **component form** of $\overrightarrow{AC}$ with $A = (x_1, y_1)$ and $C = (x_2, y_2)$ is

$$\overrightarrow{AC} = \langle x_2 - x_1, y_2 - y_1 \rangle.$$

EXAMPLE 1   Find the component form of $\overrightarrow{CF}$ if $C = (-4, -3)$ and $F = (1, 5)$.

Solution   We have

$$\overrightarrow{CF} = \langle 1 - (-4), 5 - (-3) \rangle = \langle 5, 8 \rangle.$$

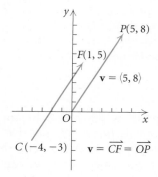

Note that vector $\overrightarrow{CF}$ is equivalent to *position vector* $\overrightarrow{OP}$ with $P = (5, 8)$ as shown in the figure at left.

Now that we know how to write vectors in component form, let's restate some definitions that we first considered in Section 4.5.

The length of a vector $\mathbf{v}$ is easy to determine when the components of the vector are known. For $\mathbf{v} = \langle v_1, v_2 \rangle$, we have

$$|\mathbf{v}|^2 = v_1^2 + v_2^2 \qquad \text{Using the Pythagorean theorem}$$
$$|\mathbf{v}| = \sqrt{v_1^2 + v_2^2}.$$

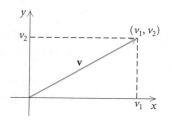

Length of a Vector

The **length**, or **magnitude**, of a vector $\mathbf{v} = \langle v_1, v_2 \rangle$ is given by

$$|\mathbf{v}| = \sqrt{v_1^2 + v_2^2}.$$

Two vectors are **equivalent** if they have the *same* magnitude and the *same* direction.

Equivalent Vectors

Let $\mathbf{u} = \langle u_1, u_2 \rangle$ and $\mathbf{v} = \langle v_1, v_2 \rangle$. Then

$$\langle u_1, u_2 \rangle = \langle v_1, v_2 \rangle \quad \text{if and only if} \quad u_1 = v_1 \quad \text{and} \quad u_2 = v_2.$$

Operations on Vectors

To multiply a vector $\mathbf{v}$ by a positive real number, we multiply its length by the number. Its direction stays the same. When a vector $\mathbf{v}$ is multiplied by 2, its length is doubled and its direction is not changed. When a vector is multiplied by 1.6, its length is increased by 60% and its direction stays the same. To multiply a vector $\mathbf{v}$ by a negative real number, we multiply its length by the number and reverse its direction. When a vector is multiplied by -2, its length is doubled and its direction is reversed. Since real numbers work like scaling factors in vector multiplication, we call them **scalars** and the products $k\mathbf{v}$ are called **scalar multiples** of $\mathbf{v}$.

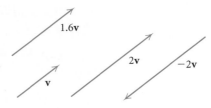

Scalar multiples of $\mathbf{v}$

Scalar Multiplication

For a real number k and a vector $\mathbf{v} = \langle v_1, v_2 \rangle$, the **scalar product** of k and $\mathbf{v}$ is

$$k\mathbf{v} = k\langle v_1, v_2 \rangle = \langle kv_1, kv_2 \rangle.$$

The vector $k\mathbf{v}$ is a **scalar multiple** of the vector $\mathbf{v}$.

EXAMPLE 2 Let $\mathbf{u} = \langle -5, 4 \rangle$ and $\mathbf{w} = \langle 1, -1 \rangle$. Find $-7\mathbf{w}$, $3\mathbf{u}$, and $-1\mathbf{w}$.

Solution

$$-7\mathbf{w} = -7\langle 1, -1 \rangle = \langle -7, 7 \rangle,$$
$$3\mathbf{u} = 3\langle -5, 4 \rangle = \langle -15, 12 \rangle,$$
$$-1\mathbf{w} = -1\langle 1, -1 \rangle = \langle -1, 1 \rangle$$

    ■

In Section 4.5, we used the parallelogram law to add two vectors, but now we can add two vectors using components. To add two vectors given in component form, we add the corresponding components. Let $\mathbf{u} = \langle u_1, u_2 \rangle$ and $\mathbf{v} = \langle v_1, v_2 \rangle$. Then

$$\mathbf{u} + \mathbf{v} = \langle u_1 + v_1, u_2 + v_2 \rangle.$$

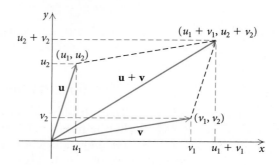

For example, if $\mathbf{v} = \langle -3, 2 \rangle$ and $\mathbf{w} = \langle 5, -9 \rangle$, then

$$\mathbf{v} + \mathbf{w} = \langle -3 + 5, 2 + (-9) \rangle = \langle 2, -7 \rangle.$$

> *Vector Addition*
>
> If $\mathbf{u} = \langle u_1, u_2 \rangle$ and $\mathbf{v} = \langle v_1, v_2 \rangle$, then
>
> $$\mathbf{u} + \mathbf{v} = \langle u_1 + v_1, u_2 + v_2 \rangle.$$

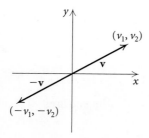

Before we define vector subtraction, we need to define $-\mathbf{v}$. The opposite of $\mathbf{v} = \langle v_1, v_2 \rangle$, shown at left, is

$$-\mathbf{v} = (-1)\mathbf{v} = (-1)\langle v_1, v_2 \rangle = \langle -v_1, -v_2 \rangle.$$

Vector subtraction such as $\mathbf{u} - \mathbf{v}$ involves subtracting corresponding components. We show this by rewriting $\mathbf{u} - \mathbf{v}$ as $\mathbf{u} + (-\mathbf{v})$. If $\mathbf{u} = \langle u_1, u_2 \rangle$ and $\mathbf{v} = \langle v_1, v_2 \rangle$, then

$$\begin{aligned}
\mathbf{u} - \mathbf{v} &= \mathbf{u} + (-\mathbf{v}) = \langle u_1, u_2 \rangle + \langle -v_1, -v_2 \rangle \\
&= \langle u_1 + (-v_1), u_2 + (-v_2) \rangle \\
&= \langle u_1 - v_1, u_2 - v_2 \rangle.
\end{aligned}$$

We can illustrate vector subtraction with parallelograms, just as we did vector addition.

Sketch **u** and **v**.

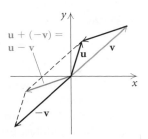

Sketch −**v**.

Sketch **u** + (−**v**), or **u** − **v**, using the parallelogram law.

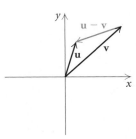

u − **v** is the vector from the terminal point of **v** to the terminal point of **u**.

Vector Subtraction

If **u** = $\langle u_1, u_2 \rangle$ and **v** = $\langle v_1, v_2 \rangle$, then

$$\mathbf{u} - \mathbf{v} = \langle u_1 - v_1, u_2 - v_2 \rangle.$$

It is interesting to compare the sum of two vectors with the difference of the same two vectors in the same parallelogram. The vectors **u** + **v** and **u** − **v** are the diagonals of the parallelogram.

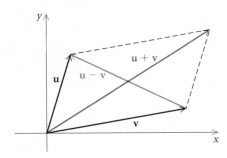

EXAMPLE 3 Do the following calculations, where **u** = $\langle 7, 2 \rangle$ and **v** = $\langle -3, 5 \rangle$.

a) **u** + **v**
b) **u** − 6**v**
c) 3**u** + 4**v**
d) $|5\mathbf{v} - 2\mathbf{u}|$

Solution

a) $\mathbf{u} + \mathbf{v} = \langle 7, 2 \rangle + \langle -3, 5 \rangle = \langle 4, 7 \rangle$

b) $\mathbf{u} - 6\mathbf{v} = \langle 7, 2 \rangle - 6\langle -3, 5 \rangle = \langle 7, 2 \rangle - \langle -18, 30 \rangle = \langle 25, -28 \rangle$

c) $3\mathbf{u} + 4\mathbf{v} = 3\langle 7, 2 \rangle + 4\langle -3, 5 \rangle = \langle 21, 6 \rangle + \langle -12, 20 \rangle = \langle 9, 26 \rangle$

d) $|5\mathbf{v} - 2\mathbf{u}| = |5\langle -3, 5\rangle - 2\langle 7, 2\rangle| = |\langle -15, 25\rangle - \langle 14, 4\rangle|$

$$= |\langle -29, 21\rangle|$$
$$= \sqrt{(-29)^2 + 21^2}$$
$$= \sqrt{1282}$$
$$\approx 35.8 \qquad \rule{1em}{0.5em}$$

Before we state the properties of vector addition and scalar multiplication, we need to define another special vector—the zero vector. The vector whose initial and terminal points are both $(0, 0)$ is the **zero vector,** denoted by **O,** or $\langle 0, 0\rangle$. Its magnitude is 0. In vector addition, the zero vector is the additive identity vector:

$$\mathbf{v} + \mathbf{O} = \mathbf{v}. \qquad \langle v_1, v_2\rangle + \langle 0, 0\rangle = \langle v_1, v_2\rangle$$

Operations on vectors share many of the same properties as operations on real numbers.

Properties of Vector Addition and Scalar Multiplication

For all vectors **u,** **v,** and **w,** and for all scalars b and c:

1. $\mathbf{u} + \mathbf{v} = \mathbf{v} + \mathbf{u}$.
2. $\mathbf{u} + (\mathbf{v} + \mathbf{w}) = (\mathbf{u} + \mathbf{v}) + \mathbf{w}$.
3. $\mathbf{v} + \mathbf{O} = \mathbf{v}$.
4. $1\mathbf{v} = \mathbf{v}; \quad 0\mathbf{v} = \mathbf{O}$.
5. $\mathbf{v} + (-\mathbf{v}) = \mathbf{O}$.
6. $b(c\mathbf{v}) = (bc)\mathbf{v}$.
7. $(b + c)\mathbf{v} = b\mathbf{v} + c\mathbf{v}$.
8. $b(\mathbf{u} + \mathbf{v}) = b\mathbf{u} + b\mathbf{v}$.

Unit Vectors

A vector of magnitude, or length, 1 is called a **unit vector.** The vector $|\mathbf{v}| = \left\langle -\frac{3}{5}, \frac{4}{5}\right\rangle$ is a unit vector because

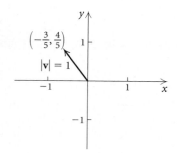

$$|\mathbf{v}| = \left|\left\langle -\tfrac{3}{5}, \tfrac{4}{5}\right\rangle\right| = \sqrt{\left(-\tfrac{3}{5}\right)^2 + \left(\tfrac{4}{5}\right)^2}$$

$$= \sqrt{\tfrac{9}{25} + \tfrac{16}{25}} = \sqrt{\tfrac{25}{25}}$$

$$= \sqrt{1} = 1.$$

EXAMPLE 4   Find a unit vector that has the same direction as the vector $\mathbf{w} = \langle -3, 5\rangle$.

Solution We first find the length of **w**:

$$|\mathbf{w}| = \sqrt{(-3)^2 + 5^2} = \sqrt{34}.$$

Thus we want a vector whose length is $1/\sqrt{34}$ of **w** and whose direction is the same as vector **w**. That vector is

$$\mathbf{u} = \frac{1}{\sqrt{34}}\mathbf{w} = \frac{1}{\sqrt{34}}\langle -3, 5 \rangle = \left\langle \frac{-3}{\sqrt{34}}, \frac{5}{\sqrt{34}} \right\rangle.$$

The vector **u** is a unit vector because

$$|\mathbf{u}| = \left| \frac{1}{\sqrt{34}}\mathbf{w} \right| = \sqrt{\left(\frac{-3}{\sqrt{34}} \right)^2 + \left(\frac{5}{\sqrt{34}} \right)^2} = \sqrt{\frac{9}{34} + \frac{25}{34}}$$

$$= \sqrt{\frac{34}{34}} = \sqrt{1} = 1.$$

> ## Unit Vector
>
> If **v** is a vector and $\mathbf{v} \neq \mathbf{O}$, then
>
> $$\frac{1}{|\mathbf{v}|} \cdot \mathbf{v}, \quad \text{or} \quad \frac{\mathbf{v}}{|\mathbf{v}|},$$
>
> is a **unit vector** in the direction of **v**.

Although unit vectors can have any direction, the unit vectors parallel to the *x*- and *y*-axes are particularly useful. They are defined as

$$\mathbf{i} = \langle 1, 0 \rangle \quad \text{and} \quad \mathbf{j} = \langle 0, 1 \rangle.$$

Any vector can be expressed as a **linear combination** of unit vectors **i** and **j**. For example, let $\mathbf{v} = \langle v_1, v_2 \rangle$. Then

$$\begin{aligned} \mathbf{v} = \langle v_1, v_2 \rangle &= \langle v_1, 0 \rangle + \langle 0, v_2 \rangle \\ &= v_1\langle 1, 0 \rangle + v_2\langle 0, 1 \rangle \\ &= v_1\mathbf{i} + v_2\mathbf{j}. \end{aligned}$$

EXAMPLE 5   Express the vector $\mathbf{r} = \langle 2, -6 \rangle$ as a linear combination of **i** and **j**.

Solution

$$\mathbf{r} = \langle 2, -6 \rangle = 2\mathbf{i} + (-6)\mathbf{j} = 2\mathbf{i} - 6\mathbf{j}$$

EXAMPLE 6   Write the vector $\mathbf{q} = -\mathbf{i} + 7\mathbf{j}$ in component form.

Solution

$$\mathbf{q} = -\mathbf{i} + 7\mathbf{j} = -1\mathbf{i} + 7\mathbf{j} = \langle -1, 7 \rangle$$

Vector operations can also be performed when vectors are written as linear combinations of **i** and **j**.

EXAMPLE 7   If $\mathbf{a} = 5\mathbf{i} - 2\mathbf{j}$ and $\mathbf{b} = -\mathbf{i} + 8\mathbf{j}$, find $3\mathbf{a} - \mathbf{b}$.

Solution

$$
\begin{aligned}
3\mathbf{a} - \mathbf{b} &= 3(5\mathbf{i} - 2\mathbf{j}) - (-\mathbf{i} + 8\mathbf{j}) \\
&= 15\mathbf{i} - 6\mathbf{j} + \mathbf{i} - 8\mathbf{j} \\
&= 16\mathbf{i} - 14\mathbf{j}
\end{aligned}
$$

Direction Angles

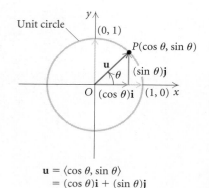

$$\mathbf{u} = \langle \cos \theta, \sin \theta \rangle$$
$$= (\cos \theta)\mathbf{i} + (\sin \theta)\mathbf{j}$$

UNIT CIRCLE
REVIEW SECTION 2.5.

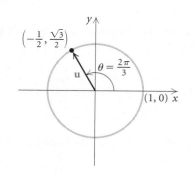

The terminal point P of a unit vector in standard position is a point on the unit circle denoted by $(\cos \theta, \sin \theta)$. Thus the unit vector can be expressed in component form,

$$\mathbf{u} = \langle \cos \theta, \sin \theta \rangle,$$

or as a linear combination of the unit vectors **i** and **j**,

$$\mathbf{u} = (\cos \theta)\mathbf{i} + (\sin \theta)\mathbf{j},$$

where the components of **u** are functions of the **direction angle** θ measured counterclockwise from the x-axis to the vector. As θ varies from 0 to 2π, the point P traces the circle $x^2 + y^2 = 1$. This takes in all possible directions for unit vectors so the equation $\mathbf{u} = (\cos \theta)\mathbf{i} + (\sin \theta)\mathbf{j}$ describes every possible unit vector in the plane.

EXAMPLE 8   Calculate and sketch the unit vector $\mathbf{u} = (\cos \theta)\mathbf{i} + (\sin \theta)\mathbf{j}$ for $\theta = 2\pi/3$. Include the unit circle in your sketch.

Solution

$$\mathbf{u} = \left(\cos \frac{2\pi}{3}\right)\mathbf{i} + \left(\sin \frac{2\pi}{3}\right)\mathbf{j} = \left(-\frac{1}{2}\right)\mathbf{i} + \left(\frac{\sqrt{3}}{2}\right)\mathbf{j}$$

Let $\mathbf{v} = \langle v_1, v_2 \rangle$ with direction angle θ. Using the definition of the tangent function, we can determine the direction angle from the components of **v**:

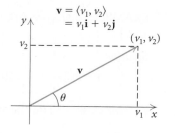

$$\tan \theta = \frac{v_2}{v_1}$$

$$\theta = \tan^{-1} \frac{v_2}{v_1}.$$

EXAMPLE 9   Determine the direction angle θ of the vector $\mathbf{w} = -4\mathbf{i} - 3\mathbf{j}$.

Solution We know that

$$\mathbf{w} = -4\mathbf{i} - 3\mathbf{j} = \langle -4, -3 \rangle.$$

Thus we have

$$\tan \theta = \frac{-3}{-4} = \frac{3}{4} \quad \text{and} \quad \theta = \tan^{-1}\frac{3}{4}.$$

Since **w** is in the third quadrant, we know that θ is a third-quadrant angle. The reference angle is

$$\tan^{-1}\frac{3}{4} \approx 37°,$$

and $\theta \approx 180° + 37°$, or $217°$.

It is convenient for work with applied problems and in subsequent courses, such as calculus, to have a way to express a vector so that both its magnitude and its direction can be determined, or read, easily. Let **v** be a vector. Then $\mathbf{v}/|\mathbf{v}|$ is a unit vector in the same direction as **v**. Thus we have

$$\frac{\mathbf{v}}{|\mathbf{v}|} = (\cos \theta)\mathbf{i} + (\sin \theta)\mathbf{j}$$

$$\mathbf{v} = |\mathbf{v}|[(\cos \theta)\mathbf{i} + (\sin \theta)\mathbf{j}] \qquad \text{Multiplying by } |\mathbf{v}|$$

$$\mathbf{v} = |\mathbf{v}|(\cos \theta)\mathbf{i} + |\mathbf{v}|(\sin \theta)\mathbf{j}.$$

Let's revisit the applied problem in Example 3 of Section 4.5 and use this new notation.

EXAMPLE 10 *Airplane Speed and Direction.* An airplane travels on a bearing of 100° at an airspeed of 190 km/h while a wind is blowing 48 km/h from 220°. Find the ground speed of the airplane and the direction of its track, or course, over the ground.

Solution We first make a drawing. The wind is represented by $\overrightarrow{OC}$ and the velocity vector of the airplane by $\overrightarrow{OA}$. The resultant velocity vector is **v**, the sum of the two vectors:

$$\mathbf{v} = \overrightarrow{OC} + \overrightarrow{OA}.$$

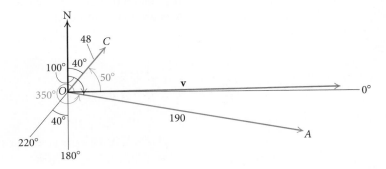

The bearing (measured from north) of the airspeed vector $\overrightarrow{OA}$ is 100°. Its *direction angle* (measured counterclockwise from the positive x-axis) is 350°. The bearing (measured from north) of the wind vector $\overrightarrow{OC}$ is 220°. Its direction angle (measured counterclockwise from the positive x-axis) is 50°. The magnitudes of $\overrightarrow{OA}$ and $\overrightarrow{OC}$ are 190 and 48, respectively. We have

$$\overrightarrow{OA} = 190(\cos 350°)\mathbf{i} + 190(\sin 350°)\mathbf{j}, \quad \text{and}$$
$$\overrightarrow{OC} = 48(\cos 50°)\mathbf{i} + 48(\sin 50°)\mathbf{j}.$$

Thus,

$$\mathbf{v} = \overrightarrow{OA} + \overrightarrow{OC}$$
$$= [190(\cos 350°)\mathbf{i} + 190(\sin 350°)\mathbf{j}] + [48(\cos 50°)\mathbf{i} + 48(\sin 50°)\mathbf{j}]$$
$$= [190(\cos 350°) + 48(\cos 50°)]\mathbf{i} + [190(\sin 350°) + 48(\sin 50°)]\mathbf{j}$$
$$\approx 217.97\mathbf{i} + 3.78\mathbf{j}.$$

From this form, we can determine the ground speed and the course:

$$\text{Ground speed} \approx \sqrt{(217.97)^2 + (3.78)^2}$$
$$\approx 218 \text{ km/h}.$$

We let α be the direction angle of $\mathbf{v}$. Then

$$\tan \alpha = \frac{3.78}{217.97}$$

$$\alpha = \tan^{-1} \frac{3.78}{217.97} \approx 1°.$$

Thus the course of the airplane (the direction from north) is $90° - 1°$, or 89°.

Angle Between Vectors

When a vector is multiplied by a scalar, the result is a vector. When two vectors are added, the result is also a vector. Thus we might expect the product of two vectors to be a vector as well, but it is not. The *dot product* of two vectors is a real number, or scalar. This product is useful in finding the angle between two vectors and in determining whether two vectors are perpendicular.

Dot Product

The **dot product** of two vectors $\mathbf{u} = \langle u_1, u_2 \rangle$ and $\mathbf{v} = \langle v_1, v_2 \rangle$ is

$$\mathbf{u} \cdot \mathbf{v} = u_1 v_1 + u_2 v_2.$$

(Note that $u_1 v_1 + u_2 v_2$ is a *scalar*, not a vector.)

EXAMPLE 11   Find the indicated dot product if

$$\mathbf{u} = \langle 2, -5 \rangle, \qquad \mathbf{v} = \langle 0, 4 \rangle, \quad \text{and} \quad \mathbf{w} = \langle -3, 1 \rangle.$$

a) $\mathbf{u} \cdot \mathbf{w}$

b) $\mathbf{w} \cdot \mathbf{v}$

Solution

a) $\mathbf{u} \cdot \mathbf{w} = 2(-3) + (-5)1 = -6 - 5 = -11$

b) $\mathbf{w} \cdot \mathbf{v} = -3(0) + 1(4) = 0 + 4 = 4$

The dot product can be used to find the angle between two vectors. The angle *between* two vectors is the smallest positive angle formed by the two directed line segments. Thus the angle θ between $\mathbf{u}$ and $\mathbf{v}$ is the same angle as between $\mathbf{v}$ and $\mathbf{u}$, and $0 \le \theta \le \pi$.

Angle Between Two Vectors

If θ is the angle between two *nonzero* vectors $\mathbf{u}$ and $\mathbf{v}$, then

$$\cos \theta = \frac{\mathbf{u} \cdot \mathbf{v}}{|\mathbf{u}|\,|\mathbf{v}|}.$$

EXAMPLE 12   Find the angle between $\mathbf{u} = \langle 3, 7 \rangle$ and $\mathbf{v} = \langle -4, 2 \rangle$.

Solution   We begin by finding $\mathbf{u} \cdot \mathbf{v}$, $|\mathbf{u}|$, and $|\mathbf{v}|$:

$$\mathbf{u} \cdot \mathbf{v} = 3(-4) + 7(2) = 2,$$
$$|\mathbf{u}| = \sqrt{3^2 + 7^2} = \sqrt{58}, \quad \text{and}$$
$$|\mathbf{v}| = \sqrt{(-4)^2 + 2^2} = \sqrt{20}.$$

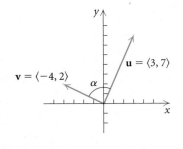

Then

$$\cos \alpha = \frac{\mathbf{u} \cdot \mathbf{v}}{|\mathbf{u}|\,|\mathbf{v}|} = \frac{2}{\sqrt{58}\,\sqrt{20}}$$

$$\alpha = \cos^{-1} \frac{2}{\sqrt{58}\,\sqrt{20}}$$

$$\alpha \approx 86.6°.$$

Forces in Equilibrium

When several forces act through the same point on an object, their vector sum must be **O** in order for a balance to occur. When a balance occurs, then the object is either stationary or moving in a straight line without acceleration. The fact that the vector sum must be **O** for a balance, and vice versa, allows us to solve many applied problems involving forces.

EXAMPLE 13   *Suspended Block.*   A 350-lb block is suspended by two cables, as shown at left. At point *A*, there are three forces acting: **W**, the block pulling down, and **R** and **S**, the two cables pulling upward and outward. Find the tension in each cable.

Solution   We draw a force diagram with the initial points of each vector at the origin. For there to be a balance, the vector sum must be the vector **O**:

$$\mathbf{R} + \mathbf{S} + \mathbf{W} = \mathbf{O}.$$

We can express each vector in terms of its magnitude and its direction angle:

$$\mathbf{R} = |\mathbf{R}|[(\cos 125°)\mathbf{i} + (\sin 125°)\mathbf{j}],$$
$$\mathbf{S} = |\mathbf{S}|[(\cos 37°)\mathbf{i} + (\sin 37°)\mathbf{j}], \quad \text{and}$$
$$\mathbf{W} = |\mathbf{W}|[(\cos 270°)\mathbf{i} + (\sin 270°)\mathbf{j}]$$
$$= 350(\cos 270°)\mathbf{i} + 350(\sin 270°)\mathbf{j}$$
$$= -350\mathbf{j}. \qquad \cos 270° = 0; \ \sin 270° = -1$$

Substituting for **R**, **S**, and **W** in $\mathbf{R} + \mathbf{S} + \mathbf{W} = \mathbf{O}$, we have

$$[|\mathbf{R}|(\cos 125°) + |\mathbf{S}|(\cos 37°)]\mathbf{i} + [|\mathbf{R}|(\sin 125°) + |\mathbf{S}|(\sin 37°) - 350]\mathbf{j}$$
$$= 0\mathbf{i} + 0\mathbf{j}.$$

This gives us a system of equations:

$$|\mathbf{R}|(\cos 125°) + |\mathbf{S}|(\cos 37°) = 0,$$
$$|\mathbf{R}|(\sin 125°) + |\mathbf{S}|(\sin 37°) - 350 = 0.$$

Solving this system, we get

$$|\mathbf{R}| \approx 280 \quad \text{and} \quad |\mathbf{S}| \approx 201.$$

The tensions in the cables are 280 lb and 201 lb.   ▬

Exercise Set **4.6**

Find the component form of the vector given the initial and terminal points. Then find the length of the vector.

1. $\overrightarrow{MN}$; $M(6, -7)$, $N(-3, -2)$

2. $\overrightarrow{CD}$; $C(1, 5)$, $D(5, 7)$

3. $\overrightarrow{FE}$; $E(8, 4)$, $F(11, -2)$

4. $\overrightarrow{BA}$; $A(9, 0)$, $B(9, 7)$

5. $\overrightarrow{KL}$; $K(4, -3)$, $L(8, -3)$

6. $\overrightarrow{GH}$; $G(-6, 10)$, $H(-3, 2)$

7. Find the magnitude of vector **u** if $\mathbf{u} = \langle -1, 6 \rangle$.

8. Find the magnitude of vector $\overrightarrow{ST}$ if $\overrightarrow{ST} = \langle -12, 5 \rangle$.

Do the indicated calculations in Exercises 9–26 for the vectors

$$\mathbf{u} = \langle 5, -2 \rangle, \quad \mathbf{v} = \langle -4, 7 \rangle, \quad \text{and} \quad \mathbf{w} = \langle -1, -3 \rangle.$$

9. $\mathbf{u} + \mathbf{w}$ 10. $\mathbf{w} + \mathbf{u}$ 11. $|3\mathbf{w} - \mathbf{v}|$

12. $6\mathbf{v} + 5\mathbf{u}$ 13. $\mathbf{v} - \mathbf{u}$ 14. $|2\mathbf{w}|$

15. $5\mathbf{u} - 4\mathbf{v}$ 16. $-5\mathbf{v}$

17. $|3\mathbf{u}| - |\mathbf{v}|$ 18. $|\mathbf{v}| + |\mathbf{u}|$

19. $\mathbf{v} + \mathbf{u} + 2\mathbf{w}$ 20. $\mathbf{w} - (\mathbf{u} + 4\mathbf{v})$

21. $2\mathbf{v} + \mathbf{O}$ 22. $10|7\mathbf{w} - 3\mathbf{u}|$

23. $\mathbf{u} \cdot \mathbf{w}$ 24. $\mathbf{w} \cdot \mathbf{u}$

25. $\mathbf{u} \cdot \mathbf{v}$ 26. $\mathbf{v} \cdot \mathbf{w}$

The vectors **u**, **v**, *and* **w** *are drawn below. Copy them on a sheet of paper. Then sketch each of the vectors in Exercises 27–30.*

27. u + v

28. u − 2v

29. u + v + w

30. $\frac{1}{2}$**u − w**

31. Vectors **u**, **v**, and **w** are determined by the sides of △*ABC* below.

   a) Find an expression for **w** in terms of **u** and **v**.
   b) Find an expression for **v** in terms of **u** and **w**.

32. In △*ABC*, vectors **u** and **w** are determined by the sides shown, where *P* is the midpoint of side *BC*. Find an expression for **v** in terms of **u** and **w**.

Find a unit vector that has the same direction as the given vector.

33. v = ⟨−5, 12⟩

34. u = ⟨3, 4⟩

35. w = ⟨1, −10⟩

36. a = ⟨6, −7⟩

37. r = ⟨−2, −8⟩

38. t = ⟨−3, −3⟩

Express the vector as a linear combination of the unit vectors **i** *and* **j**.

39. w = ⟨−4, 6⟩

40. r = ⟨−15, 9⟩

41. s = ⟨2, 5⟩

42. u = ⟨2, −1⟩

Express the vector as a linear combination of **i** *and* **j**.

43.

44.

For Exercises 45–48, use the vectors

$$\mathbf{u} = 2\mathbf{i} + \mathbf{j}, \qquad \mathbf{v} = -3\mathbf{i} - 10\mathbf{j}, \quad \text{and} \quad \mathbf{w} = \mathbf{i} - 5\mathbf{j}.$$

Perform the indicated vector operations and state the answer in two forms: (a) as a linear combination of **i** *and* **j** *and (b) in component form.*

45. 4u − 5w

46. v + 3w

47. u − (v + w)

48. (u − v) + w

Sketch (include the unit circle) and calculate the unit vector **u** = (cos θ)**i** + (sin θ)**j** *for the given direction angle.*

49. $\theta = \dfrac{\pi}{2}$

50. $\theta = \dfrac{\pi}{3}$

51. $\theta = \dfrac{4\pi}{3}$

52. $\theta = \dfrac{3\pi}{2}$

Determine the direction angle θ of the vector, to the nearest degree.

53. u = ⟨−2, −5⟩

54. w = ⟨4, −3⟩

55. q = **i** + 2**j**

56. w = 5**i** − **j**

57. t = ⟨5, 6⟩

58. b = ⟨−8, −4⟩

Find the magnitude and the direction angle θ of the vector.

59. u = 3[(cos 45°)**i** + (sin 45°)**j**]

60. w = 6[(cos 150°)**i** + (sin 150°)**j**]

61. v = $\left\langle -\dfrac{1}{2}, \dfrac{\sqrt{3}}{2} \right\rangle$

62. u = −**i** − **j**

Find the angle between the given vectors, to the nearest tenth of a degree.

63. u = ⟨2, −5⟩, **v** = ⟨1, 4⟩

64. a = ⟨−3, −3⟩, **b** = ⟨−5, 2⟩

65. w = ⟨3, 5⟩, **r** = ⟨5, 5⟩

66. v = ⟨−4, 2⟩, **t** = ⟨1, −4⟩

67. a = **i** + **j**, **b** = 2**i** − 3**j**

68. u = 3**i** + 2**j**, **v** = −**i** + 4**j**

Express each vector in Exercises 69–72 in the form a**i** + b**j** *and sketch each as an arrow in the coordinate plane.*

69. The unit vectors **u** = (cos θ)**i** + (sin θ)**j** for θ = π/6 and θ = 3π/4. Include the unit circle $x^2 + y^2 = 1$ in your sketch.

70. The unit vectors **u** = (cos θ)**i** + (sin θ)**j** for θ = −π/4 and θ = −3π/4. Include the unit circle $x^2 + y^2 = 1$ in your sketch.

71. The unit vector obtained by rotating **j** counterclockwise $3\pi/4$ radians about the origin

72. The unit vector obtained by rotating **j** clockwise $2\pi/3$ radians about the origin

For the vectors in Exercises 73 and 74, find the unit vectors $\mathbf{u} = (\cos \theta)\mathbf{i} + (\sin \theta)\mathbf{j}$ *in the same direction.*

73. $-\mathbf{i} + 3\mathbf{j}$ **74.** $6\mathbf{i} - 8\mathbf{j}$

For the vectors in Exercises 75 and 76, express each vector in terms of its magnitude and its direction.

75. $2\mathbf{i} - 3\mathbf{j}$ **76.** $5\mathbf{i} + 12\mathbf{j}$

77. Use a sketch to show that

$$\mathbf{v} = 3\mathbf{i} - 6\mathbf{j} \quad \text{and} \quad \mathbf{u} = -\mathbf{i} + 2\mathbf{j}$$

have opposite directions.

78. Use a sketch to show that

$$\mathbf{v} = 3\mathbf{i} - 6\mathbf{j} \quad \text{and} \quad \mathbf{u} = \tfrac{1}{2}\mathbf{i} - \mathbf{j}$$

have the same direction.

Exercises 79–82 appeared first in Exercise Set 4.5, where we used the law of cosines and the law of sines to solve the applied problems. For this exercise set, solve the problem using the vector form

$$\mathbf{v} = |\mathbf{v}|[(\cos \theta)\mathbf{i} + (\sin \theta)\mathbf{j}].$$

79. *Ship.*   A ship sails first N80°E for 120 nautical mi, and then S20°W for 200 nautical mi. How far is the ship, then, from the starting point, and in what direction?

80. *Boat.*   A boat heads 35°, propelled by a force of 750 lb. A wind from 320° exerts a force of 150 lb on the boat. How large is the resultant force, and in what direction is the boat moving?

81. *Airplane.*   An airplane has an airspeed of 150 km/h. It is to make a flight in a direction of 070° while there is a 25-km/h wind from 340°. What will the airplane's actual heading be?

82. *Airplane.*   An airplane flies 032° for 210 mi, and then 280° for 170 mi. How far is the airplane, then, from the starting point, and in what direction?

83. Two cables support a 1000-lb weight, as shown. Find the tension in each cable.

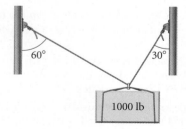

84. A 2500-kg block is suspended by two ropes, as shown. Find the tension in each rope.

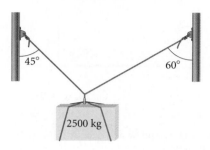

85. A 150-lb sign is hanging from the end of a hinged boom, supported by a cable inclined 42° with the horizontal. Find the tension in the cable and the compression in the boom.

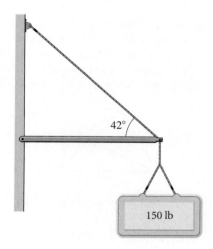

86. A weight of 200 lb is supported by a frame made of two rods and hinged at points *A*, *B*, and *C*. Find the forces exerted by the two rods.

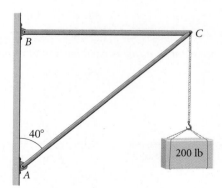

Let $\mathbf{u} = \langle u_1, u_2 \rangle$ and $\mathbf{v} = \langle v_1, v_2 \rangle$. *Prove each of the following properties.*

87. $\mathbf{u} + \mathbf{v} = \mathbf{v} + \mathbf{u}$

88. $\mathbf{u} \cdot \mathbf{v} = \mathbf{v} \cdot \mathbf{u}$

Discussion and Writing

89. Explain how unit vectors are related to the unit circle.

90. Write a vector sum problem for a classmate for which the answer is $\mathbf{v} = 5\mathbf{i} - 8\mathbf{j}$.

Skill Maintenance

Given that

$$g(x) = \frac{x + 5}{x - 1},$$

find each of the following.

91. $g(-2)$

92. $g(1)$

93. $g(-5)$

94. The domain of g

Synthesis

95. If the dot product of two nonzero vectors $\mathbf{u}$ and $\mathbf{v}$ is 0, then the vectors are perpendicular (**orthogonal**). Let $\mathbf{u} = \langle u_1, u_2 \rangle$ and $\mathbf{v} = \langle v_1, v_2 \rangle$.

  a) Prove that if $\mathbf{u} \cdot \mathbf{v} = 0$, then $\mathbf{u}$ and $\mathbf{v}$ are perpendicular.

  b) Given an example of two perpendicular vectors and show that their dot product is 0.

96. If $\overrightarrow{PQ}$ is any vector, what is $\overrightarrow{PQ} + \overrightarrow{QP}$?

97. Find all the unit vectors that are parallel to the vector $\langle 3, -4 \rangle$.

98. Find a vector of length 2 whose direction is the opposite of the direction of the vector $\mathbf{v} = -\mathbf{i} + 2\mathbf{j}$. How many such vectors are there?

99. Given the vector $\overrightarrow{AB} = 3\mathbf{i} - \mathbf{j}$ and A is the point $(2, 9)$, find the point B.

100. Find vector $\mathbf{v}$ from point A to the origin, where $\overrightarrow{AB} = 4\mathbf{i} - 2\mathbf{j}$ and B is the point $(-2, 5)$.

Chapter Summary and Review 4

Important Properties and Formulas

The Law of Sines

$$\frac{a}{\sin A} = \frac{b}{\sin B} = \frac{c}{\sin C}$$

The Law of Cosines

$$a^2 = b^2 + c^2 - 2bc \cos A,$$
$$b^2 = a^2 + c^2 - 2ac \cos B,$$
$$c^2 = a^2 + b^2 - 2ab \cos C$$

The Area of a Triangle

$$K = \frac{1}{2} bc \sin A = \frac{1}{2} ab \sin C = \frac{1}{2} ac \sin B$$

(continued)

Complex Numbers

Complex Number: $a + bi$, a, b real, $i^2 = -1$

Imaginary Number: $a + bi$, $b \neq 0$

Complex Conjugates: $a + bi$, $a - bi$

Absolute Value: $|a + bi| = \sqrt{a^2 + b^2}$

Trigonometric Notation: $a + bi = r(\cos \theta + i \sin \theta)$

Multiplication: $r_1(\cos \theta_1 + i \sin \theta_1) \cdot r_2(\cos \theta_2 + i \sin \theta_2)$

$$= r_1 r_2 [\cos (\theta_1 + \theta_2) + i \sin (\theta_1 + \theta_2)]$$

Division: $\dfrac{r_1(\cos \theta_1 + i \sin \theta_1)}{r_2(\cos \theta_2 + i \sin \theta_2)} = \dfrac{r_1}{r_2}[\cos (\theta_1 - \theta_2) + i \sin (\theta_1 - \theta_2)]$, $r_2 \neq 0$

DeMoivre's Theorem

$$[r(\cos \theta + i \sin \theta)]^n = r^n(\cos n\theta + i \sin n\theta)$$

Roots of Complex Numbers

The nth roots of $r(\cos \theta + i \sin \theta)$ are

$$r^{1/n}\left[\cos \left(\frac{\theta}{n} + k \cdot \frac{360°}{n} \right) + i \sin \left(\frac{\theta}{n} + k \cdot \frac{360°}{n} \right) \right], \quad r \neq 0, k = 0, 1, 2, \ldots, n - 1.$$

Vectors

If $\mathbf{u} = \langle u_1, u_2 \rangle$ and $\mathbf{v} = \langle v_1, v_2 \rangle$ and k is a scalar, then:

Length: $|\mathbf{v}| = \sqrt{v_1^2 + v_2^2}$

Addition: $\mathbf{u} + \mathbf{v} = \langle u_1 + v_1, u_2 + v_2 \rangle$

Subtraction: $\mathbf{u} - \mathbf{v} = \langle u_1 - v_1, u_2 - v_2 \rangle$

Scalar Multiplication: $k\mathbf{v} = \langle kv_1, kv_2 \rangle$

Dot Product: $\mathbf{u} \cdot \mathbf{v} = u_1 v_1 + u_2 v_2$

Angle Between Two Vectors: $\cos \theta = \dfrac{\mathbf{u} \cdot \mathbf{v}}{|\mathbf{u}| |\mathbf{v}|}$

REVIEW EXERCISES

Solve $\triangle ABC$, if possible.

1. $a = 23.4$ ft, $b = 15.7$ ft, $c = 8.3$ ft
2. $B = 27°$, $C = 35°$, $b = 19$ in.
3. $A = 133°28'$, $C = 31°42'$, $b = 890$ m
4. $B = 37°$, $b = 4$ yd, $c = 8$ yd

5. Find the area of $\triangle ABC$ if $b = 9.8$ m, $c = 7.3$ m, and $A = 67.3°$.

6. A parallelogram has sides of lengths 3.21 ft and 7.85 ft. One of its angles measures 147°. Find the area of the parallelogram.

7. *Sandbox.*   A child-care center has a triangular-shaped sandbox. Two of the three sides measure 15 ft and 12.5 ft and form an included angle of 42°.

To determine the amount of sand that is needed to fill the box, the director must determine the area of the floor of the box. Find the area of the floor of the box to the nearest square foot.

15 ft

42°

12.5 ft

8. *Flower Garden.* A triangular flower garden has sides of lengths 11 m, 9 m, and 6 m. Find the angles of the garden to the nearest degree.

9. In an isosceles triangle, the base angles each measure 52.3° and the base is 513 ft long. Find the lengths of the other two sides to the nearest foot.

10. *Airplanes.* Two airplanes leave an airport at the same time. The first flies 175 km/h in a direction of 305.6°. The second flies 220 km/h in a direction of 195.5°. After 2 hr, how far apart are the planes?

Express in terms of i.

11. $-\sqrt{-40}$

12. $\sqrt{-12} \cdot \sqrt{-20}$

13. $\dfrac{\sqrt{-49}}{-\sqrt{-64}}$

Simplify each of the following. Write the answer in the form a + bi, where a and b are real numbers.

14. $(6 + 2i)(-4 - 3i)$

15. $\dfrac{2 - 3i}{1 - 3i}$

16. $(3 - 5i) - (2 - i)$

17. $(6 + 2i) + (-4 - 3i)$

18. i^{23}

19. $(-3i)^{28}$

Graph the complex number and find its absolute value.

20. $2 - 5i$

21. 4

22. $2i$

23. $-3 + i$

Find trigonometric notation.

24. $1 + i$

25. $-4i$

26. $-5\sqrt{3} + 5i$

27. $\dfrac{3}{4}$

Find standard notation, a + bi.

28. $4(\cos 60° + i \sin 60°)$

29. $7(\cos 0° + i \sin 0°)$

30. $5\left(\cos \dfrac{2\pi}{3} + i \sin \dfrac{2\pi}{3}\right)$

31. $2\left[\cos\left(-\dfrac{\pi}{3}\right) + i \sin\left(-\dfrac{\pi}{3}\right)\right]$

Convert to trigonometric notation and then multiply or divide, expressing the answer in standard notation.

32. $(1 + i\sqrt{3})(1 - i)$

33. $\dfrac{2 - 2i}{2 + 2i}$

34. $\dfrac{2 + 2\sqrt{3}i}{\sqrt{3} - i}$

35. $i(3 - 3\sqrt{3}i)$

Raise the number to the given power and write trigonometric notation for the answer.

36. $[2(\cos 60° + i \sin 60°)]^3$

37. $(1 - i)^4$

Raise the number to the given power and write standard notation for the answer.

38. $(1 + i)^6$

39. $\left(\dfrac{1}{2} + \dfrac{\sqrt{3}}{2}i\right)^{10}$

40. Find the square roots of $-1 + i$.

41. Find the cube roots of $3\sqrt{3} - 3i$.

42. Find and graph the fourth roots of 81.

43. Find and graph the fifth roots of 1.

Find all the complex solutions of the equation.

44. $x^4 - i = 0$

45. $x^3 + 1 = 0$

*Magnitudes of vectors **u** and **v** and the angle θ between the vectors are given. Find the magnitude of the sum, **u** + **v**, to the nearest tenth and give the direction by specifying to the nearest degree the angle that it makes with the vector **u**.*

46. $|\mathbf{u}| = 12$, $|\mathbf{v}| = 15$, $\theta = 120°$

47. $|\mathbf{u}| = 41$, $|\mathbf{v}| = 60$, $\theta = 25°$

The vectors **u**, **v**, *and* **w** *are drawn below. Copy them on a sheet of paper. Then sketch each of the vectors in Exercises 48 and 49.*

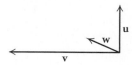

48. **u** − **v**   **49.** **u** + $\frac{1}{2}$**w**

50. Forces of 230 N and 500 N act on an object. The angle between the forces is 52°. Find the resultant, giving the angle that it makes with the smaller force.

51. *Wind.*   A wind has an easterly component of 15 km/h and a southerly component of 25 km/h. Find the magnitude and the direction of the wind.

52. *Ship.*   A ship sails N75°E for 90 nautical mi, and then S10°W for 100 nautical mi. How far is the ship, then, from the starting point, and in what direction?

Find the component form of the vector given the initial and terminal points.

53. $\overrightarrow{AB}$; $A(2, -8)$, $B(-2, -5)$

54. $\overrightarrow{TR}$; $R(0, 7)$, $T(-2, 13)$

55. Find the magnitude of vector **u** if **u** = $\langle 5, -6 \rangle$.

Do the calculations in Exercises 56–59 for the vectors **u** = $\langle 3, -4 \rangle$,   **v** = $\langle -3, 9 \rangle$,   *and*   **w** = $\langle -2, -5 \rangle$.

56. 4**u** + **w**

57. 2**w** − 6**v**

58. |**u**| + |2**w**|

59. **u** · **w**

60. Find a unit vector that has the same direction as **v** = $\langle -6, -2 \rangle$.

61. Express the vector **t** = $\langle -9, 4 \rangle$ as a linear combination of the unit vectors **i** and **j**.

62. Determine the direction angle θ of the vector **w** = $\langle -4, -1 \rangle$ to the nearest degree.

63. Find the magnitude and the direction angle θ of **u** = −5**i** − 3**j**.

64. Find the angle between **u** = $\langle 3, -7 \rangle$ and **v** = $\langle 2, 2 \rangle$ to the nearest tenth of a degree.

65. *Airplane.*   An airplane has an airspeed of 160 mph. It is to make a flight in a direction of 080° while there is a 20-mph wind from 310°. What will the airplane's actual heading be?

Do the calculations in Exercises 66–69 for the vectors **u** = 2**i** + 5**j**,   **v** = −3**i** + 10**j**,   *and*   **w** = 4**i** + 7**j**.

66. 5**u** − 8**v**   **67.** **u** − (**v** + **w**)

68. |**u** − **v**| **69.** 3|**w**| + |**v**|

70. Express the vector $\overrightarrow{PQ}$ in the form $a\mathbf{i} + b\mathbf{j}$, if P is the point $(1, -3)$ and Q is the point $(-4, 2)$.

Express each vector in Exercises 71 and 72 in the form $a\mathbf{i} + b\mathbf{j}$ *and sketch each as an arrow in the coordinate plane.*

71. The unit vectors **u** = $(\cos \theta)\mathbf{i} + (\sin \theta)\mathbf{j}$ for $\theta = \pi/4$ and $\theta = 5\pi/4$. Include the unit circle $x^2 + y^2 = 1$ in your sketch.

72. The unit vector obtained by rotating **j** counterclockwise $2\pi/3$ radians about the origin.

73. Express the vector 3**i** − **j** as a product of its magnitude and its direction.

Discussion and Writing

74. Explain why these statements are not contradictory:
The number 1 has one real cube root.
The number 1 has three complex cube roots.

75. Summarize how you can tell algebraically when solving triangles whether there is no solution, one solution, or two solutions.

76. *Golf: Distance versus Accuracy.*   It is often argued in golf that the farther you hit the ball, the more accurate it must be to stay safe. (Safe means not in the woods, water, or some other hazard.) In his book, *Golf and the Spirit* (p. 54), M. Scott Peck asserts "Deviate 5° from your aiming point on a 150-yd shot, and your ball will land approximately 20 yd to the side of where you wanted it to be. Do the same on a 300-yd shot, and it will be 40 yd off target. Twenty yards may well be in the range of safety; 40 yards probably won't. This principle not infrequently allows a mediocre, short-hitting golfer like myself to score better than the long hitter." Check the accuracy of the mathematics in this statement and comment on Peck's assertion.

Synthesis

77. Let **u** = 12**i** + 5**j**. Find a vector that has the same direction as **u** but has length 3.

78. A parallelogram has sides of lengths 3.42 and 6.97. Its area is 18.4. Find the sizes of its angles.

Analytic Geometry Topics 5

APPLICATION

Expenditures from the Social Security Trust Fund are increasing at a greater rate than the fund's income. Expenditures can be modeled with an exponential function and income can be modeled with a linear function. We can use these functions to predict that expenditures will equal income about 25 yr after 1985, or in 2010. (*Source*: Social Security Administration)

This problem appears as Exercise 65 in Section 5.4.

I n this chapter, we study *conic sections*. These curves are formed by the intersection of a cone and a plane. Conic sections and their properties were first studied by the Greeks. We also study systems of equations in which at least one equation is nonlinear. Both conic sections and nonlinear systems of equations have many applications, as we will see.

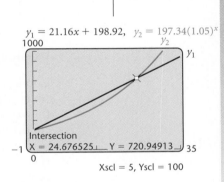

$y_1 = 21.16x + 198.92, \quad y_2 = 197.34(1.05)^x$

Intersection
X = 24.676525   Y = 720.94913

Xscl = 5, Yscl = 100

319

The Parabola
5.1

• *Given an equation of a parabola, complete the square, if necessary, and then find the vertex, the focus, and the directrix and graph the parabola.*

A **conic section** is formed when a right circular cone with two parts, called *nappes*, is intersected by a plane. One of four types of curves can be formed: a parabola, a circle, an ellipse, or a hyperbola.

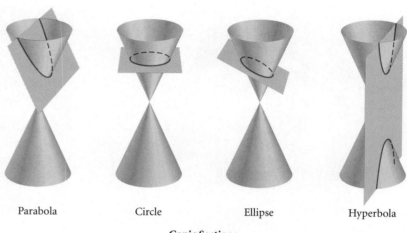

| Parabola | Circle | Ellipse | Hyperbola |

Conic Sections

Parabolas

Conic sections can be defined algebraically using second-degree equations of the form $Ax^2 + Bxy + Cy^2 + Dx + Ey + F = 0$. In addition, they can be defined geometrically as a set of points that satisfy certain conditions.

In Section 1.5, we saw that the graph of the quadratic function $f(x) = ax^2 + bx + c$, $a \neq 0$, is a parabola. A parabola can be defined geometrically.

Parabola

A **parabola** is the set of all points in a plane equidistant from a fixed line (the **directrix**) and a fixed point not on the line (the **focus**).

The line that is perpendicular to the directrix and contains the focus is the **axis of symmetry.** The **vertex** is the midpoint of the segment between the focus and the directrix. (See the figure at left.)

Let's derive the standard equation of a parabola with vertex $(0, 0)$ and directrix $y = -p$, where $p > 0$. We place the coordinate axes as shown in the figure at the top of the following page. The y-axis is the axis of symmetry and contains the focus F. The distance from the focus to the vertex is the same as the distance from the vertex to the directrix. Thus the coordinates of F are $(0, p)$.

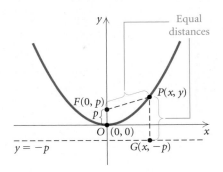

Let $P(x, y)$ be any point of the parabola and consider $\overline{PG}$ perpendicular to the line $y = -p$. The coordinates of G are $(x, -p)$. By the definition of a parabola,

$$PF = PG.$$ The distance from P to the focus is the same as the distance from P to the directrix.

Then using the distance formula, we have

$$\sqrt{(x - 0)^2 + (y - p)^2} = \sqrt{(x - x)^2 + [y - (-p)]^2}$$

$$x^2 + y^2 - 2py + p^2 = y^2 + 2py + p^2 \quad \text{Squaring both sides and squaring the binomials}$$

$$x^2 = 4py.$$

We have shown that if $P(x, y)$ is on the parabola shown above, then its coordinates satisfy this equation. The converse is also true, but we will not prove it here.

Note that if $p > 0$, as above, the graph opens up. If $p < 0$, the graph opens down.

The equation of a parabola with vertex $(0, 0)$ and directrix $x = -p$ is derived similarly. Such a parabola opens either right ($p > 0$) or left ($p < 0$).

Standard Equation of a Parabola with Vertex at the Origin

The standard equation of a parabola with vertex $(0, 0)$ and directrix $y = -p$ is

$$x^2 = 4py.$$

The focus is $(0, p)$ and the y-axis is the axis of symmetry.

The standard equation of a parabola with vertex $(0, 0)$ and directrix $x = -p$ is

$$y^2 = 4px.$$

The focus is $(p, 0)$ and the x-axis is the axis of symmetry.

EXAMPLE 1   Find the focus and the directrix of the parabola $y = -\frac{1}{12}x^2$. Then graph the parabola.

Solution   We write $y = -\frac{1}{12}x^2$ in the form $x^2 = 4py$:

$$-\frac{1}{12}x^2 = y \qquad \text{Given equation}$$

$$x^2 = -12y \qquad \text{Multiplying by } -12 \text{ on both sides}$$

$$x^2 = 4(-3)y. \qquad \text{Standard form}$$

Thus, $p = -3$, so the focus is $(0, p)$, or $(0, -3)$. The directrix is $y = -p = -(-3) = 3$.

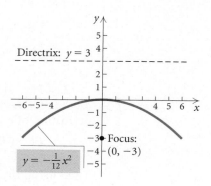

EXAMPLE 2   Find an equation of the parabola with vertex $(0, 0)$ and focus $(5, 0)$. Then graph the parabola.

Solution   The focus is on the x-axis so the line of symmetry is the x-axis. Thus the equation is of the type

$$y^2 = 4px.$$

Since the focus is 5 units to the right of the vertex, $p = 5$ and the equation is

$$y^2 = 4(5)x, \quad \text{or}$$
$$y^2 = 20x.$$

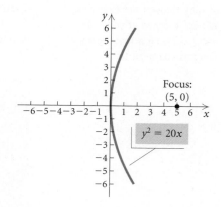

SQUARING THE WINDOW

REVIEW "INTRODUCTION TO GRAPHS AND THE GRAPHING CALCULATOR."

We can check the graph on a grapher using a squared viewing window. But first it might be necessary to solve for y:

$$y^2 = 20x$$
$$y = \pm\sqrt{20x}. \qquad \text{Using the principle of square roots}$$

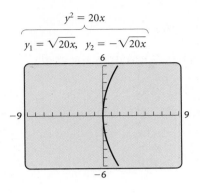

$$y^2 = 20x$$

$$y_1 = \sqrt{20x}, \quad y_2 = -\sqrt{20x}$$

We now graph $y_1 = \sqrt{20x}$ and $y_2 = -\sqrt{20x}$. On some graphers, it is possible to graph $y_1 = \sqrt{20x}$ and $y_2 = -y_1$ by using the Y-VARS menu. The hand-drawn graph appears to be correct.

Finding Standard Form by Completing the Square

If a parabola with vertex at the origin is translated horizontally $|h|$ units and vertically $|k|$ units, it has an equation as follows.

Standard Equation of a Parabola with Vertex (h, k) and Vertical Axis of Symmetry

The standard equation of a parabola with vertex (h, k) and vertical axis of symmetry is

$$(x - h)^2 = 4p(y - k),$$

where the vertex is (h, k), the focus is $(h, k + p)$, and the directrix is $y = k - p$.

(When $p < 0$, the parabola opens down.)

Standard Equation of a Parabola with Vertex (h, k) and Horizontal Axis of Symmetry

The standard equation of a parabola with vertex (h, k) and horizontal axis of symmetry is

$$(y - k)^2 = 4p(x - h),$$

where the vertex is (h, k), the focus is $(h + p, k)$, and the directrix is $x = h - p.$

(When $p < 0$, the parabola opens left.)

COMPLETING THE SQUARE

REVIEW SECTION 1.5
AND APPENDIX A.

We can complete the square on equations of the form

$$y = ax^2 + bx + c \quad \text{or} \quad x = ay^2 + by + c$$

in order to write them in standard form.

EXAMPLE 3   For the parabola

$$x^2 + 6x + 4y + 5 = 0,$$

find the vertex, the focus, and the directrix. Then draw the graph.

Solution   We first complete the square:

$x^2 + 6x + 4y + 5 = 0$

$x^2 + 6x \qquad = -4y - 5$ Subtracting $4y$ and 5 on both sides

$x^2 + 6x + 9 = -4y - 5 + 9$ Adding 9 on both sides to complete the square on the left side

$x^2 + 6x + 9 = -4y + 4$

$(x + 3)^2 = -4(y - 1)$ Factoring

$[(x - (-3)]^2 = 4(-1)(y - 1).$ Writing standard form: $(x - h)^2 = 4p(y - k)$

We now have the following:

Vertex (h, k): $(-3, 1)$;

Focus $(h, k + p)$: $(-3, 1 + (-1))$, or $(-3, 0)$;

Directrix $y = k - p$: $y = 1 - (-1)$, or $y = 2$.

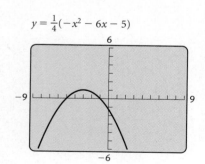

$y = \frac{1}{4}(-x^2 - 6x - 5)$

We can check the graph on a grapher using a squared viewing window. It might be necessary to solve for y first:

$x^2 + 6x + 4y + 5 = 0$

$4y = -x^2 - 6x - 5$

$y = \frac{1}{4}(-x^2 - 6x - 5).$

The hand-drawn graph appears to be correct.

EXAMPLE 4   For the parabola

$$y^2 - 2y - 8x - 31 = 0,$$

find the vertex, the focus, and the directrix. Then draw the graph.

Solution We first complete the square:

$$y^2 - 2y - 8x - 31 = 0$$

$$y^2 - 2y \qquad = 8x + 31 \qquad \text{Adding } 8x \text{ and } 31 \text{ on both sides}$$

$$y^2 - 2y + 1 = 8x + 31 + 1 \qquad \text{Adding 1 on both sides to complete the square on the left side}$$

$$y^2 - 2y + 1 = 8x + 32$$

$$(y - 1)^2 = 8(x + 4)$$

$$(y - 1)^2 = 4(2)[x - (-4)]. \qquad \text{Writing standard form: } (y - k)^2 = 4p(x - h)$$

We now have the following:

Vertex (h, k): $(-4, 1)$;

Focus $(h + p, k)$: $(-4 + 2, 1)$, or $(-2, 1)$;

Directrix $x = h - p$: $x = -4 - 2$, or $x = -6$.

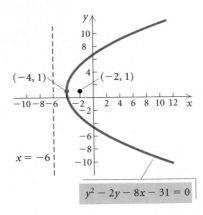

We can check the graph on a grapher using a squared window. We solve the original equation for y using the quadratic formula:

$$y^2 - 2y - 8x - 31 = 0$$

$$y^2 - 2y + (-8x - 31) = 0$$

$$a = 1, \quad b = -2, \quad c = -8x - 31$$

$$y = \frac{-(-2) \pm \sqrt{(-2)^2 - 4 \cdot 1(-8x - 31)}}{2 \cdot 1}$$

$$y = \frac{2 \pm \sqrt{32x + 128}}{2}.$$

We now graph

$$y_1 = \frac{2 + \sqrt{32x + 128}}{2} \quad \text{and} \quad y_2 = \frac{2 - \sqrt{32x + 128}}{2}.$$

The hand-drawn graph appears to be correct.

Applications

Parabolas have many applications. For example, cross sections of car headlights, flashlights, and searchlights are parabolas. The bulb is located at the focus and light from that point is reflected outward parallel to the axis of symmetry. Satellite dishes and field microphones used at sporting events often have parabolic cross sections. Incoming radio waves or sound waves parallel to the axis are reflected into the focus. Cables hung between structures in suspension bridges, such as the Golden Gate Bridge, form parabolas. When a cable supports only its own weight, however, it forms a curve called a *catenary* rather than a parabola.

Exercise Set **5.1**

In Exercises 1–6, match the equation with one of the graphs (a)–(f), which follow.

a)

b)

c)

d)

e)

f)

1. $x^2 = 8y$

2. $y^2 = -10x$

3. $(y - 2)^2 = -3(x + 4)$

4. $(x + 1)^2 = 5(y - 2)$

5. $13x^2 - 8y - 9 = 0$

6. $41x + 6y^2 = 12$

Find the vertex, the focus, and the directrix. Then draw the graph.

7. $x^2 = 20y$

8. $x^2 = 16y$

9. $y^2 = -6x$

10. $y^2 = -2x$

11. $x^2 - 4y = 0$

12. $y^2 + 4x = 0$

13. $x = 2y^2$

14. $y = \frac{1}{2}x^2$

Find an equation of a parabola satisfying the given conditions.

15. Focus $(4, 0)$, directrix $x = -4$

16. Focus $\left(0, \frac{1}{4}\right)$, directrix $y = -\frac{1}{4}$

17. Focus $(0, -\pi)$, directrix $y = \pi$

18. Focus $(-\sqrt{2}, 0)$, directrix $x = \sqrt{2}$

19. Focus $(3, 2)$, directrix $x = -4$

20. Focus $(-2, 3)$, directrix $y = -3$

Find the vertex, the focus, and the directrix. Then draw the graph.

21. $(x + 2)^2 = -6(y - 1)$

22. $(y - 3)^2 = -20(x + 2)$

23. $x^2 + 2x + 2y + 7 = 0$

24. $y^2 + 6y - x + 16 = 0$

25. $x^2 - y - 2 = 0$

26. $x^2 - 4x - 2y = 0$

27. $y = x^2 + 4x + 3$

28. $y = x^2 + 6x + 10$

29. $y^2 - y - x + 6 = 0$

30. $y^2 + y - x - 4 = 0$

31. *Satellite Dish.* An engineer designs a satellite dish with a parabolic cross section. The dish is 15 ft wide at the opening and the focus is placed 4 ft from the vertex.

a) Position a coordinate system with the origin at the vertex and the x-axis on the parabola's axis of symmetry and find an equation of the parabola.

b) Find the depth of the satellite dish at the vertex.

32. *Headlight Mirror.* A car headlight mirror has a parabolic cross section with diameter 6 in. and depth 1 in.

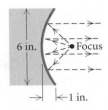

a) Position a coordinate system with the origin at the vertex and the x-axis on the parabola's axis of symmetry and find an equation of the parabola.

b) How far from the vertex should the bulb be positioned if it is to be placed at the focus?

33. *Spotlight.* A spotlight has a parabolic cross section that is 4 ft wide at the opening and 1.5 ft deep at the vertex. How far from the vertex is the focus?

34. *Field Microphone.* A field microphone used at a football game has a parabolic cross section and is 18 in. deep. The focus is 4 in. from the vertex. Find the width of the microphone at the opening.

Discussion and Writing

35. Is a parabola always the graph of a function? Why or why not?

36. Explain how the distance formula is used to find the standard equation of a parabola.

Skill Maintenance

Find each of the following exactly in radians and degrees.

37. $\arcsin\left(-\dfrac{\sqrt{2}}{2}\right)$

38. $\tan^{-1}\sqrt{3}$

Evaluate.

39. $\cos^{-1}\left[\cos\left(-\dfrac{\pi}{6}\right)\right]$

40. $\sin\left(\arccos\dfrac{\sqrt{3}}{2}\right)$

Synthesis

41. Find an equation of the parabola with a vertical axis of symmetry and vertex $(-1, 2)$ and containing the point $(-3, 1)$.

42. Find an equation of a parabola with a horizontal axis of symmetry and vertex $(-2, 1)$ and containing the point $(-3, 5)$.

Use a grapher to find the vertex, the focus, and the directrix of each of the following.

43. $4.5x^2 - 7.8x + 9.7y = 0$

44. $134.1y^2 + 43.4x - 316.6y - 122.4 = 0$

45. *Suspension Bridge.* The cables of a suspension bridge are 50 ft above the roadbed at the ends of the bridge and 10 ft above it in the center of the bridge. The roadbed is 200 ft long. Vertical cables are to be spaced every 20 ft along the bridge. Calculate the lengths of these vertical cables.

The Circle and the Ellipse

5.2

- • Given an equation of a circle, complete the square, if necessary, and then find the center and the radius and graph the circle.
- • Given an equation of an ellipse, complete the square, if necessary, and then find the center, the vertices, and the foci and graph the ellipse.

Circles

We can define a circle geometrically.

CIRCLES
SEE APPENDIX B.

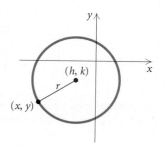

> ### Circle
> A **circle** is the set of all points in a plane that are at a fixed distance from a fixed point (the **center**) in the plane.

Recall the standard equation of a circle with center (h, k) and radius r.

> ### Standard Equation of a Circle
> The standard equation of a circle with center (h, k) and radius r is
> $$(x - h)^2 + (y - k)^2 = r^2.$$

EXAMPLE 1   For the circle
$$x^2 + y^2 - 16x + 14y + 32 = 0,$$
find the center and the radius. Then graph the circle.

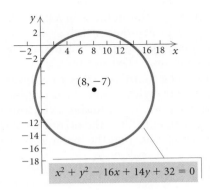

$$x^2 + y^2 - 16x + 14y + 32 = 0$$

Solution   First, we complete the square twice:

$$x^2 + y^2 - 16x + 14y + 32 = 0$$
$$x^2 - 16x \qquad + y^2 + 14y \qquad = -32$$
$$x^2 - 16x + 64 + y^2 + 14y + 49 = -32 + 64 + 49$$

> **Adding 64 and 49 on both sides to complete the square twice on the left side**

$$(x - 8)^2 + (y + 7)^2 = 81$$
$$(x - 8)^2 + [y - (-7)]^2 = 9^2. \qquad \text{Writing standard form}$$

The center is $(8, -7)$ and the radius is 9.

To use a grapher to graph the circle, it might be necessary to solve for y first. The original equation can be solved using the quadratic formula, or the standard form of the equation can be solved using the principle of square roots. The second alternative is illustrated here:

$$(x - 8)^2 + (y + 7)^2 = 81$$
$$(y + 7)^2 = 81 - (x - 8)^2$$
$$y + 7 = \pm\sqrt{81 - (x - 8)^2} \qquad \text{Using the principle of square roots}$$
$$y = -7 \pm \sqrt{81 - (x - 8)^2}.$$

$$(x - 8)^2 + (y + 7)^2 = 81$$
$$y_1 = -7 + \sqrt{81 - (x - 8)^2},$$
$$y_2 = -7 - \sqrt{81 - (x - 8)^2}$$

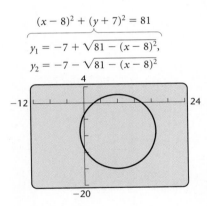

Then we graph

$$y_1 = -7 + \sqrt{81 - (x - 8)^2}$$

and

$$y_2 = -7 - \sqrt{81 - (x - 8)^2}$$

in a squared viewing window.

The hand-drawn graph appears to be correct.

> DRAW CIRCLE FEATURE
>
> SEE APPENDIX B.

Some graphers have a DRAW feature that provides a quick way to graph a circle when the center and the radius are known.

Ellipses

We have studied two conic sections, the parabola and the circle. Now we turn our attention to a third, the *ellipse*.

Ellipse

An **ellipse** is the set of all points in a plane, the sum of whose distances from two fixed points (the **foci**) is constant. The **center** of an ellipse is the midpoint of the segment between the foci.

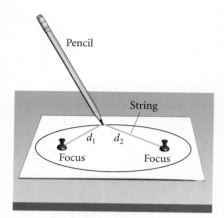

Pencil

String

d_1 d_2

Focus   Focus

We can draw an ellipse by first placing two thumbtacks in a piece of cardboard. These are the foci (singular, *focus*). We then attach a piece of string to the tacks. Its length is the constant sum of the distances $d_1 + d_2$ from the foci to any point on the ellipse. Next, we trace a curve with a pencil held tight against the string. The figure traced is an ellipse.

Let's first consider the ellipse shown below with center at the origin. The points F_1 and F_2 are the foci. The segment $\overline{A'A}$ is the **major axis,** and the points A' and A are the **vertices**. The segment $\overline{B'B}$ is the **minor axis,** and the points B and B' are the **y-intercepts**. Note that the major axis of an ellipse is longer than the minor axis.

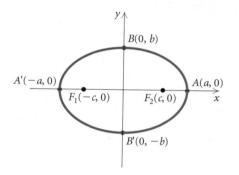

Standard Equation of an Ellipse with Center at the Origin

Major Axis Horizontal

$$\frac{x^2}{a^2} + \frac{y^2}{b^2} = 1, \quad a > b > 0$$

Vertices: $(-a, 0), (a, 0)$

y-intercepts:
$(0, -b), (0, b)$

Foci: $(-c, 0), (c, 0),$
where $c^2 = a^2 - b^2$

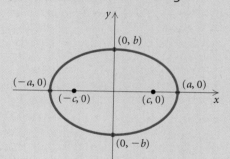

Major Axis Vertical

$$\frac{x^2}{b^2} + \frac{y^2}{a^2} = 1, \quad a > b > 0$$

Vertices: $(0, -a), (0, a)$

x-intercepts:
$(-b, 0), (b, 0)$

Foci: $(0, -c), (0, c),$
where $c^2 = a^2 - b^2$

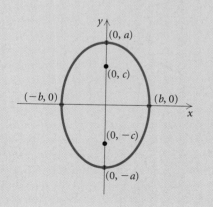

EXAMPLE 2 Find the standard equation of the ellipse with vertices $(-5, 0)$ and $(5, 0)$ and foci $(-3, 0)$ and $(3, 0)$. Then graph the ellipse.

Solution Since the foci are on the x-axis and the origin is the midpoint of the segment between them, the major axis is horizontal and $(0, 0)$ is the center of the ellipse. Thus the equation is of the form

$$\frac{x^2}{a^2} + \frac{y^2}{b^2} = 1.$$

Since the vertices are $(-5, 0)$ and $(5, 0)$ and the foci are $(-3, 0)$ and $(3, 0)$, we know that $a = 5$ and $c = 3$. These values can be used to find b^2:

$$c^2 = a^2 - b^2$$
$$3^2 = 5^2 - b^2$$
$$9 = 25 - b^2$$
$$b^2 = 16.$$

Thus the equation of the ellipse is

$$\frac{x^2}{25} + \frac{y^2}{16} = 1.$$

To graph the ellipse, we plot the vertices $(-5, 0)$ and $(5, 0)$. Since $b^2 = 16$, we know that $b = 4$ and the y-intercepts are $(0, -4)$ and $(0, 4)$. We plot these points as well and connect the four points we have plotted with a smooth curve.

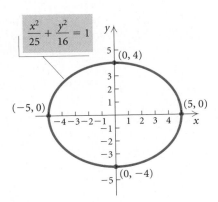

To draw the graph using a grapher, it might be necessary to solve for y first:

$$y = \pm\sqrt{\frac{400 - 16x^2}{25}}.$$

Then we graph

$$y_1 = -\sqrt{\frac{400 - 16x^2}{25}}$$

$$\frac{x^2}{25} + \frac{y^2}{16} = 1$$

$$y_1 = -\sqrt{\frac{400 - 16x^2}{25}},$$

$$y_2 = \sqrt{\frac{400 - 16x^2}{25}}$$

and

$$y_2 = \sqrt{\frac{400 - 16x^2}{25}}$$

or

$$y_1 = -\sqrt{\frac{400 - 16x^2}{25}} \quad \text{and} \quad y_2 = -y_1$$

in a squared viewing window.

EXAMPLE 3   For the ellipse

$$9x^2 + 4y^2 = 36,$$

find the vertices and the foci. Then draw the graph.

Solution   We first find standard form:

$$9x^2 + 4y^2 = 36$$

$$\frac{9x^2}{36} + \frac{4y^2}{36} = \frac{36}{36} \qquad \text{Dividing by 36 on both sides to get 1 on the right side}$$

$$\frac{x^2}{4} + \frac{y^2}{9} = 1$$

$$\frac{x^2}{2^2} + \frac{y^2}{3^2} = 1. \qquad \text{Writing standard form}$$

Thus, $a = 3$ and $b = 2$. The major axis is vertical, so the vertices are $(0, -3)$ and $(0, 3)$. Since we know that $c^2 = a^2 - b^2$, we have $c^2 = 3^2 - 2^2 = 5$, so $c = \sqrt{5}$ and the foci are $(0, -\sqrt{5})$ and $(0, \sqrt{5})$.

To graph the ellipse, we plot the vertices. Note also that since $b = 2$, the x-intercepts are $(-2, 0)$ and $(2, 0)$. We plot these points as well and connect the four points we have plotted with a smooth curve.

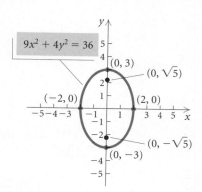

If the center of an ellipse is not at the origin but at some point (h, k), then we can think of an ellipse with center at the origin being translated left or right $|h|$ units and up or down $|k|$ units.

Standard Equation of an Ellipse with Center at (h, k)

Major Axis Horizontal

$$\frac{(x-h)^2}{a^2} + \frac{(y-k)^2}{b^2} = 1, \ a > b > 0$$

Vertices: $(h-a, k)$, $(h+a, k)$

Length of minor axis: $2b$

Foci: $(h-c, k)$, $(h+c, k)$, where $c^2 = a^2 - b^2$

Major Axis Vertical

$$\frac{(x-h)^2}{b^2} + \frac{(y-k)^2}{a^2} = 1, \ a > b > 0$$

Vertices: $(h, k-a)$, $(h, k+a)$

Length of minor axis: $2b$

Foci: $(h, k-c)$, $(h, k+c)$, where $c^2 = a^2 - b^2$

EXAMPLE 4 For the ellipse

$$4x^2 + y^2 + 24x - 2y + 21 = 0,$$

find the center, the vertices, and the foci. Then draw the graph.

Solution First, we complete the square to get standard form:

$$4x^2 + y^2 + 24x - 2y + 21 = 0$$

$$4(x^2 + 6x \quad) + (y^2 - 2y \quad) = -21$$

$$4(x^2 + 6x + 9) + (y^2 - 2y + 1) = -21 + 4 \cdot 9 + 1$$

> Completing the square twice by adding $4 \cdot 9$ and 1 on both sides

$$4(x+3)^2 + (y-1)^2 = 16$$

$$\frac{1}{16}[4(x+3)^2 + (y-1)^2] = \frac{1}{16} \cdot 16$$

$$\frac{(x+3)^2}{4} + \frac{(y-1)^2}{16} = 1$$

$$\frac{[x-(-3)]^2}{2^2} + \frac{(y-1)^2}{4^2} = 1.$$

> Writing standard form:
> $$\frac{(x-h)^2}{b^2} + \frac{(y-k)^2}{a^2} = 1$$

The center is $(-3, 1)$. Note that $a = 4$ and $b = 2$. The major axis is vertical, so the vertices are 4 units above and below the center:

$$(-3, 1 + 4) \text{ and } (-3, 1 - 4), \text{ or } (-3, 5) \text{ and } (-3, -3).$$

We know that $c^2 = a^2 - b^2$, so $c^2 = 4^2 - 2^2 = 12$ and $c = \sqrt{12}$, or $2\sqrt{3}$. Then the foci are $2\sqrt{3}$ units above and below the center:

$$(-3, 1 + 2\sqrt{3}) \text{ and } (-3, 1 - 2\sqrt{3}).$$

To graph the ellipse, we plot the vertices. Note also that since $b = 2$, two other points on the graph are the endpoints of the minor axis, 2 units right and left of the center:

$$(-3 + 2, 1) \text{ and } (-3 - 2, 1),$$

or

$$(-1, 1) \text{ and } (-5, 1).$$

We plot these points as well and connect the four points with a smooth curve.

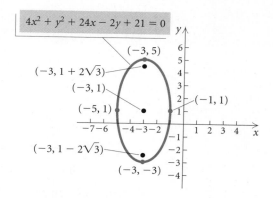

$4x^2 + y^2 + 24x - 2y + 21 = 0$

Applications

One of the most exciting recent applications of an ellipse is a medical device called a *lithotripter*. This machine uses underwater shock waves to crush kidney stones. The waves originate at one focus of an ellipse and are reflected to the kidney stone, which is positioned at the other focus. Recovery time following the use of this technique is much shorter than with conventional surgery and the mortality rate is far lower.

A room with an ellipsoidal ceiling is known as a *whispering gallery*. In such a room, a word whispered at one focus can be clearly heard at the other. Whispering galleries are found in the rotunda of the Capitol Building in Washington, D.C., and in the Mormon Tabernacle in Salt Lake City.

Ellipses have many other applications. Planets travel around the sun in elliptical orbits with the sun at one focus, for example, and satellites travel around the earth in elliptical orbits as well.

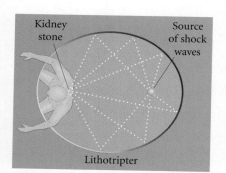

Lithotripter

Planetary orbit

Exercise Set 5.2

In Exercises 1–6, match the equation with one of the graphs (a)–(f), which follow.

a)

b)

c)

d)

e)

f)

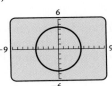

1. $x^2 + y^2 = 5$
2. $y^2 = 20 - x^2$
3. $x^2 + y^2 - 6x + 2y = 6$
4. $x^2 + y^2 + 10x - 12y = 3$
5. $x^2 + y^2 - 5x + 3y = 0$
6. $x^2 + 4x - 2 = 6y - y^2 - 6$

Find the center and the radius of the circle with the given equation. Then draw the graph.

7. $x^2 + y^2 - 14x + 4y = 11$
8. $x^2 + y^2 + 2x - 6y = -6$
9. $x^2 + y^2 + 4x - 6y - 12 = 0$
10. $x^2 + y^2 - 8x - 2y - 19 = 0$
11. $x^2 + y^2 + 6x - 10y = 0$
12. $x^2 + y^2 - 7x - 2y = 0$
13. $x^2 + y^2 - 9x = 7 - 4y$
14. $y^2 - 6y - 1 = 8x - x^2 + 3$

In Exercises 15–18, match the equation with one of the graphs (a)–(d), which follow.

a)

b)

c)

d)

15. $16x^2 + 4y^2 = 64$
16. $4x^2 + 5y^2 = 20$
17. $x^2 + 9y^2 - 6x + 90y = -225$
18. $9x^2 + 4y^2 + 18x - 16y = 11$

Find the vertices and the foci of the ellipse with the given equation. Then draw the graph.

19. $\dfrac{x^2}{4} + \dfrac{y^2}{1} = 1$
20. $\dfrac{x^2}{25} + \dfrac{y^2}{36} = 1$
21. $16x^2 + 9y^2 = 144$
22. $9x^2 + 4y^2 = 36$
23. $2x^2 + 3y^2 = 6$
24. $5x^2 + 7y^2 = 35$
25. $4x^2 + 9y^2 = 1$
26. $25x^2 + 16y^2 = 1$

Find an equation of an ellipse satisfying the given conditions.

27. Vertices: $(-7, 0)$ and $(7, 0)$;
    foci: $(-3, 0)$ and $(3, 0)$
28. Vertices: $(0, -6)$ and $(0, 6)$;
    foci: $(0, -4)$ and $(0, 4)$
29. Vertices: $(0, -8)$ and $(0, 8)$;
    length of minor axis: 10
30. Vertices: $(-5, 0)$ and $(5, 0)$;
    length of minor axis: 6
31. Foci: $(-2, 0)$ and $(2, 0)$;
    length of major axis: 6
32. Foci: $(0, -3)$ and $(0, 3)$;
    length of major axis: 10

Find the center, the vertices, and the foci of the ellipse. Then draw the graph.

33. $\dfrac{(x - 1)^2}{9} + \dfrac{(y - 2)^2}{4} = 1$

34. $\dfrac{(x - 1)^2}{1} + \dfrac{(y - 2)^2}{4} = 1$

35. $\dfrac{(x+3)^2}{25} + \dfrac{(y-5)^2}{36} = 1$

36. $\dfrac{(x-2)^2}{16} + \dfrac{(y+3)^2}{25} = 1$

37. $3(x+2)^2 + 4(y-1)^2 = 192$

38. $4(x-5)^2 + 3(y-4)^2 = 48$

39. $4x^2 + 9y^2 - 16x + 18y - 11 = 0$

40. $x^2 + 2y^2 - 10x + 8y + 29 = 0$

41. $4x^2 + y^2 - 8x - 2y + 1 = 0$

42. $9x^2 + 4y^2 + 54x - 8y + 49 = 0$

The **eccentricity** of an ellipse is defined as $e = c/a$. For an ellipse, $0 < c < a$, so $0 < e < 1$. When e is close to 0, an ellipse appears to be nearly circular. When e is close to 1, an ellipse is very flat.

43. Observe the shapes of the ellipses in Examples 2 and 4. Which ellipse has the smaller eccentricity? Confirm your answer by computing the eccentricity of each ellipse.

44. Which ellipse below has the smaller eccentricity? (Assume that the coordinate systems have the same scale.)

a)

b)

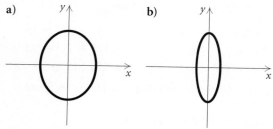

45. Find an equation of an ellipse with vertices $(0, -4)$ and $(0, 4)$ and $e = \frac{1}{4}$.

46. Find an equation of an ellipse with vertices $(-3, 0)$ and $(3, 0)$ and $e = \frac{7}{10}$.

47. *Bridge Supports.* The bridge support shown in the figure below is the top half of an ellipse. Assuming that a coordinate system is superimposed on the drawing in such a way that the center of the ellipse is at point Q, find an equation of the ellipse.

48. *The Ellipse.* In Washington, D.C., there is a large grassy area south of the White House known as the Ellipse. It is actually an ellipse with major axis of length 1048 ft and minor axis of length 898 ft. Assuming that a coordinate system is superimposed on the area in such a way that the center is at the origin and the major and minor axes are on the x- and y-axes of the coordinate system, respectively, find an equation of the ellipse.

49. *The Earth's Orbit.* The maximum distance of the earth from the sun is 9.3×10^7 miles. The minimum distance is 9.1×10^7 miles. The sun is at one focus of the elliptical orbit. Find the distance from the sun to the other focus.

50. *Carpentry.* A carpenter is cutting a 3-ft by 4-ft elliptical sign from a 3-ft by 4-ft piece of plywood. The ellipse will be drawn using a string attached to the board at the foci of the ellipse.

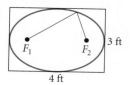

a) How far from the ends of the board should the string be attached?

b) How long should the string be?

Discussion and Writing

51. Explain why function notation is not used in this section.

52. Would you prefer to graph the ellipse

$$\dfrac{(x-3)^2}{4} + \dfrac{(y+5)^2}{9} = 1$$

by hand or using a grapher? Explain why you answered as you did.

Skill Maintenance

In Exercises 53–56, given the function:

a) *Determine whether it is one-to-one.*

b) *If it is one-to-one, find a formula for the inverse.*

53. $f(x) = 2x - 3$

54. $f(x) = x^3 + 2$

55. $f(x) = \dfrac{5}{x-1}$

56. $f(x) = \sqrt{x+4}$

Synthesis

Find an equation of an ellipse satisfying the given conditions.

57. Vertices: $(3, -4)$, $(3, 6)$;
endpoints of minor axis: $(1, 1)$, $(5, 1)$

58. Vertices: $(-1, -1)$, $(-1, 5)$;
endpoints of minor axis: $(-3, 2)$, $(1, 2)$

59. Vertices: $(-3, 0)$ and $(3, 0)$; passing through $\left(2, \frac{22}{3}\right)$

60. Center: $(-2, 3)$; major axis vertical;
length of major axis: 4;
length of minor axis: 1

Use a grapher to find the center and the vertices of each of the following.

61. $4x^2 + 9y^2 - 16.025x + 18.0927y - 11.346 = 0$

62. $9x^2 + 4y^2 + 54.063x - 8.016y + 49.872 = 0$

63. *Bridge Arch.* A bridge with a semielliptical arch spans a river as shown below. What is the clearance 6 ft from the riverbank?

14 ft

50 ft

The Hyperbola

5.3

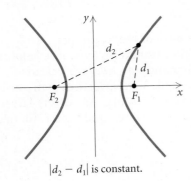

$|d_2 - d_1|$ is constant.

• *Given an equation of a hyperbola, complete the square, if necessary, and then find the center, the vertices, and the foci and graph the hyperbola.*

The last type of conic section that we will study is the *hyperbola*.

> ### Hyperbola
> A **hyperbola** is the set of all points in a plane for which the absolute value of the difference of the distances from two fixed points (the **foci**) is constant. The midpoint of the segment between the foci is the **center** of the hyperbola.

Standard Equations of Hyperbolas

We first consider the equation of a hyperbola with center at the origin. In the figure below, F_1 and F_2 are the foci. The segment $\overline{V_2 V_1}$ is the **transverse axis** and the points V_2 and V_1 are the **vertices**.

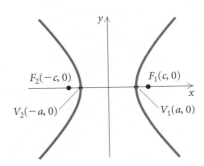

$F_2(-c, 0)$ $F_1(c, 0)$

$V_2(-a, 0)$ $V_1(a, 0)$

Standard Equation of a Hyperbola with Center at the Origin

Transverse Axis Horizontal

$$\frac{x^2}{a^2} - \frac{y^2}{b^2} = 1$$

Vertices: $(-a, 0)$, $(a, 0)$

Foci: $(-c, 0)$, $(c, 0)$,
where $c^2 = a^2 + b^2$

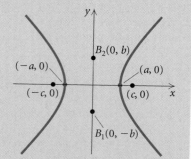

Transverse Axis Vertical

$$\frac{y^2}{a^2} - \frac{x^2}{b^2} = 1$$

Vertices: $(0, -a)$, $(0, a)$

Foci: $(0, -c)$, $(0, c)$,
where $c^2 = a^2 + b^2$

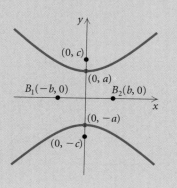

The segment $\overline{B_1 B_2}$ is the **conjugate axis** of the hyperbola.

To graph a hyperbola with a horizontal transverse axis, it is helpful to begin by graphing the lines $y = -(b/a)x$ and $y = (b/a)x$. These are the **asymptotes** of the hyperbola. For a hyperbola with a vertical transverse axis, the asymptotes are $y = -(a/b)x$ and $y = (a/b)x$. As $|x|$ gets larger and larger, the graph of the hyperbola gets closer and closer to the asymptotes.

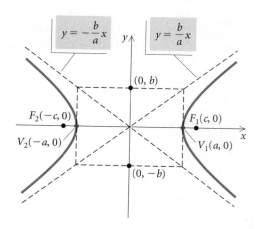

EXAMPLE 1 Find an equation of the hyperbola with vertices $(0, -4)$ and $(0, 4)$ and foci $(0, -6)$ and $(0, 6)$.

Solution We know that $a = 4$ and $c = 6$. We find b^2.

$$c^2 = a^2 + b^2$$
$$6^2 = 4^2 + b^2$$
$$36 = 16 + b^2$$
$$20 = b^2$$

Since the vertices and the foci are on the y-axis, we know that the transverse axis is vertical. We can now write the equation of the hyperbola:

$$\frac{y^2}{a^2} - \frac{x^2}{b^2} = 1$$
$$\frac{y^2}{16} - \frac{x^2}{20} = 1.$$

STUDY TIP

Take the time to include all the steps when working your homework problems. Doing so will help you organize your thinking and avoid computational errors. It will also give you complete, step-by-step solutions of the exercises that can be used to study for an exam.

EXAMPLE 2 For the hyperbola given by

$$9x^2 - 16y^2 = 144,$$

find the vertices, the foci, and the asymptotes. Then graph the hyperbola.

Solution First, we find standard form:

$$9x^2 - 16y^2 = 144$$

$$\frac{1}{144}(9x^2 - 16y^2) = \frac{1}{144} \cdot 144 \qquad \text{Multiplying by } \tfrac{1}{144} \text{ to get 1 on the right side}$$

$$\frac{x^2}{16} - \frac{y^2}{9} = 1$$

$$\frac{x^2}{4^2} - \frac{y^2}{3^2} = 1. \qquad \text{Writing standard form}$$

The hyperbola has a horizontal transverse axis, so the vertices are $(-a, 0)$ and $(a, 0)$, or $(-4, 0)$ and $(4, 0)$. From the standard form of the equation, we know that $a^2 = 4^2$, or 16, and $b^2 = 3^2$, or 9. We find the foci:

$$c^2 = a^2 + b^2$$
$$c^2 = 16 + 9$$
$$c^2 = 25$$
$$c = 5.$$

Thus the foci are $(-5, 0)$ and $(5, 0)$.

Next, we find the asymptotes:

$$y = \frac{b}{a}x = \frac{3}{4}x$$

and

$$y = -\frac{b}{a}x = -\frac{3}{4}x.$$

To draw the graph, we sketch the asymptotes first. This is easily done by drawing the rectangle with horizontal sides passing through (0, 3) and (0, −3) and vertical sides through (4, 0) and (−4, 0). Then we draw and extend the diagonals of this rectangle. The two extended diagonals are the asymptotes of the hyperbola. Next, we plot the vertices and draw the branches of the hyperbola outward from the vertices toward the asymptotes.

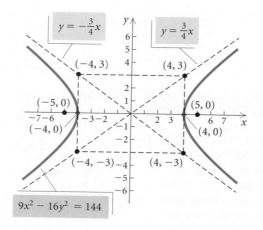

We can use a grapher as a check. It might be necessary to solve for y first and then graph the top and bottom halves of the hyperbola in the same squared viewing window.

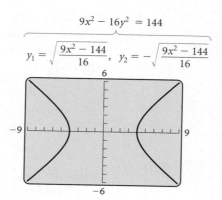

If a hyperbola with center at the origin is translated left or right $|h|$ units and up or down $|k|$ units, the center is at the point (h, k).

Standard Equation of a Hyperbola with Center (h, k)

Transverse Axis Horizontal

$$\frac{(x-h)^2}{a^2} - \frac{(y-k)^2}{b^2} = 1$$

Vertices: $(h-a, k)$, $(h+a, k)$

Asymptotes: $y - k = \frac{b}{a}(x-h)$, $y - k = -\frac{b}{a}(x-h)$

Foci: $(h-c, k)$, $(h+c, k)$, where $c^2 = a^2 + b^2$

Transverse Axis Vertical

$$\frac{(y-k)^2}{a^2} - \frac{(x-h)^2}{b^2} = 1$$

Vertices: $(h, k-a)$, $(h, k+a)$

Asymptotes: $y - k = \frac{a}{b}(x-h)$, $y - k = -\frac{a}{b}(x-h)$

Foci: $(h, k-c)$, $(h, k+c)$, where $c^2 = a^2 + b^2$

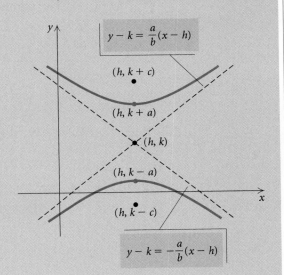

EXAMPLE 3 For the hyperbola given by

$$4y^2 - x^2 + 24y + 4x + 28 = 0,$$

find the center, the vertices, and the foci. Then draw the graph.

Solution First, we complete the square to get standard form:

$$4y^2 - x^2 + 24y + 4x + 28 = 0$$

$$4(y^2 + 6y \qquad\quad) - (x^2 - 4x \qquad\quad) = -28$$

$$4(y^2 + 6y + 9 - 9) - (x^2 - 4x + 4 - 4) = -28$$

$$4(y^2 + 6y + 9) - (x^2 - 4x + 4) = -28 + 36 - 4.$$

Then

$$4(y + 3)^2 - (x - 2)^2 = 4$$

$$\frac{(y + 3)^2}{1} - \frac{(x - 2)^2}{4} = 1 \qquad \text{Dividing by 4}$$

$$\frac{[y - (-3)]^2}{1^2} - \frac{(x - 2)^2}{2^2} = 1. \qquad \text{Standard form}$$

The center is $(2, -3)$. Note that $a = 1$ and $b = 2$. The transverse axis is vertical, so the vertices are 1 unit below and above the center:

$$(2, -3 - 1) \text{ and } (2, -3 + 1), \quad \text{or} \quad (2, -4) \text{ and } (2, -2).$$

We know that $c^2 = a^2 + b^2$, so $c^2 = 1^2 + 2^2 = 1 + 4 = 5$ and $c = \sqrt{5}$. Thus the foci are $\sqrt{5}$ units below and above the center:

$$(2, -3 - \sqrt{5}) \quad \text{and} \quad (2, -3 + \sqrt{5}).$$

The asymptotes are

$$y - (-3) = \frac{1}{2}(x - 2) \quad \text{and} \quad y - (-3) = -\frac{1}{2}(x - 2),$$

or

$$y + 3 = \frac{1}{2}(x - 2) \quad \text{and} \quad y + 3 = -\frac{1}{2}(x - 2).$$

We sketch the asymptotes, plot the vertices, and draw the graph.

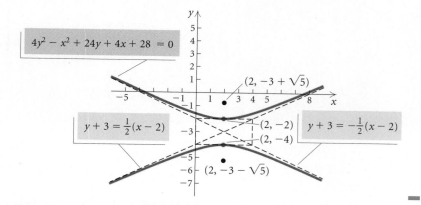

CONNECTING THE CONCEPTS

CLASSIFYING EQUATIONS OF CONIC SECTIONS

EQUATION	TYPE OF CONIC SECTION	GRAPH
$x - 4 + 4y = y^2$	Only one variable is squared, so this cannot be a circle, an ellipse, or a hyperbola. Find an equivalent equation: $$x = (y - 2)^2.$$ This is an equation of a parabola.	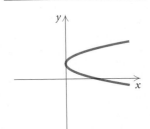
$3x^2 + 3y^2 = 75$	Both variables are squared, so this cannot be a parabola. The squared terms are added, so this cannot be a hyperbola. Divide by 3 on both sides to find an equivalent equation: $$x^2 + y^2 = 25.$$ This is an equation of a circle.	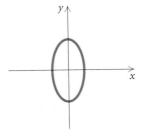
$y^2 = 16 - 4x^2$	Both variables are squared, so this cannot be a parabola. Add $4x^2$ on both sides to find an equivalent equation: $4x^2 + y^2 = 16$. The squared terms are added, so this cannot be a hyperbola. The coefficients of x^2 and y^2 are not the same, so this is not a circle. Divide by 16 on both sides to find an equivalent equation: $$\frac{x^2}{4} + \frac{y^2}{16} = 1.$$ This is an equation of an ellipse.	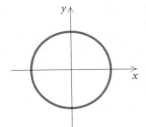
$x^2 = 4y^2 + 36$	Both variables are squared, so this cannot be a parabola. Subtract $4y^2$ on both sides to find an equivalent equation: $x^2 - 4y^2 = 36$. The squared terms are not added, so this cannot be a circle or an ellipse. Divide by 36 on both sides to find an equivalent equation: $$\frac{x^2}{36} - \frac{y^2}{9} = 1.$$ This is an equation of a hyperbola.	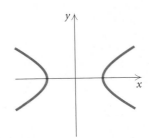

Applications

Some comets travel in hyperbolic paths with the sun at one focus. Such comets pass by the sun only one time unlike those with elliptical orbits, which reappear at intervals. A cross section of an amphitheater might be one branch of a hyperbola. A cross section of a nuclear cooling tower might also be a hyperbola.

One other application of hyperbolas is in the long range navigation system LORAN. This system uses transmitting stations in three locations to send out simultaneous signals to a ship or aircraft. The difference in the arrival times of the signals from one pair of transmitters is recorded on the ship or aircraft. This difference is also recorded for signals from another pair of transmitters. For each pair, a computation is performed to determine the difference in the distances from each member of the pair to the ship or aircraft. If each pair of differences is kept constant, two hyperbolas can be drawn. Each has one of the pairs of transmitters as foci and the ship or aircraft lies on the intersection of two of their branches.

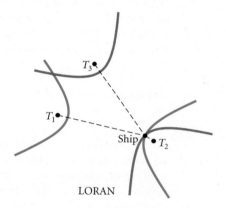

LORAN

Exercise Set 5.3

In Exercises 1–6, match the equation with one of the graphs (a)–(f), which follow.

a)

b)

c)

d)

e)

f)

1. $\dfrac{x^2}{25} - \dfrac{y^2}{9} = 1$

2. $\dfrac{y^2}{4} - \dfrac{x^2}{36} = 1$

3. $\dfrac{(y-1)^2}{16} - \dfrac{(x+3)^2}{1} = 1$

4. $\dfrac{(x+4)^2}{100} - \dfrac{(y-2)^2}{81} = 1$

5. $25x^2 - 16y^2 = 400$

6. $y^2 - x^2 = 9$

Find an equation of a hyperbola satisfying the given conditions.

7. Vertices at $(0, 3)$ and $(0, -3)$; foci at $(0, 5)$ and $(0, -5)$

8. Vertices at $(1, 0)$ and $(-1, 0)$; foci at $(2, 0)$ and $(-2, 0)$

9. Asymptotes $y = \frac{3}{2}x$, $y = -\frac{3}{2}x$; one vertex $(2, 0)$

10. Asymptotes $y = \frac{5}{4}x$, $y = -\frac{5}{4}x$; one vertex $(0, 3)$

Find the center, the vertices, the foci, and the asymptotes. Then draw the graph.

11. $\dfrac{x^2}{4} - \dfrac{y^2}{4} = 1$

12. $\dfrac{x^2}{1} - \dfrac{y^2}{9} = 1$

13. $\dfrac{(x-2)^2}{9} - \dfrac{(y+5)^2}{1} = 1$

14. $\dfrac{(x-5)^2}{16} - \dfrac{(y+2)^2}{9} = 1$

15. $\dfrac{(y+3)^2}{4} - \dfrac{(x+1)^2}{16} = 1$

16. $\dfrac{(y+4)^2}{25} - \dfrac{(x+2)^2}{16} = 1$

17. $x^2 - 4y^2 = 4$

18. $4x^2 - y^2 = 16$

19. $9y^2 - x^2 = 81$

20. $y^2 - 4x^2 = 4$

21. $x^2 - y^2 = 2$

22. $x^2 - y^2 = 3$

23. $y^2 - x^2 = \frac{1}{4}$

24. $y^2 - x^2 = \frac{1}{9}$

Find the center, the vertices, the foci, and the asymptotes of the hyperbola. Then draw the graph.

25. $x^2 - y^2 - 2x - 4y - 4 = 0$

26. $4x^2 - y^2 + 8x - 4y - 4 = 0$

27. $36x^2 - y^2 - 24x + 6y - 41 = 0$

28. $9x^2 - 4y^2 + 54x + 8y + 41 = 0$

29. $9y^2 - 4x^2 - 18y + 24x - 63 = 0$

30. $x^2 - 25y^2 + 6x - 50y = 41$

31. $x^2 - y^2 - 2x - 4y = 4$

32. $9y^2 - 4x^2 - 54y - 8x + 41 = 0$

33. $y^2 - x^2 - 6x - 8y - 29 = 0$

34. $x^2 - y^2 = 8x - 2y - 13$

The **eccentricity** of a hyperbola is defined as $e = c/a$. For a hyperbola, $c > a > 0$, so $e > 1$. When e is close to 1, a hyperbola appears to be very narrow. As the eccentricity increases, the hyperbola becomes "wider."

35. Observe the shapes of the hyperbolas in Examples 2 and 3. Which hyperbola has the larger eccentricity? Confirm your answer by computing the eccentricity of each hyperbola.

36. Which hyperbola below has the larger eccentricity? (Assume that the coordinate systems have the same scale.)

a)

b)

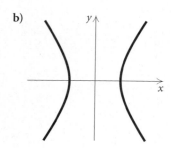

37. Find an equation of a hyperbola with vertices $(3, 7)$ and $(-3, 7)$ and $e = \frac{5}{3}$.

38. Find an equation of a hyperbola with vertices $(-1, 3)$ and $(-1, 7)$ and $e = 4$.

39. *Hyperbolic Mirror.*   Certain telescopes contain both a parabolic and a hyperbolic mirror. In the telescope shown in the figure below, the parabola and the hyperbola share focus F_1, which is 14 m above the vertex of the parabola. The hyperbola's second focus F_2 is 2 m above the parabola's vertex. The vertex of the hyperbolic mirror is 1 m below F_1. Position a coordinate system with the origin at the center of the hyperbola and with the foci on the y-axis. Then find the equation of the hyperbola.

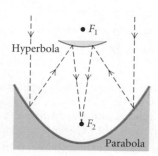

40. *Nuclear Cooling Tower.*   A cross section of a

nuclear cooling tower is a hyperbola with equation

$$\frac{x^2}{90^2} - \frac{y^2}{130^2} = 1.$$

The tower is 450 ft tall and the distance from the top of the tower to the center of the hyperbola is half the distance from the base of the tower to the center of the hyperbola. Find the diameter of the top and the base of the tower.

Discussion and Writing

41. How does the graph of a parabola differ from the graph of one branch of a hyperbola?

42. Would you prefer to graph the hyperbola

$$\frac{(y + 5)^2}{16} + \frac{(x - 4)^2}{36} = 1$$

by hand or using a grapher? Explain why you answered as you did.

Skill Maintenance

Solve the triangle, if possible.

43. $A = 125°$, $C = 31°$, $b = 22$

44. $B = 143°$, $a = 34$, $c = 30$

Find the area of the triangle.

45. $C = 72°29'$, $a = 2.5$ yd, $b = 4.2$ yd

46. $A = 115°$, $b = 6$ cm, $c = 8$ cm

Synthesis

Find an equation of a hyperbola satisfying the given conditions.

47. Vertices at $(3, -8)$ and $(3, -2)$; asymptotes $y = 3x - 14$, $y = -3x + 4$

48. Vertices at $(-9, 4)$ and $(-5, 4)$; asymptotes $y = 3x + 25$, $y = -3x - 17$

Use a grapher to find the center, the vertices, and the asymptotes.

49. $5x^2 - 3.5y^2 + 14.6x - 6.7y + 3.4 = 0$

50. $x^2 - y^2 - 2.046x - 4.088y - 4.228 = 0$

51. *Navigation.* Two radio transmitters positioned 300 mi apart send simultaneous signals to a ship that is 200 mi offshore, sailing parallel to the shoreline. The signal from transmitter S reaches the ship 200 microseconds later than the signal from transmitter T. The signals travel at a speed of 186,000 miles per second, or 0.186 mile per microsecond. Find the equation of the hyperbola with foci S and T on which the ship is located.

(*Hint*: For any point on the hyperbola, the absolute value of the difference of its distances from the foci is $2a$.)

Nonlinear Systems of Equations

5.4

- Solve a nonlinear system of equations.
- Use nonlinear systems of equations to solve applied problems.

The systems of equations that we have studied so far have been composed of linear equations. Now we consider systems of two equations in two variables in which at least one equation is not linear.

Nonlinear Systems of Equations

The graph of a nonlinear system of equations can have no point of intersection or one or more points of intersection. The coordinates of each point of intersection are a solution of the system of equations. When no point of intersection exists, the system of equations has no real-number solution.

Real-number solutions of nonlinear systems of equations can be found algebraically, using the substitution or elimination method, or graphically. However, since a graph shows *only* real-number solutions, solutions involving imaginary numbers must always be found algebraically.

When we are solving algebraically, the substitution method is preferable for a system consisting of one linear and one nonlinear equation. The elimination method is preferable in most, but not all, cases when both equations are nonlinear.

EXAMPLE 1 Solve the following system of equations:

$$x^2 + y^2 = 25, \qquad (1) \qquad \text{(The graph is a circle.)}$$
$$3x - 4y = 0. \qquad (2) \qquad \text{(The graph is a line.)}$$

Algebraic Solution

We use the substitution method. First, we solve equation (2) for x:

$$x = \tfrac{4}{3}y. \quad (3) \qquad \text{We could have solved for } y \text{ instead.}$$

Next, we substitute $\tfrac{4}{3}y$ for x in equation (1) and solve for y:

$$\left(\tfrac{4}{3}y\right)^2 + y^2 = 25$$
$$\tfrac{16}{9}y^2 + y^2 = 25$$
$$\tfrac{25}{9}y^2 = 25$$
$$y^2 = 9 \qquad \text{Multiplying by } \tfrac{9}{25}$$
$$y = \pm 3.$$

Now we substitute these numbers for y in equation (3) and solve for x:

$$x = \tfrac{4}{3}(3) = 4, \qquad \text{The pair } (4, 3) \text{ appears to be a solution.}$$
$$x = \tfrac{4}{3}(-3) = -4. \qquad \text{The pair } (-4, -3) \text{ appears to be a solution.}$$

CHECK: For (4, 3):

$$\begin{array}{c|c} x^2 + y^2 = 25 \\ \hline 4^2 + 3^2 \;?\; 25 \\ 16 + 9 \\ 25 & 25 \quad \text{TRUE} \end{array} \qquad \begin{array}{c|c} 3x - 4y = 0 \\ \hline 3(4) - 4(3) \;?\; 0 \\ 12 - 12 \\ 0 & 0 \quad \text{TRUE} \end{array}$$

For (−4, −3):

$$\begin{array}{c|c} x^2 + y^2 = 25 \\ \hline (-4)^2 + (-3)^2 \;?\; 25 \\ 16 + 9 \\ 25 & 25 \quad \text{TRUE} \end{array} \qquad \begin{array}{c|c} 3x - 4y = 0 \\ \hline 3(-4) - 4(-3) \;?\; 0 \\ -12 + 12 \\ 0 & 0 \quad \text{TRUE} \end{array}$$

The pairs $(4, 3)$ and $(-4, -3)$ check, so they are the solutions.

Graphical Solution

We graph both equations in the same viewing window and note that there are two points of intersection. We can find their coordinates using the INTERSECT feature.

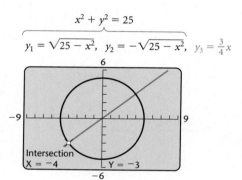

$x^2 + y^2 = 25$

$y_1 = \sqrt{25 - x^2}, \quad y_2 = -\sqrt{25 - x^2}, \quad y_3 = \tfrac{3}{4}x$

$x^2 + y^2 = 25$

$y_1 = \sqrt{25 - x^2}, \quad y_2 = -\sqrt{25 - x^2}, \quad y_3 = \tfrac{3}{4}x$

The solutions are $(4, 3)$ and $(-4, -3)$.

The graph can also be used to check the solutions that we found algebraically.

In the algebraic solution in Example 1, suppose that to find x, we had substituted 3 and −3 in equation (1) rather than equation (3). If $y = 3$, $y^2 = 9$, and if $y = -3$, $y^2 = 9$, so both substitutions can be performed at the same time:

$$x^2 + y^2 = 25 \quad (1)$$
$$x^2 + (\pm 3)^2 = 25$$
$$x^2 + 9 = 25$$
$$x^2 = 16$$
$$x = \pm 4.$$

Thus, if $y = 3$, $x = 4$ or $x = -4$, and if $y = -3$, $x = 4$ or $x = -4$. The possible solutions are $(4, 3)$, $(-4, 3)$, $(4, -3)$, and $(-4, -3)$. A check reveals that $(4, -3)$ and $(-4, 3)$ are not solutions of equation (2). Had we graphed the system of equations before solving algebraically, it would have been clear that there are only two real-number solutions.

EXAMPLE 2 Solve the following system of equations:

$$x + y = 5, \quad (1) \qquad \text{(The graph is a line.)}$$
$$y = 3 - x^2. \quad (2) \qquad \text{(The graph is a parabola.)}$$

Algebraic Solution

We use the substitution method, substituting $3 - x^2$ for y in equation (1):

$$x + 3 - x^2 = 5$$
$$-x^2 + x - 2 = 0 \qquad \text{Subtracting 5 and rearranging}$$
$$x^2 - x + 2 = 0. \qquad \text{Multiplying by } -1$$

Next, we use the quadratic formula:

$$x = \frac{-b \pm \sqrt{b^2 - 4ac}}{2a} = \frac{-(-1) \pm \sqrt{(-1)^2 - 4(1)(2)}}{2(1)}$$
$$= \frac{1 \pm \sqrt{1 - 8}}{2} = \frac{1 \pm \sqrt{-7}}{2} = \frac{1 \pm i\sqrt{7}}{2} = \frac{1}{2} \pm \frac{\sqrt{7}}{2}i.$$

Now, we substitute these values for x in equation (1) and solve for y:

$$\frac{1}{2} + \frac{\sqrt{7}}{2}i + y = 5$$
$$y = 5 - \frac{1}{2} - \frac{\sqrt{7}}{2}i = \frac{9}{2} - \frac{\sqrt{7}}{2}i$$

and $\quad \dfrac{1}{2} - \dfrac{\sqrt{7}}{2}i + y = 5$

$$y = 5 - \frac{1}{2} + \frac{\sqrt{7}}{2}i = \frac{9}{2} + \frac{\sqrt{7}}{2}i.$$

The solutions are

$$\left(\frac{1}{2} + \frac{\sqrt{7}}{2}i, \frac{9}{2} - \frac{\sqrt{7}}{2}i\right) \quad \text{and} \quad \left(\frac{1}{2} - \frac{\sqrt{7}}{2}i, \frac{9}{2} + \frac{\sqrt{7}}{2}i\right).$$

There are no real-number solutions.

Graphical Solution

We graph both equations in the same viewing window.

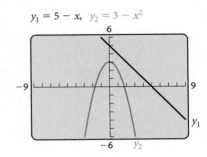

$y_1 = 5 - x, \quad y_2 = 3 - x^2$

Note that there are no points of intersection. This indicates that there are no real-number solutions. Algebra must be used, as at left, to find the imaginary-number solutions.

EXAMPLE 3   Solve the following system of equations:

$$2x^2 + 5y^2 = 39, \quad (1) \qquad \text{(The graph is an ellipse.)}$$
$$3x^2 - y^2 = -1. \quad (2) \qquad \text{(The graph is a hyperbola.)}$$

Algebraic Solution

We use the elimination method. First, we multiply equation (2) by 5 and add to eliminate the y^2-term:

$$2x^2 + 5y^2 = 39 \quad (1)$$
$$\underline{15x^2 - 5y^2 = -5} \quad \text{Multiplying (2) by 5}$$
$$17x^2 \qquad\quad = 34 \quad \text{Adding}$$
$$x^2 = 2$$
$$x = \pm\sqrt{2}.$$

If $x = \sqrt{2}$, $x^2 = 2$, and if $x = -\sqrt{2}$, $x^2 = 2$. Thus substituting $\sqrt{2}$ or $-\sqrt{2}$ for x in equation (2) gives us

$$3(\pm\sqrt{2})^2 - y^2 = -1$$
$$3 \cdot 2 - y^2 = -1$$
$$6 - y^2 = -1$$
$$-y^2 = -7$$
$$y^2 = 7$$
$$y = \pm\sqrt{7}.$$

Thus, for $x = \sqrt{2}$, we have $y = \sqrt{7}$ or $y = -\sqrt{7}$, and for $x = -\sqrt{2}$, we have $y = \sqrt{7}$ or $y = -\sqrt{7}$. The possible solutions are $(\sqrt{2}, \sqrt{7})$, $(\sqrt{2}, -\sqrt{7})$, $(-\sqrt{2}, \sqrt{7})$, and $(-\sqrt{2}, -\sqrt{7})$. All four pairs check, so they are the solutions.

Graphical Solution

We graph both equations in the same viewing window and note that there are four points of intersection. We can use the INTERSECT feature to find their coordinates.

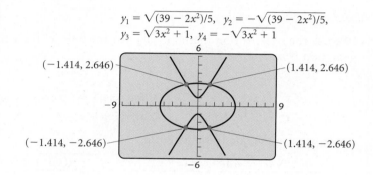

$$y_1 = \sqrt{(39 - 2x^2)/5}, \quad y_2 = -\sqrt{(39 - 2x^2)/5},$$
$$y_3 = \sqrt{3x^2 + 1}, \quad y_4 = -\sqrt{3x^2 + 1}$$

$(-1.414, 2.646)$ $(1.414, 2.646)$
$(-1.414, -2.646)$ $(1.414, -2.646)$

Note that the algebraic method yields exact solutions whereas the graphical method yields decimal approximations of the solutions.

The solutions are approximately $(1.414, 2.646)$, $(1.414, -2.646)$, $(-1.414, 2.646)$, and $(-1.414, -2.646)$.

EXAMPLE 4 Solve the following system of equations:

$$x^2 - 3y^2 = 6, \quad (1)$$
$$xy = 3. \quad (2)$$

Algebraic Solution

We use the substitution method. First, we solve equation (2) for y:

$$xy = 3$$
$$y = \frac{3}{x}. \quad (3)$$

Next, we substitute $3/x$ for y in equation (1) and solve for x:

$$x^2 - 3\left(\frac{3}{x}\right)^2 = 6$$
$$x^2 - 3 \cdot \frac{9}{x^2} = 6$$
$$x^2 - \frac{27}{x^2} = 6$$

$$x^4 - 27 = 6x^2 \qquad \text{Multiplying by } x^2$$
$$x^4 - 6x^2 - 27 = 0$$
$$u^2 - 6u - 27 = 0 \qquad \text{Letting } u = x^2$$
$$(u - 9)(u + 3) = 0 \qquad \text{Factoring}$$
$$u = 9 \quad \text{or} \quad u = -3 \qquad \text{Principle of zero products}$$
$$x^2 = 9 \quad \text{or} \quad x^2 = -3 \qquad \text{Substituting } x^2 \text{ for } u$$
$$x = \pm 3 \quad \text{or} \quad x = \pm i\sqrt{3}.$$

Since $y = 3/x$,

when $x = 3$, $\quad y = \frac{3}{3} = 1;$

when $x = -3$, $\quad y = \frac{3}{-3} = -1;$

when $x = i\sqrt{3}$, $\quad y = \frac{3}{i\sqrt{3}} = \frac{3}{i\sqrt{3}} \cdot \frac{-i\sqrt{3}}{-i\sqrt{3}} = -i\sqrt{3};$

when $x = -i\sqrt{3}$, $\quad y = \frac{3}{-i\sqrt{3}} = \frac{3}{-i\sqrt{3}} \cdot \frac{i\sqrt{3}}{i\sqrt{3}} = i\sqrt{3}.$

The pairs $(3, 1)$, $(-3, -1)$, $(i\sqrt{3}, -i\sqrt{3})$, and $(-i\sqrt{3}, i\sqrt{3})$ check, so they are the solutions.

Graphical Solution

We graph both equations in the same viewing window and find the coordinates of their points of intersection.

Again, note that the graphical method yields only the real-number solutions of the system of equations. The algebraic method must be used to find *all* the solutions.

Modeling and Problem Solving

EXAMPLE 5 *Dimensions of a Piece of Land.* **For** a student recreation building at Southport Community College, an architect wants to lay out a rectangular piece of land that has a perimeter of 204 m and an area of 2565 m². Find the dimensions of the piece of land.

Solution

1. **Familiarize.** We make a drawing and label it, letting l = the length of the piece of land, in meters, and w = the width, in meters.

2. **Translate.** We now have the following:

   Perimeter: $2w + 2l = 204,$ (1)

   Area: $lw = 2565.$ (2)

3. **Carry out.** We solve the system of equations both algebraically and graphically.

Algebraic Solution

We solve the system of equations

$$2w + 2l = 204,$$
$$lw = 2565.$$

Solving the second equation for l gives us $l = 2565/w$. We then substitute $2565/w$ for l in equation (1) and solve for w:

$$2w + 2\left(\frac{2565}{w}\right) = 204$$

$$2w^2 + 2(2565) = 204w \qquad \text{Multiplying by } w$$

$$2w^2 - 204w + 2(2565) = 0$$

$$w^2 - 102w + 2565 = 0 \qquad \text{Multiplying by } \tfrac{1}{2}$$

$$(w - 57)(w - 45) = 0$$

$$w = 57 \quad or \quad w = 45.$$

Principle of zero products

If $w = 57$, then $l = 2565/w = 2565/57 = 45$. If $w = 45$, then $l = 2565/w = 2565/45 = 57$. Since length is generally considered to be longer than width, we have the solution $l = 57$ and $w = 45$, or $(57, 45)$.

Area = lw
= 2565 m²

w

l

Perimeter = $2w + 2l$
= 204 m

Graphical Solution

We replace l with x and w with y, graph $y_1 = (204 - 2x)/2$ and $y_2 = 2565/x$, and find the point(s) of intersection of the graphs.

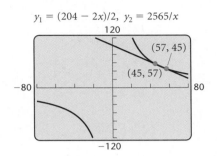

$$y_1 = (204 - 2x)/2, \ y_2 = 2565/x$$

As in the algebraic solution, we have two possible solutions: $(45, 57)$ and $(57, 45)$. Since length, x, is generally considered to be longer than width, y, we have the solution $(57, 45)$.

4. Check. If $l = 57$ and $w = 45$, the perimeter is $2 \cdot 57 + 2 \cdot 45$, or 204. The area is $57 \cdot 45$, or 2565. The numbers check.

5. State. The length of the piece of land is 57 m and the width is 45 m.

Exercise Set **5.4**

In Exercises 1–6, match the system of equations with one of the graphs (a)–(f), which follow.

a)

b)

c)

d)

e)

f)

1. $x^2 + y^2 = 16,,$
  $x + y = 3$

2. $16x^2 + 9y^2 = 144,$
  $x - y = 4$

3. $y = x^2 - 4x - 2,$
  $2y - x = 1$

4. $4x^2 - 9y^2 = 36,$
  $x^2 + y^2 = 25$

5. $y = x^2 - 3,$
  $x^2 + 4y^2 = 16$

6. $y^2 - 2y = x + 3,$
  $xy = 4$

Solve.

7. $x^2 + y^2 = 25$,
$y - x = 1$

8. $x^2 + y^2 = 100$,
$y - x = 2$

9. $4x^2 + 9y^2 = 36$,
$3y + 2x = 6$

10. $9x^2 + 4y^2 = 36$,
$3x + 2y = 6$

11. $x^2 + y^2 = 25$,
$y^2 = x + 5$

12. $y = x^2$,
$x = y^2$

13. $x^2 + y^2 = 9$,
$x^2 - y^2 = 9$

14. $y^2 - 4x^2 = 4$,
$4x^2 + y^2 = 4$

15. $y^2 - x^2 = 9$,
$2x - 3 = y$

16. $x + y = -6$,
$xy = -7$

17. $y^2 = x + 3$,
$2y = x + 4$

18. $y = x^2$,
$3x = y + 2$

19. $x^2 + y^2 = 25$,
$xy = 12$

20. $x^2 - y^2 = 16$,
$x + y^2 = 4$

21. $x^2 + y^2 = 4$,
$16x^2 + 9y^2 = 144$

22. $x^2 + y^2 = 25$,
$25x^2 + 16y^2 = 400$

23. $x^2 + 4y^2 = 25$,
$x + 2y = 7$

24. $y^2 - x^2 = 16$,
$2x - y = 1$

25. $x^2 - xy + 3y^2 = 27$,
$x - y = 2$

26. $2y^2 + xy + x^2 = 7$,
$x - 2y = 5$

27. $x^2 + y^2 = 16$,
$y^2 - 2x^2 = 10$

28. $x^2 + y^2 = 14$,
$x^2 - y^2 = 4$

29. $x^2 + y^2 = 5$,
$xy = 2$

30. $x^2 + y^2 = 20$,
$xy = 8$

31. $3x + y = 7$,
$4x^2 + 5y = 56$

32. $2y^2 + xy = 5$,
$4y + x = 7$

33. $a + b = 7$,
$ab = 4$

34. $p + q = -4$,
$pq = -5$

35. $x^2 + y^2 = 13$,
$xy = 6$

36. $x^2 + 4y^2 = 20$,
$xy = 4$

37. $x^2 + y^2 + 6y + 5 = 0$,
$x^2 + y^2 - 2x - 8 = 0$

38. $2xy + 3y^2 = 7$,
$3xy - 2y^2 = 4$

39. $2a + b = 1$,
$b = 4 - a^2$

40. $4x^2 + 9y^2 = 36$,
$x + 3y = 3$

41. $a^2 + b^2 = 89$,
$a - b = 3$

42. $xy = 4$,
$x + y = 5$

43. $xy - y^2 = 2$,
$2xy - 3y^2 = 0$

44. $4a^2 - 25b^2 = 0$,
$2a^2 - 10b^2 = 3b + 4$

45. $m^2 - 3mn + n^2 + 1 = 0$,
$3m^2 - mn + 3n^2 = 13$

46. $ab - b^2 = -4$,
$ab - 2b^2 = -6$

47. $x^2 + y^2 = 5$,
$x - y = 8$

48. $4x^2 + 9y^2 = 36$,
$y - x = 8$

49. $a^2 + b^2 = 14$,
$ab = 3\sqrt{5}$

50. $x^2 + xy = 5$,
$2x^2 + xy = 2$

51. $x^2 + y^2 = 25$,
$9x^2 + 4y^2 = 36$

52. $x^2 + y^2 = 1$,
$9x^2 - 16y^2 = 144$

53. $5y^2 - x^2 = 1$,
$xy = 2$

54. $x^2 - 7y^2 = 6$,
$xy = 1$

55. *Picture Frame Dimensions.* Frank's Frame Shop is building a frame for a rectangular oil painting with a perimeter of 28 cm and a diagonal of 10 cm. Find the dimensions of the painting.

10 cm

56. *Landscaping.* Green Leaf Landscaping is planting a rectangular wildflower garden with a perimeter of 6 m and a diagonal of $\sqrt{5}$ m. Find the dimensions of the garden.

57. *Pamphlet Design.* A graphic artist is designing a rectangular advertising pamphlet that is to have an area of 20 in^2 and a perimeter of 18 in. Find the dimensions of the pamphlet.

58. *Sign Dimensions.* Peden's Advertising is building a rectangular sign with an area of 2 yd^2 and a perimeter of 6 yd. Find the dimensions of the sign.

59. *Banner Design.* A rectangular banner with an area of $\sqrt{3}$ m^2 is being designed to advertise an exhibit at the Madison Art League. The length of a diagonal is 2 m. Find the dimensions of the banner.

Area = $\sqrt{3}$ m^2

60. *Carpentry.* A carpenter wants to make a rectangular tabletop with an area of $\sqrt{2}$ m² and a diagonal of $\sqrt{3}$ m. Find the dimensions of the tabletop.

61. *Fencing.* It will take 210 yd of fencing to enclose a rectangular dog run. The area of the run is 2250 yd². What are the dimensions of the run?

62. *Office Dimensions.* The diagonal of the floor of a rectangular office cubicle is 1 ft longer than the length of the cubicle and 3 ft longer than twice the width. Find the dimensions of the cubicle.

63. *Seed Test Plots.* The Burton Seed Company has two square test plots. The sum of their areas is 832 ft² and the difference of their areas is 320 ft². Find the length of a side of each plot.

64. *Investment.* Jenna made an investment for 1 yr that earned $7.50 simple interest. If the principal had been $25 more and the interest rate 1% less, the interest would have been the same. Find the principal and the rate.

65. *Social Security Trust Fund.* Income and expenditures of the Social Security Trust Fund are shown in the table below.

YEAR	INCOME (IN BILLIONS)	EXPENDITURES (IN BILLIONS)
1985	$203.5	$190.6
1990	315.4	253.1
1991	329.7	274.2
1992	342.6	291.9
1993	355.6	308.8
1994	381.1	323.0
1995	399.7	339.8
1996	424.5	353.6
1997	457.7	369.1
1998	484.3	382.9
1999	503.7	396.3

Source: Social Security Administration

a) Find a linear function $I(x)$ that models the income data. Let x represent the number of years after 1985.
b) Find an exponential function $E(x)$ that models the expenditure data. Let x represent the number of years after 1985.
c) Use the functions found in parts (a) and (b) to determine when income will equal expenditures.

66. *Higher Education.* The percent of associate's, bachelor's, master's, first professional, and doctor's degrees conferred on men and women in the United States in recent years are shown in the table below.

YEAR	PERCENT MALE	PERCENT FEMALE
1960	65.8	34.2
1965	61.5	38.5
1970	59.2	40.8
1975	56.0	44.0
1980	51.1	48.9
1985	49.3	50.7
1990	46.6	53.4
1995	44.9	55.1

Source: U.S. National Center for Education Statistics

a) Find an exponential function $M(x)$ that models the percent of degrees that were earned by men. Let x represent the number of years after 1960.
b) Find a linear function $F(x)$ that models the percent of degrees that were earned by women. Let x represent the number of years after 1960.
c) Use the functions found in parts (a) and (b) to determine when equal percents of men and women earned degrees.

Discussion and Writing

67. What would you say to a classmate who tells you that any nonlinear system of equations can be solved graphically?

68. Write a problem that can be translated to a nonlinear system of equations, and ask a classmate to solve it. Devise the problem so that the solution is "The dimensions of the rectangle are 6 ft by 8 ft."

Skill Maintenance

Simplify. Write answers in the form a + bi, where a and b are real numbers.

69. $(4 - 7i)(3 + 8i)$

70. $\dfrac{i}{3 - i}$

Find standard notation, a + bi.

71. $2(\cos 60° + i \sin 60°)$

72. $\sqrt{3}\left[\cos\left(-\dfrac{\pi}{3}\right) + i \sin\left(-\dfrac{\pi}{3}\right)\right]$

Synthesis

73. Find an equation of the circle that passes through the points (2, 4) and (3, 3) and whose center is on the line $3x - y = 3$.

74. Find an equation of the circle that passes through the points $(-2, 3)$ and $(-4, 1)$ and whose center is on the line $5x + 8y = -2$.

75. Find an equation of an ellipse centered at the origin that passes through the points $(1, \sqrt{3}/2)$ and $(\sqrt{3}, 1/2)$.

76. Find an equation of a hyperbola of the type

$$\frac{x^2}{b^2} - \frac{y^2}{a^2} = 1$$

that passes through the points $(-3, -3\sqrt{5}/2)$ and $(-3/2, 0)$.

77. Find an equation of the circle that passes through the points (4, 6), $(-6, 2)$, and $(1, -3)$.

78. Find an equation of the circle that passes through the points (2, 3), (4, 5), and $(0, -3)$.

79. Show that a hyperbola does not intersect its asymptotes. That is, solve the system of equations

$$\frac{x^2}{a^2} - \frac{y^2}{b^2} = 1,$$

$$y = \frac{b}{a}x \quad \left(\text{or } y = -\frac{b}{a}x \right).$$

80. *Numerical Relationship.* Find two numbers whose product is 2 and the sum of whose reciprocals is $\frac{33}{8}$.

81. *Numerical Relationship.* The square of a number exceeds twice the square of another number by $\frac{1}{8}$. The sum of their squares is $\frac{5}{16}$. Find the numbers.

82. *Box Dimensions.* Four squares with sides 5 in. long are cut from the corners of a rectangular metal sheet that has an area of 340 in². The edges are bent up to form an open box with a volume of 350 in³. Find the dimensions of the box.

83. *Numerical Relationship.* The sum of two numbers is 1, and their product is 1. Find the sum of their cubes. There is a method to solve this problem that is easier than solving a nonlinear system of equations. Can you discover it?

84. Solve for x and y:

$$x^2 - y^2 = a^2 - b^2,$$
$$x - y = a - b.$$

Solve.

85. $x^3 + y^3 = 72,$
$\quad x + y = 6$

86. $a + b = \dfrac{5}{6},$
$\quad \dfrac{a}{b} + \dfrac{b}{a} = \dfrac{13}{6}$

87. $p^2 + q^2 = 13,$
$\quad \dfrac{1}{pq} = -\dfrac{1}{6}$

88. $x^2 + y^2 = 4,$
$\quad (x - 1)^2 + y^2 = 4$

Solve using a grapher.

89. $x^2 + y^2 = 19{,}380{,}510.36,$
$\quad 27{,}942.25x - 6.125y = 0$

90. $2x + 2y = 1660,$
$\quad xy = 35{,}325$

91. $14.5x^2 - 13.5y^2 - 64.5 = 0,$
$\quad 5.5x - 6.3y - 12.3 = 0$

92. $13.5xy + 15.6 = 0,$
$\quad 5.6x - 6.7y - 42.3 = 0$

93. $0.319x^2 + 2688.7y^2 = 56{,}548,$
$\quad 0.306x^2 - 2688.7y^2 = 43{,}452$

94. $18.465x^2 + 788.723y^2 = 6408,$
$\quad 106.535x^2 - 788.723y^2 = 2692$

Rotation of Axes

5.5

CONIC SECTIONS

REVIEW SECTION 5.1.

- *Use rotation of axes to graph conic sections.*
- *Use the discriminant to determine the type of conic represented by a given equation.*

In Section 5.1, we saw that conic sections can be defined algebraically using a second-degree equation of the form $Ax^2 + Bxy + Cy^2 + Dx + Ey + F = 0$. Up to this point, we have considered only equations of this form for which $B = 0$. Now we turn our attention to equations of conics that contain an xy-term.

Rotation of Axes

When B is nonzero, the graph of $Ax^2 + Bxy + Cy^2 + Dx + Ey + F = 0$ is a conic section with an axis that is not parallel to the x- or y-axis. We use a technique called **rotation of axes** when we graph such an equation. The goal is to rotate the x- and y-axes through a positive angle θ to yield an $x'y'$-coordinate system, as shown at left. For the appropriate choice of θ, the graph of any conic section with an xy-term will have its axis parallel to the x'-axis or the y'-axis.

Algebraically we want to rewrite an equation

$$Ax^2 + Bxy + Cy^2 + Dx + Ey + F = 0$$

in the xy-coordinate system in the form

$$A'(x')^2 + C'(y')^2 + D'x' + E'y' + F' = 0$$

in the $x'y'$-coordinate system. Equations of this second type were graphed in Sections 5.1–5.3.

To achieve our goal, we find formulas relating the xy-coordinates of a point and the $x'y'$-coordinates of the same point. We begin by letting P be a point with coordinates (x, y) in the xy-coordinate system and (x', y') in the $x'y'$-coordinate system.

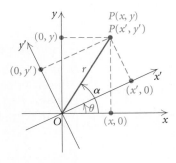

We let r represent the distance OP, and we let α represent the angle from the x-axis to OP. Then

$$\cos \alpha = \frac{x}{r} \quad \text{and} \quad \sin \alpha = \frac{y}{r},$$

so

$$x = r \cos \alpha \quad \text{and} \quad y = r \sin \alpha.$$

We also see from the figure above that

$$\cos (\alpha - \theta) = \frac{x'}{r} \quad \text{and} \quad \sin (\alpha - \theta) = \frac{y'}{r},$$

so

$$x' = r \cos (\alpha - \theta) \quad \text{and} \quad y' = r \sin (\alpha - \theta).$$

Then

$$x' = r \cos \alpha \cos \theta + r \sin \alpha \sin \theta$$

and

$$y' = r \sin \alpha \cos \theta - r \cos \alpha \sin \theta.$$

Substituting x for $r \cos \alpha$ and y for $r \sin \alpha$ gives us

$$x' = x \cos \theta + y \sin \theta \qquad (1)$$

and

$$y' = y \cos \theta - x \sin \theta. \qquad (2)$$

We can use these formulas to find the $x'y'$-coordinates of any point given that point's xy-coordinates and an angle of rotation θ. To express xy-coordinates in terms of $x'y'$-coordinates and an angle of rotation θ, we solve the system composed of equations (1) and (2) above for x and y. (See Exercise 45.) We get

$$x = x' \cos \theta - y' \sin \theta$$

and

$$y = x' \sin \theta + y' \cos \theta.$$

Rotation of Axes Formulas

If the x- and y-axes are rotated about the origin through a positive acute angle θ, then the coordinates (x, y) and (x', y') of a point P in the xy- and $x'y'$-coordinate systems are related by the following formulas:

$$x' = x \cos \theta + y \sin \theta, \quad y' = -x \sin \theta + y \cos \theta;$$
$$x = x' \cos \theta - y' \sin \theta, \quad y = x' \sin \theta + y' \cos \theta.$$

EXAMPLE 1   Suppose that the xy-axes are rotated through an angle of $45°$. Write the equation $xy = 1$ in the $x'y'$-coordinate system.

Solution   We substitute 45° for θ in the rotation of axes formulas for x and y:

$$x = x' \cos 45° - y' \sin 45°,$$
$$y = x' \sin 45° + y' \cos 45°.$$

Then we have

$$x = x'\left(\frac{\sqrt{2}}{2}\right) - y'\left(\frac{\sqrt{2}}{2}\right) = \frac{\sqrt{2}}{2}(x' - y')$$

and

$$y = x'\left(\frac{\sqrt{2}}{2}\right) + y'\left(\frac{\sqrt{2}}{2}\right) = \frac{\sqrt{2}}{2}(x' + y').$$

Next, we substitute these expressions for x and y in the equation $xy = 1$:

$$\frac{\sqrt{2}}{2}(x' - y') \cdot \frac{\sqrt{2}}{2}(x' + y') = 1$$

$$\frac{1}{2}[(x')^2 - (y')^2] = 1$$

$$\frac{(x')^2}{2} - \frac{(y')^2}{2} = 1, \quad \text{or} \quad \frac{(x')^2}{(\sqrt{2})^2} - \frac{(y')^2}{(\sqrt{2})^2} = 1.$$

We have the equation of a hyperbola in the $x'y'$-coordinate system with its axis on the x'-axis and with vertices $(-\sqrt{2}, 0)$ and $(\sqrt{2}, 0)$. Its asymptotes are $y' = -x'$ and $y' = x'$. These correspond to the axes of the xy-coordinate system.

Now let's substitute the rotation of axes formulas for x and y in the equation

$$Ax^2 + Bxy + Cy^2 + Dx + Ey + F = 0.$$

We have

$$A(x' \cos \theta - y' \sin \theta)^2 + B(x' \cos \theta - y' \sin \theta)(x' \sin \theta + y' \cos \theta)$$
$$+ C(x' \sin \theta + y' \cos \theta)^2 + D(x' \cos \theta - y' \sin \theta)$$
$$+ E(x' \sin \theta + y' \cos \theta) + F = 0.$$

Performing the operations indicated and collecting like terms yields the equation

$$A'(x')^2 + B'x'y' + C'(y')^2 + D'x' + E'y' + F' = 0, \qquad (3)$$

where

$$A' = A \cos^2 \theta + B \sin \theta \cos \theta + C \sin^2 \theta,$$
$$B' = 2(C - A) \sin \theta \cos \theta + B(\cos^2 \theta - \sin^2 \theta),$$
$$C' = A \sin^2 \theta - B \sin \theta \cos \theta + C \cos^2 \theta,$$
$$D' = D \cos \theta + E \sin \theta,$$
$$E' = -D \sin \theta + E \cos \theta, \quad \text{and}$$
$$F' = F.$$

Recall that our goal is to produce an equation without an $x'y'$-term, or with $B' = 0$. Then we must have

$$2(C - A) \sin \theta \cos \theta + B(\cos^2 \theta - \sin^2 \theta) = 0$$

$$(C - A) \sin 2\theta + B \cos 2\theta = 0 \qquad \text{Using double-angle formulas}$$

$$B \cos 2\theta = (A - C) \sin 2\theta$$

$$\frac{\cos 2\theta}{\sin 2\theta} = \frac{A - C}{B}$$

$$\cot 2\theta = \frac{A - C}{B}.$$

Thus, when θ is chosen so that

$$\cot 2\theta = \frac{A - C}{B},$$

equation (3) will have no $x'y'$-term. Although we will not do so here, it can be shown that we can always find θ such that $0° < 2\theta < 180°$, or $0° < \theta < 90°$.

Eliminating the xy-Term

To eliminate the xy-term from the equation

$$Ax^2 + Bxy + Cy^2 + Dx + Ey + F = 0, \quad B \neq 0,$$

select an angle θ such that

$$\cot 2\theta = \frac{A - C}{B}, \quad 0° < 2\theta < 180°,$$

and use the rotation of axes formulas.

EXAMPLE 2 Graph the equation
$$3x^2 - 2\sqrt{3}xy + y^2 + 2x + 2\sqrt{3}y = 0.$$

Solution We have
$$A = 3, \quad B = -2\sqrt{3}, \quad C = 1, \quad D = 2, \quad E = 2\sqrt{3}, \quad \text{and} \quad F = 0.$$

To select the angle of rotation θ, we must have
$$\cot 2\theta = \frac{A - C}{B} = \frac{3 - 1}{-2\sqrt{3}} = -\frac{1}{\sqrt{3}}.$$

Thus, $2\theta = 120°$, and $\theta = 60°$. We substitute this value for θ in the rotation of axes formulas for x and y:
$$x = x' \cos 60° - y' \sin 60°,$$
$$y = x' \sin 60° + y' \cos 60°.$$

This gives us
$$x = x' \cdot \frac{1}{2} - y' \cdot \frac{\sqrt{3}}{2} = \frac{x'}{2} - \frac{y'\sqrt{3}}{2}$$

and
$$y = x' \cdot \frac{\sqrt{3}}{2} + y' \cdot \frac{1}{2} = \frac{x'\sqrt{3}}{2} + \frac{y'}{2}.$$

Now we substitute these expressions for x and y in the given equation:
$$3\left(\frac{x'}{2} - \frac{y'\sqrt{3}}{2}\right)^2 - 2\sqrt{3}\left(\frac{x'}{2} - \frac{y'\sqrt{3}}{2}\right)\left(\frac{x'\sqrt{3}}{2} + \frac{y'}{2}\right) +$$
$$\left(\frac{x'\sqrt{3}}{2} + \frac{y'}{2}\right)^2 + 2\left(\frac{x'}{2} - \frac{y'\sqrt{3}}{2}\right) + 2\sqrt{3}\left(\frac{x'\sqrt{3}}{2} + \frac{y'}{2}\right) = 0.$$

After simplifying, we get
$$4(y')^2 + 4x' = 0, \quad \text{or}$$
$$(y')^2 = -x'.$$

This is the equation of a parabola with its vertex at $(0, 0)$ of the $x'y'$-coordinate system and axis of symmetry $y' = 0$. We sketch the graph.

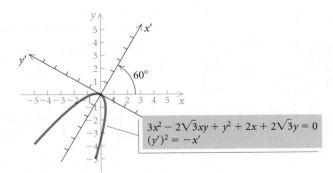

The Discriminant

It is possible to determine the type of conic represented by the equation $Ax^2 + Bxy + Cy^2 + Dx + Ey + F = 0$ before rotating the axes. Using the expressions for A', B', and C' in terms of A, B, C, and θ developed earlier, it can be shown that

$$(B')^2 - 4A'C' = B^2 - 4AC.$$

Now when θ is chosen so that

$$\cot 2\theta = \frac{A - C}{B},$$

rotation of axes gives us an equation

$$A'(x')^2 + C'(y')^2 + D'x' + E'y' + F' = 0.$$

If A' and C' have the same sign, or $A'C' > 0$, then the graph of this equation is an ellipse or a circle. If A' and C' have different signs, or $A'C' < 0$, then the graph is a hyperbola. And, if either $A' = 0$ or $C' = 0$, or $A'C' = 0$, the graph is a parabola.

Since $B' = 0$ and $(B')^2 - 4A'C' = B^2 - 4AC$, it follows that $B^2 - 4AC = -4A'C'$. Then the graph is an ellipse or a circle if $B^2 - 4AC < 0$, a hyperbola if $B^2 - 4AC > 0$, or a parabola if $B^2 - 4AC = 0$. (There are certain special cases, called *degenerate conics*, where these statements do not hold, but we will not concern ourselves with these here.) The expression $B^2 - 4AC$ is the **discriminant** of the equation $Ax^2 + Bxy + Cy^2 + Dx + Ey + F = 0$.

The graph of the equation

$$Ax^2 + Bxy + Cy^2 + Dx + Ey + F = 0$$

is, except in degenerate cases,

1. an ellipse or a circle if $B^2 - 4AC < 0$,

2. a hyperbola if $B^2 - 4AC > 0$, and

3. a parabola if $B^2 - 4AC = 0$.

EXAMPLE 3   Graph the equation $3x^2 + 2xy + 3y^2 = 16$.

Solution   We have

$$A = 3, \qquad B = 2, \quad \text{and} \quad C = 3, \qquad \text{so}$$
$$B^2 - 4AC = 2^2 - 4 \cdot 3 \cdot 3 = 4 - 36 = -32.$$

Since the discriminant is negative, the graph is an ellipse or a circle. Now, to rotate the axes, we begin by determining θ:

$$\cot 2\theta = \frac{A - C}{B} = \frac{3 - 3}{2} = \frac{0}{2} = 0.$$

```
2²−4*3*3
                    −32
```

Then $2\theta = 90°$ and $\theta = 45°$, so

$$\sin \theta = \frac{\sqrt{2}}{2} \quad \text{and} \quad \cos \theta = \frac{\sqrt{2}}{2}.$$

Substituting in the rotation of axes formulas gives

$$x = x' \cos \theta - y' \sin \theta = x'\left(\frac{\sqrt{2}}{2}\right) - y'\left(\frac{\sqrt{2}}{2}\right) = \frac{\sqrt{2}}{2}(x' - y')$$

and

$$y = x' \sin \theta + y' \cos \theta = x'\left(\frac{\sqrt{2}}{2}\right) + y'\left(\frac{\sqrt{2}}{2}\right) = \frac{\sqrt{2}}{2}(x' + y').$$

Now we substitute for x and y in the given equation:

$$3\left[\frac{\sqrt{2}}{2}(x' - y')\right]^2 + 2\left[\frac{\sqrt{2}}{2}(x' - y')\right]\left[\frac{\sqrt{2}}{2}(x' + y')\right] +$$
$$3\left[\frac{\sqrt{2}}{2}(x' + y')\right]^2 = 16.$$

After simplifying, we have

$$4(x')^2 + 2(y')^2 = 16, \quad \text{or}$$
$$\frac{(x')^2}{4} + \frac{(y')^2}{8} = 1.$$

This is the equation of an ellipse with vertices $(0, -\sqrt{8})$ and $(0, \sqrt{8})$, or $(0, -2\sqrt{2})$ and $(0, 2\sqrt{2})$, on the y'-axis. The x'-intercepts are $(-2, 0)$ and $(2, 0)$. We sketch the graph.

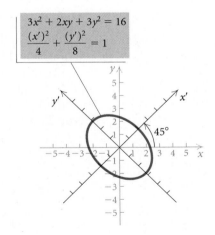

$$3x^2 + 2xy + 3y^2 = 16$$
$$\frac{(x')^2}{4} + \frac{(y')^2}{8} = 1$$

EXAMPLE 4 Graph the equation $4x^2 - 24xy - 3y^2 - 156 = 0$.

Solution We have

$$A = 4, \quad B = -24, \quad \text{and} \quad C = -3, \quad \text{so}$$
$$B^2 - 4AC = (-24)^2 - 4 \cdot 4(-3) = 576 + 48 = 624.$$

Since the discriminant is positive, the graph is a hyperbola. To rotate the axes, we begin by determining θ:

$$\cot 2\theta = \frac{A - C}{B} = \frac{4 - (-3)}{-24} = -\frac{7}{24}.$$

Since $\cot 2\theta < 0$, we have $90° < 2\theta < 180°$. From the triangle at left, we see that $\cos 2\theta = -\frac{7}{25}$.

Using half-angle formulas, we have

$$\sin \theta = \sqrt{\frac{1 - \cos 2\theta}{2}} = \sqrt{\frac{1 - \left(-\frac{7}{25}\right)}{2}} = \frac{4}{5}$$

and

$$\cos \theta = \sqrt{\frac{1 + \cos 2\theta}{2}} = \sqrt{\frac{1 + \left(-\frac{7}{25}\right)}{2}} = \frac{3}{5}.$$

Substituting in the rotation of axes formulas gives us

$$x = x' \cos \theta - y' \sin \theta = \tfrac{3}{5}x' - \tfrac{4}{5}y'$$

and

$$y = x' \sin \theta + y' \cos \theta = \tfrac{4}{5}x' + \tfrac{3}{5}y'.$$

Now we substitute for x and y in the given equation:

$$4\left(\tfrac{3}{5}x' - \tfrac{4}{5}y'\right)^2 - 24\left(\tfrac{3}{5}x' - \tfrac{4}{5}y'\right)\left(\tfrac{4}{5}x' + \tfrac{3}{5}y'\right) - 3\left(\tfrac{4}{5}x' + \tfrac{3}{5}y'\right)^2 - 156 = 0.$$

After simplifying, we have

$$13(y')^2 - 12(x')^2 - 156 = 0$$
$$13(y')^2 - 12(x')^2 = 156$$
$$\frac{(y')^2}{12} - \frac{(x')^2}{13} = 1.$$

The graph of this equation is a hyperbola with vertices $(0, -\sqrt{12})$ and $(0, \sqrt{12})$, or $(0, -2\sqrt{3})$ and $(0, 2\sqrt{3})$, on the y'-axis. Since we know that $\sin \theta = \frac{4}{5}$ and $0° < \theta < 90°$, we can use a grapher to find that $\theta \approx 53.1°$. Thus the xy-axes are rotated through an angle of about $53.1°$ in order to obtain the $x'y'$-axes. We sketch the graph.

Exercise Set 5.5

For the given angle of rotation and coordinates of a point in the xy-coordinate system, find the coordinates of the point in the x'y'-coordinate system.

1. $\theta = 45°$, $(\sqrt{2}, -\sqrt{2})$ **2.** $\theta = 45°$, $(-1, 3)$

3. $\theta = 30°$, $(0, 2)$ **4.** $\theta = 60°$, $(0, \sqrt{3})$

For the given angle of rotation and coordinates of a point in the x'y'-coordinate system, find the coordinates of the point in the xy-coordinate system.

5. $\theta = 45°$, $(1, -1)$ **6.** $\theta = 45°$, $(-3\sqrt{2}, \sqrt{2})$

7. $\theta = 30°$, $(2, 0)$ **8.** $\theta = 60°$, $(-1, -\sqrt{3})$

Use the discriminant to determine whether the graph of the equation is an ellipse (or a circle), a hyperbola, or a parabola.

9. $3x^2 - 5xy + 3y^2 - 2x + 7y = 0$

10. $5x^2 + 6xy - 4y^2 + x - 3y + 4 = 0$

11. $x^2 - 3xy - 2y^2 + 12 = 0$

12. $4x^2 + 7xy + 2y^2 - 3x + y = 0$

13. $4x^2 - 12xy + 9y^2 - 3x + y = 0$

14. $6x^2 + 5xy + 6y^2 + 15 = 0$

15. $2x^2 - 8xy + 7y^2 + x - 2y + 1 = 0$

16. $x^2 + 6xy + 9y^2 - 3x + 4y = 0$

17. $8x^2 - 7xy + 5y^2 - 17 = 0$

18. $x^2 + xy - y^2 - 4x + 3y - 2 = 0$

Graph the equation.

19. $3x^2 + 2xy + 3y^2 = 16$

20. $3x^2 + 10xy + 3y^2 + 8 = 0$

21. $x^2 - 10xy + y^2 + 36 = 0$

22. $x^2 + 2xy + y^2 + 4\sqrt{2}x - 4\sqrt{2}y = 0$

23. $x^2 - 2\sqrt{3}xy + 3y^2 - 12\sqrt{3}x - 12y = 0$

24. $13x^2 + 6\sqrt{3}xy + 7y^2 - 16 = 0$

25. $7x^2 + 6\sqrt{3}xy + 13y^2 - 32 = 0$

26. $x^2 + 4xy + y^2 - 9 = 0$

27. $11x^2 + 10\sqrt{3}xy + y^2 = 32$

28. $5x^2 - 8xy + 5y^2 = 81$

29. $\sqrt{2}x^2 + 2\sqrt{2}xy + \sqrt{2}y^2 - 8x + 8y = 0$

30. $x^2 + 2\sqrt{3}xy + 3y^2 - 8x + 8\sqrt{3}y = 0$

31. $x^2 + 6\sqrt{3}xy - 5y^2 + 8x - 8\sqrt{3}y - 48 = 0$

32. $3x^2 - 2xy + 3y^2 - 6\sqrt{2}x + 2\sqrt{2}y - 26 = 0$

33. $x^2 + xy + y^2 = 24$

34. $4x^2 + 3\sqrt{3}xy + y^2 = 55$

35. $4x^2 - 4xy + y^2 - 8\sqrt{5}x - 16\sqrt{5}y = 0$

36. $9x^2 - 24xy + 16y^2 - 400x - 300y = 0$

37. $11x^2 + 7xy - 13y^2 = 621$

38. $3x^2 + 4xy + 6y^2 = 28$

Discussion and Writing

39. Explain how the procedure you would follow for graphing an equation of the form $Ax^2 + Bxy + Cy^2 + Dx + Ey + F = 0$ when $B \neq 0$ differs from the procedure you would follow when $B = 0$.

40. Discuss some circumstances under which you might use rotation of axes.

Skill Maintenance

Convert to radian measure.

41. $120°$ **42.** $-315°$

Convert to degree measure.

43. $\dfrac{\pi}{3}$ **44.** $\dfrac{3\pi}{4}$

Synthesis

45. Solve this system of equations for x and y:

$$x' = x \cos \theta + y \sin \theta,$$
$$y' = y \cos \theta - x \sin \theta.$$

Show your work.

46. Show that substituting $x' \cos \theta - y' \sin \theta$ for x and $x' \sin \theta + y' \cos \theta$ for y in the equation

$$Ax^2 + Bxy + Cy^2 + Dx + Ey + F = 0$$

yields the equation

$$A'(x')^2 + B'x'y' + C'(y')^2 + D'x' + E'y' + F' = 0,$$

where

$A' = A \cos^2 \theta + B \sin \theta \cos \theta + C \sin^2 \theta,$

$B' = 2(C - A) \sin \theta \cos \theta + B(\cos^2 \theta - \sin^2 \theta),$

$C' = A \sin^2 \theta - B \sin \theta \cos \theta + C \cos^2 \theta,$

$D' = D \cos \theta + E \sin \theta,$

$E' = -D \sin \theta + E \cos \theta,$ and

$F' = F.$

47. Show that $A + C = A' + C'$.

48. Show that for any angle θ, the equation $x^2 + y^2 = r^2$ becomes $(x')^2 + (y')^2 = r^2$ when the rotation of axes formulas are applied.

Polar Coordinates and Graphs

5.6

- *Graph points given their polar coordinates.*
- *Convert from rectangular to polar coordinates and from polar to rectangular coordinates.*
- *Convert from rectangular to polar equations and from polar to rectangular equations.*
- *Graph polar equations.*

Polar Coordinates

All graphing throughout this text has been done with rectangular coordinates, (x, y), in the Cartesian coordinate system. We now introduce the polar coordinate system. As shown in the diagram at left, any point P has rectangular coordinates (x, y) and polar coordinates (r, θ). On a polar graph, the origin is called the **pole** and the positive half of the x-axis is called the **polar axis**. The point P can be plotted given the directed angle θ from the polar axis to the ray OP and the directed distance r from the pole to the point. The angle θ can be expressed in degrees or radians.

To plot points on a polar graph:

1. Locate the directed angle θ.
2. Move a directed distance r from the pole. If $r > 0$, move along ray OP. If $r < 0$, move in the opposite direction of ray OP.

Polar graph paper, shown below, facilitates plotting. Points B and G illustrate that θ may be in radians. Points E and F illustrate that the polar coordinates of a point are not unique.

EXAMPLE 1 Graph each of the following points.

a) $A(3, 60°)$ **b)** $B(0, 10°)$

c) $C(-5, 120°)$ **d)** $D(1, -60°)$

e) $E\left(2, \dfrac{3\pi}{2}\right)$ **f)** $F\left(-4, \dfrac{\pi}{3}\right)$

Solution

To convert from rectangular to polar coordinates and from polar to rectangular coordinates, we need to recall the following relationships.

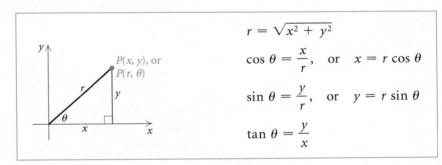

$$r = \sqrt{x^2 + y^2}$$

$$\cos\theta = \frac{x}{r}, \quad \text{or} \quad x = r\cos\theta$$

$$\sin\theta = \frac{y}{r}, \quad \text{or} \quad y = r\sin\theta$$

$$\tan\theta = \frac{y}{x}$$

EXAMPLE 2 Convert each of the following to polar coordinates.

a) $(3, 3)$ **b)** $(2\sqrt{3}, -2)$

Solution

a) We first find r:

$$r = \sqrt{3^2 + 3^2} = \sqrt{18} = 3\sqrt{2}.$$

Then we determine θ:

$$\tan\theta = \frac{3}{3} = 1; \quad \text{therefore,} \quad \theta = 45°, \text{ or } \frac{\pi}{4}.$$

We know that $\theta = \pi/4$ and not $5\pi/4$ since $(3, 3)$ is in quadrant I. Thus, $(r, \theta) = (3\sqrt{2}, 45°)$, or $(3\sqrt{2}, \pi/4)$. Other possibilities for polar coordinates include $(3\sqrt{2}, -315°)$ and $(-3\sqrt{2}, 5\pi/4)$.

b) We first find r:

$$r = \sqrt{(2\sqrt{3})^2 + (-2)^2} = \sqrt{12 + 4} = \sqrt{16} = 4.$$

Then we determine θ:

$$\tan \theta = \frac{-2}{2\sqrt{3}} = -\frac{1}{\sqrt{3}}; \quad \text{therefore,} \quad \theta = 330°, \text{ or } \frac{11\pi}{6}.$$

Thus, $(r, \theta) = (4, 330°)$, or $(4, 11\pi/6)$. Other possibilities for polar coordinates for this point include $(4, -\pi/6)$ and $(-4, 150°)$.   ▬

It is easier to convert from polar to rectangular coordinates than from rectangular to polar coordinates.

EXAMPLE 3 Convert each of the following to rectangular coordinates.

a) $\left(10, \dfrac{\pi}{3} \right)$ **b)** $(-5, 135°)$

Solution

a) The ordered pair $(10, \pi/3)$ gives us $r = 10$ and $\theta = \pi/3$. We now find x and y:

$$x = r \cos \theta = 10 \cos \frac{\pi}{3} = 10 \cdot \frac{1}{2} = 5$$

and

$$y = r \sin \theta = 10 \sin \frac{\pi}{3} = 10 \cdot \frac{\sqrt{3}}{2} = 5\sqrt{3}.$$

Thus, $(x, y) = (5, 5\sqrt{3})$.

b) From the ordered pair $(-5, 135°)$, we know that $r = -5$ and $\theta = 135°$. We now find x and y:

$$x = -5 \cos 135° = -5 \cdot \left(-\frac{\sqrt{2}}{2} \right) = \frac{5\sqrt{2}}{2}$$

and

$$y = -5 \sin 135° = -5 \cdot \left(\frac{\sqrt{2}}{2} \right) = -\frac{5\sqrt{2}}{2}.$$

Thus, $(x, y) = (5\sqrt{2}/2, -5\sqrt{2}/2)$.   ▬

The above conversions can be easily made with some graphers.

Polar and Rectangular Equations

Some curves have simpler equations in polar coordinates than in rectangular coordinates. For others, the reverse is true.

EXAMPLE 4   Convert each of the following to a polar equation.

a) $x^2 + y^2 = 25$

b) $2x - y = 5$

Solution

a) We have

$$x^2 + y^2 = 25$$
$$(r \cos \theta)^2 + (r \sin \theta)^2 = 25 \qquad \text{Substituting for } x \text{ and } y$$
$$r^2 \cos^2 \theta + r^2 \sin^2 \theta = 25$$
$$r^2(\cos^2 \theta + \sin^2 \theta) = 25$$
$$r^2 = 25 \qquad \cos^2 \theta + \sin^2 \theta = 1$$
$$r = 5.$$

This example illustrates that the polar equation of a circle centered at the origin is much simpler than the rectangular equation.

b) We have

$$2x - y = 5$$
$$2(r \cos \theta) - (r \sin \theta) = 5$$
$$r(2 \cos \theta - \sin \theta) = 5.$$

In this example, we see that the rectangular equation is simpler than the polar equation.   ▬

EXAMPLE 5   Convert each of the following to a rectangular equation.

a) $r = 4$

b) $r \cos \theta = 6$

c) $r = 2 \cos \theta + 3 \sin \theta$

Solution

a) We have

$$r = 4$$
$$\sqrt{x^2 + y^2} = 4 \qquad \text{Substituting for } r$$
$$x^2 + y^2 = 16. \qquad \text{Squaring}$$

In squaring, we must be careful not to introduce solutions of the equation that are not already present. In this case, we did not, because the graph of either equation is a circle of radius 4 centered at the origin.

b) We have

$$r \cos \theta = 6$$
$$x = 6. \qquad x = r \cos \theta$$

The graph of $r \cos \theta = 6$, or $x = 6$, is a vertical line.

c) We have

$$r = 2 \cos \theta + 3 \sin \theta$$
$$r^2 = 2r \cos \theta + 3r \sin \theta \qquad \text{Multiplying by } r \text{ on both sides}$$
$$x^2 + y^2 = 2x + 3y. \qquad \text{Substituting } x^2 + y^2 \text{ for } r^2, x$$
$$\text{for } r \cos \theta, \text{ and } y \text{ for } r \sin \theta$$

Graphing Polar Equations

To graph a polar equation, we can make a table of values, choosing values of θ and calculating corresponding values of r. We plot the points and complete the graph, as we do when graphing a rectangular equation. A difference occurs in the case of a polar equation, because as θ increases sufficiently, points may begin to repeat and the curve will be traced again and again. When this happens, the curve is complete.

EXAMPLE 6   Make a hand-drawn graph of $r = 1 - \sin \theta$.

Solution   We first make a table of values. The TABLE feature on a grapher is the most efficient way to create this list. Note that the points begin to repeat at $\theta = 360°$. We plot these points and draw the curve, as shown below.

θ	r
0°	1
15°	0.74118
30°	0.5
45°	0.29289
60°	0.13397
75°	0.03407
90°	0
105°	0.03407
120°	0.13397
135°	0.29289
150°	0.5
165°	0.74118
180°	1

θ	r
195°	1.2588
210°	1.5
225°	1.7071
240°	1.866
255°	1.9659
270°	2
285°	1.9659
300°	1.866
315°	1.7071
330°	1.5
345°	1.2588
360°	1
375°	0.74118
390°	0.5

$r = 1 - \sin \theta$

Because of its heart shape, this curve is called a *cardioid*.

We plotted points in Example 6 because we feel that it is important to understand how these curves are developed. However, for the most part, we will graph polar equations using a grapher. The equation must usually be written first in the form $r = f(\theta)$. When graphing polar equations, it is necessary to decide on not only the best window dimensions but also the range of values for θ. Typically, we begin with a range of 0 to 2π for θ in radians and 0° to 360° for θ in degrees. Because most polar graphs are curved, it is important to square the window to minimize distortion.

Graph $r = 4 \sin 3\theta$. Begin by setting the grapher in POLAR mode and use either of the following windows:

Window
(Radians)
  θmin = 0
  θmax = 2π
  θstep = $\pi/24$
  Xmin = −9
  Xmax = 9
  Xscl = 1
  Ymin = −6
  Ymax = 6
  Yscl = 1

Window
(Degrees)
  θmin = 0
  θmax = 360
  θstep = 1
  Xmin = −9
  Xmax = 9
  Xscl = 1
  Ymin = −6
  Ymax = 6
  Yscl = 1

$r = 4 \sin 3\theta$

We observe the same graph in both windows. The grapher allows us to view the curve as it is formed.

Now graph each of the following equations and observe the effect of changing the coefficient of $\sin 3\theta$ and the coefficient of θ:

$$r = 2 \sin 3\theta, \qquad r = 6 \sin 3\theta, \qquad r = 4 \sin \theta,$$
$$r = 4 \sin 5\theta, \qquad r = 4 \sin 2\theta, \qquad r = 4 \sin 4\theta.$$

Polar equations of the form $r = a \cos n\theta$ and $r = a \sin n\theta$ have rose-shaped curves. The number a determines the length of the petals, and the number n determines the number of petals. If n is odd, there are n petals. If n is even, there are $2n$ petals.

EXAMPLE 7   Graph each of the following polar equations. Try to visualize the shape of the curve before graphing it.

a) $r = 3$ **b)** $r = 5 \sin \theta$

c) $r = \cos \theta$ **d)** $r = 2 \csc \theta$

Solution

a)

$r = 3$

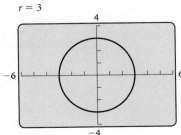

b)

$r = 5 \sin \theta$

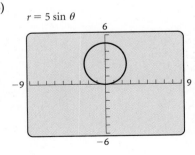

c)

$r = \cos \theta$

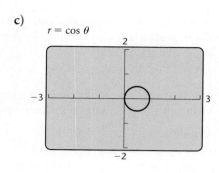

d)

$r = 2 \csc \theta = 2/\sin \theta$

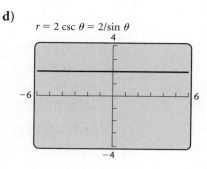

We can verify our graphs in parts (a)–(d) by converting the polar equation to the equivalent rectangular equation and visualizing the graph. For $r = 3$, we substitute $\sqrt{x^2 + y^2}$ for r and square. The resulting equation, $x^2 + y^2 = 3^2$, is the equation of a circle with radius 3 centered at the origin.

In part (d), we have

$$r = 2 \csc \theta$$

$$r = \frac{2}{\sin \theta}$$

$$r \sin \theta = 2$$

$$y = 2. \qquad \text{Substituting } y \text{ for } r \sin \theta$$

The graph of $y = 2$ is a horizontal line passing through $(0, 2)$.

Parts (b) and (c) can also be checked by rewriting the equations in rectangular form.

EXAMPLE 8   Graph: $r + 1 = 2 \cos 2\theta$.

Solution   We first solve for r:

$$r = 2 \cos 2\theta - 1.$$

We then obtain the following graph.

$r = 2 \cos 2\theta - 1$

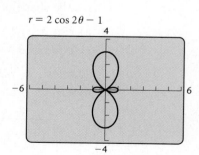

Exercise Set 5.6

Graph the point on a polar grid.

1. $(2, 45°)$ **2.** $(4, \pi)$ **3.** $(3.5, 210°)$

4. $(-3, 135°)$ **5.** $\left(1, \dfrac{\pi}{6}\right)$ **6.** $(2.75, 150°)$

7. $\left(-5, \dfrac{\pi}{2}\right)$ **8.** $(0, 15°)$ **9.** $(3, -315°)$

10. $\left(1.2, -\dfrac{2\pi}{3}\right)$ **11.** $(4.3, -60°)$ **12.** $(3, 405°)$

Find polar coordinates of points A, B, C, and D. Give three answers for each point.

13.

14.

Find the polar coordinates of the point. Express the angle in degrees and then in radians, using the smallest possible positive angle.

15. $(0, -3)$ **16.** $(-4, 4)$

17. $(3, -3\sqrt{3})$ **18.** $(-\sqrt{3}, 1)$

19. $(4\sqrt{3}, -4)$ **20.** $(2\sqrt{3}, 2)$

21. $(-\sqrt{2}, -\sqrt{2})$ **22.** $(-3, 3\sqrt{3})$

Use a grapher to convert from rectangular to polar coordinates. Express the answer in both degrees and radians, using the smallest possible positive angle.

23. $(3, 7)$ **24.** $(-2, -\sqrt{5})$

25. $(-\sqrt{10}, 3.4)$ **26.** $(0.9, -6)$

Find the rectangular coordinates of the point.

27. $(5, 60°)$ **28.** $(0, -23°)$

29. $(-3, 45°)$ **30.** $(6, 30°)$

31. $(3, -120°)$ **32.** $\left(7, \dfrac{\pi}{6}\right)$

33. $\left(-2, \dfrac{5\pi}{3}\right)$ **34.** $(1.4, 225°)$

Use a grapher to convert from polar to rectangular coordinates. Round the coordinates to the nearest hundredth.

35. $(3, -43°)$ **36.** $\left(-5, \dfrac{\pi}{7}\right)$

37. $\left(-4.2, \dfrac{3\pi}{5}\right)$ **38.** $(2.8, 166°)$

Convert to a polar equation.

39. $3x + 4y = 5$ **40.** $5x + 3y = 4$

41. $x = 5$ **42.** $y = 4$

43. $x^2 + y^2 = 36$ **44.** $x^2 - 4y^2 = 4$

45. $x^2 = 25y$ **46.** $2x - 9y + 3 = 0$

47. $y^2 - 5x - 25 = 0$ **48.** $x^2 + y^2 = 8y$

Convert to a rectangular equation.

49. $r = 5$ **50.** $\theta = \dfrac{3\pi}{4}$

51. $r \sin \theta = 2$ **52.** $r = -3 \sin \theta$

53. $r + r \cos \theta = 3$ **54.** $r = \dfrac{2}{1 - \sin \theta}$

55. $r - 9 \cos \theta = 7 \sin \theta$ **56.** $r + 5 \sin \theta = 7 \cos \theta$

57. $r = 5 \sec \theta$ **58.** $r = 3 \cos \theta$

Graph the equation by plotting points. Then check your work using a grapher.

59. $r = \sin \theta$ **60.** $r = 1 - \cos \theta$

61. $r = 4 \cos 2\theta$ **62.** $r = 1 - 2 \sin \theta$

In Exercises 63–74, use a grapher to match the equation with one of figures (a)–(l), which follow. Try matching the graphs mentally before using a grapher.

a)

b)

c)

d)

e)

f)

g)

h)

i)

j)

k)

l)

63. $r = 3 \sin 2\theta$

64. $r = 4 \cos \theta$

65. $r = \theta$

66. $r^2 = \sin 2\theta$

67. $r = \dfrac{5}{1 + \cos \theta}$

68. $r = 1 + 2 \sin \theta$

69. $r = 3 \cos 2\theta$

70. $r = 3 \sec \theta$

71. $r = 3 \sin \theta$

72. $r = 4 \cos 5\theta$

73. $r = 2 \sin 3\theta$

74. $r \sin \theta = 6$

Discussion and Writing

75. Explain why the rectangular coordinates of a point are unique and the polar coordinates of a point are not unique.

76. Give an example of an equation that is easier to graph on a grapher in polar notation than in rectangular notation and explain why.

Skill Maintenance

Solve.

77. $4 \sin^2 x = 3$

78. $2 \cos^2 x + \cos x = 1$

Graph.

79. $y = 2x - 5$

80. $4x - y = 6$

81. $x = -3$

82. $y = 0$

Synthesis

83. Convert to a rectangular equation:
$$r = \sec^2 \frac{\theta}{2}.$$

84. The center of a regular hexagon is at the origin, and one vertex is the point $(4, 0°)$. Find the coordinates of the other vertices.

Graph.

85. $r = \sin \theta \tan \theta$ (Cissoid)

86. $r = 3\theta$ (Spiral of Archimedes)

87. $r = e^{\theta/10}$ (Logarithmic spiral)

88. $r = 10^{2\theta}$ (Logarithmic spiral)

89. $r = \cos 2\theta \sec \theta$ (Strophoid)

90. $r = \cos 2\theta - 2$ (Peanut)

91. $r = \frac{1}{4} \tan^2 \theta \sec \theta$ (Semicubical parabola)

92. $r = \sin 2\theta + \cos \theta$ (Twisted sister)

Polar Equations of Conics

5.7

CONIC SECTIONS

REVIEW SECTIONS 5.1–5.3.

• Graph polar equations of conics.
• Convert from polar to rectangular equations of conics.
• Find polar equations of conics.

In Sections 5.1–5.3, we saw that the parabola, the ellipse, and the hyperbola have different definitions in rectangular coordinates. When polar coordinates are used, we can give a single definition that applies to all three conics.

An Alternate Definition of Conics

Let L be a fixed line (the **directrix**); let F be a fixed point (the **focus**), not on L; and let e be a positive constant (the **eccentricity**). A **conic** is the set of all points P in the plane such that

$$\frac{PF}{PL} = e,$$

where PF is the distance from P to F and PL is the distance from P to L. The conic is a parabola if $e = 1$, an ellipse if $e < 1$, and a hyperbola if $e > 1$.

Note that if $e = 1$, then $PF = PL$ and the alternate definition of a parabola is identical to the definition presented in Section 5.1.

Polar Equations of Conics

To derive equations for the conics in polar coordinates, we position the focus F at the pole and position the directrix L either perpendicular to the polar axis or parallel to it. In the figure below, we place L perpendicular to the polar axis and p units to the right of the focus, or pole.

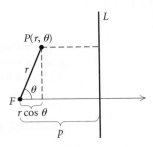

Note that $PL = p - r \cos \theta$. Then if P is any point on the conic, we have

$$\frac{PF}{PL} = e$$

$$\frac{r}{p - r \cos \theta} = e$$

$$r = ep - er \cos \theta$$

$$r + er \cos \theta = ep$$

$$r(1 + e \cos \theta) = ep$$

$$r = \frac{ep}{1 + e \cos \theta}.$$

Thus we see that the polar equation of a conic with focus at the pole and directrix perpendicular to the polar axis and p units to the right of the pole is

$$r = \frac{ep}{1 + e \cos \theta},$$

where e is the eccentricity of the conic.

For an ellipse and a hyperbola, we can make the following statement regarding eccentricity.

For an ellipse and a hyperbola, the **eccentricity e** is given by

$$e = \frac{c}{a},$$

where c is the distance from the center to a focus and a is the distance from the center to a vertex.

Interactive Discovery

Graph each of the following equations in POLAR mode in the square window $[-12, 12, -8, 8]$. Graph the third equation in DOT mode.

$$r = \frac{3}{1 + \cos \theta}, \qquad r = \frac{5}{1 + 0.4 \cos \theta}, \qquad r = \frac{8}{1 + 4 \cos \theta}$$

Describe each curve and use its equation to find its eccentricity. Do your graphs confirm the relationship between the graph of a conic and its eccentricity?

EXAMPLE 1 Describe and graph the conic $r = \dfrac{18}{6 + 3 \cos \theta}$.

Solution We begin by dividing the numerator and the denominator by 6 to obtain a constant term of 1 in the denominator:

$$r = \frac{3}{1 + 0.5 \cos \theta}.$$

This equation is in the form

$$r = \frac{ep}{1 + e \cos \theta}$$

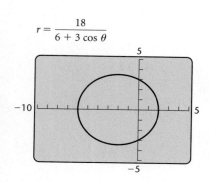

$$r = \frac{18}{6 + 3 \cos \theta}$$

with $e = 0.5$. Since $e < 1$, the graph is an ellipse. Also, since $e = 0.5$ and $ep = 0.5p = 3$, we have $p = 6$. Thus the ellipse has a vertical directrix that lies 6 units to the right of the pole. We graph the equation in the square window $[-10, 5, -5, 5]$.

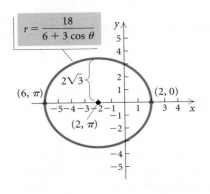

$$r = \frac{18}{6 + 3 \cos \theta}$$

Note that the major axis is horizontal and lies on the polar axis. The vertices are found by letting $\theta = 0$ and $\theta = \pi$. They are $(2, 0)$ and $(6, \pi)$. The center of the ellipse is at the midpoint of the segment connecting the vertices, or at $(2, \pi)$.

The length of the major axis is 8, so we have $2a = 8$, or $a = 4$. From the equation of the conic, we know that $e = 0.5$. Using the equation $e = c/a$, we can find that $c = 2$. Finally, using $a = 4$ and $c = 2$ in $b^2 = a^2 - c^2$ gives us

$$b^2 = 4^2 - 2^2 = 16 - 4 = 12$$
$$b = \sqrt{12}, \text{ or } 2\sqrt{3},$$

so the length of the minor axis is $\sqrt{12}$, or $2\sqrt{3}$. This is useful to know when sketching a hand-drawn graph of the conic.

Other derivations similar to the one on p. 375 lead to the following result.

Polar Equations of Conics

A polar equation of any of the four forms

$$r = \frac{ep}{1 \pm e \cos \theta}, \qquad r = \frac{ep}{1 \pm e \sin \theta}$$

is a conic section. The conic is a parabola if $e = 1$, an ellipse if $0 < e < 1$, and a hyperbola if $e > 1$.

The table below describes the polar equations of conics with a focus at the pole and the directrix either perpendicular to or parallel to the polar axis.

EQUATION	DESCRIPTION
$r = \dfrac{ep}{1 + e \cos \theta}$	Vertical directrix p units to the right of the pole (or focus)
$r = \dfrac{ep}{1 - e \cos \theta}$	Vertical directrix p units to the left of the pole (or focus)
$r = \dfrac{ep}{1 + e \sin \theta}$	Horizontal directrix p units above the pole (or focus)
$r = \dfrac{ep}{1 - e \sin \theta}$	Horizontal directrix p units below the pole (or focus)

EXAMPLE 2   Describe and graph the conic $r = \dfrac{10}{5 - 5 \sin \theta}$.

Solution   We first divide the numerator and the denominator by 5:

$$r = \frac{2}{1 - \sin \theta}.$$

This equation is in the form

$$r = \frac{ep}{1 - e \sin \theta}$$

with $e = 1$, so the graph is a parabola. Since $e = 1$ and $ep = 1 \cdot p = 2$, we have $p = 2$. Thus the parabola has a horizontal directrix 2 units below the pole. We graph the equation in the square window $[-9, 9, -4, 8]$.

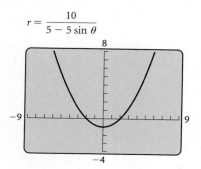

$$r = \frac{10}{5 - 5 \sin \theta}$$

The parabola has a vertical axis of symmetry. We find the vertex of the parabola by letting $\theta = 3\pi/2$. It is $(1, 3\pi/2)$.

EXAMPLE 3   Describe and graph the conic $r = \dfrac{4}{2 + 6 \sin \theta}$.

Solution   We first divide the numerator and the denominator by 2:

$$r = \frac{2}{1 + 3 \sin \theta}.$$

This equation is in the form

$$r = \frac{ep}{1 + e \sin \theta}$$

with $e = 3$. Since $e > 1$, the graph is a hyperbola. We have $e = 3$ and $ep = 3p = 2$, so $p = \frac{2}{3}$. Thus the hyperbola has a horizontal directrix that lies $\frac{2}{3}$ unit above the pole. We graph the equation in the square window $[-6, 6, -3, 5]$ using DOT mode.

Note that the transverse axis is vertical. To find the vertices, we let $\theta = \pi/2$ and $\theta = 3\pi/2$. The vertices are $(1/2, \pi/2)$ and $(-1, 3\pi/2)$.

$$r = \frac{4}{2 + 6 \sin \theta}$$

The center of the hyperbola is the midpoint of the segment connecting the vertices, or $(3/4, \pi/2)$. Thus the distance c from the center to a focus is $3/4$. Using $c = 3/4$, $e = 3$, and $e = c/a$, we have $a = 1/4$. Then since $c^2 = a^2 + b^2$, we have

$$b^2 = \left(\frac{3}{4}\right)^2 - \left(\frac{1}{4}\right)^2 = \frac{9}{16} - \frac{1}{16} = \frac{1}{2}$$

$$b = \frac{1}{\sqrt{2}}, \text{ or } \frac{\sqrt{2}}{2}.$$

Knowing the values of a and b allows us to sketch the asymptotes if we are graphing the hyperbola by hand. We can also easily plot the points $(2, 0)$ and $(2, \pi)$ on the polar axis.

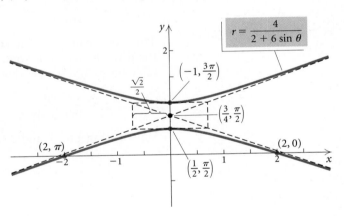

Converting from Polar to Rectangular Equations

We can use the relationships between polar and rectangular coordinates that were developed in Section 5.6 to convert polar equations of conics to rectangular equations.

EXAMPLE 4   Convert to a rectangular equation: $r = \dfrac{2}{1 - \sin \theta}$.

Solution   We have

$$r = \frac{2}{1 - \sin\theta}$$

$$r - r\sin\theta = 2$$

$$r = r\sin\theta + 2$$

$$\sqrt{x^2 + y^2} = y + 2 \qquad \text{Substituting } \sqrt{x^2 + y^2} \text{ for } r \text{ and } y \text{ for } r\sin\theta$$

$$x^2 + y^2 = y^2 + 4y + 4 \qquad \text{Squaring both sides}$$

$$x^2 = 4y + 4, \quad \text{or}$$

$$x^2 - 4y - 4 = 0.$$

This is the equation of a parabola, as we should have anticipated, since $e = 1$.

Finding Polar Equations of Conics

We can find the polar equation of a conic with a focus at the pole if we know its eccentricity and the equation of the directrix.

EXAMPLE 5   Find a polar equation of the conic with a focus at the pole, eccentricity $\frac{1}{3}$, and directrix $r = 2\csc\theta$.

Solution   The equation of the directrix can be written

$$r = \frac{2}{\sin\theta}, \quad \text{or} \quad r\sin\theta = 2.$$

This corresponds to the equation $y = 2$ in rectangular coordinates, so the directrix is a horizontal line 2 units above the polar axis. Using the table on p. 377, we see that the equation is of the form

$$r = \frac{ep}{1 + e\sin\theta}.$$

Substituting $\frac{1}{3}$ for e and 2 for p gives us

$$r = \frac{\frac{1}{3} \cdot 2}{1 + \frac{1}{3}\sin\theta} = \frac{\frac{2}{3}}{1 + \frac{1}{3}\sin\theta} = \frac{2}{3 + \sin\theta}.$$

Exercise Set 5.7

In Exercises 1–6, match the equation with one of the graphs (a)–(f), which follow.

a)

b)

c)

d)

e)

f)

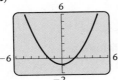

1. $r = \dfrac{3}{1 + \cos \theta}$

2. $r = \dfrac{4}{1 + 2 \sin \theta}$

3. $r = \dfrac{8}{4 - 2 \cos \theta}$

4. $r = \dfrac{12}{4 + 6 \sin \theta}$

5. $r = \dfrac{5}{3 - 3 \sin \theta}$

6. $r = \dfrac{6}{3 + 2 \cos \theta}$

For each equation:

a) *Tell whether the equation describes a parabola, an ellipse, or a hyperbola.*

b) *State whether the directrix is vertical or horizontal and give its location in relation to the pole.*

c) *Find the vertex or vertices.*

d) *Graph the equation.*

7. $r = \dfrac{1}{1 + \cos \theta}$

8. $r = \dfrac{4}{2 + \cos \theta}$

9. $r = \dfrac{15}{5 - 10 \sin \theta}$

10. $r = \dfrac{12}{4 + 8 \sin \theta}$

11. $r = \dfrac{8}{6 - 3 \cos \theta}$

12. $r = \dfrac{6}{2 + 2 \sin \theta}$

13. $r = \dfrac{20}{10 + 15 \sin \theta}$

14. $r = \dfrac{10}{8 - 2 \cos \theta}$

15. $r = \dfrac{9}{6 + 3 \cos \theta}$

16. $r = \dfrac{4}{3 - 9 \sin \theta}$

17. $r = \dfrac{3}{2 - 2 \sin \theta}$

18. $r = \dfrac{12}{3 + 9 \cos \theta}$

19. $r = \dfrac{4}{2 - \cos \theta}$

20. $r = \dfrac{5}{1 - \sin \theta}$

21. $r = \dfrac{7}{2 + 10 \sin \theta}$

22. $r = \dfrac{3}{8 - 4 \cos \theta}$

23.–38. Convert the equations in Exercises 7–22 to rectangular equations.

Find a polar equation of the conic with a focus at the pole and the given eccentricity and directrix.

39. $e = 2$, $r = 3 \csc \theta$

40. $e = \frac{2}{3}$, $r = -\sec \theta$

41. $e = 1$, $r = 4 \sec \theta$

42. $e = 3$, $r = 2 \csc \theta$

43. $e = \frac{1}{2}$, $r = -2 \sec \theta$

44. $e = 1$, $r = 4 \csc \theta$

45. $e = \frac{3}{4}$, $r = 5 \csc \theta$

46. $e = \frac{4}{5}$, $r = 2 \sec \theta$

47. $e = 4$, $r = -2 \csc \theta$

48. $e = 3$, $r = 3 \csc \theta$

Discussion and Writing

49. Consider the graphs of

$$r = \frac{e}{1 - e \sin \theta}$$

for $e = 0.2$, 0.4, 0.6, and 0.8. Explain the effect of the value of e on the graph.

50. When using a grapher, would you prefer to graph a conic in rectangular form or in polar form? Why?

Skill Maintenance

For $f(x) = (x - 3)^2 + 4$, find each of the following.

51. $f(t)$

52. $f(2t)$

53. $f(t - 1)$

54. $f(t + 2)$

Synthesis

Parabolic Orbit. *Suppose that a comet travels in a parabolic orbit with the sun as its focus. Position a polar coordinate system with the pole at the sun and the axis of the orbit perpendicular to the polar axis. When the comet is the given distance from the sun, the segment from the comet to the sun makes the given angle with the polar axis. Find a polar equation of the orbit, assuming that the directrix lies above the pole.*

55. 100 million miles, $\dfrac{\pi}{6}$

56. 120 million miles, $\dfrac{\pi}{4}$

Parametric Equations and Graphs

5.8

- *Graph parametric equations and determine an equivalent rectangular equation.*
- *Determine parametric equations for a rectangular equation.*
- *Determine the location of a moving object at a specific time.*

Graphing Parametric Equations

Much of our graphing of curves in this text has been with rectangular equations involving only two variables, x and y. We graphed sets of ordered pairs, (x, y), where y is a function of x. In this section, we introduce a third variable, t, such that x and y are each a function of t.

Consider a point P with coordinates (x, y) in the rectangular coordinate plane. As P moves in the plane, its location at time t is given by two functions

$$x = f(t) \quad \text{and} \quad y = g(t).$$

For example, let the equations be

$$x = \tfrac{1}{2}t \quad \text{and} \quad y = t^2 - 3, \quad t \geq 0.$$

We restrict t to nonnegative values since t represents time. When $t = 2$, we have

$$x = \tfrac{1}{2} \cdot 2 = 1 \quad \text{and} \quad y = 2^2 - 3 = 1.$$

Thus after 2 sec, the coordinates of P are $(1, 1)$. The table below lists other ordered pairs. We plot them and draw the curve for $t \geq 0$.

t	x	y	(x, y)
0	0	-3	$(0, -3)$
1	$\tfrac{1}{2}$	-2	$\left(\tfrac{1}{2}, -2\right)$
2	1	1	$(1, 1)$
3	$\tfrac{3}{2}$	6	$\left(\tfrac{3}{2}, 6\right)$

$x = \tfrac{1}{2}t, y = t^2 - 3, t \geq 0$

This curve appears to be part of a parabola. Let's verify this by finding the equivalent rectangular equation. Solving $x = \tfrac{1}{2}t$ for t, we get $t = 2x$. Substituting $2x$ for t in $y = t^2 - 3$, we have

$$y = (2x)^2 - 3 = 4x^2 - 3.$$

We know from Section 1.5 that this is a quadratic equation and that the graph is a parabola. Thus the curve is part of the parabola $y = 4x^2 - 3$.

The equations $x = \tfrac{1}{2}t$ and $y = t^2 - 3$ are called the **parametric equations** for the curve. The variable t is called the **parameter**. The equation $y = 4x^2 - 3$, with the restriction $x \geq 0$, is the equivalent rectangular equation for the curve. Since $t \geq 0$ and $x = \frac{1}{2}t$, the restriction $x \geq 0$ must be included. In this example, t represents time, but t, in general, can represent any real number in a specified interval. One advantage of parametric equations is that the restriction on the parameter t allows us to investigate a portion of the curve.

Parametric Equations

If f and g are continuous functions of t on an interval I, then the set of ordered pairs (x, y) such that $x = f(t)$ and $y = g(t)$ is a **plane curve**. The equations $x = f(t)$ and $y = g(t)$ are **parametric equations** for the curve. The variable t is the **parameter**.

Plane curves described with parametric equations can also be graphed on graphers. Consult your grapher's manual or the *Graphing Calculator Manual* that accompanies this text for specific instructions.

EXAMPLE 1 Using a grapher, graph each of the following plane curves given their respective parametric equations and the restriction for the parameter. Then find the equivalent rectangular equation.

a) $x = t^2$, $y = t - 1$; $-1 \leq t \leq 4$

b) $x = \sqrt{t}$, $y = 2t + 3$; $0 \leq t \leq 3$

Solution

a) When using a grapher set in PARAMETRIC mode, we must set minimum and maximum values for x, y, and t.

WINDOW
Tmin = −1
Tmax = 4
Tstep = .1
Xmin = −2
Xmax = 18
Xscl = 1
Ymin = −4
Ymax = 4
Yscl = 1

$x = t^2$, $y = t - 1$; $-1 \leq t \leq 4$

Using the TRACE feature, we can see the T, X, and Y values along the curve. To find an equivalent rectangular equation, we can solve either equation for t. Let's select the simpler equation $y = t - 1$ and solve for t:

$$y = t - 1$$
$$y + 1 = t.$$

We then substitute $y + 1$ for t in $x = t^2$:

$$x = t^2$$
$$x = (y + 1)^2. \quad \text{Parabola}$$

This is an equation of a parabola that opens to the right. Given that $-1 \le t \le 4$, we have the corresponding restrictions on x and y: $0 \le x \le 16$ and $-2 \le y \le 3$. Thus the equivalent rectangular equation is

$$x = (y + 1)^2; \quad 0 \le x \le 16, \quad -2 \le y \le 3.$$

b) Using a grapher set in PARAMETRIC mode, we have

WINDOW
  Tmin = 0
  Tmax = 3
  Tstep = .1
  Xmin = −3
  Xmax = 3
  Xscl = 1
  Ymin = −2
  Ymax = 10
  Yscl = 1

To find an equivalent rectangular equation, we first solve $x = \sqrt{t}$ for t:

$$x = \sqrt{t}$$
$$x^2 = t.$$

Then we substitute x^2 for t in $y = 2t + 3$:

$$y = 2t + 3$$
$$y = 2x^2 + 3. \quad \text{Parabola}$$

When $0 \le t \le 3$, $0 \le x \le \sqrt{3}$ and $3 \le y \le 9$. The equivalent rectangular equation is

$$y = 2x^2 + 3; \quad 0 \le x \le \sqrt{3}, \quad 3 \le y \le 9. \quad \text{—}$$

We first graphed in PARAMETRIC mode in Section 2.5 (p. 166). There we used an angle measure, in degrees, as the parameter.

EXAMPLE 2 Graph the plane curve described by $x = \cos t$ and $y = \sin t$, with t in $[0, 2\pi]$. Then determine an equivalent rectangular equation.

Solution Using a squared window and a Tstep of $\pi/48$, we obtain the graph at left. It appears to be the unit circle.

The equivalent rectangular equation can be obtained by squaring each parametric equation:

$$x^2 = \cos^2 t \quad \text{and} \quad y^2 = \sin^2 t.$$

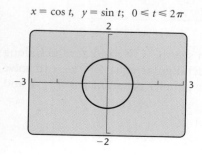

$x = \cos t, \ y = \sin t; \ 0 \le t \le 2\pi$

This allows us to use the trigonometric identity $\sin^2 \theta + \cos^2 \theta = 1$. Substituting, we get

$$x^2 + y^2 = 1.$$

As expected, this is an equation of the unit circle.　　　▬

One advantage of graphing the unit circle parametrically is that it provides a method of finding trigonometric function values.

EXAMPLE 3　Using the VALUE feature from the CALC menu and the parametric graph of the unit circle, find each of the following function values.

a) $\cos \dfrac{7\pi}{6}$　　　　　　　　　　　　　　**b)** $\sin 4.13$

Solution

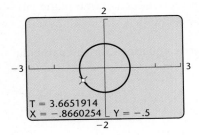

T = 3.6651914
X = −.8660254　Y = −.5

a) Using the VALUE feature from the CALC menu, we enter $7\pi/6$ for t. The value of x, which is $\cos t$, appears, as shown at left. Thus,

$$\cos \frac{7\pi}{6} \approx -0.8660.$$

b) We enter 4.13 for t. The value of y, which is $\sin t$, will appear on the screen. Thus $\sin 4.13 \approx -0.8352$.　　　▬

EXAMPLE 4　Graph the plane curve represented by

$$x = 5 \cos t \quad \text{and} \quad y = 3 \sin t, \quad 0 \le t \le 2\pi.$$

Then eliminate the parameter to find the rectangular equation.

Solution

$x = 5 \cos t, \; y = 3 \sin t; \; 0 \le t \le 2\pi$

This appears to be the graph of an ellipse. To find the rectangular equation, we first solve for $\cos t$ and $\sin t$ in the parametric equations:

$$x = 5 \cos t \qquad y = 3 \sin t$$

$$\frac{x}{5} = \cos t, \qquad \frac{y}{3} = \sin t.$$

Using the identity $\sin^2 \theta + \cos^2 \theta = 1$, we can substitute to eliminate the parameter:

$$\sin^2 t + \cos^2 t = 1$$

$$\left(\frac{y}{3}\right)^2 + \left(\frac{x}{5}\right)^2 = 1 \qquad \text{Substituting}$$

$$\frac{x^2}{25} + \frac{y^2}{9} = 1. \qquad \text{Ellipse}$$

The rectangular form of the equation confirms that the graph is an ellipse centered at the origin with vertices at $(5, 0)$ and $(-5, 0)$.   ▬

Determining Parametric Equations for a Given Rectangular Equation

Many sets of parametric equations can represent the same plane curve. In fact, there are infinitely many such equations.

EXAMPLE 5   Find four sets of parametric equations for the parabola

$$y = 4 - (x + 3)^2.$$

SOLUTION

If $x = t$, then $y = 4 - (t + 3)^2$.

If $x = t - 3$, then $y = 4 - (t - 3 + 3)^2$, or $4 - t^2$.

If $x = \dfrac{t}{3}$, then $y = 4 - \left(\dfrac{t}{3} + 3\right)^2$, or $-\dfrac{t^2}{9} - 2t - 5$.

In each of the above three sets, t is any real number. The following equations restrict t to any number greater than or equal to 0.

If $x = \sqrt{t} - 3$, then $y = 4 - (\sqrt{t} - 3 + 3)^2$, or $4 - t$; $t \geq 0$.   ▬

Applications

The motion of an object can be described with parametric equations.

EXAMPLE 6   *Motion of a Baseball.* A baseball player can throw a ball from a height of 5.9 ft with an initial velocity of 100 ft/sec at an angle of 41° with the horizontal. Neglecting air resistance, we know that the parametric equations for the path of the ball are

$$x = (100 \cos 41°)t \quad \text{and} \quad y = 5.9 + (100 \sin 41°)t - 16t^2,$$

where x and y are in feet and t is in seconds. Since t represents the time that the ball is in the air, we must have t greater than or equal to 0.

a) Graph the plane curve.

b) Determine the location of the ball after 0.5 sec, 1 sec, and 2 sec.

c) Find a rectangular equation that describes the curve.

Solution

a) We set the grapher to DEGREE mode and enter the parametric equations.

$$x = (100 \cos 41°)t, \quad y = 5.9 + (100 \sin 41°)t - 16t^2$$

b) Using the VALUE feature, we find that

when $T = 0.5$, $(x, y) = (37.735479, 34.702951)$,

when $T = 1$, $(x, y) = (75.470958, 55.505903)$, and

when $T = 2$, $(x, y) = (150.94192, 73.111806)$.

In each ordered pair, the x-coordinate represents the horizontal distance and the y-coordinate represents the corresponding vertical height.

c) To find the equivalent rectangular equation, we solve $x = (100 \cos 41°)t$ for t to obtain

$$t = \frac{x}{100 \cos 41°}.$$

Then we substitute $x/(100 \cos 41°)$ for t in the equation for y:

$$y = 5.9 + (100 \sin 41°)\frac{x}{100 \cos 41°} - 16\left(\frac{x}{100 \cos 41°}\right)^2$$

$$y = 5.9 + x \tan 41° - \frac{16}{100^2 \cos^2 41°}x^2. \qquad \textbf{Rectangular equation}$$

Note that this is a quadratic equation, as expected, since the graph is a parabola. ▬

The path of a fixed point on the circumference of a circle as it rolls along a line is called a **cycloid**. For example, a point on the rim of a bicycle wheel traces out a cycloid curve.

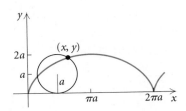

EXAMPLE 7   Graph the cycloid described by the parametric equations

$$x = 3(t - \sin t) \quad \text{and} \quad y = 3(1 - \cos t); \quad 0 \le t \le 6\pi.$$

Solution

$x = 3(t - \sin t), \; y = 3(1 - \cos t); \; 0 \le t \le 6\pi$

Exercise Set **5.8**

Using a grapher, graph the plane curve given by the set of parametric equations and the restriction for the parameter. Then find the equivalent rectangular equation.

1. $x = \frac{1}{2}t, \; y = 6t - 7; \; -1 \le t \le 6$

2. $x = t, \; y = 5 - t; \; -2 \le t \le 3$

3. $x = t^3, \; y = t - 4; \; -1 \le t \le 10$

4. $x = \sqrt{t}, \; y = 2t + 3; \; 0 \le t \le 8$

5. $x = t^2, \; y = \sqrt{t}; \; 0 \le t \le 4$

6. $x = t^3 + 1, \; y = t; \; -3 \le t \le 3$

7. $x = t + 3, \; y = \dfrac{1}{t + 3}; \; -2 \le t \le 2$

8. $x = 2t^3 + 1, \; y = 2t^3 - 1; \; -4 \le t \le 4$

9. $x = 2t - 1, \; y = t^2; \; -3 \le t \le 3$

10. $x = \frac{1}{3}t, \; y = t; \; -5 \le t \le 5$

11. $x = e^{-t}, \; y = e^t; \; -\infty < t < \infty$

12. $x = 2 \ln t, \; y = t^2; \; 0 < t < \infty$

13. $x = 3 \cos t, \; y = 3 \sin t; \; 0 \le t \le 2\pi$

14. $x = 2 \cos t, \; y = 4 \sin t; \; 0 \le t \le 2\pi$

15. $x = \cos t, \; y = 2 \sin t; \; 0 \le t \le 2\pi$

16. $x = 2 \cos t, \; y = 2 \sin t; \; 0 \le t \le 2\pi$

17. $x = \sec t, \; y = \cos t; \; -\dfrac{\pi}{2} < t < \dfrac{\pi}{2}$

18. $x = \sin t, \; y = \csc t; \; 0 < t < \pi$

19. $x = 1 + 2 \cos t, \; y = 2 + 2 \sin t; \; 0 \le t \le 2\pi$

20. $x = 2 + \sec t, \; y = 1 + 3 \tan t; \; 0 < t < \dfrac{\pi}{2}$

Using a parametric graph of the unit circle, find the function value.

21. $\sin \dfrac{\pi}{4}$

22. $\cos \dfrac{2\pi}{3}$

23. $\cos \dfrac{17\pi}{12}$

24. $\sin \dfrac{4\pi}{5}$

25. $\tan \dfrac{\pi}{5}$

26. $\tan \dfrac{2\pi}{7}$

27. $\cos 5.29$

28. $\sin 1.83$

Find two sets of parametric equations for the rectangular equation.

29. $y = 4x - 3$

30. $y = x^2 - 1$

31. $y = (x - 2)^2 - 6x$

32. $y = x^3 + 3$

33. *Motion of a Ball.*   A ball is thrown from a height of 6.25 ft with an initial velocity of 80 ft/sec at an angle of 35° with the horizontal. The parametric equations for the path of the ball are

$$x = (80 \cos 35°)t \quad \text{and}$$
$$y = 6.25 + (80 \sin 35°)t - 16t^2,$$

where x and y are in feet and t is in seconds.

a) Find the height of the ball after 1 sec.

b) Find the total horizontal distance traveled by the ball.

c) How long does it take the ball to hit the ground?

34. For Exercise 33, determine the rectangular equation describing the path of the ball.

35. In Example 6, use a grapher to determine the maximum height that the ball reaches.

36. The path of a projectile ejected with an initial velocity of 200 ft/sec at an angle of 55° with the horizontal is described with the equations

$$x = (210 \cos 55°)t \quad \text{and} \quad y = (210 \sin 55°)t - 16t^2,$$

where $t \geq 0$.

a) Find the height after 4 sec.

b) Determine the equivalent rectangular equation for this curve.

Graph each cycloid for t in the indicated interval.

37. $x = 2(t - \sin t), \ y = 2(1 - \cos t); \ 0 \leq t \leq 4\pi$

38. $x = 4t - 4 \sin t, \ y = 4 - 4 \cos t; \ 0 \leq t \leq 6\pi$

39. $x = t - \sin t, \ y = 1 - \cos t; \ -2\pi \leq t \leq 2\pi$

40. $x = 5(t - \sin t), \ y = 5(1 - \cos t); \ -4\pi \leq t \leq 4\pi$

Discussion and Writing

41. Show that $x = a \cos t + h$ and $y = b \sin t + k$, $0 \leq t \leq 2\pi$, are parametric equations of an ellipse with center (h, k).

42. Consider the graph in Example 3. Explain how the values of T, X, and Y displayed relate to both the parametric and rectangular equations of the unit circle.

Skill Maintenance

Determine the direction angle θ of the vector, to the nearest degree.

43. $\mathbf{w} = \langle 3, -5 \rangle$ **44.** $\mathbf{q} = \mathbf{i} - 3\mathbf{j}$

Find the angle between the given vectors, to the nearest tenth of a degree.

45. $\mathbf{u} = \langle 6, -2 \rangle, \mathbf{v} = \langle -3, -1 \rangle$

46. $\mathbf{a} = \mathbf{i} - \mathbf{j}, \mathbf{b} = 4\mathbf{i} + 3\mathbf{j}$

Synthesis

47. Graph the curve described by

$$x = 3 \cos t \quad \text{and} \quad y = 3 \sin t, 0 \leq t \leq 2\pi.$$

As t increases, the path of the curve is generated counterclockwise. How can this set of equations be changed so that the curve is generated clockwise?

48. Graph the plane curve described by

$$x = \cos^3 t \quad \text{and} \quad y = \sin^3 t, 0 \leq t \leq 2\pi.$$

Then find the equivalent rectangular equation.

Chapter Summary and Review   5

Important Properties and Formulas

Standard Equation of a Parabola with Vertex at the Origin

The standard equation of a parabola with vertex $(0, 0)$ and directrix $y = -p$ is

$$x^2 = 4py.$$

The focus is $(0, p)$ and the y-axis is the axis of symmetry.

The standard equation of a parabola with vertex $(0, 0)$ and directrix $x = -p$ is

$$y^2 = 4px.$$

The focus is $(p, 0)$ and the x-axis is the axis of symmetry.

Standard Equation of a Parabola with Vertex (h, k) and Vertical Axis of Symmetry

The standard equation of a parabola with vertex (h, k) and vertical axis of symmetry is

$$(x - h)^2 = 4p(y - k),$$

where the vertex is (h, k), the focus is $(h, k + p)$, and the directrix is $y = k - p$.

Standard Equation of a Parabola with Vertex (h, k) and Horizontal Axis of Symmetry

The standard equation of a parabola with vertex (h, k) and horizontal axis of symmetry is

$$(y - k)^2 = 4p(x - h),$$

where the vertex is (h, k), the focus is $(h + p, k)$, and the directrix is $x = h - p$.

Standard Equation of a Circle

The standard equation of a circle with center (h, k) and radius r is

$$(x - h)^2 + (y - k)^2 = r^2.$$

Standard Equation of an Ellipse with Center at the Origin

Major axis horizontal

$$\frac{x^2}{a^2} + \frac{y^2}{b^2} = 1, \ a > b > 0$$

Vertices: $(-a, 0), (a, 0)$

y-intercepts: $(0, -b), (0, b)$

Foci: $(-c, 0), (c, 0)$, where $c^2 = a^2 - b^2$

Major axis vertical

$$\frac{x^2}{b^2} + \frac{y^2}{a^2} = 1, \ a > b > 0$$

Vertices: $(0, -a), (0, a)$

x-intercepts: $(-b, 0), (b, 0)$

Foci: $(0, -c), (0, c)$, where $c^2 = a^2 - b^2$

Standard Equation of an Ellipse with Center at (h, k)

Major axis horizontal

$$\frac{(x - h)^2}{a^2} + \frac{(y - k)^2}{b^2} = 1, \ a > b > 0$$

Vertices: $(h - a, k), (h + a, k)$

Length of minor axis: $2b$

Foci: $(h - c, k), (h + c, k)$, where $c^2 = a^2 - b^2$

Major axis vertical

$$\frac{(x - h)^2}{b^2} + \frac{(y - k)^2}{a^2} = 1, \ a > b > 0$$

Vertices: $(h, k - a), (h, k + a)$

Length of minor axis: $2b$

Foci: $(h, k - c), (h, k + c)$, where $c^2 = a^2 - b^2$

Standard Equation of a Hyperbola with Center at the Origin

Transverse axis horizontal

$$\frac{x^2}{a^2} - \frac{y^2}{b^2} = 1$$

Vertices: $(-a, 0)$, $(a, 0)$

Asymptotes: $y = -\frac{b}{a}x$, $y = \frac{b}{a}x$

Foci: $(-c, 0)$, $(c, 0)$, where $c^2 = a^2 + b^2$

Transverse axis vertical

$$\frac{y^2}{a^2} - \frac{x^2}{b^2} = 1$$

Vertices: $(0, -a)$, $(0, a)$

Asymptotes: $y = -\frac{a}{b}x$, $y = \frac{a}{b}x$

Foci: $(0, -c)$, $(0, c)$, where $c^2 = a^2 + b^2$

Standard Equation of a Hyperbola with Center at (h, k)

Transverse axis horizontal

$$\frac{(x - h)^2}{a^2} - \frac{(y - k)^2}{b^2} = 1$$

Vertices: $(h - a, k)$, $(h + a, k)$

Asymptotes: $y - k = \frac{b}{a}(x - h)$,

$$y - k = -\frac{b}{a}(x - h)$$

Foci: $(h - c, k)$, $(h + c, k)$, where $c^2 = a^2 + b^2$

Transverse axis vertical

$$\frac{(y - k)^2}{a^2} - \frac{(x - h)^2}{b^2} = 1$$

Vertices: $(h, k - a)$, $(h, k + a)$

Asymptotes: $y - k = \frac{a}{b}(x - h)$,

$$y - k = -\frac{a}{b}(x - h)$$

Foci: $(h, k - c)$, $(h, k + c)$, where $c^2 = a^2 + b^2$

Rotation of Axes Formulas

$$x' = x \cos \theta + y \sin \theta,$$
$$y' = -x \sin \theta + y \cos \theta$$
$$x = x' \cos \theta - y' \sin \theta,$$
$$y = x' \sin \theta + y' \cos \theta$$

Eliminating the xy-Term

To eliminate the xy-term from the equation

$$Ax^2 + Bxy + Cy^2 + Dx + Ey + F = 0,$$
$$B \neq 0,$$

select an angle θ such that

$$\cot 2\theta = \frac{A - C}{B}, \quad 0 < 2\theta < 180°,$$

and use the rotation of axes formulas.

The Discriminant

The graph of the equation $Ax^2 + Bxy + Cy^2 + Dx + Ey + F = 0$ is, except in degenerate cases,

1. an ellipse or a circle if $B^2 - 4AC < 0$,
2. a hyperbola if $B^2 - 4AC > 0$,   and
3. a parabola if $B^2 - 4AC = 0$.

Polar Equations of Conics

A polar equation of any of the four forms

$$r = \frac{ep}{1 \pm e \cos \theta}, \qquad r = \frac{ep}{1 \pm e \sin \theta}$$

is a conic section. The conic is a parabola if $e = 1$, an ellipse if $0 < e < 1$, and a hyperbola if $e > 1$.

REVIEW EXERCISES

In Exercises 1–8, match the equation with one of the graphs (a)–(h), which follow.

a)

b)

c)

d)

e)

f)

g)

h)

1. $y^2 = 5x$

2. $y^2 = 9 - x^2$

3. $3x^2 + 4y^2 = 12$

4. $9y^2 - 4x^2 = 36$

5. $x^2 + y^2 + 2x - 3y = 8$

6. $4x^2 + y^2 - 16x - 6y = 15$

7. $x^2 - 8x + 6y = 0$

8. $\dfrac{(x+3)^2}{16} - \dfrac{(y-1)^2}{25} = 1$

9. Find an equation of the parabola with directrix $y = \frac{3}{2}$ and focus $\left(0, -\frac{3}{2}\right)$.

10. Find the focus, the vertex, and the directrix of the parabola given by
$$y^2 = -12x.$$

11. Find the vertex, the focus, and the directrix of the parabola given by
$$x^2 + 10x + 2y + 9 = 0.$$

12. Find the center, the vertices, and the foci of the ellipse given by
$$16x^2 + 25y^2 - 64x + 50y - 311 = 0.$$
Then draw the graph.

13. Find an equation of the ellipse having vertices $(0, -4)$ and $(0, 4)$ with minor axis of length 6.

14. Find the center, the vertices, the foci, and the asymptotes of the hyperbola given by
$$x^2 - 2y^2 + 4x + y - \tfrac{1}{8} = 0.$$

15. *Spotlight.* A spotlight has a parabolic cross section that is 2 ft wide at the opening and 1.5 ft deep at the vertex. How far from the vertex is the focus?

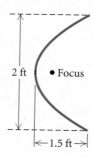

Solve.

16. $x^2 - 16y = 0,$
$\quad x^2 - y^2 = 64$

17. $4x^2 + 4y^2 = 65,$
$\quad 6x^2 - 4y^2 = 25$

18. $x^2 - y^2 = 33,$
$\quad x + y = 11$

19. $x^2 - 2x + 2y^2 = 8,$
$\quad 2x + y = 6$

20. $x^2 - y = 3,$
$\quad 2x - y = 3$

21. $x^2 + y^2 = 25,$
$\quad x^2 - y^2 = 7$

22. $x^2 - y^2 = 3,$
$\quad y = x^2 - 3$

23. $x^2 + y^2 = 18,$
$\quad 2x + y = 3$

24. $x^2 + y^2 = 100,$
$\quad 2x^2 - 3y^2 = -120$

25. $x^2 + 2y^2 = 12,$
$\quad xy = 4$

26. *Numerical Relationship.* The sum of two numbers is 11 and the sum of their squares is 65. Find the numbers.

27. *Dimensions of a Rectangle.* A rectangle has a perimeter of 38 m and an area of 84 m². What are the dimensions of the rectangle?

28. *Numerical Relationship.* Find two positive integers whose sum is 12 and the sum of whose reciprocals is $\frac{3}{8}$.

29. *Perimeter.* The perimeter of a square is 12 cm more than the perimeter of another square. The area of the first square exceeds the area of the other by 39 cm². Find the perimeter of each square.

30. *Radius of a Circle.* The sum of the areas of two circles is 130π ft². The difference of the areas is 112π ft². Find the radius of each circle.

Graph the equation.

31. $5x^2 - 2xy + 5y^2 - 24 = 0$

32. $x^2 - 10xy + y^2 + 12 = 0$

33. $5x^2 + 6\sqrt{3}xy - y^2 = 16$

34. $x^2 + 2xy + y^2 - \sqrt{2}x + \sqrt{2}y = 0$

35. Find the polar coordinates of each of these points. Give three answers for each point.

Find the polar coordinates of the point. Express the answer in degrees and then in radians.

36. $(-4\sqrt{2}, 4\sqrt{2})$

37. $(0, -5)$

Use a grapher to convert from rectangular to polar coordinates. Express the answer in degrees and then in radians.

38. $(-2, 5)$

39. $(-4.2, \sqrt{7})$

Find the rectangular coordinates of the point.

40. $\left(3, \dfrac{\pi}{4}\right)$

41. $(-6, -120°)$

Use a grapher to convert from polar to rectangular coordinates. Round the coordinates to the nearest hundredth.

42. $(2, -15°)$

43. $\left(-2.3, \dfrac{\pi}{5}\right)$

Convert to a polar equation.

44. $5x - 2y = 6$

45. $y = 3$

46. $x^2 + y^2 = 9$

47. $y^2 - 4x - 16 = 0$

Convert to a rectangular equation.

48. $r = 6$

49. $r + r \sin \theta = 1$

50. $r = \dfrac{3}{1 - \cos \theta}$

51. $r - 2 \cos \theta = 3 \sin \theta$

In Exercises 52–55, use a grapher to match the equation with each of figures (a)–(d), which follow. Try matching the graphs mentally before using a grapher.

a)

b)

c)

d)

52. $r = 2 \sin \theta$

53. $r^2 = \cos 2\theta$

54. $r = 1 + 3 \cos \theta$

55. $r \sin \theta = 4$

Graph the equation. State whether the directrix is vertical or horizontal, describe its location in relation to the pole, and find the vertex or vertices.

56. $r = \dfrac{6}{3 - 3 \sin \theta}$

57. $r = \dfrac{8}{2 + 4 \cos \theta}$

58. $r = \dfrac{4}{2 - \cos \theta}$

59. $r = \dfrac{18}{9 + 6 \sin \theta}$

60.–63. Convert the equations in Exercises 56–59 to rectangular equations.

Find a polar equation of the conic with a focus at the pole and the given eccentricity and directrix.

64. $e = \frac{1}{2}, r = 2 \sec \theta$

65. $e = 3, r = -6 \csc \theta$

66. $e = 1, r = -4 \sec \theta$

67. $e = 2, r = 3 \csc \theta$

Using a grapher, graph the plane curve given by the set of parametric equations and the restrictions for the parameter. Then find the equivalent rectangular equation.

68. $x = t, y = 2 + t, \quad -3 \le t \le 3$

69. $x = \sqrt{t}, y = t - 1, \quad 0 \le t \le 9$

70. $x = 2 \cos t, y = 2 \sin t, \quad 0 \le t \le 2\pi$

71. $x = 3 \sin t, y = \cos t, \quad 0 \le t \le 2\pi$

Find two sets of parametric equations for the given rectangular equation.

72. $y = 2x - 3$

73. $y = x^2 + 4$

Discussion and Writing

74. What would you say to a classmate who tells you that an algebraic solution of a nonlinear system of equations is always preferable to a graphical solution?

75. Is a circle a special type of ellipse? Why or why not?

Synthesis

76. Find an equation of the ellipse containing the point $(-1/2, 3\sqrt{3}/2)$ and with vertices $(0, -3)$ and $(0, 3)$.

77. Find two numbers whose product is 4 and the sum of whose reciprocals is $\frac{65}{56}$.

78. Find an equation of the circle that passes through the points $(10, 7)$, $(-6, 7)$, and $(-8, 1)$.

79. *Navigation.* Two radio transmitters positioned 400 mi apart send simultaneous signals to a ship that is 250 mi offshore, sailing parallel to the shoreline. The signal from transmitter A reaches the ship 300 microseconds before the signal from transmitter B. The signals travel at a speed of 186,000 miles per second, or 0.186 mile per microsecond. Find the equation of the hyperbola with foci A and B on which the ship is located. (*Hint*: For any point on the hyperbola, the absolute value of the difference of its distances from the foci is $2a$.)

80. Convert to a rectangular equation:

$$r = \csc^2 \frac{\theta}{2}.$$

Exponential and Logarithmic Functions 6

The total credit card volume for Visa, MasterCard, American Express, and Discover has dramatically increased in recent years (*Source*: CardWeb Inc.'s CardData). The exponential function

$$f(x) = 249.7267077(1.50881332)^x$$

can be used to predict the total credit card volume for these four cards x years after 1988.

This problem appears as Example 6 in Section 6.5.

I n this chapter, we will consider two kinds of closely related functions. The first type, *exponential functions,* has a variable in the exponent. Such functions have many applications to the growth of populations, commodities, and investments. The inverses of exponential functions, called *logarithmic functions,* or *logarithm functions,* are also important in many applications such as earthquake magnitude, sound level, and chemical pH.

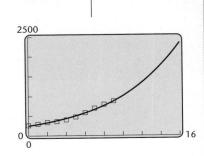

Exponential Functions and Graphs

6.1

- *Graph exponential equations and functions.*
- *Solve applied problems involving exponential functions and their graphs.*

We now turn our attention to the study of a set of functions that are very rich in application. Consider the following graphs. Each one illustrates an *exponential function*. In this section, we consider such functions, their inverses, and some important applications of exponential functions.

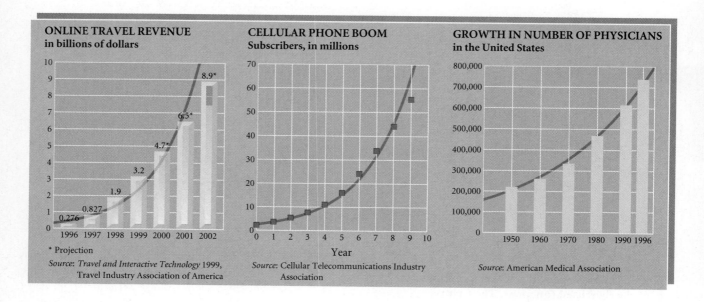

ONLINE TRAVEL REVENUE
in billions of dollars

* Projection

Source: *Travel and Interactive Technology* 1999, Travel Industry Association of America

CELLULAR PHONE BOOM
Subscribers, in millions

Year

Source: Cellular Telecommunications Industry Association

GROWTH IN NUMBER OF PHYSICIANS
in the United States

Source: American Medical Association

Graphing Exponential Functions

We now define exponential functions. We assume that a^x has meaning for any real number x and any positive real number a and that the laws of exponents still hold, though we will not prove them here.

Exponential Function

The function $f(x) = a^x$, where x is a real number, $a > 0$ and $a \neq 1$, is called the **exponential function,** base a.

We require the **base** to be positive in order to avoid the complex numbers that would occur by taking even roots of negative numbers—an example is $(-1)^{1/2}$, the square root of -1, which is not a real number. The restriction $a \neq 1$ is made to exclude the constant function $f(x) = 1^x = 1$, which does not have an inverse because it is not one-to-one.

The following are examples of exponential functions:

$$f(x) = 2^x, \qquad f(x) = \left(\frac{1}{2}\right)^x, \qquad f(x) = (3.57)^x.$$

Note that, in contrast to functions like $f(x) = x^5$ and $f(x) = x^{1/2}$ in which the variable is the base of an exponential expression, the variable in an exponential function is *in the exponent*. Let's now consider graphs of exponential functions.

EXAMPLE 1   Graph the exponential function

$$y = f(x) = 2^x.$$

Solution   We compute some function values and list the results in a table (at left).

	y	
x	$y = f(x) = 2^x$	(x, y)
0	1	$(0, 1)$
1	2	$(1, 2)$
2	4	$(2, 4)$
3	8	$(3, 8)$
-1	$\frac{1}{2}$	$\left(-1, \frac{1}{2}\right)$
-2	$\frac{1}{4}$	$\left(-2, \frac{1}{4}\right)$
-3	$\frac{1}{8}$	$\left(-3, \frac{1}{8}\right)$

$$f(0) = 2^0 = 1; \qquad\qquad f(-1) = 2^{-1} = \frac{1}{2^1} = \frac{1}{2};$$

$$f(1) = 2^1 = 2; \qquad\qquad f(-2) = 2^{-2} = \frac{1}{2^2} = \frac{1}{4};$$

$$f(2) = 2^2 = 4; \qquad\qquad f(-3) = 2^{-3} = \frac{1}{2^3} = \frac{1}{8}$$

$$f(3) = 2^3 = 8;$$

Next, we plot these points and connect them with a smooth curve. Be sure to plot enough points to determine how steeply the curve rises.

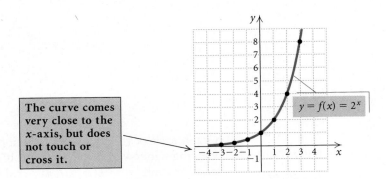

The curve comes very close to the x-axis, but does not touch or cross it.

$y = f(x) = 2^x$

Note that as x increases, the function values increase without bound. As x decreases, the function values decrease, getting close to 0. That is, as $x \to -\infty$, $y \to 0$. The x-axis, or the line $y = 0$, is a horizontal asymptote. As the x-inputs decrease, the curve gets closer and closer to this line, but does not cross it.

Check the graph of $y = f(x) = 2^x$ using a grapher. Then graph $y = f(x) = 3^x$. Use the TRACE and TABLE features to confirm that the graph never crosses the x-axis.

x	$y = f(x) = 2^{-x}$	(x, y)
0	1	$(0, 1)$
1	$\frac{1}{2}$	$(1, \frac{1}{2})$
2	$\frac{1}{4}$	$(2, \frac{1}{4})$
3	$\frac{1}{8}$	$(3, \frac{1}{8})$
−1	2	$(-1, 2)$
−2	4	$(-2, 4)$
−3	8	$(-3, 8)$

The top of the table has a header y spanning the middle column.

EXAMPLE 2   Graph the exponential function $y = f(x) = \left(\frac{1}{2}\right)^x$.

Solution   We compute some function values and list the results in a table (at left). Before we plot these points and draw the curve, note that

$$y = f(x) = \left(\frac{1}{2}\right)^x = (2^{-1})^x = 2^{-x}.$$

This tells us, before we begin graphing, that this graph is a reflection of the graph of $y = 2^x$ across the y-axis.

$$f(0) = 2^{-0} = 1; \qquad f(-1) = 2^{-(-1)} = 2^1 = 2;$$

$$f(1) = 2^{-1} = \frac{1}{2^1} = \frac{1}{2}; \qquad f(-2) = 2^{-(-2)} = 2^2 = 4;$$

$$f(2) = 2^{-2} = \frac{1}{2^2} = \frac{1}{4}; \qquad f(-3) = 2^{-(-3)} = 2^3 = 8$$

$$f(3) = 2^{-3} = \frac{1}{2^3} = \frac{1}{8};$$

Next, we plot these points and connect them with a smooth curve.

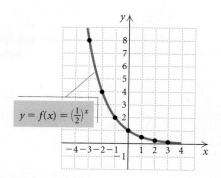

$y = f(x) = \left(\frac{1}{2}\right)^x$

Interactive Discovery

Graph each of the following functions using the same set of axes and a viewing window of $[-10, 10, -10, 10]$. Try entering these functions with one equation: $y_1 = \{1.1, 2.7, 4\} \wedge x$. Look for patterns in the graphs.

$$f(x) = 1.1^x, \qquad f(x) = 2.7^x, \qquad f(x) = 4^x$$

Next, clear the screen and graph each of the following functions using the same set of axes and viewing window. Again, look for patterns in the graphs.

$$f(x) = 0.1^x, \qquad f(x) = 0.6^x, \qquad f(x) = 0.23^x$$

What relationship do you see between the base a and the shape of the resulting graph of a^x? What do all the graphs have in common? How do they differ?

PROPERTIES OF EXPONENTIAL FUNCTIONS

Let's list and compare some characteristics of exponential functions, keeping in mind that the definition of an exponential function, $f(x) = a^x$, requires that a be positive and different from 1.

$f(x) = a^x$ for $0 < a < 1$ or
$f(x) = a^{-x}$, $a > 1$:

Continuous

One-to-one

Domain: $(-\infty, \infty)$

Range: $(0, \infty)$

Decreasing

Horizontal asymptote is x-axis

y-intercept: $(0, 1)$

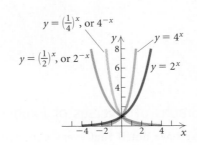

$f(x) = a^x$, $a > 1$:

Continuous

One-to-one

Domain: $(-\infty, \infty)$

Range: $(0, \infty)$

Increasing

Horizontal asymptote is x-axis

y-intercept: $(0, 1)$

TRANSFORMATIONS
OF FUNCTIONS

REVIEW SECTION 1.4.

To graph other types of exponential functions, keep in mind the ideas of translation, stretching, and reflection. All these concepts allow us to visualize the graph before drawing it.

EXAMPLE 3 Graph each of the following. Before doing so, describe how each graph can be obtained from $f(x) = 2^x$.

a) $f(x) = 2^{x-2}$

b) $f(x) = 2^x - 4$

c) $f(x) = 5 - 2^{-x}$

Solution

a) The graph of this function is the graph of $y = 2^x$ shifted *right* 2 units.

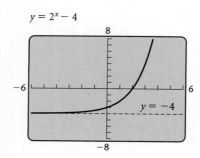

$y = 2^x - 4$

b) The graph is the graph of $y = 2^x$ shifted *down* 4 units (see the graph at left).

c) The graph is a reflection of the graph of $y = 2^x$ across the y-axis, followed by a reflection across the x-axis and then a shift *up* 5 units.

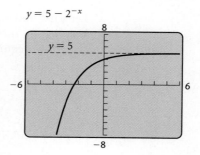

$y = 5 - 2^{-x}$

Graphs of Inverses of Exponential Functions

We have noted that every exponential function (with $a > 0$ and $a \neq 1$) is one-to-one. Thus each such function has an inverse that is a function. In the next section, we will name these inverse functions and use them in applications. For now, we draw their graphs by interchanging x and y.

EXAMPLE 4   Graph: $x = 2^y$.

Solution   Note that x is alone on one side of the equation. We can find ordered pairs that are solutions by choosing values for y and then computing the corresponding x-values.

For $y = 0$, $x = 2^0 = 1$.

For $y = 1$, $x = 2^1 = 2$.

For $y = 2$, $x = 2^2 = 4$.

For $y = 3$, $x = 2^3 = 8$.

For $y = -1$, $x = 2^{-1} = \dfrac{1}{2^1} = \dfrac{1}{2}$.

For $y = -2$, $x = 2^{-2} = \dfrac{1}{2^2} = \dfrac{1}{4}$.

For $y = -3$, $x = 2^{-3} = \dfrac{1}{2^3} = \dfrac{1}{8}$.

x		
$x = 2^y$	y	(x, y)
1	0	$(1, 0)$
2	1	$(2, 1)$
4	2	$(4, 2)$
8	3	$(8, 3)$
$\dfrac{1}{2}$	-1	$\left(\dfrac{1}{2}, -1\right)$
$\dfrac{1}{4}$	-2	$\left(\dfrac{1}{4}, -2\right)$
$\dfrac{1}{8}$	-3	$\left(\dfrac{1}{8}, -3\right)$

(1) Choose values for y.
(2) Compute values for x.

We plot the points and connect them with a smooth curve. Note that the curve does not touch or cross the *y*-axis. The *y*-axis is a vertical asymptote.

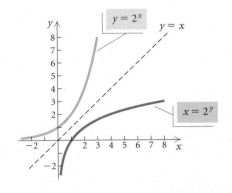

Note too that this curve looks just like the graph of $y = 2^x$, except that it is reflected across the line $y = x$, as we would expect for an inverse. The inverse of $y = 2^x$ is $x = 2^y$.

Applications

Graphers are especially helpful when working with exponential functions. They not only facilitate computations but they also allow us to visualize the functions.

EXAMPLE 5 *Compound Interest.* The amount of money A that a principal P will be worth after t years at interest rate i, compounded n times per year, is given by the formula

$$A = P\left(1 + \frac{i}{n}\right)^{nt}.$$

Suppose that $100,000 is invested at 8% interest, compounded semiannually.

a) Find a function for the amount of money after t years.

b) Graph the function.

c) Find the amount of money in the account at $t = 0$, 4, 8, and 10 yr.

d) When will the amount of money in the account reach $400,000?

Solution

a) Since $P = 100,000$, $i = 8\% = 0.08$, and $n = 2$, we can substitute these values and write the following function:

$$A(t) = 100,000\left(1 + \frac{0.08}{2}\right)^{2 \cdot t} = 100,000(1.04)^{2t}.$$

b) For the graph, we use the viewing window [0, 20, 0, 500,000] because of the large numbers and the fact that negative time values and amounts of money have no meaning in this application.

$y = 100{,}000(1.04)^{2x}$

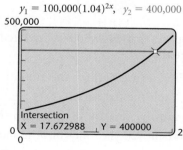

$y = 100{,}000(1.04)^{2x}$

c) We can compute function values using function notation on the home screen of a grapher. (See the window at left.) We can also calculate the values directly on a grapher by substituting in the expression for $A(t)$:

$$A(0) = 100{,}000(1.04)^{2 \cdot 0} = 100{,}000;$$
$$A(4) = 100{,}000(1.04)^{2 \cdot 4} \approx 136{,}856.91;$$
$$A(8) = 100{,}000(1.04)^{2 \cdot 8} \approx 187{,}298.12;$$
$$A(10) = 100{,}000(1.04)^{2 \cdot 10} \approx 219{,}112.31.$$

d) To find the amount of time it takes for the account to grow to $400,000, we set

$$100{,}000(1.04)^{2t} = 400{,}000$$

and solve. One way we can do this is by graphing the equations

$$y_1 = 100{,}000(1.04)^{2x} \quad \text{and} \quad y_2 = 400{,}000.$$

INTERSECT METHOD

REVIEW SECTION 3.5.

Then we can use the Intersect method to estimate the first coordinate of the point of intersection. (See Fig. 1 below.)

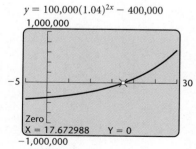

FIGURE 1

FIGURE 2

ZERO METHOD

REVIEW SECTION 3.5.

We can also use the Zero method to estimate the zero of the function $y = 100{,}000(1.04)^{2x} - 400{,}000$. (See Fig. 2 above.)

Regardless of the method we use, we see that the account grows to $400,000 after about 17.67 yr, or about 17 yr, 8 mo, and 2 days. ▬

*Interactive
Discovery*

Take a sheet of $8\frac{1}{2}$-in. by 11-in. paper and cut it into two equal pieces. Then cut each of these in half. Next, cut each of the four pieces in half again. Continue this process.

	t	$f(t) = 0.004 \cdot 2^t$
Start	0	$0.004 \cdot 2^0$, or 0.004
Step 1	1	$0.004 \cdot 2^1$, or 0.008
Step 2	2	$0.004 \cdot 2^2$, or 0.016
Step 3	3	
Step 4	4	
Step 5	5	

a) Place all the pieces in a stack and measure the thickness with a micrometer or other measuring device such as a dial caliper.

b) A piece of paper is typically 0.004 in. thick. Check the measurement in part (a) by completing the table shown here.

c) Graph the function $f(t) = 0.004(2)^t$.

d) Compute the thickness of the paper (in miles) after 25 steps.

The Number e

We now consider a very special number in mathematics. In 1741, Leonhard Euler named this number e. Though you may not have encountered it before, you will see here and in future mathematics that it has many important applications. To explain this number, we use the compound interest formula $A = P(1 + i/n)^{nt}$ in Example 5. Suppose that \$1 is an initial investment at 100% interest for 1 yr. (No bank would pay this.) The formula above becomes a function A defined in terms of the number of compounding periods n:

$$A(n) = \left(1 + \frac{1}{n}\right)^n.$$

Let's visualize the function

$$A(n) = \left(1 + \frac{1}{n}\right)^n$$

using a grapher and explore the values of $A(n)$ as $n \to \infty$.

Interactive Discovery

Graph the function $y = (1 + 1/x)^x$. Consider the graph for larger and larger values of x. Does this function have a horizontal asymptote? Explore with a grapher to determine the asymptote, if it exists. You might try to use the TABLE feature.

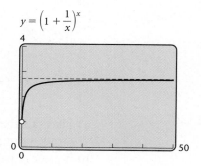

$$y = \left(1 + \frac{1}{x}\right)^x$$

Let's find some function values using the TABLE feature on a grapher.

n	$A(n) = \left(1 + \dfrac{1}{n}\right)^n$
1 (compounded annually)	$2.00
2 (compounded semiannually)	$2.25
3	$2.3704
4 (compounded quarterly)	$2.4414
5	$2.4883
100	$2.7048
365 (compounded daily)	$2.7146
8760 (compounded hourly)	$2.7181

As the values of n get larger and larger, the function values get closer and closer to the number Euler named e. Its decimal representation does not terminate or repeat; it is irrational.

$$e = 2.7182818284\ldots$$

EXAMPLE 6   Find each value of e^x, to four decimal places, using the $\boxed{e^x}$ key on a grapher.

a) e^3 **b)** $e^{-0.23}$

c) e^0 **d)** e^1

Solution

FUNCTION VALUE	READOUT	ROUNDED
a) e^3	e^(3) 20.08553692	20.0855
b) $e^{-0.23}$	e^(−.23) .7945336025	0.7945
c) e^0	e^(0) 1	1
d) e^1	e^(1) 2.718281828	2.7183

Graphs of Exponential Functions, Base e

We demonstrate ways in which to graph exponential functions.

EXAMPLE 7   Graph $f(x) = e^x$ and $g(x) = e^{-x}$.

Solution   We can compute points for each equation using the $\boxed{e^x}$ key on a grapher. Then we plot these points and draw the graphs of the functions.

$y_1 = e^x, \quad y_2 = e^{-x}$

X	Y₁	Y₂
−3	.04979	20.086
−2	.13534	7.3891
−1	.36788	2.7183
0	1	1
1	2.7183	.36788
2	7.3891	.13534
3	20.086	.04979

X = 3

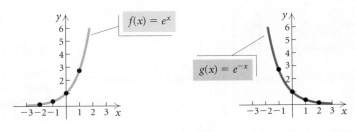

Note that the graph of g is a reflection of the graph of f across the y-axis.

EXAMPLE 8   Graph each of the following using a grapher. Briefly describe how each graph can be obtained from $y = e^x$.

a) $f(x) = e^{-0.5x}$

b) $f(x) = 1 - e^{-2x}$

c) $f(x) = e^{x+3}$

Solution

a) We note that the graph of this function is a horizontal stretching of the graph of $y = e^x$ followed by a reflection across the y-axis. (See Fig. 1.)

b) The graph is a horizontal shrinking of the graph of $y = e^x$, followed by a reflection across the y-axis, then across the x-axis, followed by a translation up 1 unit. (See Fig. 2.)

$y = e^{-0.5x}$

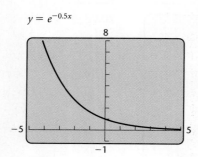

FIGURE 1

$y = 1 - e^{-2x}$

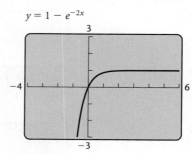

FIGURE 2

$y = e^{x+3}$

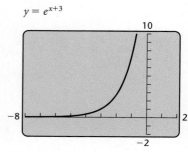

FIGURE 3

c) The graph is a translation of the graph of $y = e^x$ to the left 3 units. (See Fig. 3.)

Exercise Set 6.1

Find each of the following, to four decimal places, using a grapher.

1. e^4

2. e^{10}

3. $e^{-2.458}$

4. $\left(\dfrac{1}{e^3}\right)^2$

Make a hand-drawn graph of the function. Then check your work using a grapher, if possible.

5. $f(x) = 3^x$

6. $f(x) = 5^x$

7. $f(x) = 6^x$

8. $f(x) = 3^{-x}$

9. $f(x) = \left(\dfrac{1}{4}\right)^x$

10. $f(x) = \left(\dfrac{2}{3}\right)^x$

11. $x = 3^y$

12. $x = 4^y$

13. $x = \left(\dfrac{1}{2}\right)^y$

14. $x = \left(\dfrac{4}{3}\right)^y$

15. $y = \dfrac{1}{4}e^x$

16. $y = 2e^{-x}$

17. $f(x) = 1 - e^{-x}$

18. $f(x) = e^x - 2$

Graph the function using a grapher. Describe how each graph can be obtained from a basic exponential function.

19. $f(x) = 2^{x+1}$

20. $f(x) = 2^{x-1}$

21. $f(x) = 2^x - 3$

22. $f(x) = 2^x + 1$

23. $f(x) = 4 - 3^{-x}$

24. $f(x) = 2^{x-1} - 3$

25. $f(x) = \left(\dfrac{3}{2}\right)^{x-1}$

26. $f(x) = 3^{4-x}$

27. $f(x) = 2^{x+3} - 5$

28. $f(x) = -3^{x-2}$

29. $f(x) = e^{2x}$

30. $f(x) = e^{-0.2x}$

31. $y = e^{-x+1}$

32. $y = e^{2x} + 1$

33. $f(x) = 2(1 - e^{-x})$

34. $f(x) = 1 - e^{-0.01x}$

35. *Growth of AIDS.* The total number of AIDS cases reported in the United States is approximated by the exponential function

$$N(t) = 274,390(1.195)^t,$$

where $t = 0$ corresponds to 1992 (*Source*: U.S. Centers for Disease Control and Prevention, HIV/AID Surveillance Reports).

a) According to this function, how many AIDS cases had been reported by 1998?

b) Predict the number of AIDS cases that will be reported by 2001.

c) Graph the function.

d) Using a grapher and either the Intersect or the Zero method (see Example 5), estimate the length of time it will take for the number of AIDS cases to reach 3 million.

36. *Growth of Bacteria* Escherichia coli. The bacteria *Escherichia coli* are commonly found in the human bladder. Suppose that 3000 of the bacteria are present at time $t = 0$. Then under certain conditions, t minutes later, the number of bacteria present is

$$N(t) = 3000(2)^{t/20}.$$

a) How many bacteria will be present after 10 min? 20 min? 30 min? 40 min? 60 min?

b) Graph the function.

c) This bacteria can cause bladder infections in humans when the number of bacteria reaches 100,000,000. Using a grapher and either the Intersect or the Zero method (see Example 5), find the length of time it takes for a bladder infection to be possible.

37. *Recycling Aluminum Cans.* It is estimated that two thirds of all aluminum cans distributed will be recycled each year (*Source*: Alcoa Corporation). A beverage company distributes 350,000 cans. The number still in use after time t, in years, is given by the exponential function

$$N(t) = 350,000\left(\tfrac{2}{3}\right)^t.$$

a) How many cans are still in use after 0 yr? 1 yr? 4 yr? 10 yr?

b) Graph the function.

c) After how long will 2000 of the cans still be in use?

38. *Interest in a College Trust Fund.* Following the birth of a child, Juan deposits $10,000 in a college trust fund where interest is 6.4%, compounded semiannually.

a) Find a function for the amount in the account after t years.

b) Find the amount of money in the account at $t = 0, 4, 8, 10,$ and 18 yr.

c) Graph the function.

d) After how long will the account contain $100,000?

39. *Salvage Value.* A top-quality fax–copying machine is purchased for $5800. Its value each year is about 80% of the value of the preceding year. After t years, its value, in dollars, is given by the exponential function

$$V(t) = 5800(0.8)^t.$$

a) Graph the function.

b) Find the value of the machine after 0 yr, 1 yr, 2 yr, 5 yr, and 10 yr.

c) The company decides to replace the machine when its value has declined to $500. After how long will the machine be replaced?

40. *Cellular Phone Boom.* The number of people using cellular phones has grown exponentially in recent years. The total number of subscribers C, in millions, is given by

$$C(t) = 0.4703477987(1.523963441)^x,$$

where t is the number of years since 1985.

CELL PHONE BOOM

Source: Cellular Telecommunications Industry Association

a) Find the total number of subscribers in 1995, 1999, 2000, and 2004.

b) Graph the function with a scatterplot of the data.

c) After how long will the number of subscribers be 500,000,000?

41. *Timber Demand.* World demand for timber is increasing exponentially. The demand N, in billions

of cubic feet, purchased is given by

$$N(t) = 46.6(1.018)^t,$$

where t is the number of years since 1981 (*Sources*: U.N. Food and Agricultural Organization, American Forest and Paper Association).

a) Graph the function.
b) Find the demand for timber in 2000 and 2010.
c) After how many years will the demand for timber be 93.4 billion cubic feet?

42. *Typing Speed.* Sarah is taking keyboarding at a community college. After she practices for t hours, her speed, in words per minute, is given by the function

$$S(t) = 200[1 - (0.86)^t].$$

a) Graph the function.
b) What is Sarah's speed after practicing for 10 hr? 20 hr? 40 hr? 100 hr?
c) How much time passes before Sarah's speed is 100 words per minute?
d) Does this graph have an asymptote? If so, what is it, and what is its significance to Sarah's learning?

43. *Advertising.* A company begins a radio advertising campaign in New York City to market a new CD-ROM video game. The percentage of the target market that buys a game is generally a function of the length of the advertising campaign. The estimated percentage is given by

$$f(t) = 100(1 - e^{-0.04t}),$$

where t is the number of days of the campaign.

a) Graph the function.
b) Find $f(25)$, the percentage of the target market that has bought the product after a 25-day advertising campaign.
c) After how long will 90% of the target market have bought the product?

44. *Growth of a Stock.* The value of a stock is given by the function

$$V(t) = 58(1 - e^{-1.1t}) + 20,$$

where V is the value of the stock after time t, in months.

a) Graph the function.
b) Find $V(1)$, $V(2)$, $V(4)$, $V(6)$, and $V(12)$.
c) After how long will the value of the stock be $75?

In Exercises 45–58, use a grapher to match the equation with one of figures (a)–(n), which follow.

a)

b)

c)

d)

e)

f)

g)

h)

i)

j)

k)

l)

m)

n)

45. $y = 3^x - 3^{-x}$

46. $y = 3^{-(x+1)^2}$

47. $f(x) = -2.3^x$

48. $f(x) = 30,000(1.4)^x$

49. $y = 2^{-|x|}$

50. $y = 2^{-(x-1)}$

51. $f(x) = (0.58)^x - 1$

52. $y = 2^x + 2^{-x}$

53. $g(x) = e^{|x|}$

54. $f(x) = |2^x - 1|$

55. $y = 2^{-x^2}$

56. $y = |2^{x^2} - 8|$

57. $g(x) = \dfrac{e^x - e^{-x}}{2}$

58. $f(x) = \dfrac{e^x + e^{-x}}{2}$

Use a grapher to find the point(s) of intersection of the graphs of each of the following pairs of equations.

59. $y = |1 - 3^x|$,
$\quad y = 4 + 3^{-x^2}$

60. $y = 4^x + 4^{-x}$,
$\quad y = 8 - 2x - x^2$

61. $y = 2e^x - 3$, $\; y = \dfrac{e^x}{x}$

62. $y = \dfrac{1}{e^x + 1}$, $\; y = 0.3x + \dfrac{7}{9}$

Use a grapher. Solve graphically.

63. $5.3^x - 4.2^x = 1073$

64. $e^x = x^3$

65. $2^x > 1$

66. $3^x \le 1$

67. $2^x + 3^x = x^2 + x^3$

68. $31{,}245e^{-3x} = 523{,}467$

Discussion and Writing

69. Describe the differences between the graphs of $f(x) = x^3$ and $g(x) = 3^x$.

70. Suppose that $10,000 is invested for 8 yr at 6.4% interest, compounded annually. In what year will the most interest be earned? Why?

71. Graph each pair of equations using the same set of axes. Then compare the results in parts (a) and (b).

a) $y = 3^x$, $\; x = 3^y$ **b)** $y = 1^x$, $\; x = 1^y$

Skill Maintenance

Solve, finding all solutions in $[0, 2\pi)$ or $[\infty, 360°)$.

72. $\csc^2 - 2 = 0$

73. $6 \sin^2 x - 5 \sin x = 4$

74. $2 \csc x \tan x - 2 \csc x + \tan x - 1 = 0$

75. $5 \tan x - 3 = 7$

Synthesis

76. Which is larger, 7^π or π^7? 70^{80} or 80^{70}?

77. Graph $f(x) = x^{1/(x-1)}$ for $x > 0$. Use a grapher and the TABLE feature to identify the horizontal asymptote.

In Exercises 78 and 79:

a) *Graph using a grapher.*
b) *Approximate the zeros.*
c) *Approximate the relative maximum and minimum values. If your grapher has a MAX–MIN feature, use it.*

78. $f(x) = x^2 e^{-x}$

79. $f(x) = e^{-x^2}$

Logarithmic Functions and Graphs

6.2

- Graph logarithmic functions.
- Convert between exponential and logarithmic equations.
- Find common and natural logarithms using a grapher.

We now consider *logarithmic*, or *logarithm*, *functions*. These functions are inverses of exponential functions and have many applications.

Logarithmic Functions

Consider the function $f(x) = 2^x$ graphed at right. We see from the graph that this function passes the horizontal-line test and is therefore one-to-one. Thus, f has an inverse that is a function.

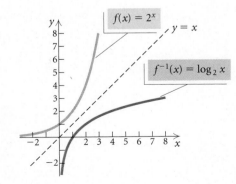

To find a formula for f^{-1} when $f(x) = 2^x$, we try to use the method of Section 1.6:

1. Replace $f(x)$ with y: $f(x) = 2^x$
$$y = 2^x$$

2. Interchange x and y: $x = 2^y$

3. Solve for y: $y =$ the power to which we raise 2 to get x

4. Replace y with $f^{-1}(x)$: $f^{-1}(x) =$ the power to which we raise 2 to get x.

Mathematicians have defined a new symbol to replace the words "the power to which we raise 2 to get x." That symbol is "$\log_2 x$," read "the logarithm, base 2, of x."

Logarithmic Function, Base 2

"$\log_2 x$," read "the logarithm, base 2, of x," means "the power to which we raise 2 to get x."

Thus if $f(x) = 2^x$, then $f^{-1}(x) = \log_2 x$. For example,

$$f^{-1}(8) = \log_2 8 = 3,$$

because

3 is the power to which we raise 2 to get 8.

Similarly, $\log_2 13$ is the power to which we raise 2 to get 13. As yet, we have no simpler way to say this other than

"$\log_2 13$ is the power to which we raise 2 to get 13."

Later, however, we will learn how to approximate this expression using a grapher.

For any exponential function $f(x) = a^x$, its inverse is called a **logarithmic function, base a.** The graph of the inverse can be obtained by reflecting the graph of $y = a^x$ across the line $y = x$, to obtain $x = a^y$. Then $x = a^y$ is equivalent to $y = \log_a x$. We read $\log_a x$ as "the logarithm, base a, of x."

The inverse of $f(x) = a^x$ is given by $f^{-1}(x) = \log_a x$.

Logarithmic Function, Base a

We define $y = \log_a x$ as that number y such that $x = a^y$, where $x > 0$ and a is a positive constant other than 1.

Let's look at the graphs of $f(x) = a^x$ and $f^{-1}(x) = \log_a x$ for $a > 1$ and $0 < a < 1$.

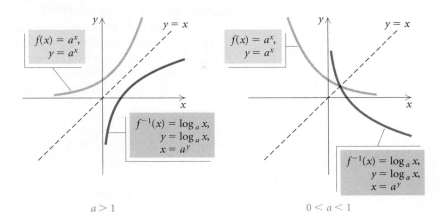

$$a > 1 \qquad\qquad\qquad 0 < a < 1$$

CONNECTING THE CONCEPTS

COMPARING EXPONENTIAL AND LOGARITHMIC FUNCTIONS

Usually, we use a number that is greater than 1 for the logarithmic base. In the following table, we compare exponential and logarithmic functions with bases a greater than 1. Similar statements would be made for a, where $0 < a < 1$. It is helpful to visualize the differences by carefully observing the graphs.

EXPONENTIAL FUNCTION	LOGARITHMIC FUNCTION
$y = a^x$ $f(x) = a^x$ $a > 1$ Continuous One-to-one Domain: All real numbers, $(-\infty, \infty)$ Range: All positive real numbers, $(0, \infty)$ Increasing Horizontal asymptote is x-axis: $\quad (a^x \to 0 \text{ as } x \to -\infty)$ y-intercept: $(0, 1)$ There is no x-intercept.	$x = a^y$ $f^{-1}(x) = \log_a x$ $a > 1$ Continuous One-to-one Domain: All positive real numbers, $(0, \infty)$ Range: All real numbers, $(-\infty, \infty)$ Increasing Vertical asymptote is y-axis: $\quad (\log_a x \to -\infty \text{ as } x \to 0^+)$ x-intercept: $(1, 0)$ There is no y-intercept.

Converting Between Exponential and Logarithmic Equations

It is helpful in dealing with logarithmic functions to remember that a logarithm of a number is an *exponent*. It is the exponent y in $x = a^y$. You might think to yourself, "the logarithm, base a, of a number x is the power to which a must be raised to get x."

We are led to the following. (The symbol $\longleftrightarrow$ means that the two statements are equivalent; that is, when one is true, the other is true. The words "if and only if" can be used in place of $\longleftrightarrow$.)

$$\log_a x = y \longleftrightarrow x = a^y \qquad \text{A logarithm is an exponent!}$$

EXAMPLE 1   Convert each of the following to a logarithmic equation.

a) $16 = 2^x$ b) $10^{-3} = 0.001$ c) $e^t = 70$

Solution

                                       — The exponent is the logarithm.

a) $16 = 2^x \qquad \log_2 16 = x$

                                       — The base remains the same.

b) $10^{-3} = 0.001 \rightarrow \log_{10} 0.001 = -3$

c) $e^t = 70 \rightarrow \log_e 70 = t$

EXAMPLE 2   Convert each of the following to an exponential equation.

a) $\log_2 32 = 5$ b) $\log_a Q = 8$ c) $x = \log_t M$

Solution

                                       — The logarithm is the exponent.

a) $\log_2 32 = 5 \qquad 2^5 = 32$

                                       — The base remains the same.

b) $\log_a Q = 8 \rightarrow a^8 = Q$

c) $x = \log_t M \rightarrow t^x = M$

Finding Certain Logarithms

Let's use the definition of logarithms to find some logarithmic values.

EXAMPLE 3   Find each of the following logarithms.

a) $\log_{10} 10{,}000$ b) $\log_{10} 0.01$

c) $\log_2 8$ d) $\log_9 3$

e) $\log_6 1$ f) $\log_8 8$

Solution

RESULT	REASON
a) $\log_{10} 10{,}000 = 4$	Think of the meaning of $\log_{10} 10{,}000$. It is the exponent to which we raise 10 to get 10,000. That exponent is 4.
b) $\log_{10} 0.01 = -2$	We have $0.01 = \dfrac{1}{100} = \dfrac{1}{10^2} = 10^{-2}$. The exponent to which we raise 10 to get 0.01 is -2.
c) $\log_2 8 = 3$	$8 = 2^3$. The exponent to which we raise 2 to get 8 is 3.
d) $\log_9 3 = \frac{1}{2}$	$3 = \sqrt{9} = 9^{1/2}$. The exponent to which we raise 9 to get 3 is $\frac{1}{2}$.
e) $\log_6 1 = 0$	$1 = 6^0$. The exponent to which we raise 6 to get 1 is 0.
f) $\log_8 8 = 1$	$8 = 8^1$. The exponent to which we raise 8 to get 8 is 1.

Examples 3(e) and 3(f) illustrate two important properties of logarithms. The property $\log_a 1 = 0$ follows from the fact that $a^0 = 1$. Thus, $\log_5 1 = 0$, $\log_{10} 1 = 0$, and so on. The property $\log_a a = 1$ follows from the fact that $a^1 = a$. Thus, $\log_5 5 = 1$, $\log_{10} 10 = 1$, and so on.

$$\log_a 1 = 0 \quad \text{and} \quad \log_a a = 1, \quad \text{for any logarithmic base } a.$$

Finding Logarithms on a Grapher

Before calculators became so widely available, base-10, or **common logarithms,** were used extensively to simplify complicated calculations. In fact, that is why logarithms were invented. The abbreviation **log**, with no base written, is used to represent the common logarithms, or base-10 logarithms. Thus,

$$\log 29 \quad \text{means} \quad \log_{10} 29.$$

Let's compare $\log 29$ with $\log 10$ and $\log 100$:

$$\left.\begin{array}{l} \log 10 = \log_{10} 10 = 1 \\ \qquad\quad \log 29 = \ ? \\ \log 100 = \log_{10} 100 = 2 \end{array}\right\} \quad \begin{array}{l} \text{Since 29 is between 10 and 100, it seems} \\ \text{reasonable that } \log 29 \text{ is between 1 and 2.} \end{array}$$

On a grapher, the scientific key for common logarithms is generally marked ⎡LOG⎤. Using that key, we find that

$$\log 29 \approx 1.462397998 \approx 1.4624$$

rounded to four decimal places. Since $1 < 1.4624 < 2$, our answer seems reasonable. This also tells us that $10^{1.4624} \approx 29$.

Interactive Discovery

Find log 1000 and log 10,000 without using a grapher. Between what two whole numbers is log 9874? Now approximate log 9874, rounded to four decimal places, on a grapher.

EXAMPLE 4   Find each of the following common logarithms on a grapher set in REAL mode. Round to four decimal places.

a) log 645,778
b) log 0.0000239
c) log (−3)

Solution

Function Value	Readout	Rounded
a) log 645,778	log(645778) 5.810083246	5.8101
b) log 0.0000239	log(0.0000239) −4.621602099	−4.6216
c) log (−3)	ERR:NONREAL ANS *	Does not exist

A check for part (a) is shown below. Since 5.810083246 is the power to which we raise 10 to get 645,778, we can check by finding $10^{5.810083246}$. In part (c), log (−3) does not exist as a real number because there is no real-number power to which we can raise 10 to get −3. The number 10 raised to any real-number power is positive. The common logarithm of a negative number does not exist as a real number.

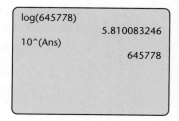

---

*If the grapher is set in $a + bi$ mode, the readout is $.4771212547 + 1.364376354i$.

Natural Logarithms

Logarithms, base e, are called **natural logarithms.** The abbreviation "ln" is generally used for natural logarithms. Thus,

ln 53   means $\log_e 53$.

On a grapher, the scientific key for natural logarithms is generally marked [LN]. Using that key, we find that

$$\ln 53 \approx 3.970291914$$
$$\approx 3.9703$$

rounded to four decimal places.

EXAMPLE 5 Find each of the following natural logarithms on a grapher set in REAL mode. Round to four decimal places.

a) ln 645,778   **b)** ln 0.0000239
c) ln (−5)   **d)** ln e
e) ln 1

Solution

FUNCTION VALUE	READOUT	ROUNDED
a) ln 645,778	ln(645778) 13.37821107	13.3782
b) ln 0.0000239	ln(0.0000239) −10.6416321	−10.6416
c) ln (−5)	ERR:NONREAL ANS *	Does not exist
d) ln e	ln(e) 1	1
e) ln 1	ln(1) 0	0

Note that $\ln e = \log_e e = 1$ and $\ln 1 = \log_e 1 = 0$. A check for part (b) is shown at left.

---

*If the grapher is set in $a + bi$ mode, the readout is $1.609437912 + 3.141592654i$.

Changing Logarithmic Bases

Most graphers give the values of both common logarithms and natural logarithms. To find a logarithm with a base other than 10 or e, we can use the following conversion formula.

The Change-of-Base Formula

For any logarithmic bases a and b, and any positive number M,

$$\log_b M = \frac{\log_a M}{\log_a b}.$$

We will prove this result in the next section.

EXAMPLE 6 Find $\log_5 8$ using common logarithms.

Solution First, we let $a = 10$, $b = 5$, and $M = 8$. Then we substitute into the change-of-base formula:

$$\log_5 8 = \frac{\log_{10} 8}{\log_{10} 5} \qquad \text{Substituting}$$

$$\approx 1.2920. \qquad \text{Using a grapher}$$

The check is shown in Fig. 1.

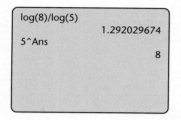

FIGURE 1

```
log(8)/log(5)
                1.292029674
5^Ans
                          8
```

We can also use base e for a conversion.

EXAMPLE 7 Find $\log_5 8$ using natural logarithms.

Solution Substituting e for a, 5 for b, and 8 for M, we have

$$\log_5 8 = \frac{\log_e 8}{\log_e 5}$$

$$= \frac{\ln 8}{\ln 5} \approx 1.2920.$$

The check is shown in Fig. 2.

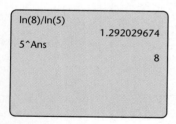

FIGURE 2

```
ln(8)/ln(5)
                1.292029674
5^Ans
                          8
```

Graphs of Logarithmic Functions

We demonstrate several ways to graph logarithmic functions.

EXAMPLE 8 Graph: $y = f(x) = \log_5 x$.

Solution

Method 1. The equation $y = \log_5 x$ is equivalent to $x = 5^y$. We can find ordered pairs that are solutions by choosing values for y and computing the x-values. We then plot points, remembering that x still is the first coordinate.

For $y = 0$, $x = 5^0 = 1$.
For $y = 1$, $x = 5^1 = 5$.
For $y = 2$, $x = 5^2 = 25$.
For $y = 3$, $x = 5^3 = 125$.

For $y = -1$, $x = 5^{-1} = \dfrac{1}{5}$.

For $y = -2$, $x = 5^{-2} = \dfrac{1}{25}$.

x, or 5^y	y
1	0
5	1
25	2
125	3
$\dfrac{1}{5}$	-1
$\dfrac{1}{25}$	-2

$y = \log_5 x$

(1) Select y.
(2) Compute x.

Method 2. To use a grapher, we must first change the base. Here we change from base 5 to base e:

$$y = \log_5 x = \frac{\ln x}{\ln 5}. \qquad \text{Using } \log_b M = \frac{\log_a M}{\log_a b}$$

The graph is as shown in Fig. 3 below.

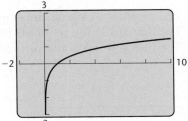

$y = \log_5 x = \dfrac{\ln x}{\ln 5}$

$y_1 = 5^x, \quad y_2 = \log_5 x$

FIGURE 3

FIGURE 4

Method 3. Some graphers have a feature that graphs inverses automatically. If we begin with $y_1 = 5^x$, the graphs of both y_1 and its inverse $y_2 = \log_5 x$ will be drawn. (See Fig. 4 above.)

EXAMPLE 9 Graph: $g(x) = \ln x$.

Solution To graph $y = g(x) = \ln x$, we first write its equivalent exponential equation, $x = e^y$. Next, we select values for y and use a grapher to find the corresponding values of e^y. We then plot points and draw the curve.

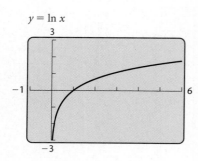

$y = \ln x$

x, or e^y	y
1	0
2.7	1
7.4	2
20.1	3
0.1	-2
0.4	-1

$g(x) = \ln x$
or $x = e^y$

We could also use the $\boxed{\text{LN}}$ key on a grapher to find function values. The easiest way is to graph the function $y = \ln x$ using a grapher. (See the window at left.)

Recall that the graph of $f(x) = \log_a x$, for any base a, has the x-intercept (1, 0). The domain is the set of positive real numbers. The range is the set of all real numbers.

EXAMPLE 10   Graph each of the following using a grapher. Describe how each graph can be obtained from $y = \ln x$. Give the domain of each function and discuss the vertical asymptotes.

a) $f(x) = \ln (x + 3)$
b) $f(x) = 3 - \frac{1}{2} \ln x$
c) $f(x) = |\ln (x - 1)|$

Solution

a) The graph is a shift of the graph of $y = \ln x$ left 3 units (see Fig. 5). The domain is the set of all real numbers greater than -3, $(-3, \infty)$. The line $x = -3$ is the vertical asymptote.

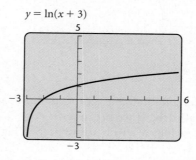

$y = \ln(x + 3)$

FIGURE 5

$y = 3 - \frac{1}{2} \ln x$

FIGURE 6

b) The graph is a shrinking of the graph of $y = \ln x$ in the y-direction, followed by a reflection across the x-axis, and then a translation up 3 units (see Fig. 6). The domain is the set of all positive real numbers, $(0, \infty)$. The y-axis is the vertical asymptote.

$y = |\ln(x - 1)|$

FIGURE 7

c) The graph is a translation of the graph of $y = \ln x$, 1 unit to the right. Then the absolute value has the effect of reflecting negative outputs across the x-axis (see Fig. 7). The domain is the set of all real numbers greater than 1, $(1, \infty)$. The line $x = 1$ is the vertical asymptote.   ▬

Applications

EXAMPLE 11   *Walking Speed.* In a study by psychologists Bornstein and Bornstein, it was found that the average walking speed w, in feet per second, of a person living in a city of population P, in thousands, is given by the function

$$w(P) = 0.37 \ln P + 0.05$$

(*Source: International Journal of Psychology*).

a) The population of San Diego, California, is 1,171,000. Find the average walking speed of people living in San Diego.

b) Graph the function.

c) A sociologist computes the average walking speed in a city to be approximately 2.0 ft/sec. Use this information to estimate the population of the city.

Solution

a) We substitute 1171 for P, since P is in thousands:

$$w(1171) = 0.37 \ln 1171 + 0.05 \quad \text{Substituting}$$
$$\approx 2.7 \text{ ft/sec.} \quad \text{Finding the natural logarithm and simplifying}$$

The average walking speed of people living in San Diego is about 2.7 ft/sec.

b) We graph with a viewing window of $[0, 600, 0, 4]$ because inputs are very large and outputs are very small by comparison.

$y = 0.37 \ln x + 0.05$

c) To find the population for which the average walking speed is 2.0 ft/sec, we set

$$2.0 = 0.37 \ln P + 0.05$$

and solve.

$y_1 = 0.37 \ln x + 0.05, \quad y_2 = 2$

We will use the Intersect method. We graph the equations $y_1 = 0.37 \ln x + 0.05$ and $y_2 = 2$ and use the INTERSECT feature to approximate the point of intersection (see the window at left). We see that in a city in which the average walking speed is 2.0 ft/sec, the population is about 194.5 thousand, or 194,500.   ▬

EXAMPLE 12   *Earthquake Magnitude.* The magnitude R, measured on the Richter scale, of an earthquake of intensity I is defined as

$$R = \log \frac{I}{I_0},$$

where I_0 is a minimum intensity used for comparison. We can think of I_0 as a threshold intensity that is the weakest earthquake that can be recorded on a seismograph. If one earthquake is 10 times as intense as another, its magnitude on the Richter scale is 1 greater than that of the other. If one earthquake is 100 times as intense as another, its magnitude on the Richter scale is 2 higher, and so on. Thus an earthquake whose magnitude is 7 on the Richter scale is 10 times as intense as an earthquake whose magnitude is 6. Earthquakes can be interpreted as multiples of the minimum intensity I_0.

The earthquake in western Turkey on August 17, 1999, had an intensity of $10^{7.4} \cdot I_0$. What was its magnitude on the Richter scale?

Solution   We substitute into the formula:

$$R = \log \frac{I}{I_0} = \log \frac{10^{7.4} I_0}{I_0}$$
$$= \log 10^{7.4} = 7.4.$$

The magnitude of the earthquake was 7.4 on the Richter scale.   ▬

Exercise Set 6.2

Make a hand-drawn graph of each of the following. Then check your work using a grapher.

1. $y = \log_3 x$ **2.** $y = \log_4 x$

3. $f(x) = \log x$ **4.** $f(x) = \ln x$

Find each of the following. Do not use a grapher.

5. $\log_2 16$ **6.** $\log_3 9$

7. $\log_5 125$ **8.** $\log_2 64$

9. $\log 0.001$ **10.** $\log 100$

11. $\log_2 \frac{1}{4}$ **12.** $\log_8 2$

13. $\ln 1$ **14.** $\ln e$

15. $\log 10$ **16.** $\log 1$

Convert to a logarithmic equation.

17. $10^3 = 1000$ **18.** $5^{-3} = \frac{1}{125}$

19. $8^{1/3} = 2$ **20.** $10^{0.3010} = 2$

21. $e^3 = t$ **22.** $Q^t = x$

23. $e^2 = 7.3891$ **24.** $e^{-1} = 0.3679$

25. $p^k = 3$ **26.** $e^{-t} = 4000$

Convert to an exponential equation.

27. $\log_5 5 = 1$ **28.** $t = \log_4 7$

29. $\log 0.01 = -2$ **30.** $\log 7 = 0.845$

31. $\ln 30 = 3.4012$ **32.** $\ln 0.38 = -0.9676$

33. $\log_a M = -x$

34. $\log_t Q = k$

35. $\log_a T^3 = x$

36. $\ln W^5 = t$

Find each of the following using a grapher. Round to four decimal places.

37. log 3   **38.** log 8

39. log 532   **40.** log 93,100

41. log 0.57   **42.** log 0.082

43. log (−2)   **44.** ln 50

45. ln 2   **46.** ln (−4)

47. ln 809.3   **48.** ln 0.00037

49. ln (−1.32)   **50.** ln 0

Find the logarithm using the change-of-base formula.

51. $\log_4 100$ **52.** $\log_3 20$

53. $\log_{100} 0.3$ **54.** $\log_\pi 100$

55. $\log_{200} 50$ **56.** $\log_{5.3} 1700$

For each of the following functions, briefly describe how the graph can be obtained from a basic logarithmic function. Then graph the function using a grapher. Give the domain of the function and discuss the vertical asymptotes.

57. $f(x) = \log_2 (x + 3)$ **58.** $f(x) = \log_3 (x - 2)$

59. $y = \log_3 x - 1$ **60.** $y = 3 + \log_2 x$

61. $f(x) = 4 \ln x$ **62.** $f(x) = \frac{1}{2} \ln x$

63. $y = 2 - \ln x$ **64.** $y = \ln (x + 1)$

Graph the function and its inverse using the same set of axes. Use any method.

65. $f(x) = 3^x,\ f^{-1}(x) = \log_3 x$

66. $f(x) = \log_4 x,\ f^{-1}(x) = 4^x$

67. $f(x) = \log x,\ f^{-1}(x) = 10^x$

68. $f(x) = e^x,\ f^{-1}(x) = \ln x$

69. *Walking Speed.* Refer to Example 11. Various cities and their populations are given below. Find the average walking speed in each city.

  a) Albuquerque, New Mexico: 419,681
  b) Chicago, Illinois: 2,721,547
  c) Pittsburgh, Pennsylvania: 350,363
  d) Durham, North Carolina: 149,799
  e) Green Bay, Wisconsin: 94,466

70. *Earthquake Magnitude.* Refer to Example 12. Various locations of earthquakes and their intensities are given below. What was the magnitude on the Richter scale?

  a) Mexico City, 1978: $10^{7.85} \cdot I_0$
  b) San Francisco, 1906: $10^{8.25} \cdot I_0$
  c) Chile, 1960: $10^{9.6} \cdot I_0$
  d) Italy, 1980: $10^{7.85} \cdot I_0$
  e) San Francisco, 1989: $10^{6.9} \cdot I_0$

71. *Forgetting.* Students in an accounting class took a final exam. They took equivalent forms of the exam in monthly intervals thereafter. The average score $S(t)$, in percent, after t months was found to be given by the function

$$S(t) = 78 - 15 \log (t + 1), \quad t \geq 0.$$

  a) What was the average score when they initially took the test, $t = 0$?
  b) What was the average score after 4 months? 24 months?
  c) Graph the function.
  d) After what time t was the average score 50?

72. *pH of Substances in Chemistry.* In chemistry, the pH of a substance is defined as

$$pH = -\log [H^+],$$

where H^+ is the hydrogen ion concentration, in moles per liter. Find the pH of each substance.

Litmus paper is used to test pH.

SUBSTANCE	HYDROGEN ION CONCENTRATION
a) Pineapple juice	1.6×10^{-4}
b) Hair rinse	0.0013
c) Mouthwash	6.3×10^{-7}
d) Eggs	1.6×10^{-8}
e) Tomatoes	6.3×10^{-5}

73. Find the hydrogen ion concentration of each substance, given the pH (see Exercise 72). Express the answer in scientific notation.

SUBSTANCE	pH
a) Tap water	7
b) Rainwater	5.4
c) Orange juice	3.2
d) Wine	4.8

74. *Advertising.* A model for advertising response is given by the function

$$N(a) = 1000 + 200 \ln a, \quad a \geq 1,$$

where $N(a)$ is the number of units sold and a is the amount spent on advertising, in thousands of dollars.

a) How many units were sold after spending $1000 ($a = 1$) on advertising?
b) How many units were sold after spending $5000?
c) Graph the function.
d) How much would have to be spent in order to sell 2000 units?

75. *Loudness of Sound.* The **loudness L**, in bels (after Alexander Graham Bell), of a sound of intensity I is defined to be

$$L = \log \frac{I}{I_0},$$

where I_0 is the minimum intensity detectable by the human ear (such as the tick of a watch at 20 ft under quiet conditions). If a sound is 10 times as intense as another, its loudness is 1 bel greater than that of the other. If a sound is 100 times as intense as another, its loudness is 2 bels greater, and so on. The bel is a large unit, so a subunit, the **decibel**, is generally used. For L, in decibels, the formula is

$$L = 10 \log \frac{I}{I_0}.$$

Find the loudness, in decibels, of each sound with the given intensity.

SOUND	INTENSITY
a) Library	$2510 \cdot I_0$
b) Dishwasher	$2,500,000 \cdot I_0$
c) Conversational speech	$10^6 \cdot I_0$
d) Heavy truck	$10^9 \cdot I_0$

Discussion and Writing

76. Explain how the graph of $f(x) = \ln x$ can be used to obtain the graph of $g(x) = e^{x-2}$.
77. If $\log b < 0$, what can you say about b?

Skill Maintenance

Find an equation of a parabola satisfying the given conditions.

78. Focus $(2, -8)$, directrix $y = -12$
79. Focus $(0, -6)$, directrix $x = 6$

Find an equation of an ellipse satisfying the given conditions.

80. Vertices: $(-10, 0)$ and $(10, 0)$; foci: $(-8, 0)$ and $(8, 0)$
81. Vertices: $(0, -7)$ and $(0, 7)$; foci: $(0, -6)$ and $(0, 6)$

Synthesis

Simplify.

82. $\dfrac{\log_3 64}{\log_3 16}$

83. $\dfrac{\log_5 8}{\log_5 2}$

Find the domain of the function.

84. $f(x) = \log_4 x^2$
85. $f(x) = \log_5 x^3$
86. $f(x) = \log (3x - 4)$
87. $f(x) = \ln |x|$

Solve.

88. $\log_2 (x - 3) \geq 4$
89. $\log_2 (2x + 5) < 0$

90. Using a grapher, find the point(s) of intersection of the graphs.

$$y = 4 \ln x, \qquad y = \frac{4}{e^x + 1}$$

In Exercises 91–94, match the equation with one of figures (a)–(d), which follow. If needed, use a grapher.

a)

b)

c)

d)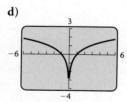

91. $f(x) = \ln |x|$
92. $f(x) = |\ln x|$
93. $f(x) = \ln x^2$
94. $g(x) = |\ln (x - 1)|$

For Exercises 95–98:
a) Graph the function.
b) Estimate the zeros.
c) Estimate the relative maximum and the minimum values.

95. $f(x) = x \ln x$
96. $f(x) = x^2 \ln x$
97. $f(x) = \dfrac{\ln x}{x^2}$
98. $f(x) = e^{-x} \ln x$

Properties of Logarithmic Functions

6.3

- Convert from logarithms of products, powers, and quotients to expressions in terms of individual logarithms, and conversely.
- Simplify expressions of the type $\log_a a^x$ and $a^{\log_a x}$.

We now establish some properties of logarithmic functions. These properties are based on corresponding rules for exponents.

Logarithms of Products

Interactive Discovery

Graph each of the following functions. Use the graphs to discover a pair of equivalent equations.

$$y_1 = \log (4x^2 \cdot x^3), \qquad y_2 = \log (4x^2) + \log (x^3),$$
$$y_3 = \log (4x^2 + x^3), \qquad y_4 = [\log (4x^2)][\log (x^3)]$$

Test the result by using the TABLE feature to complete the following tables and compare the y-values.

X	Y₁	Y₂
1		
10		
20		
300		

X = 1

X	Y₃	Y₄
1		
10		
20		
300		

X = 1

The first property of logarithms corresponds to the rule of exponents: $a^m \cdot a^n = a^{m+n}$.

The Product Rule

For any positive numbers M and N and any logarithmic base a,

$$\log_a MN = \log_a M + \log_a N.$$

(The logarithm of a product is the sum of the logarithms of the factors.)

EXAMPLE 1   Express as a sum of logarithms: $\log_3 (9 \cdot 27)$.

Solution   We have

$$\log_3 (9 \cdot 27) = \log_3 9 + \log_3 27. \qquad \text{Using the product rule}$$

As a check, note that

$$\log_3 (9 \cdot 27) = \log_3 243 = 5$$

and $\log_3 9 + \log_3 27 = 2 + 3 = 5.$

EXAMPLE 2 Express as a single logarithm: $\log_2 p^3 + \log_2 q$.

Solution We have

$$\log_2 p^3 + \log_2 q = \log_2 (p^3 q).$$

A PROOF OF THE PRODUCT RULE: Let $\log_a M = x$ and $\log_a N = y$. Converting to exponential equations, we have $a^x = M$ and $a^y = N$. Then

$$MN = a^x \cdot a^y = a^{x+y}.$$

Converting back to a logarithmic equation, we get

$$\log_a MN = x + y.$$

Remembering what x and y represent, we know it follows that

$$\log_a MN = \log_a M + \log_a N.$$

Logarithms of Powers

Interactive Discovery

Graph each of the following functions. Use the graphs to discover a pair of equivalent equations.

$$y_1 = \log x^5, \qquad y_2 = \log x + \log 5,$$
$$y_3 = 5 \log x, \qquad y_4 = \log (x + 5)$$

Test the result by using the TABLE feature to complete the following tables and compare the y-values.

X	Y₁	Y₂
1		
10		
20		
300		

X = 1

X	Y₃	Y₄
1		
10		
20		
300		

X = 1

The second property of logarithms corresponds to the rule of exponents: $(a^m)^n = a^{mn}$.

The Power Rule

For any positive number M, any logarithmic base a, and any real number p,

$$\log_a M^p = p \log_a M.$$

(The logarithm of a power of M is the exponent times the logarithm of M.)

EXAMPLE 3 Express each of the following as a product.

a) $\log_a 11^{-3}$

b) $\log_a \sqrt[4]{7}$

Solution

a) $\log_a 11^{-3} = -3 \log_a 11$ Using the power rule

b) $\log_a \sqrt[4]{7} = \log_a 7^{1/4}$ Writing exponential notation

$\phantom{\log_a \sqrt[4]{7}} = \frac{1}{4} \log_a 7$ Using the power rule

━

A PROOF OF THE POWER RULE: Let $x = \log_a M$. The equivalent exponential equation is $a^x = M$. Raising both sides to the power p, we obtain

$$(a^x)^p = M^p, \quad \text{or} \quad a^{xp} = M^p.$$

Converting back to a logarithmic equation, we get

$$\log_a M^p = xp.$$

But $x = \log_a M$, so substituting gives us

$$\log_a M^p = (\log_a M)p$$
$$= p \log_a M.$$

Logarithms of Quotients

Interactive Discovery

Graph each of the following functions. Use the graphs to discover a pair of equivalent equations.

$$y_1 = \log\left(\frac{4x^5}{x^2}\right),$$

$$y_2 = \log(4x^5) + \log(x^2),$$

$$y_3 = \log(4x^5) - \log(x^2),$$

$$y_4 = \log(4x + 5x^2)$$

Test the result by using the TABLE feature to complete the following tables and compare the y-values.

X	Y₁	Y₂
1		
10		
20		
300		

X = 1

X	Y₃	Y₄
1		
10		
20		
300		

X = 1

The third property of logarithms corresponds to the rule of exponents: $a^m/a^n = a^{m-n}$.

The Quotient Rule

For any positive numbers M and N, and any logarithmic base a,

$$\log_a \frac{M}{N} = \log_a M - \log_a N.$$

(The logarithm of a quotient is the logarithm of the numerator minus the logarithm of the denominator.)

EXAMPLE 4   Express as a difference of logarithms: $\log_t \frac{8}{w}$.

Solution

$$\log_t \frac{8}{w} = \log_t 8 - \log_t w \qquad \text{Using the quotient rule}$$

EXAMPLE 5   Express as a single logarithm: $\log_b 64 - \log_b 16$.

Solution

$$\log_b 64 - \log_b 16 = \log_b \frac{64}{16} = \log_b 4$$

A PROOF OF THE QUOTIENT RULE: The proof follows from both the product and the power rules:

$$\log_a \frac{M}{N} = \log_a MN^{-1}$$
$$= \log_a M + \log_a N^{-1} \qquad \text{Using the product rule}$$
$$= \log_a M + (-1) \log_a N \qquad \text{Using the power rule}$$
$$= \log_a M - \log_a N.$$

Note the following.

Common Errors

$\log_a MN \neq (\log_a M)(\log_a N)$ — The logarithm of a product is *not* the product of the logarithms.

$\log_a (M + N) \neq \log_a M + \log_a N$ — The logarithm of a sum is *not* the sum of the logarithms.

$\log_a \frac{M}{N} \neq \frac{\log_a M}{\log_a N}$ — The logarithm of a quotient is *not* the quotient of the logarithms.

Using the Properties Together

EXAMPLE 6 Express each of the following in terms of sums and differences of logarithms.

a) $\log_a \dfrac{x^2 y^5}{z^4}$ 　　　　**b)** $\log_a \sqrt[3]{\dfrac{a^2 b}{c^5}}$ 　　　　**c)** $\log_b \dfrac{ay^5}{m^3 n^4}$

Solution

a) $\log_a \dfrac{x^2 y^5}{z^4} = \log_a (x^2 y^5) - \log_a z^4$ 　　　　　Using the quotient rule

$\qquad\qquad = \log_a x^2 + \log_a y^5 - \log_a z^4$ 　　　Using the product rule

$\qquad\qquad = 2 \log_a x + 5 \log_a y - 4 \log_a z$ 　　　Using the power rule

b) $\log_a \sqrt[3]{\dfrac{a^2 b}{c^5}} = \log_a \left(\dfrac{a^2 b}{c^5} \right)^{1/3}$ 　　　　Writing exponential notation

$\qquad\qquad = \dfrac{1}{3} \log_a \dfrac{a^2 b}{c^5}$ 　　　　Using the power rule

$\qquad\qquad = \dfrac{1}{3} (\log_a a^2 b - \log_a c^5)$ 　　Using the quotient rule

$\qquad\qquad = \dfrac{1}{3} (2 \log_a a + \log_a b - 5 \log_a c)$ 　　Using the product and power rules. The parentheses are important.

$\qquad\qquad = \dfrac{1}{3} (2 + \log_a b - 5 \log_a c)$ 　　$\log_a a = 1$

$\qquad\qquad = \dfrac{2}{3} + \dfrac{1}{3} \log_a b - \dfrac{5}{3} \log_a c$ 　　Multiplying to remove parentheses

c) $\log_b \dfrac{ay^5}{m^3 n^4} = \log_b ay^5 - \log_b m^3 n^4$ 　　Using the quotient rule

$\qquad\qquad = (\log_b a + \log_b y^5) - (\log_b m^3 + \log_b n^4)$ 　　Using the product rule

$\qquad\qquad = \log_b a + \log_b y^5 - \log_b m^3 - \log_b n^4$ 　　Removing parentheses

$\qquad\qquad = \log_b a + 5 \log_b y - 3 \log_b m - 4 \log_b n$ 　　Using the power rule

EXAMPLE 7 Express as a single logarithm:

$$5 \log_b x - \log_b y + \dfrac{1}{4} \log_b z.$$

Solution

$$5 \log_b x - \log_b y + \frac{1}{4} \log_b z = \log_b x^5 - \log_b y + \log_b z^{1/4}$$

Using the power rule

$$= \log_b \frac{x^5}{y} + \log_b z^{1/4}$$

Using the quotient rule

$$= \log_b \frac{x^5 z^{1/4}}{y}, \text{ or } \log_b \frac{x^5 \sqrt[4]{z}}{y}$$

Using the product rule

EXAMPLE 8   Given that $\log_a 2 \approx 0.301$ and $\log_a 3 \approx 0.477$, find each of the following.

a) $\log_a 6$ **b)** $\log_a \dfrac{2}{3}$ **c)** $\log_a 81$

d) $\log_a \sqrt{a}$ **e)** $\log_a 5$ **f)** $\dfrac{\log_a 3}{\log_a 2}$

Solution

a) $\log_a 6 = \log_a (2 \cdot 3) = \log_a 2 + \log_a 3$ Using the product rule
$$\approx 0.301 + 0.477$$
$$\approx 0.778$$

b) $\log_a \frac{2}{3} = \log_a 2 - \log_a 3$ Using the quotient rule
$$\approx 0.301 - 0.477 \approx -0.176$$

c) $\log_a 81 = \log_a 3^4 = 4 \log_a 3$ Using the power rule
$$\approx 4(0.477) \approx 1.908$$

d) $\log_a \sqrt{a} = \log_a a^{1/2} = \frac{1}{2} \log_a a$ Using the power rule
$$= \frac{1}{2} \cdot 1 = \frac{1}{2}$$

e) $\log_a 5$ *cannot be found using these properties and the given information.*
$$(\log_a 5 \neq \log_a 2 + \log_a 3)$$

f) $\dfrac{\log_a 3}{\log_a 2} \approx \dfrac{0.477}{0.301} \approx 1.585$ We simply divided, not using any of the properties.

In Example 8, a is actually 10 so we have common logarithms. Check as many results as possible using a grapher.

Simplifying Expressions of the Type $\log_a a^x$ and $a^{\log_a x}$

We have two final properties to consider. The first follows from the product rule: Since $\log_a a^x = x \log_a a = x \cdot 1 = x$, we have $\log_a a^x = x$. This property also follows from the definition of a logarithm: x is the power to which we raise a in order to get a^x.

The Logarithm of a Base to a Power

For any base a and any real number x,

$$\log_a a^x = x.$$

(The logarithm, base a, of a to a power is the power.)

EXAMPLE 9 Simplify each of the following.

a) $\log_a a^8$ b) $\ln e^{-t}$ c) $\log 10^{3k}$

Solution

a) $\log_a a^8 = 8$ 8 is the power to which we raise a in order to get a^8.

b) $\ln e^{-t} = \log_e e^{-t} = -t$

c) $\log 10^{3k} = \log_{10} 10^{3k} = 3k$

Let $M = \log_a x$. Then $a^M = x$. Substituting $\log_a x$ for M, we obtain $a^{\log_a x} = x$. This also follows from the definition of a logarithm: $\log_a x$ is the power to which a is raised in order to get x.

A Base to a Logarithmic Power

For any base a and any positive real number x,

$$a^{\log_a x} = x.$$

(The number a raised to the power $\log_a x$ is x.)

EXAMPLE 10 Simplify each of the following.

a) $4^{\log_4 k}$ b) $e^{\ln 5}$ c) $10^{\log 7t}$

Solution

a) $4^{\log_4 k} = k$

b) $e^{\ln 5} = e^{\log_e 5} = 5$

c) $10^{\log 7t} = 10^{\log_{10} 7t} = 7t$

A PROOF OF THE CHANGE-OF-BASE FORMULA: We close this section by proving the change-of-base formula and summarizing the properties of logarithms considered thus far in this chapter. In Section 6.2, we used the change-of-base formula,

$$\log_b M = \frac{\log_a M}{\log_a b},$$

to make base conversions in order to graph logarithmic functions on a grapher. Let $x = \log_b M$. Then

$$b^x = M \qquad \text{Definition of logarithm}$$
$$\log_a b^x = \log_a M \qquad \text{Taking the log on both sides}$$
$$x \log_a b = \log_a M \qquad \text{Using the power rule}$$
$$x = \frac{\log_a M}{\log_a b}, \qquad \text{Dividing by } \log_a b$$

so

$$x = \log_b M = \frac{\log_a M}{\log_a b}.$$

Following is a summary of the properties of logarithms.

Summary of the Properties of Logarithms

The Product Rule:	$\log_a MN = \log_a M + \log_a N$
The Power Rule:	$\log_a M^p = p \log_a M$
The Quotient Rule:	$\log_a \dfrac{M}{N} = \log_a M - \log_a N$
The Change-of-Base Formula:	$\log_b M = \dfrac{\log_a M}{\log_a b}$
Other Properties:	$\log_a a = 1, \qquad \log_a 1 = 0,$
	$\log_a a^x = x, \qquad a^{\log_a x} = x$

Exercise Set 6.3

Express as a sum of logarithms.

1. $\log_3 (81 \cdot 27)$

2. $\log_2 (8 \cdot 64)$

3. $\log_5 (5 \cdot 125)$

4. $\log_4 (64 \cdot 32)$

5. $\log_t 8Y$

6. $\log_e Qx$

Express as a product.

7. $\log_b t^3$

8. $\log_a x^4$

9. $\log y^8$

10. $\ln y^5$

11. $\log_c K^{-6}$

12. $\log_b Q^{-8}$

Express as a difference of logarithms.

13. $\log_t \dfrac{M}{8}$

14. $\log_a \dfrac{76}{13}$

15. $\log_a \dfrac{x}{y}$

16. $\log_b \dfrac{3}{w}$

Express in terms of sums and differences of logarithms.

17. $\log_a 6xy^5z^4$

18. $\log_a x^3y^2z$

19. $\log_b \dfrac{p^2q^5}{m^4b^9}$

20. $\log_b \dfrac{x^2y}{b^3}$

21. $\log_a \sqrt{\dfrac{x^6}{p^5q^8}}$

22. $\log_c \sqrt[3]{\dfrac{y^3z^2}{x^4}}$

23. $\log_a \sqrt[4]{\dfrac{m^8n^{12}}{a^3b^5}}$

24. $\log_a \sqrt{\dfrac{a^6b^8}{a^2b^5}}$

Express as a single logarithm and, if possible, simplify.

25. $\log_a 75 + \log_a 2$

26. $\log 0.01 + \log 1000$

27. $\log 10,000 - \log 100$

28. $\ln 54 - \ln 6$

29. $\frac{1}{2} \log_a x + 4 \log_a y - 3 \log_a x$

30. $\frac{2}{5} \log_a x - \frac{1}{3} \log_a y$

31. $\ln x^2 - 2 \ln \sqrt{x}$

32. $\ln 2x + 3(\ln x - \ln y)$

33. $\ln (x^2 - 4) - \ln (x + 2)$

34. $\log_a \dfrac{a}{\sqrt{x}} - \log_a \sqrt{ax}$

35. $\ln x - 3[\ln (x - 5) + \ln (x + 5)]$

36. $\frac{2}{3}[\ln (x^2 - 9) - \ln (x + 3)] + \ln (x + y)$

37. $\frac{3}{2} \ln 4x^6 - \frac{4}{5} \ln 2y^{10}$

38. $120(\ln \sqrt[5]{x^3} + \ln \sqrt[3]{y^2} - \ln \sqrt[4]{16z^5})$

Given that $\log_b 3 = 1.0986$ *and* $\log_b 5 = 1.6094$, *find each of the following.*

39. $\log_b \frac{3}{5}$

40. $\log_b 15$

41. $\log_b \frac{1}{5}$

42. $\log_b \frac{5}{3}$

43. $\log_b \sqrt{b}$

44. $\log_b \sqrt{b^3}$

45. $\log_b 5b$

46. $\log_b 9$

47. $\log_b 75$

48. $\log_b \dfrac{1}{b}$

Simplify.

49. $\log_p p^3$

50. $\log_t t^{2713}$

51. $\log_e e^{|x-4|}$

52. $\log_q q^{\sqrt{3}}$

53. $3^{\log_3 4x}$

54. $5^{\log_5 (4x-3)}$

55. $10^{\log w}$

56. $e^{\ln x^3}$

57. $\ln e^{8t}$

58. $\log 10^{-k}$

Discussion and Writing

59. Given that $f(x) = a^x$ and $g(x) = \log_a x$, find $(f \circ g)(x)$ and $(g \circ f)(x)$. These results are alternative proofs of what properties of logarithms already proven in this section? Explain.

60. Explain the errors, if any, in the following:
$$\log_a ab^3 = (\log_a a)(\log_a b^3) = 3 \log_a b.$$

Skill Maintenance

Use a grapher to find each of the following in radians, rounded to three decimal places, and in degrees, rounded to the nearest tenth of a degree.

61. $\sin^{-1} 0.8337$

62. $\cot^{-1} 2.344$

63. $\arctan 0.6527$

64. $\arccos (-0.5003)$

Synthesis

Solve for x. Do an algebraic solution.

65. $5^{\log_5 8} = 2x$

66. $\ln e^{3x-5} = -8$

Express as a single logarithm and, if possible, simplify.

67. $\log_a (x^2 + xy + y^2) + \log_a (x - y)$

68. $\log_a (a^{10} - b^{10}) - \log_a (a + b)$

Express as a sum or a difference of logarithms.

69. $\log_a \dfrac{x - y}{\sqrt{x^2 - y^2}}$

70. $\log_a \sqrt{9 - x^2}$

71. Given that $\log_a x = 2$, $\log_a y = 3$, and $\log_a z = 4$, find
$$\log_a \dfrac{\sqrt[4]{y^2 z^5}}{\sqrt[4]{x^3 z^{-2}}}.$$

Determine whether each of the following is true. Assume that a, x, M, and N are positive.

72. $\log_a M + \log_a N = \log_a (M + N)$

73. $\log_a M - \log_a N = \log_a \dfrac{M}{N}$

74. $\dfrac{\log_a M}{\log_a N} = \log_a M - \log_a N$

75. $\dfrac{\log_a M}{x} = \log_a M^{1/x}$

76. $\log_a x^3 = 3 \log_a x$

77. $\log_a 8x = \log_a x + \log_a 8$

78. $\log_N (MN)^x = x \log_N M + x$

Suppose that $\log_a x = 2$. *Find each of the following.*

79. $\log_a \left(\dfrac{1}{x}\right)$

80. $\log_{1/a} x$

81. Simplify:
$$\log_{10} 11 \cdot \log_{11} 12 \cdot \log_{12} 13 \cdots \log_{998} 999 \cdot \log_{999} 1000.$$

Prove each of the following for any base a and any positive number x.

82. $\log_a \left(\dfrac{1}{x}\right) = -\log_a x = \log_{1/a} x$

83. $\log_a \left(\dfrac{x + \sqrt{x^2 - 5}}{5}\right) = -\log_a (x - \sqrt{x^2 - 5})$

Solving Exponential and Logarithmic Equations

6.4

• *Solve exponential and logarithmic equations.*

Solving Exponential Equations

Equations with variables in the exponents, such as

$$3^x = 20 \quad \text{and} \quad 2^{5x} = 64,$$

are called **exponential equations.** We now consider solving exponential equations.

Sometimes, as is the case with the equation $2^{5x} = 64$, we can write each side as a power of the same number:

$$2^{5x} = 2^6.$$

We can then set the exponents equal and solve:

$$5x = 6$$
$$x = \frac{6}{5}$$
$$x = 1.2.$$

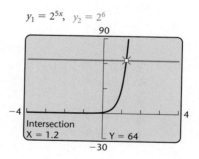

$$y_1 = 2^{5x}, \quad y_2 = 2^6$$

Intersection
X = 1.2   Y = 64

We use the following property.

> **Base–Exponent Property**
> For any $a > 0$, $a \neq 1$,
> $$a^x = a^y \longleftrightarrow x = y.$$

ONE-TO-ONE FUNCTIONS

REVIEW SECTION 1.6.

This property follows from the fact that for any $a > 0$, $a \neq 1$, $f(x) = a^x$ is a one-to-one function. If $a^x = a^y$, then $f(x) = f(y)$. Then since f is one-to-one, it follows that $x = y$. Conversely, if $x = y$, it follows that $a^x = a^y$, since we are raising a to the same power.

EXAMPLE 1 Solve: $2^{3x-7} = 32$.

Algebraic Solution

Note that $32 = 2^5$. Thus we can write each side as a power of the same number:

$$2^{3x-7} = 2^5.$$

Since the bases are the same number, 2, we can use the base–exponent property and set the exponents equal:

$$3x - 7 = 5$$
$$3x = 12$$
$$x = 4.$$

CHECK: $\dfrac{2^{3x-7} = 32}{}$

$2^{3(4)-7}$? 32
2^{12-7}
2^5
32 | 32 TRUE

The solution is 4.

Graphical Solution

We will use the Intersect method. We graph

$$y_1 = 2^{3x-7} \quad \text{and} \quad y_2 = 32$$

to find the coordinates of the point of intersection.

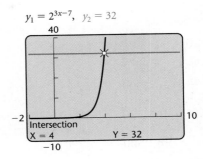

$y_1 = 2^{3x-7}, \; y_2 = 32$

The first coordinate of this point is the solution of the equation $y_1 = y_2$, or $2^{3x-7} = 32$. The solution is 4.

We could also write the equation in the form $2^{3x-7} - 32 = 0$ and use the Zero method.   ▬

When it does not seem possible to write each side as a power of the same base, we can take the common or natural logarithm on each side and use the power rule for logarithms.

EXAMPLE 2   Solve: $3^x = 20$.

Algebraic Solution

We have

$$3^x = 20$$
$$\log 3^x = \log 20 \qquad \text{Taking the common logarithm on both sides}$$
$$x \log 3 = \log 20 \qquad \text{Using the power rule}$$
$$x = \frac{\log 20}{\log 3}. \qquad \text{Dividing by } \log 3$$

This is an exact answer. We cannot simplify further, but we can approximate using a grapher:

$$x = \frac{\log 20}{\log 3} \approx 2.7268.$$

We can check this by finding $3^{2.7268}$. The solution is about 2.7268.

Graphical Solution

We will use the Intersect method. We graph

$$y_1 = 3^x \quad \text{and} \quad y_2 = 20$$

to find the x-coordinate of the point of intersection.

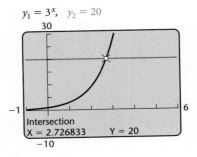

$y_1 = 3^x, \; y_2 = 20$

The solution is about 2.7268.   ▬

It will make our work easier if we take the natural logarithm when working with equations that have e as a base.

EXAMPLE 3   Solve: $e^{0.08t} = 2500$.

Algebraic Solution

We have

$e^{0.08t} = 2500$

$\ln e^{0.08t} = \ln 2500$ — Taking the natural logarithm on both sides

$0.08t = \ln 2500$ — Finding the logarithm of a base to a power: $\log_a a^x = x$

$t = \dfrac{\ln 2500}{0.08}$ — Dividing by 0.08

≈ 97.8.

The solution is about 97.8.

Graphical Solution

Using the Intersect method, we graph the equations

$$y_1 = e^{0.08x} \quad \text{and} \quad y_2 = 2500$$

and determine the point of intersection.

$y_1 = e^{0.08x}, \quad y_2 = 2500$

Intersection
X = 97.800575   Y = 2500

The solution is about 97.8.

EXAMPLE 4   Solve: $e^x + e^{-x} - 6 = 0$.

Algebraic Solution

In this case, we have more than one term with x in the exponent. To get a single expression with x in the exponent, we do the following:

$e^x + e^{-x} - 6 = 0$

$e^x + \dfrac{1}{e^x} - 6 = 0$ — Rewriting e^{-x} with a positive exponent

$e^{2x} + 1 - 6e^x = 0$. — Multiplying by e^x on both sides

This equation is reducible to quadratic with $u = e^x$:

$u^2 - 6u + 1 = 0$.

The coefficients of the reduced quadratic equation are $a = 1$, $b = -6$, and $c = 1$. Using the quadratic formula, we obtain

$u = e^x = 3 \pm \sqrt{8}$.

We now take the natural logarithm on both sides:

$$\ln e^x = \ln (3 \pm \sqrt{8})$$
$$x = \ln (3 \pm \sqrt{8}). \qquad \text{Using } \ln e^x = x$$

Approximating each of the solutions, we obtain 1.76 and -1.76.

Graphical Solution

Using the Zero method, we begin by graphing the function

$$y = e^x + e^{-x} - 6.$$

Then we find the zeros of the function.

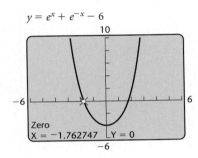

$y = e^x + e^{-x} - 6$

Zero
X = -1.762747 Y = 0

The leftmost zero is about -1.76. Using the ZERO feature one more time, we find that the other zero is about 1.76.
The solutions are about -1.76 and 1.76.   ▬

It is possible that when encountering an equation like the one in Example 4, you might not recognize that it could be solved in the algebraic manner shown. This points out the value of the graphical solution.

Solving Logarithmic Equations

Equations containing variables in logarithmic expressions, such as $\log_2 x = 4$ and $\log x + \log (x + 3) = 1$, are called **logarithmic equations.**

> To solve logarithmic equations algebraically, first try to obtain a single logarithmic expression on one side and then write an equivalent exponential equation.

EXAMPLE 5   Solve: $\log_3 x = -2$.

Algebraic Solution

We have

$\log_3 x = -2$

$3^{-2} = x$ Converting to an exponential equation

$\dfrac{1}{3^2} = x$

$\dfrac{1}{9} = x.$

CHECK: $\dfrac{\log_3 x = -2}{}$

$\log_3 \dfrac{1}{9}$? -2

$\log_3 3^{-2}$

-2 | -2 TRUE

The solution is $\frac{1}{9}$.

Graphical Solution

We use the change-of-base formula and graph the equations

$$y_1 = \log_3 x = \frac{\ln x}{\ln 3} \quad \text{and} \quad y_2 = -2.$$

Then we use the Intersect method.

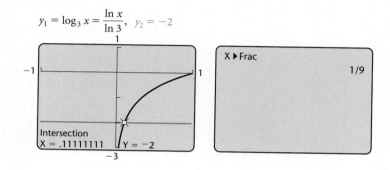

$y_1 = \log_3 x = \dfrac{\ln x}{\ln 3}, \quad y_2 = -2$

If the solution is a rational number, we usually can find fractional notation for the exact solution by using the FRAC feature from the MATH submenu of the MATH menu.

The solution is $\frac{1}{9}$.

EXAMPLE 6   Solve: $\log x + \log (x + 3) = 1$.

Algebraic Solution

In this case, we have common logarithms. Including the base 10's will help us understand the problem:

$$\log_{10} x + \log_{10} (x + 3) = 1$$

$$\log_{10} [x(x + 3)] = 1 \qquad \text{Using the product rule to obtain a single logarithm}$$

$$x(x + 3) = 10^1 \qquad \text{Writing an equivalent exponential equation}$$

$$x^2 + 3x = 10$$

$$x^2 + 3x - 10 = 0$$

$$(x - 2)(x + 5) = 0 \qquad \text{Factoring}$$

$$x - 2 = 0 \quad or \quad x + 5 = 0$$

$$x = 2 \quad or \qquad x = -5.$$

CHECK: For 2:

$$\log x + \log (x + 3) = 1$$

$$\log 2 + \log (2 + 3) \ ? \ 1$$
$$\log 2 + \log 5$$
$$\log 10$$
$$1 \ \big| \ 1 \quad \text{TRUE}$$

For -5:

$$\log x + \log (x + 3) = 1$$

$$\log (-5) + \log (-5 + 3) \ ? \ 1 \quad \text{FALSE}$$

The number -5 is not a solution because negative numbers do not have real-number logarithms. The solution is 2.

Graphical Solution

We can graph the equations

$$y_1 = \log x + \log (x + 3)$$

and

$$y_2 = 1$$

and use the Intersect method. The first coordinate of the point of intersection is the solution of the equation.

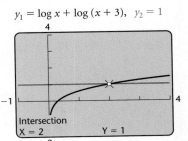

$y_1 = \log x + \log (x + 3), \ y_2 = 1$

We could also graph the function

$$y = \log x + \log (x + 3) - 1$$

and use the Zero method. The zero of the function is the solution of the equation.

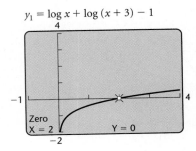

$y_1 = \log x + \log (x + 3) - 1$

With either method, we see that the solution is 2. Note that the graphical solution gives only the one *true* solution.

EXAMPLE 7 Solve: $\log_3 (2x - 1) - \log_3 (x - 4) = 2$.

Algebraic Solution

We have

$$\log_3 (2x - 1) - \log_3 (x - 4) = 2$$

$$\log_3 \frac{2x - 1}{x - 4} = 2 \qquad \text{Using the quotient rule}$$

$$\frac{2x - 1}{x - 4} = 3^2 \qquad \begin{array}{l}\text{Writing an equivalent}\\\text{exponential equation}\end{array}$$

$$\frac{2x - 1}{x - 4} = 9$$

$$2x - 1 = 9(x - 4) \qquad \begin{array}{l}\text{Multiplying by}\\\text{the LCD, } x - 4\end{array}$$

$$2x - 1 = 9x - 36$$

$$35 = 7x$$

$$5 = x.$$

CHECK: $\log_3 (2x - 1) - \log_3 (x - 4) = 2$

$\overline{\log_3 (2 \cdot 5 - 1) - \log_3 (5 - 4) \ ? \ 2}$

$\log_3 9 - \log_3 1 \ \bigg|$

$2 - 0 \ \bigg|$

$2 \ \bigg| \ 2 \quad$ TRUE

The solution is 5.

Graphical Solution

Here we use the Intersect method to find the solutions of the equation. We use the change-of-base formula and graph the equations

$$y_1 = \frac{\ln (2x - 1)}{\ln 3} - \frac{\ln (x - 4)}{\ln 3}$$

and

$$y_2 = 2.$$

$$y_1 = \frac{\ln (2x - 1)}{\ln 3} - \frac{\ln (x - 4)}{\ln 3}, \quad y_2 = 2$$

Intersection
X = 5   Y = 2

The solution is 5.

Sometimes we encounter equations for which an algebraic solution seems difficult or impossible.

EXAMPLE 8 Solve: $e^{0.5x} - 7.3 = 2.08x + 6.2$.

$y_1 = e^{0.5x} - 7.3, \quad y_2 = 2.08x + 6.2$

$x = -6.471, y = -7.261$
$x = 6.610, y = 19.949$

Graphical Solution

We graph the equations

$$y_1 = e^{0.5x} - 7.3 \quad \text{and} \quad y_2 = 2.08x + 6.2$$

and use the Intersect method. (See the window at left.)
We can also consider the equation

$$y = e^{0.5x} - 7.3 - 2.08x - 6.2, \quad \text{or} \quad y = e^{0.5x} - 2.08x - 13.5,$$

and use the Zero method. The approximate solutions are -6.471 and 6.610.

Exercise Set 6.4

Solve the exponential equation algebraically. Then check using a grapher.

1. $3^x = 81$

2. $2^x = 32$

3. $2^{2x} = 8$

4. $3^{7x} = 27$

5. $2^x = 33$

6. $2^x = 40$

7. $5^{4x-7} = 125$

8. $4^{3x-5} = 16$

9. $27 = 3^{5x} \cdot 9^{x^2}$

10. $3^{x^2+4x} = \frac{1}{27}$

11. $84^x = 70$

12. $28^x = 10^{-3x}$

13. $e^t = 1000$

14. $e^{-t} = 0.04$

15. $e^{-0.03t} = 0.08$

16. $1000e^{0.09t} = 5000$

17. $3^x = 2^{x-1}$

18. $5^{x+2} = 4^{1-x}$

19. $(3.9)^x = 48$

20. $250 - (1.87)^x = 0$

21. $e^x + e^{-x} = 5$

22. $e^x - 6e^{-x} = 1$

23. $\dfrac{e^x + e^{-x}}{e^x - e^{-x}} = 3$

24. $\dfrac{5^x - 5^{-x}}{5^x + 5^{-x}} = 8$

Solve the logarithmic equation algebraically. Then check using a grapher.

25. $\log_5 x = 4$

26. $\log_2 x = -3$

27. $\log x = -4$

28. $\log x = 1$

29. $\ln x = 1$

30. $\ln x = -2$

31. $\log_2 (10 + 3x) = 5$

32. $\log_5 (8 - 7x) = 3$

33. $\log x + \log (x - 9) = 1$

34. $\log_2 (x + 1) + \log_2 (x - 1) = 3$

35. $\log_8 (x + 1) - \log_8 x = 2$

36. $\log x - \log (x + 3) = -1$

37. $\log_4 (x + 3) + \log_4 (x - 3) = 2$

38. $\ln (x + 1) - \ln x = \ln 4$

39. $\log (2x + 1) - \log (x - 2) = 1$

40. $\log_5 (x + 4) + \log_5 (x - 4) = 2$

Use only a grapher. Find approximate solutions of the equation or approximate the point(s) of intersection of the pair of equations.

41. $e^{7.2x} = 14.009$

42. $0.082e^{0.05x} = 0.034$

43. $xe^{3x} - 1 = 3$

44. $5e^{5x} + 10 = 3x + 40$

45. $4 \ln (x + 3.4) = 2.5$

46. $\ln x^2 = -x^2$

47. $\log_8 x + \log_8 (x + 2) = 2$

48. $\log_3 x + 7 = 4 - \log_5 x$

49. $\log_5 (x + 7) - \log_5 (2x - 3) = 1$

50. $y = \ln 3x, \quad y = 3x - 8$

51. $2.3x + 3.8y = 12.4, \quad y = 1.1 \ln (x - 2.05)$

52. $y = 2.3 \ln (x + 10.7), \quad y = 10e^{-0.07x^2}$

53. $y = 2.3 \ln (x + 10.7), \quad y = 10e^{-0.007x^2}$

Discussion and Writing

54. In Example 3, we took the natural logarithm on both sides of the equation. What would have happened had we taken the common logarithm? Explain which approach seems better to you and why.

55. Explain how Exercises 29 and 30 could be solved using the graph of $f(x) = \ln x$.

Skill Maintenance

In Exercises 56–59:

a) Find the vertex.

b) Find the line of symmetry.

c) Determine whether there is a maximum or minimum value and find that value.

56. $f(x) = -x^2 + 6x - 8$

57. $g(x) = x^2 - 6$

58. $H(x) = 3x^2 - 12x + 16$

59. $G(x) = -2x^2 - 4x - 7$

Synthesis

Solve using any method.

60. $\ln (\ln x) = 2$

61. $\ln (\log x) = 0$

62. $\ln \sqrt[4]{x} = \sqrt{\ln x}$

63. $\sqrt{\ln x} = \ln \sqrt{x}$

64. $\log_3 (\log_4 x) = 0$

65. $(\log_3 x)^2 - \log_3 x^2 = 3$

66. $(\log x)^2 - \log x^2 = 3$

67. $\ln x^2 = (\ln x)^2$

68. $e^{2x} - 9 \cdot e^x + 14 = 0$

69. $5^{2x} - 3 \cdot 5^x + 2 = 0$

70. $x \left(\ln \frac{1}{6} \right) = \ln 6$

71. $\log_3 |x| = 2$

72. $x^{\log x} = \dfrac{x^3}{100}$

73. $\ln x^{\ln x} = 4$

74. $\dfrac{(e^{3x+1})^2}{e^4} = e^{10x}$

75. $\dfrac{\sqrt{(e^{2x} \cdot e^{-5x})^{-4}}}{e^x \div e^{-x}} = e^7$

76. $|\log_a x| = \log_a |x|$, $a > 1$

77. $\ln (x - 2) > 4$

78. $e^x < \dfrac{4}{5}$

79. $|\log_5 x| + 3 \log_5 |x| = 4$

80. $|2^{x^2} - 8| = 3$

81. Given that $a = \log_8 225$ and $b = \log_2 15$, express a as a function of b.

82. Given that $a = (\log_{125} 5)^{\log_5 125}$, find the value of $\log_3 a$.

83. Given that
$$\log_2 [\log_3 (\log_4 x)] = \log_3 [\log_2 (\log_4 y)]$$
$$= \log_4 [\log_3 (\log_2 z)]$$
$$= 0,$$
find $x + y + z$.

84. Given that $f(x) = e^x - e^{-x}$, find $f^{-1}(x)$ if it exists.

Applications and Models: Growth and Decay

6.5

• Solve applied problems involving exponential growth and decay.
• Find models involving exponential and logarithmic functions.

Exponential and logarithmic functions with base e are rich in applications to many fields such as business, science, psychology, and sociology. In this section, we consider some basic applications and then use curve fitting to do others.

Population Growth

The function

$$P(t) = P_0 e^{kt}, \quad k > 0$$

is a model of many kinds of population growth, whether it be a population of people, bacteria, cellular phones, or money. In this function, P_0 is the population at time 0, P is the population after time t, and k is called the **exponential growth rate.** The graph of such an equation is shown at left.

EXAMPLE 1 *Population Growth of India.* In 1998, the population of India was about 984 million and the exponential growth rate was 1.8% per year (*Source: Statistical Abstract of the United States*).

a) Find the exponential growth function.

b) Graph the exponential growth function.

c) What will the population be in 2005?

d) After how long will the population be double what it was in 1998?

Solution

a) At $t = 0$ (1998), the population was 984 million. We substitute 984 for P_0 and 1.8%, or 0.018, for k to obtain the exponential growth function

$$P(t) = 984e^{0.018t}.$$

b) Using a grapher, we obtain the graph of the exponential growth function, shown at left.

$y = 984e^{0.018x}$

c) In 2005, $t = 7$; that is, 7 yr have passed since 1998. To find the population in 2005, we substitute 7 for t:

$$P(7) = 984e^{0.018(7)} = 984e^{0.126} \approx 1116.$$

We can also use the VALUE feature on a grapher. (See the window at left.) In 2005, the population of India will be about 1116 million, or 1,116,000,000.

d) We are looking for the time T for which $P(T) = 2 \cdot 984$, or 1968. The number T is called the **doubling time.** To find T, we solve the equation

$$1968 = 984e^{0.018T}.$$

Algebraic Solution

We have

$1968 = 984e^{0.018T}$	Substituting 1968 for $P(T)$
$2 = e^{0.018T}$	Dividing by 984
$\ln 2 = \ln e^{0.018T}$	
$\ln 2 = 0.018T$	$\ln e^x = x$
$\dfrac{\ln 2}{0.018} = T$	Dividing by 0.018
$39 \approx T.$	

The population of India will be double what it was in 1998 about 39 yr after 1998.

Graphical Solution

Using the Intersect method, we graph the equations

$$y_1 = 984e^{0.018x} \quad \text{and} \quad y_2 = 1968$$

and find the first coordinate of their point of intersection.

$y_1 = 984e^{0.018x}, \quad y_2 = 1968$

The solution is about 39 yr, so the population of India will be double that of 1998 about 39 yr after 1998.

Interest Compounded Continuously

Here we explore the mathematics behind the concept of **interest compounded continuously.** Suppose that an amount P_0 is invested in a savings account at interest rate k *compounded continuously.* The amount $P(t)$ in the account after t years is given by the exponential function

$$P(t) = P_0 e^{kt}.$$

EXAMPLE 2 *Interest Compounded Continuously.* Suppose that $2000 is invested at interest rate k, compounded continuously, and grows to $2983.65 in 5 yr.

a) What is the interest rate?

b) Find the exponential growth function.

c) What will the balance be after 10 yr?

d) After how long will the $2000 have doubled?

Solution

a) At $t = 0$, $P(0) = P_0 = \$2000$. Thus the exponential growth function is

$$P(t) = 2000 e^{kt}.$$

We know that $P(5) = \$2983.65$. We substitute and solve for k, as shown below.

Algebraic Solution

We have

$$2983.65 = 2000 e^{k(5)}$$
     Substituting 2983.65 for $P(t)$ and 5 for t

$$2983.65 = 2000 e^{5k}$$

$$\frac{2983.65}{2000} = e^{5k}$$
     Dividing by 2000

$$\ln \frac{2983.65}{2000} = \ln e^{5k}$$
     Taking the natural logarithm

$$\ln \frac{2983.65}{2000} = 5k$$
     Using $\ln e^x = x$

$$\frac{\ln \dfrac{2983.65}{2000}}{5} = k$$
     Dividing by 5

$$0.08 \approx k.$$

The interest rate is about 0.08, or 8%.

Graphical Solution

We use the Intersect method. We graph the equations

$$y_1 = 2000 e^{5x} \quad \text{and} \quad y_2 = 2983.65$$

and use the INTERSECT feature to approximate the first coordinate of the point of intersection.

$y_1 = 2000 e^{5x}, \quad y_2 = 2983.65$

The solution is about 0.08, or 8%.

b) The exponential growth function is
$$P(t) = 2000e^{0.08t}.$$

c) The balance after 10 yr is
$$P(10) = 2000e^{0.08(10)}$$
$$= 2000e^{0.8}$$
$$\approx \$4451.08.$$

d) We solve using both an algebraic method and a graphical method.

Algebraic Solution

To find the doubling time T, we set $P(T) = 2 \cdot P_0 = \$4000$ and solve for T:

$$4000 = 2000e^{0.08T}$$

$2 = e^{0.08T}$	Dividing by 2000
$\ln 2 = \ln e^{0.08T}$	Taking the natural logarithm
$\ln 2 = 0.08T$	$\ln e^x = x$
$\dfrac{\ln 2}{0.08} = T$	Dividing by 0.08
$8.7 \approx T.$	

Thus the original investment of $2000 will double in about 8.7 yr.

Graphical Solution

The money will have doubled when $P(t) = 2 \cdot P_0 = 4000$, or when $2000e^{0.08t} = 4000$. We use the Zero method. We graph the equation

$$y = 2000e^{0.08x} - 4000$$

and find the zero of the function. The zero of the function is the solution of the equation.

$y = 2000e^{0.08x} - 4000$

The solution is about 8.7, so the original investment of $2000 will double in about 8.7 yr.

We can find a general expression relating the growth rate k and the doubling time T by solving the following equation:

$2P_0 = P_0e^{kT}$	Substituting $2P_0$ for P and T for t
$2 = e^{kT}$	Dividing by P_0
$\ln 2 = \ln e^{kt}$	Taking the natural logarithm
$\ln 2 = kT$	Using $\ln e^x = x$
$\dfrac{\ln 2}{k} = T.$	

Growth Rate and Doubling Time
The **growth rate** k and the **doubling time** T are related by

$$kT = \ln 2, \quad \text{or} \quad k = \frac{\ln 2}{T}, \quad \text{or} \quad T = \frac{\ln 2}{k}.$$

Note that the relationship between k and T does not depend on P_0.

EXAMPLE 3 *World Population Growth.* The population of the world is now doubling every 54.6 yr. What is the exponential growth rate?

Solution We have

$$k = \frac{\ln 2}{T} = \frac{\ln 2}{54.6} \approx 1.3\%.$$

The growth rate of the world population is about 1.3% per year. ▬

Models of Limited Growth

The model $P(t) = P_0 e^{kt}$ has many applications involving unlimited population growth. However, in some populations, there can be factors that prevent a population from exceeding some limiting value—perhaps a limitation on food, living space, or other natural resources. One model of such growth is

$$P(t) = \frac{a}{1 + be^{-kt}},$$

which is called a **logistic function.** This function increases toward a *limiting value a* as $t \to \infty$.

EXAMPLE 4 *Limited Population Growth.* A ship carrying 1000 passengers has the misfortune to be shipwrecked on a small island from which the passengers are never rescued. The natural resources of the island limit the population to 5780. The population gets closer and closer to this limiting value, but never reaches it. The population of the island after time t, in years, is given by the logistic equation

$$P(t) = \frac{5780}{1 + 4.78e^{-0.4t}}.$$

a) Graph the function.

b) Find the population after 0, 1, 2, 5, 10, and 20 yr.

Solution

a) We use a grapher to graph the function. The graph is the S-shaped curve shown at left. Note that this function increases toward a limiting value of 5780.

$$y_1 = \frac{5780}{1 + 4.78e^{-0.4x}}$$

b) We can use the TABLE feature on a grapher set in ASK mode to find the function values. (See the window on the left below.) The VALUE feature can also be used. (See the window on the right below.)

X	Y₁	
0	1000	
1	1374.8	
2	1836.2	
5	3509.6	
10	5314.7	
20	5770.7	

X =

Thus the population will be about 1000 after 0 yr, 1375 after 1 yr, 1836 after 2 yr, 3510 after 5 yr, 5315 after 10 yr, and 5771 after 20 yr.

Another model of limited growth is provided by the function

$$P(t) = L(1 - e^{-kt}), \quad k > 0,$$

which is shown graphed at right. This function also increases toward a limiting value L, as $x \to \infty$.

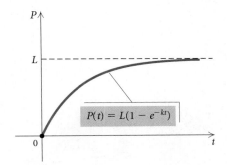

How can scientists determine that an animal bone has lost 30% of its carbon-14? The assumption is that the percentage of carbon-14 in the atmosphere and in living plants and animals is the same. When a plant or an animal dies, the amount of carbon-14 decays exponentially. The scientist burns the animal bone and uses a Geiger counter to determine the percentage of the smoke that is carbon-14. It is the amount that this varies from the percentage in the atmosphere that tells how much carbon-14 has been lost.

The process of carbon-14 dating was developed by the American chemist Willard E. Libby in 1952. It is known that the radioactivity in a living plant is 16 disintegrations per gram per minute. Since the half-life of carbon-14 is 5750 years, an object with an activity of 8 disintegrations per gram per minute is 5750 years old, one with an activity of 4 disintegrations per gram per minute is 11,500 years old, and so on. Carbon-14 dating can be used to measure the age of objects up to 40,000 years old. Beyond such an age, it is too difficult to measure the radioactivity and some other method would have to be used.

Carbon-14 was indeed used to find the age of the Dead Sea Scrolls. It was used recently to refute the authenticity of the Shroud of Turin, presumed to have covered the body of Christ.

Exponential Decay

The function

$$P(t) = P_0 e^{-kt}, \quad k > 0$$

is an effective model of the decline, or decay, of a population. An example is the decay of a radioactive substance. In this case, P_0 is the amount of the substance at time $t = 0$, and $P(t)$ is the amount of the substance left after time t, where k is a positive constant that depends on the situation. The constant k is called the **decay rate.**

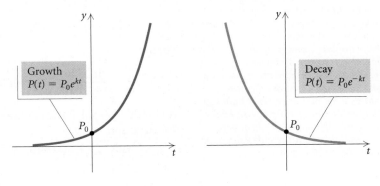

The **half-life** of bismuth is 5 days. This means that half of an amount of bismuth will cease to be radioactive in 5 days. The effect of half-life T is shown in the graph below for nonnegative inputs. The exponential function gets close to 0, but never reaches 0, as t gets very large. Thus, according to an exponential decay model, a radioactive substance never completely decays.

Radioactive decay curve

In 1947, a Bedouin youth looking for a stray goat climbed into a cave at Kirbet Qumran on the shores of the Dead Sea near Jericho and came upon earthenware jars containing an incalculable treasure of ancient manuscripts. Shown here are fragments of those so-called Dead Sea Scrolls, a portion of some 600 or so texts found so far and which concern the Jewish books of the Bible. Officials date them before 70 A.D., making them the oldest Biblical manuscripts by 1000 years.

EXAMPLE 5 *Carbon Dating.* The radioactive element carbon-14 has a half-life of 5750 yr. The percentage of carbon-14 present in the remains of organic matter can be used to determine the age of that organic matter. Archaeologists discovered that the linen wrapping from one of the Dead Sea Scrolls had lost 22.3% of its carbon-14 at the time it was found. How old was the linen wrapping?

Solution We first find k. When $t = 5750$ (the half-life), $P(t)$ will be half of P_0. We substitute $\frac{1}{2}P_0$ for $P(t)$ and 5750 for t and solve for k. Then

$$\tfrac{1}{2}P_0 = P_0 e^{-k(5750)}$$

or

$$\tfrac{1}{2} = e^{-5750k}.$$

We take the natural logarithm on both sides:

$$\ln \tfrac{1}{2} = \ln e^{-5750k}$$
$$\ln 0.5 = -5750k.$$

Then

$$k = \frac{\ln 0.5}{-5750} \approx 0.00012.$$

We could also solve the equation $\frac{1}{2} = e^{-5750k}$ using a grapher. Now we have the function

$$P(t) = P_0 e^{-0.00012t}.$$

(This equation can be used for any subsequent carbon-dating problem.) If the linen wrapping has lost 22.3% of its carbon-14 from an initial amount P_0, then 77.7%P_0 is the amount present. To find the age t of the wrapping, we solve the following equation for t:

$$77.7\%P_0 = P_0e^{-0.00012t} \qquad \text{Substituting } 77.7\%P_0 \text{ for } P$$
$$0.777 = e^{-0.00012t}$$
$$\ln 0.777 = \ln e^{-0.00012t}$$
$$\ln 0.777 = -0.00012t \qquad \ln e^x = x$$
$$\frac{\ln 0.777}{-0.00012} = t$$
$$2103 \approx t.$$

Thus the linen wrapping on the Dead Sea Scrolls was about 2103 yr old when it was found. ▬

Exponential and Logarithmic Curve Fitting

We have added several new functions that can be considered when we fit curves to data. Let's review some of them.

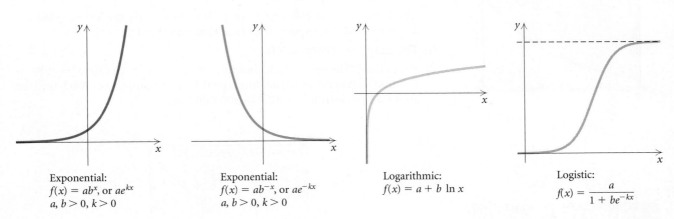

Exponential:
$f(x) = ab^x$, or ae^{kx}
$a, b > 0, k > 0$

Exponential:
$f(x) = ab^{-x}$, or ae^{-kx}
$a, b > 0, k > 0$

Logarithmic:
$f(x) = a + b \ln x$

Logistic:
$$f(x) = \frac{a}{1 + be^{-kx}}$$

Now, when we analyze a set of data for curve fitting, these models can be considered as well as polynomial functions (such as linear, quadratic, cubic, and quartic functions) and rational functions.

EXAMPLE 6 *Credit Card Volume.* The total credit card volume for Visa, MasterCard, American Express, and Discover has increased dramatically in recent years, as shown in the table on the next page.

a) Use a grapher to fit an exponential function to the data.

b) Graph the function with the scatterplot of the data.

c) Predict the total credit card volume for Visa, MasterCard, American Express, and Discover in 2003.

YEAR, x	CREDIT CARD VOLUME, y (IN BILLIONS)
1988, 0	$261.0
1989, 1	296.3
1990, 2	338.4
1991, 3	361.0
1992, 4	403.1
1993, 5	476.7
1994, 6	584.8
1995, 7	701.2
1996, 8	798.3
1997, 9	885.2

Source: CardWeb Inc.'s CardData

```
ExpReg
y=a*b^x
a=249.7267077
b=1.150881332
r²=.9877691696
r=.9938657704
```

$y = 249.7267077(1.150881332)^x$

Solution

a) We will fit an equation of the type $y = a \cdot b^x$ to the data, where x is the number of years since 1988. Entering the data into the grapher and carrying out the regression procedure, we find that the equation is

$$y = 249.7267077(1.150881332)^x.$$

The correlation coefficient is very close to 1. This gives us a good indication that the exponential function fits the data well.

b) The graph is shown at left.

c) We evaluate the function found in part (a) for $x = 15$ (2003 − 1988 = 15) and estimate that the total credit card volume in 2003 will be about $2056 billion, or $2,056,000,000,000.

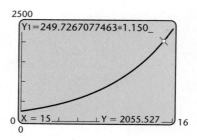

On some graphers, there may be a REGRESSION feature that yields an exponential function, base e. If not, and you wish to find such a function, a conversion can be done using the following.

Converting from Base b to Base e

$$b^x = e^{x(\ln b)}$$

Then, for the equation in Example 6, we have

$$y = 249.7267077(1.150881332)^x$$
$$= 249.7267077e^{x(\ln 1.150881332)}$$
$$= 249.7267077e^{0.1405280246x}.$$

We can prove this conversion formula using properties of logarithms, as follows:

$$e^{x(\ln b)} = e^{\ln b^x} = b^x.$$

Exercise Set 6.5

1. *World Population Growth.* In 1999, the world population was 6.0 billion. The exponential growth rate was 1.3% per year.

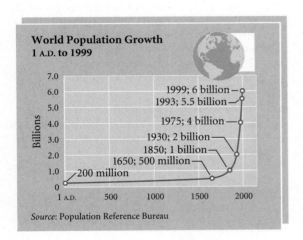

World Population Growth
1 A.D. to 1999

Source: Population Reference Bureau

a) Find the exponential growth function.
b) Predict the population of the world in 2005 and 2010.
c) When will the world population be 8 billion?
d) Find the doubling time.

2. *Population Growth of Rabbits.* Under ideal conditions, a population of rabbits has an exponential growth rate of 11.7% per day. Consider an initial population of 100 rabbits.

a) Find the exponential growth function.
b) Graph the function.
c) What will the population be after 7 days?
d) Find the doubling time.

3. *Population Growth.* Complete the following table.

POPULATION	GROWTH RATE, k	DOUBLING TIME, T
a) Mexico	1.9% per year	
b) Japan		346 yr
c) Mozambique	3.3% per year	
d) Norway	0.5% per year	
e) Syria		20.4 yr
f) Philippines		31.5 yr

4. *Female Olympic Athletes.* In 1985, the number of female athletes participating in Summer

Number of female athletes at Summer Olympic-type Games

Olympic-Type Games was 500. In 1996, about 3600 participated in the Summer Olympics in Atlanta. Assuming the exponential model applies:

a) Find the value of k ($P_0 = 500$), and write the function.

b) Estimate the number of female athletes in the Summer Olympics of 2000 and 2004.

5. *Population Growth of Israel.*   The population of Israel has a growth rate of 2.6% per year. In 1998, the population was 5,644,000. The land area of Israel is 24,313,062,400 square yards. (*Source: Statistical Abstract of the United States*) Assuming this growth rate continues and is exponential, after how long will there be one person for every square yard of land?

6. *Value of Manhattan Island.*   In 1626, Peter Minuit of the Dutch West India Company purchased Manhattan Island from the Indians for $24. Assuming an exponential rate of inflation of 8% per year, how much will Manhattan be worth in 2003?

7. *Interest Compounded Continuously.*   Suppose that $10,000 is invested at an interest rate of 5.4% per year, compounded continuously.

a) Find the exponential function that describes the amount in the account after time t, in years.

b) What is the balance after 1 yr? 2 yr? 5 yr? 10 yr?

c) What is the doubling time?

8. *Interest Compounded Continuously.*   Complete the following table.

INITIAL INVESTMENT AT $t = 0$, P_0	INTEREST RATE, k	DOUBLING TIME, T	AMOUNT AFTER 5 YR
a) $35,000	6.2%		
b) $5000			$ 7,130.90
c)	8.4%		$11,414.71
d)		11 yr	$17,539.32

9. *Carbon Dating.*   A mummy discovered in the pyramid Khufu in Egypt has lost 46% of its carbon-14. Determine its age.

10. *Carbon Dating.*   The statue of Zeus at Olympia in Greece is one of the Seven Wonders of the World. It is made of gold and ivory. The ivory was found to

have lost 35% of its carbon-14. Determine the age of the statue.

11. *Radioactive Decay.*   Complete the following table.

RADIOACTIVE SUBSTANCE	DECAY RATE, k	HALF-LIFE, T
a) Polonium		3 min
b) Lead		22 yr
c) Iodine-131	9.6% per day	
d) Krypton-85	6.3% per year	
e) Strontium-90		25 yr
f) Uranium-238		4560 yr
g) Plutonium		23,105 yr

12. *Number of Farms.*   The number N of farms in the United States has declined continually since 1950. In 1950, there were 5,647,800 farms, and in 1995 that number had decreased to 2,071,520 (*Source: U.S. Department of Agriculture*). Assuming the number of farms decreased according to the exponential model:

a) Find the value of k, and write an exponential function that describes the number of farms after time t, in years, where t is the number of years since 1950.

b) Estimate the number of farms in 2000, 2005, and 2010.

c) At this decay rate, in what year will only 100,000 farms remain?

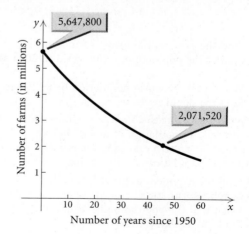

Number of years since 1950

13. *Decline in Beef Consumption.* In 1985, the average annual consumption of beef B was about 80 lb per person. In 1996, it was about 67 lb per person. Assuming consumption is decreasing according to the exponential-decay model:

a) Find the value k, and write an equation that describes beef consumption after time t, in years.

b) Estimate the consumption of beef in 2002.

c) After how many years (theoretically) will the average annual consumption of beef be 20 lb per person?

14. *The Value of Mark McGwire's Baseball Card.* The collecting of baseball cards has become a popular hobby. The card shown here is a photograph of Mark McGwire when he was a member of the Olympic USA Baseball Team in 1984.

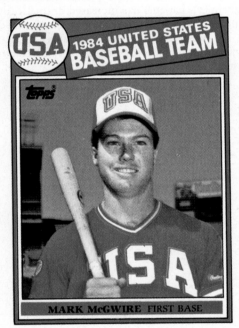

In 1987, the value of the card was $8 and in 1997, its value was $20 (*Source: SportsCards*, Joe Clemens, Price Guide Coordinator). Assuming the value of the card has grown exponentially:

a) Find the value k, and determine the exponential growth function, assuming $V_0 = 8$.

b) Using the function found in part (a), estimate the value of the card in 2000.

c) What is the doubling time for the value of the card?

d) After how long will the value of the card be $2000, assuming there is no change in the growth rate?

e) In the Baseball Price Guide in *SportsCards*, April 2000, this card was valued at $200. How does your estimate in part (b) compare with the $200 value? Explain why they differ.

15. *Spread of an Epidemic.* In a town whose population is 2000, a disease creates an epidemic. The number of people N infected t days after the disease has begun is given by the function

$$N(t) = \frac{2000}{1 + 19.9e^{-0.6t}}.$$

a) Graph the function.

b) How many are initially infected with the disease $(t = 0)$?

c) Find the number infected after 2 days, 5 days, 8 days, 12 days, and 16 days.

16. *Acceptance of Seat Belt Laws.* In recent years, many states have passed mandatory seat belt laws. The total number of states N that have passed a seat belt law t years after 1984 is given by the function

$$N(t) = \frac{50}{1 + 22e^{-0.6t}}$$

(*Source*: National Highway Traffic Safety Administration).

a) Graph the function.

b) How many states had passed the law in 1984? ($t = 0$ corresponds to 1984.)

c) Find the number of states that had passed the law by 1996 and 2002.

d) If the function were to continue to be appropriate, would all 50 states ever pass the law? Explain.

17. *Limited Population Growth in a Lake.* A lake is stocked with 400 fish of a new variety. The size of the lake, the availability of food, and the number of other fish restrict growth in the lake to a *limiting value* of 2500. The population of fish in the lake after time t, in months, is given by the function

$$P(t) = \frac{2500}{1 + 5.25e^{-0.32t}}.$$

a) Graph the function.

b) Find the population after 0, 1, 5, 10, 15, and 20 months.

In Exercises 18–23, determine which, if any, of these functions might be used as a model for the data in the scatterplot.

a) *Quadratic,* $f(x) = ax^2 + bx + c$

b) *Polynomial, not quadratic*

c) *Exponential,* $f(x) = ab^x$, or Be^{kx}, $k > 0$

d) *Exponential,* $f(x) = ab^x$, *or* Be^{-kx}, $k > 0$
e) *Logarithmic,* $f(x) = a + b \ln x$
f) *Logistic,* $f(x) = \dfrac{a}{1 + be^{-kx}}$

18.

19.

20.

21.

22.

23.

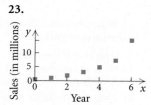

24. *Number of Physicians.* The following table contains data regarding the number of physicians in the United States in selected years.

YEAR	TOTAL NUMBER OF PHYSICIANS
1950	219,997
1955	241,711
1960	260,484
1965	292,088
1970	334,028
1975	393,742
1980	467,679
1985	552,716
1990	615,421
1994	684,414
1995	720,325
1996	737,764

Source: American Medical Association

a) Create a scatterplot of these data. Determine whether an exponential function appears to fit the data.
b) Use a grapher to fit the data with an exponential function and determine whether the function is a good fit.
c) Graph the function found in part (b) with a scatterplot of the data.
d) Estimate the number of physicians in 2000 and 2025.

25. *Projected Number of Alzheimer's Patients.* German psychiatrist Alois Alzheimer first described the disease, later called Alzheimer's disease, in 1906. Since life expectancy has significantly increased in the last century, the number of Alzheimer's patients has increased dramatically. The number of patients in the United States reached 4 million in 2000. The following table lists projected data regarding the number of Alzheimer's patients in years beyond 2000.

YEAR	PROJECTED NUMBER OF ALZHEIMER'S PATIENTS IN THE UNITED STATES (IN MILLIONS)
2000	4.0
2010	5.8
2020	6.8
2030	8.7
2040	11.8
2050	14.3

Source: "Alzheimer's, Unlocking the Mystery," by Geoffrey Cowley, *Newsweek*, January 31, 2000

a) Use a grapher to fit the data with an exponential function and determine whether the function is a good fit.
b) Graph the function found in part (a) with a scatterplot of the data.
c) Estimate the number of Alzheimer's patients in 2005, 2025, and 2100.

26. *Forgetting.* In an art class, students were tested at the end of the course on a final exam. Then they were retested with an equivalent test at subsequent time intervals. Their scores after time t, in months, are given in the table at the top of the next page.

TIME, t (IN MONTHS)	SCORE, y
1	84.9%
2	84.6%
3	84.4%
4	84.2%
5	84.1%
6	83.9%

a) Use a grapher to fit a logarithmic function
$y = a + b \ln x$ to the data.
b) Use the function to predict test scores after 8, 10, 24, and 36 months.
c) After how long will the test scores fall below 82%?

27. *On-line Travel Revenue.* With the explosion of increased Internet use, more and more travelers are booking their travel reservations on-line. The following table lists the total on-line revenue for recent years. Most of the revenue is from airline tickets.

YEAR	ON-LINE TRAVEL REVENUE (IN MILLIONS)
1996	$ 276
1997	827
1998	1900
1999	3200
2000*	4700
2001*	6500
2002*	8900

*Projections
Source: Travel and Interactive Technology
1999; Travel Industry Association of America

a) Create a scatterplot of the data. Let $x = $ the number of years since 1996.
b) Use a grapher to fit the data with linear, quadratic, and exponential functions. Determine which function has the best fit.
c) Graph all three functions found in part (b) with the scatterplot in part (a).
d) Use the functions found in part (b) to estimate the on-line travel revenue in 2010. Which function provides the most realistic prediction?

28. *Effect of Advertising.* A company introduces a new software product on a trial run in a city. They advertised the product on television and found the following data relating the percent P of people who bought the product after x ads were run.

NUMBER OF ADS, x	PERCENTAGE WHO BOUGHT, P
0	0.2
10	0.7
20	2.7
30	9.2
40	27
50	57.6
60	83.3
70	94.8
80	98.5
90	99.6

a) Use a grapher to fit a logistic function
$$P(x) = \frac{a}{1 + be^{-kx}}$$
to the data.
b) What percent will buy the product when 55 ads are run? 100 ads?
c) Find an asymptote for the graph. Interpret the asymptote in terms of the advertising situation.

Discussion and Writing

29. Browse through some newspapers or magazines until you find some data and/or a graph that seem as though they can be fit to an exponential function. Make a case for why such a fit is appropriate. Then fit an exponential function to the data and make some predictions.

30. *Atmospheric Pressure.* Atmospheric pressure P at an altitude a is given by
$$P = P_0 e^{-0.00005a},$$
where P_0 is the pressure at sea level ≈ 14.7 lb/in^2 (pounds per square inch). Explain how a barometer, or some device for measuring atmospheric pressure, can be used to find the height of a skyscraper.

Skill Maintenance

Solve the triangle, if possible.

31. $A = 131.2°$, $a = 23.6$, $c = 17.9$

32. $a = 10$ in., $b = 13$ in., $c = 7$ in.

Find each of the following.

33. $(1 - i)^7$ **34.** $(3 + 2i)^3$

Synthesis

35. *Present Value.* Following the birth of a child, a parent wants to make an initial investment P_0 that will grow to $50,000 for the child's education at age 18. Interest is compounded continuously at 7%. What should the initial investment be? Such an amount is called the **present value** of $50,000 due 18 yr from now.

36. *Present Value.* Referring to Exercise 35:
   a) Solve $P = P_0e^{kt}$ for P_0.
   b) Find the present value of $50,000 due 18 yr from now at interest rate 6.4%.

37. *Supply and Demand.* The supply and demand for the sale of a certain type of VCR are given by
$$S(p) = 480e^{-0.003p} \quad \text{and} \quad D(p) = 150e^{0.004p},$$
where $S(p)$ is the number of VCRs that the company is willing to sell at price p and $D(p)$ is the quantity that the public is willing to buy at price p. Find p, called the **equilibrium price,** such that $D(p) = S(p)$.

38. *Carbon Dating.* Recently, while digging in Chaco Canyon, New Mexico, archeologists found corn pollen that was 4000 yr old (*Source: American Anthropologist*). This was evidence that Native Americans had been cultivating crops in the Southwest centuries earlier than scientists had thought. What percent of the carbon-14 had been lost from the pollen?

39. *Newton's Law of Cooling.* Suppose that a body with temperature T_1 is placed in surroundings with temperature T_0 different from that of T_1. The body will either cool or warm to temperature $T(t)$ after time t, in minutes, where
$$T(t) = T_0 + |T_1 - T_0|e^{-kt}.$$
A cup of coffee with temperature 105°F is placed in a freezer with temperature 0°F. After 5 min, the temperature of the coffee is 70°F. What will its temperature be after 10 min?

40. *When Was the Murder Committed?* The police discover the body of a math professor. Critical to solving the crime is determining when the murder was committed. The coroner arrives at the murder scene at 12:00 P.M. She immediately takes the temperature of the body and finds it to be 94.6°. She then takes the temperature 1 hr later and finds it to be 93.4°. The temperature of the room is 70°. When was the murder committed? (Use Newton's law of cooling in Exercise 39.)

41. *Electricity.* The formula
$$i = \frac{V}{R}[1 - e^{-(R/L)t}]$$
occurs in the theory of electricity. Solve for t.

42. *The Beer–Lambert Law.* A beam of light enters a medium such as water or smog with initial intensity I_0. Its intensity decreases depending on the thickness (or concentration) of the medium. The intensity I at a depth (or concentration) of x units is given by
$$I = I_0e^{-\mu x}.$$
The constant μ (the Greek letter "mu") is called the **coefficient of absorption,** and it varies with the medium. For sea water, $\mu = 1.4$.
   a) What percentage of light intensity I_0 remains at a depth of sea water that is 1 m? 3 m? 5 m? 50 m?
   b) Plant life cannot exist below 10 m. What percentage of I_0 remains at 10 m?

43. Given that $y = ae^x$, take the natural logarithm on both sides. Let $Y = \ln y$. Consider Y as a function of x. What kind of function is Y?

44. Given that $y = ax^b$, take the natural logarithm on both sides. Let $Y = \ln y$ and $X = \ln x$. Consider Y as a function of X. What kind of function is Y?

Chapter Summary and Review

Important Properties and Formulas

Exponential Function:	$f(x) = a^x$
The Number $e = 2.7182818284\ldots$	
Logarithmic Function:	$f(x) = \log_a x$
A Logarithm is an Exponent:	$\log_a x = y \longleftrightarrow x = a^y$
The Change-of-Base Formula:	$\log_b M = \dfrac{\log_a M}{\log_a b}$
The Product Rule:	$\log_a MN = \log_a M + \log_a N$
The Power Rule:	$\log_a M^p = p \log_a M$
The Quotient Rule:	$\log_a \dfrac{M}{N} = \log_a M - \log_a N$
Other Properties:	$\log_a a = 1, \qquad \log_a 1 = 0,$
	$\log_a a^x = x, \qquad a^{\log_a x} = x$
Base–Exponent Property:	$a^x = a^y \longleftrightarrow x = y,$ for $a > 0,\ a \neq 1$
Exponential Growth Model:	$P(t) = P_0 e^{kt}$
Exponential Decay Model:	$P(t) = P_0 e^{-kt}$
Interest Compounded Continuously:	$P(t) = P_0 e^{kt}$
Limited Growth:	$P(t) = \dfrac{a}{1 + be^{-kt}}$

REVIEW EXERCISES

In Exercises 1–6, match the equation with one of figures (a)–(f), which follow. If needed, use a grapher.

a)

b)

c)

d)

e)

f)

1. $f(x) = e^{x-3}$

2. $f(x) = \log_3 x$

3. $y = -\log_3 (x + 1)$

4. $y = \left(\frac{1}{2}\right)^x$

5. $f(x) = 3(1 - e^{-x})$, $x \geq 0$

6. $f(x) = |\ln (x - 4)|$

7. Convert to an exponential equation:
$$\log_4 x = 2.$$

8. Convert to a logarithmic equation:
$$e^x = 80.$$

Solve. Use any method.

9. $\log_4 x = 2$

10. $3^{1-x} = 9^{2x}$

11. $e^x = 80$

12. $4^{2x-1} - 3 = 61$

13. $\log_{16} 4 = x$

14. $\log_x 125 = 3$

15. $\log_2 x + \log_2 (x - 2) = 3$

16. $\log (x^2 - 1) - \log (x - 1) = 1$

17. $\log x^2 = \log x$

18. $e^{-x} = 0.02$

Express as a single logarithm and, if possible, simplify.

19. $3 \log_b x - 4 \log_b y + \frac{1}{2} \log_b z$

20. $\ln (x^3 - 8) - \ln (x^2 + 2x + 4) + \ln (x + 2)$

Express in terms of sums and differences of logarithms.

21. $\ln \sqrt[4]{wr^2}$

22. $\log \sqrt[3]{\dfrac{M^2}{N}}$

Given that $\log_a 2 = 0.301$, $\log_a 5 = 0.699$, and $\log_a 6 = 0.778$, find each of the following.

23. $\log_a 3$

24. $\log_a 50$

25. $\log_a \frac{1}{5}$

26. $\log_a \sqrt[3]{5}$

Simplify.

27. $\ln e^{-5k}$

28. $\log_5 5^{-6t}$

29. How long will it take an investment to double itself if it is invested at 8.6%, compounded continuously?

30. The population of a city doubled in 30 yr. What was the exponential growth rate?

31. How old is a skeleton that has lost 27% of its carbon-14?

32. The hydrogen ion concentration of milk is 2.3×10^{-6}. What is the pH?

33. What is the loudness, in decibels, of a sound whose intensity is $1000I_0$?

34. *The Population of Zimbabwe.*   The population of Zimbabwe was 11 million in 1998, and the exponential growth rate was 1.2% per year (*Source:* U.S. Bureau of the Census, World Population Profile).

a) Find the exponential growth function.
b) What will the population be in 2004? in 2020?
c) When will the population be 25 million?
d) What is the doubling time?

35. *Toll-free Numbers.*   The use of toll-free numbers has grown exponentially. In 1967, there were 7 million such calls, and in 1991, there were 10.2 billion such calls (*Source:* Federal Communication Commission).

a) Find the exponential growth rate k.
b) Find the exponential growth function.
c) Graph the exponential growth function.
d) How many toll-free number calls will be placed in 2002? in 2005?
e) In what year will 1 trillion such calls be placed?

36. *Walking Speed.*   The average walking speed w, in feet per second, of a person living in a city of population P, in thousands, is given by the function
$$w(P) = 0.37 \ln P + 0.05.$$

a) The population of Phoenix, Arizona, is 1,200,000. Find the average walking speed.
b) A city's population has an average walking speed of 3.4 ft/sec. Find the population.

37. *Cholesterol Level and the Risk of Heart Attack.* The data in the following table show the relationship of cholesterol level in men to the risk of a heart attack.

CHOLESTEROL LEVEL, x	MEN, PER 10,000, WHO SUFFER A HEART ATTACK, y
100	30
200	65
250	100
275	130
300	180

Source: Nutrition Action Healthletter

a) Use the grapher to fit an exponential function to the data.

b) Graph the function with a scatterplot of the data.

c) Predict the heart attack rate for men with cholesterol levels of 150, 350, and 400.

d) Compare your answers in part (c) with the answer to part (c) in Example 4 of Section 3.1. Which function, power or exponential, is a better fit to the given data?

Discussion and Writing

38. Suppose that you were trying to convince a fellow student that

$$\log_2 (x + 5) \neq \log_2 x + \log_2 5.$$

Give as many explanations as you can.

39. Explain how the graph of $f(x) = e^x$ could be used to graph the function $g(x) = \ln x - 1$.

Synthesis

Solve.

40. $|\log_4 x| = 3$

41. $\log x = \ln x$

42. $5^{\sqrt{x}} = 625$

43. a) Use a grapher to graph $f(x) = 5e^{-x} \ln x$ in the viewing window $[-1, 10, -5, 5]$.

  b) Estimate the relative maximum and the minimum values.

44. Find the domain: $f(x) = \log_3 (\ln x)$.

45. Find the points of intersection of the graphs of the following equations:

$$y = 5x^2 e^{-x}, \qquad y = 2 - e^{-x^2}.$$

Appendixes

Solving Equations

A

- *Solve linear, quadratic, rational, and radical equations and equations with absolute value.*
- *Solve a system of two linear equations in two variables by graphing.*
- *Solve a system of two linear equations in two variables using the substitution and the elimination methods.*
- *Solve a formula for a given variable.*

An **equation** is a statement that two expressions are equal. To **solve** an equation in one variable is to find all the values of the variable that make the equation true. Each of these numbers is a **solution** of the equation. The set of all solutions of an equation is its **solution set**. Equations that have the same solution set are called **equivalent equations**.

Linear and Quadratic Equations

A **linear equation in one variable** is an equation that is equivalent to one of the form $ax + b = 0$, where a and b are real numbers and $a \neq 0$.

A **quadratic equation** is an equation that is equivalent to one of the form $ax^2 + bx + c = 0$, where a, b, and c are real numbers and $a \neq 0$.

The following principles allow us to solve many linear and quadratic equations.

Equation-Solving Principles

For any real numbers a, b, and c,

The Addition Principle: If $a = b$ is true, then $a + c = b + c$ is true.

The Multiplication Principle: If $a = b$ is true, then $ac = bc$ is true.

The Principle of Zero Products: If $ab = 0$ is true, then $a = 0$ or $b = 0$, and if $a = 0$ or $b = 0$, then $ab = 0$.

The Principle of Square Roots: If $x^2 = k$, then $x = \sqrt{k}$ or $x = -\sqrt{k}$.

EXAMPLE 1 Solve: $2(5 - 3x) = 8 - 3(x + 2)$.

Solution We solve this both algebraically and graphically, as shown below.

Algebraic Solution

We have

$$2(5 - 3x) = 8 - 3(x + 2)$$

$10 - 6x = 8 - 3x - 6$	Using the distributive property
$10 - 6x = 2 - 3x$	Combining like terms
$10 = 2 + 3x$	Using the addition principle to add $6x$ on both sides
$8 = 3x$	Using the addition principle to add -2, or subtract 2, on both sides
$\frac{8}{3} = x.$	Using the multiplication principle to multiply by $\frac{1}{3}$, or divide by 3, on both sides

CHECK:
$$\frac{2(5 - 3x) = 8 - 3(x + 2)}{}$$

$2\left(5 - 3 \cdot \frac{8}{3}\right)$? $8 - 3\left(\frac{8}{3} + 2\right)$ Substituting $\frac{8}{3}$ for x

$\quad 2(5 - 8) \mid 8 - 3\left(\frac{14}{3}\right)$

$\quad\quad 2(-3) \mid 8 - 14$

$\quad\quad\quad -6 \mid -6$ TRUE

The solution is $\frac{8}{3}$.

We can also use the TABLE feature on a grapher, set in ASK mode, to check the solution. We let $y_1 = 2(5 - 3x)$ and $y_2 = 8 - 3(x + 2)$.

When $\frac{8}{3}$ is entered for x, we see that y_1 and y_2 are both -6. (Note that the grapher converts $\frac{8}{3}$ to decimal notation.)

X	Y₁	Y₂
2.6667	−6	−6

X =

Graphical Solution

We graph $y_1 = 2(5 - 3x)$ and $y_2 = 8 - 3(x + 2)$. The value of x for which $y_1 = y_2$ is the solution of the original equation. This x-value is the first coordinate of the point of intersection of the graphs of y_1 and y_2. Using the INTERSECT feature (see "Introduction to Graphs and the Graphing Calculator"), we find that the first coordinate of the point of intersection is 2.6666667. This is a decimal approximation of the solution.

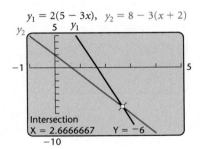

$y_1 = 2(5 - 3x), \quad y_2 = 8 - 3(x + 2)$

If the solution is a rational number, we can find fractional notation for the exact solution by using the ▶FRAC feature.

X ▶ Frac

8/3

The solution is $\frac{8}{3}$.

EXAMPLE 2 Solve: $2x^2 - x = 3$.

Algebraic Solution

We have

$$2x^2 - x = 3$$
$$2x^2 - x - 3 = 0 \qquad \text{Subtracting 3 on both sides}$$
$$(x + 1)(2x - 3) = 0 \qquad \text{Factoring}$$
$$x + 1 = 0 \quad or \quad 2x - 3 = 0 \qquad \begin{array}{l}\text{Using the principle of}\\ \text{zero products}\end{array}$$
$$x = -1 \quad or \quad 2x = 3$$
$$x = -1 \quad or \quad x = \tfrac{3}{2}.$$

Both numbers check. We can use the TABLE feature on a grapher, set in ASK mode, to confirm this. We let $y = 2x^2 - x$. The y-values in the table should be 3 for both $x = -1$ and $x = \tfrac{3}{2}$.

$y_1 = 2x^2 - x$

X	Y1	
−1	3	
1.5	3	
X =		

The solutions are -1 and $\tfrac{3}{2}$.

Graphical Solution

We graph $y_1 = 2x^2 - x$ and $y_2 = 3$ and use the INTERSECT feature twice.

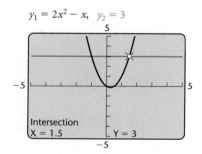

The solutions are -1 and 1.5. ▬

EXAMPLE 3 Solve: $2x^2 - 10 = 0$.

Solution We solve algebraically:

$$2x^2 - 10 = 0$$
$$2x^2 = 10 \qquad \text{Adding 10 on both sides}$$
$$x^2 = 5 \qquad \text{Dividing by 2 on both sides}$$
$$x = \sqrt{5} \quad or \quad x = -\sqrt{5}. \qquad \text{Using the principle of square roots}$$

$y_1 = 2x^2 - 10$

X	Y1	
2.2361	0	
−2.236	0	
X =		

Both numbers check. The solutions are $\sqrt{5}$ and $-\sqrt{5}$, or $\pm\sqrt{5}$. We can use the TABLE feature on a grapher, set in ASK mode, to confirm this. Note that the grapher converts $\sqrt{5}$ and $-\sqrt{5}$ to decimal notation.

If we were to use a grapher to solve this equation, we would find that $x \approx 2.236$ or $x \approx -2.236$. Since the exact solutions are not rational numbers, the ▶FRAC operation will not yield exact answers. ▬

Neither of the preceding methods would yield the *exact* solutions to a quadratic equation like $x^2 - 6x - 10 = 0$. If we wish to do that, we can use a procedure called **completing the square** and then use the principle of square roots.

EXAMPLE 4 Find the zeros of $f(x) = x^2 - 6x - 10$ by completing the square.

Solution We find the values of x for which $f(x) = 0$. That is, we solve the associated equation $x^2 - 6x - 10 = 0$. Our goal is to find an equivalent equation of the form $x^2 + bx + c = d$ in which $x^2 + bx + c$ is a perfect square. Since

$$x^2 + bx + \left(\frac{b}{2}\right)^2 = \left(x + \frac{b}{2}\right)^2,$$

the number c is found by taking half the coefficient of the x-term and squaring it. Then for the equation $x^2 - 6x - 10 = 0$, we have

$$
\begin{aligned}
x^2 - 6x - 10 &= 0 \\
x^2 - 6x &= 10 \qquad && \text{Adding 10} \\
x^2 - 6x + 9 &= 10 + 9 \qquad && \text{Adding 9 to complete the square:} \\
&&& \left(\frac{b}{2}\right)^2 = \left(\frac{-6}{2}\right)^2 = 9 \\
x^2 - 6x + 9 &= 19.
\end{aligned}
$$

Because $x^2 - 6x + 9$ is a perfect square, we are able to write it as the square of a binomial. We can then use the principle of square roots to finish the solution:

$$
\begin{aligned}
(x - 3)^2 &= 19 \qquad && \text{Factoring} \\
x - 3 &= \pm\sqrt{19} \qquad && \text{Using the principle of square roots} \\
x &= 3 \pm \sqrt{19}. \qquad && \text{Adding 3}
\end{aligned}
$$

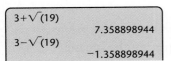

Thus the solutions of the equation are $3 + \sqrt{19}$ and $3 - \sqrt{19}$, or simply $3 \pm \sqrt{19}$. The zeros of $f(x) = x^2 - 6x - 10$ are also $3 + \sqrt{19}$ and $3 - \sqrt{19}$, or $3 \pm \sqrt{19}$.

Decimal approximations for $3 \pm \sqrt{19}$ can be found using a grapher:

$$3 + \sqrt{19} \approx 7.359 \quad \text{and} \quad 3 - \sqrt{19} \approx -1.359.$$

Approximations for the zeros can also be found by using the ZERO feature to find the zeros of the quadratic function $f(x) = x^2 - 6x - 10$, as shown below.

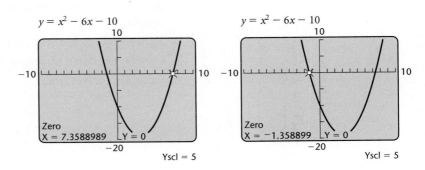

The zeros are approximately 7.359 and −1.359.

EXAMPLE 5 Solve: $2x^2 - 1 = 3x$.

Solution We have

$$2x^2 - 1 = 3x$$

$$2x^2 - 3x - 1 = 0 \qquad \text{Subtracting } 3x. \text{ We are unable to factor the result.}$$

$$2x^2 - 3x = 1 \qquad \text{Adding 1}$$

$$x^2 - \frac{3}{2}x = \frac{1}{2} \qquad \text{Dividing by 2 to make the } x^2\text{-coefficient 1}$$

$$x^2 - \frac{3}{2}x + \frac{9}{16} = \frac{1}{2} + \frac{9}{16} \qquad \text{Completing the square: } \frac{1}{2}\left(-\frac{3}{2}\right) = -\frac{3}{4} \text{ and } \left(-\frac{3}{4}\right)^2 = \frac{9}{16}; \text{ adding } \frac{9}{16}$$

$$\left(x - \frac{3}{4}\right)^2 = \frac{17}{16} \qquad \text{Factoring and simplifying}$$

$$x - \frac{3}{4} = \pm\frac{\sqrt{17}}{4} \qquad \text{Using the principle of square roots and the quotient rule for radicals}$$

$$x = \frac{3 \pm \sqrt{17}}{4}. \qquad \text{Adding } \frac{3}{4}$$

The solutions are

$$\frac{3 + \sqrt{17}}{4} \quad \text{and} \quad \frac{3 - \sqrt{17}}{4}, \quad \text{or} \quad \frac{3 \pm \sqrt{17}}{4}.$$

We can calculate approximate solutions using a grapher and check the solutions graphically by graphing the associated function $f(x) = 2x^2 - 3x - 1$.

———

> To solve a quadratic equation by completing the square:
>
> 1. Isolate the terms with variables on one side of the equation and arrange them in descending order.
> 2. Divide by the coefficient of the squared term if that coefficient is not 1.
> 3. Complete the square by taking half the coefficient of the first-degree term and adding its square on both sides of the equation.
> 4. Express one side of the equation as the square of a binomial.
> 5. Use the principle of square roots.
> 6. Solve for the variable.

Using the Quadratic Formula

Because completing the square works for *any* quadratic equation, it can be used to solve the general quadratic equation $ax^2 + bx + c = 0$ for x. The result will be a formula that can be used to solve any quadratic equation quickly.

Consider any quadratic equation in standard form:

$$ax^2 + bx + c = 0, \quad a \neq 0.$$

For now, we assume that $a > 0$ and solve by completing the square. As the steps are carried out, compare them with those of Example 5.

$$
\begin{aligned}
ax^2 + bx + c &= 0 && \text{Standard form} \\
ax^2 + bx &= -c && \text{Adding } -c \\
x^2 + \frac{b}{a}x &= -\frac{c}{a} && \text{Dividing by } a
\end{aligned}
$$

Half of $\dfrac{b}{a}$ is $\dfrac{b}{2a}$ and $\left(\dfrac{b}{2a}\right)^2 = \dfrac{b^2}{4a^2}$. Thus we add $\dfrac{b^2}{4a^2}$:

$$
x^2 + \frac{b}{a}x + \frac{b^2}{4a^2} = -\frac{c}{a} + \frac{b^2}{4a^2} \qquad \text{Adding } \frac{b^2}{4a^2} \text{ to complete the square}
$$

$$
\left(x + \frac{b}{2a}\right)^2 = -\frac{4ac}{4a^2} + \frac{b^2}{4a^2} \qquad
\begin{array}{l}
\text{Factoring and finding a common} \\
\text{denominator:} \\
-\dfrac{c}{a} = -\dfrac{4a}{4a} \cdot \dfrac{c}{a} = -\dfrac{4ac}{4a^2}
\end{array}
$$

$$
\left(x + \frac{b}{2a}\right)^2 = \frac{b^2 - 4ac}{4a^2}
$$

$$
x + \frac{b}{2a} = \pm\frac{\sqrt{b^2 - 4ac}}{2a} \qquad
\begin{array}{l}
\text{Using the principle of square roots} \\
\text{and the quotient rule for radicals.} \\
\text{Since } a > 0, \sqrt{4a^2} = 2a.
\end{array}
$$

$$
x = \frac{-b \pm \sqrt{b^2 - 4ac}}{2a}. \qquad \text{Adding } -\frac{b}{2a}
$$

It can also be shown that this result holds if $a < 0$.

The Quadratic Formula

The solutions of $ax^2 + bx + c = 0$, $a \neq 0$, are given by

$$
x = \frac{-b \pm \sqrt{b^2 - 4ac}}{2a}.
$$

EXAMPLE 6 Solve $3x^2 + 2x = 7$. Find exact solutions and approximate solutions rounded to the nearest thousandth.

Solution We show both algebraic and graphical solutions.

Algebraic Solution

After finding standard form, we are unable to factor, so we identify a, b, and c in order to use the quadratic formula:

$$3x^2 + 2x - 7 = 0;$$
$$a = 3, \quad b = 2, \quad c = -7.$$

We then use the quadratic formula:

$$x = \frac{-b \pm \sqrt{b^2 - 4ac}}{2a}$$

$$= \frac{-2 \pm \sqrt{2^2 - 4(3)(-7)}}{2(3)} \quad \text{Substituting}$$

$$= \frac{-2 \pm \sqrt{4 + 84}}{6} = \frac{-2 \pm \sqrt{88}}{6}$$

$$= \frac{-2 \pm \sqrt{4 \cdot 22}}{6} = \frac{-2 \pm 2\sqrt{22}}{6}$$

$$= \frac{2}{2} \cdot \frac{-1 \pm \sqrt{22}}{3} = \frac{-1 \pm \sqrt{22}}{3}.$$

The exact solutions are

$$\frac{-1 - \sqrt{22}}{3} \quad \text{and} \quad \frac{-1 + \sqrt{22}}{3}.$$

Using a grapher, we approximate the solutions to be -1.897 and 1.230.

A table set in ASK mode can be used to check the solutions. If the exact solutions are entered, the grapher converts them to decimal approximations in the table.

X	Y1	
−1.897	0	
1.2301	0	
X =		

A graph can also provide a visual check of the algebraic solution.

Graphical Solution

We graph $y_1 = 3x^2 + 2x$ and $y_2 = 7$ and use the INTERSECT feature to find the coordinates of the points of intersection. The first coordinates of these points are the solutions of the equation $y_1 = y_2$, or $3x^2 + 2x = 7$.

The solutions are approximately -1.897 and 1.230. We could also write the equation in standard form, $3x^2 + 2x - 7 = 0$, and use the ZERO feature.

Rational Equations

Equations containing rational expressions are called **rational equations.** Solving such equations algebraically involves multiplying on both sides by the least common denominator (LCD).

EXAMPLE 7   Solve: $\dfrac{x-8}{3} + \dfrac{x-3}{2} = 0$.

Algebraic Solution

We have

$$\frac{x-8}{3} + \frac{x-3}{2} = 0 \qquad \text{The LCD is } 3 \cdot 2, \text{ or } 6.$$

$$6\left(\frac{x-8}{3} + \frac{x-3}{2}\right) = 6 \cdot 0 \qquad \begin{array}{l}\text{Multiplying by}\\ \text{the LCD on}\\ \text{both sides}\end{array}$$

$$6 \cdot \frac{x-8}{3} + 6 \cdot \frac{x-3}{2} = 0$$

$$2(x-8) + 3(x-3) = 0$$

$$2x - 16 + 3x - 9 = 0$$

$$5x - 25 = 0$$

$$5x = 25$$

$$x = 5.$$

The possible solution is 5. We check using a table in ASK mode.

$$y = \frac{x-8}{3} + \frac{x-3}{2}$$

X	Y₁	
5	0	

X =

Since the value of

$$\frac{x-8}{3} + \frac{x-3}{2}$$

is 0 when $x = 5$, the number 5 is the solution.

Graphical Solution

We use the ZERO feature. The solutions of the given equation are the zeros of the function

$$f(x) = \frac{x-8}{3} + \frac{x-3}{2}.$$

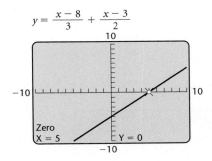

$$y = \frac{x-8}{3} + \frac{x-3}{2}$$

The solution is 5.

When we use the multiplication principle to multiply (or divide) both sides of an equation by an expression with a variable, we might not obtain an equivalent equation. We must check possible solutions obtained in this manner by substituting in the original equation. The next example illustrates this.

EXAMPLE 8 Solve: $\dfrac{x^2}{x-3} = \dfrac{9}{x-3}$.

Solution The LCD is $x - 3$.

$$(x-3) \cdot \frac{x^2}{x-3} = (x-3) \cdot \frac{9}{x-3}$$

$$x^2 = 9$$

$$x = -3 \quad or \quad x = 3 \qquad \text{Using the principle of square roots}$$

The possible solutions are -3 and 3. We check.

$y_1 = \dfrac{x^2}{x-3},\ \ y_2 = \dfrac{9}{x-3}$

X	Y1	Y2
−3	−1.5	−1.5
3	ERROR	ERROR

X =

CHECK: For -3:

$$\frac{x^2}{x-3} = \frac{9}{x-3}$$

$$\frac{(-3)^2}{-3-3} \ ? \ \frac{9}{-3-3}$$

$$\frac{9}{-6} \ \bigg|\ \frac{9}{-6} \qquad \text{TRUE}$$

For 3:

$$\frac{x^2}{x-3} = \frac{9}{x-3}$$

$$\frac{3^2}{3-3} \ ? \ \frac{9}{3-3}$$

$$\frac{9}{0} \ \bigg|\ \frac{9}{0} \qquad \text{UNDEFINED}$$

The number -3 checks, so it is a solution. Since division by 0 is undefined, 3 is not a solution. Note that 3 is not in the domain of $x^2/(x-3)$ or $9/(x-3)$. (See the table at left.) ▬

Radical Equations

A **radical equation** is an equation in which variables appear in one or more radicands. For example,

$$\sqrt{2x-5} - \sqrt{x-3} = 1$$

is a radical equation. The following principle is used to solve such equations.

The Principle of Powers

For any positive integer n:

If $a = b$ is true, then $a^n = b^n$ is true.

EXAMPLE 9 Solve: $5 + \sqrt{x + 7} = x$.

Algebraic Solution

We first isolate the radical and then use the principle of powers.

$$5 + \sqrt{x + 7} = x$$

$$\sqrt{x + 7} = x - 5 \qquad \text{Subtracting 5 on both sides}$$

$$(\sqrt{x + 7})^2 = (x - 5)^2 \qquad \begin{array}{l}\text{Using the principle of powers;}\\ \text{squaring both sides}\end{array}$$

$$x + 7 = x^2 - 10x + 25$$

$$0 = x^2 - 11x + 18$$ $\qquad$ Subtracting x and 7

$$0 = (x - 9)(x - 2) \qquad \text{Factoring}$$

$$x - 9 = 0 \quad or \quad x - 2 = 0$$

$$x = 9 \quad or \qquad x = 2$$

The possible solutions are 9 and 2.

CHECK: For 9:

$$\dfrac{5 + \sqrt{x + 7} = x}{5 + \sqrt{9 + 7} \; ? \; 9}$$

$$5 + \sqrt{16}$$

$$5 + 4$$

$$9 \;\big|\; 9 \qquad \text{TRUE}$$

For 2:

$$\dfrac{5 + \sqrt{x + 7} = x}{5 + \sqrt{2 + 7} \; ? \; 2}$$

$$5 + \sqrt{9}$$

$$5 + 3$$

$$8 \;\big|\; 2 \qquad \text{FALSE}$$

Since 9 checks but 2 does not, the only solution is 9.

Graphical Solution

We graph $y_1 = 5 + \sqrt{x + 7}$ and $y_2 = x$. Using the INTERSECT feature, we see that the solution is 9.

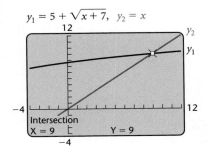

$$y_1 = 5 + \sqrt{x + 7}, \quad y_2 = x$$

We can also use the ZERO feature to get this result. To do so, we first write the equivalent equation

$$5 + \sqrt{x + 7} - x = 0.$$

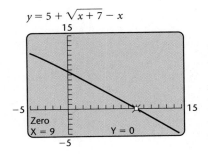

$$y = 5 + \sqrt{x + 7} - x$$

The zero of the function is 9, so the solution of the original equation is 9. Note that graphs show that the equation has only one solution. ▬

When we raise both sides of an equation to an even power, the resulting equation can have solutions that the original equation does not. This is because the converse of the principle of powers is not necessarily true. That is, if $a^n = b^n$ is true, we do not know that $a = b$ is true. For example, $(-2)^2 = 2^2$, but $-2 \neq 2$. Thus, as we see in Example 9, it is necessary to check the possible solutions in the original equation when the principle of powers is used to raise both sides of an equation to an even power.

When a radical equation has two radical terms on one side, we isolate one of them and then use the principle of powers. If, after doing so, a radical term remains, we repeat these steps.

EXAMPLE 10   Solve: $\sqrt{x-3} + \sqrt{x+5} = 4$.

Solution   We have

$$\sqrt{x-3} = 4 - \sqrt{x+5}$$ Isolating one radical

$$(\sqrt{x-3})^2 = (4 - \sqrt{x+5})^2$$ Using the principle of powers; squaring both sides

$$x - 3 = 16 - 8\sqrt{x+5} + (x+5)$$

$$x - 3 = 21 - 8\sqrt{x+5} + x$$ Combining like terms

$$-24 = -8\sqrt{x+5}$$ Isolating the remaining radical; subtracting x and 21 on both sides

$$3 = \sqrt{x+5}$$ Dividing by -8 on both sides

$$3^2 = (\sqrt{x+5})^2$$ Using the principle of powers; squaring both sides

$$9 = x + 5$$

$$4 = x.$$

The number 4 checks and is the solution. Graphs or the TABLE feature on a grapher can be used to confirm this.

Equations with Absolute Value

Recall that the absolute value of a number is its distance from 0 on the number line. We use this concept to solve equations with absolute value.

For $a > 0$:

$$|X| = a \text{ is equivalent to } X = -a \text{ or } X = a.$$

EXAMPLE 11   Solve each of the following.

a) $|x| = 5$

b) $|x - 3| = 2$

Solution

a) We solve $|x| = 5$ both algebraically and graphically.

Algebraic Solution

We have

$$|x| = 5$$

$x = -5 \quad or \quad x = 5.$

Writing an equivalent statement

The solutions are -5 and 5.
We can use the TABLE feature on a grapher to confirm this.

$y = |x|$

X	Y1	
-5	5	
5	5	

X =

Graphical Solution

Using the INTERSECT feature, we graph $y_1 = |x|$ and $y_2 = 5$ and find the first coordinates of the points of intersection.

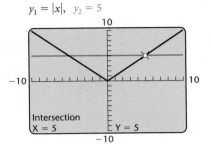

The solutions are -5 and 5.
We could also have gotten this result by graphing $y = |x| - 5$ and using the ZERO feature twice.

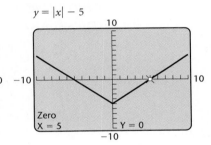

The zeros of $f(x) = |x| - 5$ are -5 and 5, so the solutions of the original equation are -5 and 5.

b) We solve $|x - 3| = 2$ algebraically:

$$|x - 3| = 2$$

$x - 3 = -2 \quad or \quad x - 3 = 2$ Writing an equivalent statement

$x = 1 \quad\quad or \quad\quad\quad x = 5.$ Adding 3

The solutions are 1 and 5.

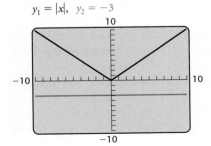

When $a = 0$, $|X| = a$ is equivalent to $x = 0$. Note that for $a < 0$, $|X| = a$ has no solution, because the absolute value of an expression is never negative. We can use a graph to illustrate the last statement for a specific value of a. For example, we let $a = -3$ and graph $y_1 = |x|$ and $y_2 = -3$. The graphs do not intersect, as shown at left. Thus the equation $|x| = -3$ has no solution. The solution set is the **empty set,** denoted $\varnothing$.

A **system of equations** is composed of two or more equations considered simultaneously. For example,

$$x + y = 11,$$
$$3x - y = 5$$

is a system of two linear equations in two variables. The solution set of this system consists of all ordered pairs that make *both* equations true. The ordered pair $(4, 7)$ is a solution of the system of equations above. We can verify this by substituting 4 for x and 7 for y in *each* equation.

$x + y = 11$		$3x - y = 5$	
$4 + 7$? 11		$3 \cdot 4 - 7$? 5	
11 $\vert$ 11	TRUE	$12 - 7$	
		5 $\vert$ 5	TRUE

Interactive Discovery

Graph $y_1 = 2x - 5$ and $y_2 = -x + 1$ in the same viewing window. Use the grapher's INTERSECT feature to find the coordinates of the point of intersection of the graphs. Do the same for $y_1 = x + 4$ and $y_2 = -1.25x - 5$ and for $y_1 = 3x - 7$ and $y_2 = -2x + \frac{1}{2}$. What does each of these ordered pairs represent?

Solving Systems of Equations Graphically

Recall that the graph of a linear equation is a line that contains all the ordered pairs in the solution set of the equation. When we graph a system of linear equations, each point at which the equations intersect is a solution of *both* equations and therefore a solution of the system of equations.

Consider the system

$$x + y = 11,$$
$$3x - y = 5.$$

$y_1 = -x + 11, \quad y_2 = 3x - 5$

When we graph these equations on the same set of axes, we see that they intersect at a single point, $(4, 7)$, so $(4, 7)$ is the solution of the system of equations. (Note that when a grapher is used, it might be necessary to write each equation in "$Y = \cdots$" form. If so, we would graph $y_1 = -x + 11$ and $y_2 = 3x - 5$.)

To check this result, we can substitute 4 for x and 7 for y in both equations, as we did above. We can also use the TABLE feature on a grapher to do a check, observing that when $x = 4$, $y_1 = 7$ and $y_2 = 7$.

The graphs of most of the systems of equations that we use to model applications intersect at a single point, like the system above. However, it is possible that the graphs will have no points in common or infinitely many points in common. Each of these possibilities is illustrated below.

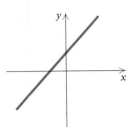

Exactly one common point
One solution

Consistent
Independent

Parallel lines
No common points
No solution

Inconsistent
Independent

Lines are identical
Infinitely many common points
Infinitely many solutions

Consistent
Dependent

If a system of two linear equations in two variables has at least one solution, it is **consistent**. If the system has no solutions, it is **inconsistent**. In addition, if a system of two linear equations in two variables has an infinite number of solutions, the equations are **dependent**. Otherwise, they are **independent**.

The Substitution Method

Solving a system of equations graphically is not always accurate when the solutions are not integers. A pair like $\left(\frac{43}{27}, -\frac{19}{27}\right)$, for instance, will be difficult to determine from a hand-drawn graph. Even when a grapher is used, decimal approximations for the solution of a system of equations are not always successfully converted to fractions.

Algebraic methods for solving systems of equations, when used correctly, always give accurate results. One such technique is the **substitution method.** It is used most often when a variable is alone on one side of an equation or when it is easy to solve for a variable. To apply the substitution method, we begin by using one of the equations to express one variable in terms of the other; then we substitute that expression in the other equation of the system.

EXAMPLE 12 Use the substitution method to solve the system

$$x + y = 11, \quad (1)$$
$$3x - y = 5. \quad (2)$$

Solution First, we solve equation (1) for y. (We could just as well solve for x.)

$$y = 11 - x$$

Then we substitute $11 - x$ for y in equation (2). This gives an equation in one variable, which we know how to solve.

$$3x - (11 - x) = 5 \qquad \text{The parentheses are necessary.}$$
$$3x - 11 + x = 5 \qquad \text{Removing parentheses}$$
$$4x - 11 = 5$$
$$4x = 16$$
$$x = 4$$

Now we substitute 4 for x in either of the original equations (this is called **back-substitution**) and solve for y. We choose equation (1):

$$4 + y = 11$$
$$y = 7.$$

We have previously checked the pair $(4, 7)$ in both equations both algebraically and graphically. The solution of the system of equations is $(4, 7)$.

The Elimination Method

Another algebraic technique for solving systems of equations is the **elimination method.** With this method, we eliminate a variable by adding two equations. Before we add, it might be necessary to multiply one or both equations by suitable constants in order to find two equations in which coefficients are opposites.

EXAMPLE 13   Solve each of the following systems using the elimination method.

a) $2x + y = 2,$ (1)   **b)** $4x + 3y = 11,$ (1)
     $x - y = 7$ (2) $-5x + 2y = 15$ (2)

Solution

a) Since the y-coefficients are opposites, we can eliminate y by adding the equations:

$$\begin{array}{ll} 2x + y = 2 & (1) \\ \underline{x - y = 7} & (2) \\ 3x \quad\;\; = 9 & \text{Adding} \\ x = 3. & \end{array}$$

We then back-substitute 3 for x in either equation and solve for y. We choose equation (1):

$$2 \cdot 3 + y = 2$$
$$6 + y = 2$$
$$y = -4.$$

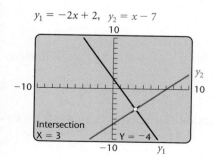

$y_1 = -2x + 2, \;\; y_2 = x - 7$

We can check the pair $(3, -4)$ either algebraically by substituting in both equations or graphically using the INTERSECT feature, as shown at left. The solution is $(3, -4)$.

b) We can obtain x-coefficients that are opposites by multiplying the first equation by 5 and the second equation by 4:

$$\begin{array}{ll} 20x + 15y = 55 & \text{Multiplying equation (1) by 5} \\ \underline{-20x + 8y = 60} & \text{Multiplying equation (2) by 4} \\ 23y = 115 & \text{Adding} \\ y = 5. \end{array}$$

We then back-substitute 5 for y in either equation (1) or (2) and solve for x. We choose equation (1):

$$\begin{array}{ll} 4x + 3 \cdot 5 = 11 & \text{Substituting 5 for } y \text{ in equation (1)} \\ 4x + 15 = 11 \\ 4x = -4 \\ x = -1. \end{array}$$

We can check the pair $(-1, 5)$ either algebraically by substituting in both equations or graphically, as shown at left. The solution is $(-1, 5)$.

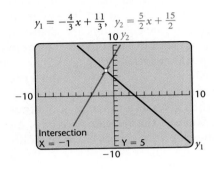

In Example 13(b), the two systems

$$\begin{array}{cc} 4x + 3y = 11, & 20x + 15y = 55, \\ -5x + 2y = 15 \quad \text{and} & -20x + 8y = 60 \end{array}$$

are **equivalent** because they have exactly the same solutions. When we use the elimination method, we often multiply one or both equations by constants to find equivalent equations that allow us to eliminate a variable by adding.

EXAMPLE 14 Solve each of the following systems using the elimination method.

a) $\begin{array}{ll} x - 3y = 1, & (1) \\ -2x + 6y = 5 & (2) \end{array}$ **b)** $\begin{array}{ll} 2x + 3y = 6, & (1) \\ 4x + 6y = 12 & (2) \end{array}$

Solution

a) We multiply equation (1) by 2 and add:

$$\begin{array}{ll} 2x - 6y = 2 & \text{Multiplying equation (1) by 2} \\ \underline{-2x + 6y = 5} & (2) \\ 0 = 7. & \text{Adding} \end{array}$$

There are no values of x and y for which $0 = 7$ is true, so the system has no solution. The solution set is $\varnothing$. The system of equations is inconsistent.

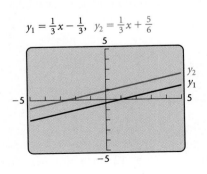

The graphs of the equations are parallel lines. In fact, we see this when we write the equations in simplified "$y = \cdots$" form to enter them on a grapher. The slopes are the same, $\frac{1}{3}$, but the y-intercepts are different, $\left(0, -\frac{1}{3}\right)$ and $\left(0, \frac{5}{6}\right)$.

b) We multiply equation (1) by -2 and add:

$$-4x - 6y = -12 \qquad \text{Multiplying equation (1) by } -2$$
$$\underline{4x + 6y = 12} \qquad \text{(2)}$$
$$0 = 0 \ . \qquad \text{Adding}$$

We obtain the equation $0 = 0$, which is true for all values of x and y. This tells us that the equations are dependent, so there are infinitely many solutions. That is, any solution of one equation of the system is also a solution of the other. The graphs of the equations are identical. (In fact, when we write the equations in simplified "$y = \cdots$" form to enter them on a grapher, we get identical equations.)

Using either equation, we can write $y = -\frac{2}{3}x + 2$, so we can write the solutions of the system as ordered pairs of the form $\left(x, -\frac{2}{3}x + 2\right)$. Any real value that we choose for x then gives us a value for y and thus an ordered pair in the solution set. Some of the solutions are $(-3, 4)$, $(0, 2)$, and $(6, -2)$. Similarly, we can write $x = -\frac{3}{2}y + 3$, so the solutions can also be expressed as $\left(-\frac{3}{2}y + 3, y\right)$.

$y_1 = -\frac{2}{3}x + 2, \ \ y_2 = -\frac{2}{3}x + 2$

Formulas

A **formula** is an equation that can be used to *model* a situation. For example, the formula $P = 2l + 2w$ gives the perimeter of a rectangle with length l and width w. The equation-solving principles presented earlier can be used to solve a formula for a given variable.

EXAMPLE 15 Solve $P = 2l + 2w$ for l.

Solution

$$P = 2l + 2w$$
$$P - 2w = 2l \qquad \text{Subtracting } 2w \text{ on both sides}$$
$$\frac{P - 2w}{2} = l \qquad \text{Dividing by 2 on both sides}$$

The formula $l = \dfrac{P - 2w}{2}$ can be used to determine a rectangle's length if we are given its perimeter and its width.

On many graphers, an equation must be solved for y before it can be entered. Solving for y is the same as solving a formula for a given variable.

EXAMPLE 16 Solve $3x + 4y = 12$ for y.

Solution We have

$$3x + 4y = 12$$
$$4y = -3x + 12 \qquad \text{Subtracting } 3x \text{ on both sides}$$
$$y = \frac{-3x + 12}{4}. \qquad \text{Dividing by 4 on both sides}$$

If we were to graph the equation $3x + 4y = 12$ on a grapher, we would enter it as $y = (-3x + 12)/4$.

Exercise Set A

Solve.

1. $4x + 5 = 21$
2. $2y - 1 = 3$
3. $y + 1 = 2y - 7$
4. $5 - 4x = x - 13$
5. $5x - 2 + 3x = 2x + 6 - 4x$
6. $5x - 17 - 2x = 6x - 1 - x$
7. $7(3x + 6) = 11 - (x + 2)$
8. $4(5y + 3) = 3(2y - 5)$
9. $(2x - 3)(3x - 2) = 0$
10. $(5x - 2)(2x + 3) = 0$
11. $3x^2 + x - 2 = 0$
12. $10x^2 - 16x + 6 = 0$
13. $4x^2 - 12 = 0$
14. $6x^2 = 36$
15. $2x^2 = 6x$
16. $18x + 9x^2 = 0$
17. $3y^3 - 5y^2 - 2y = 0$
18. $3t^3 + 2t = 5t^2$
19. $7x^3 + x^2 - 7x - 1 = 0$
    (*Hint*: Factor by grouping.)
20. $3x^3 + x^2 - 12x - 4 = 0$
    (*Hint*: Factor by grouping.)

Solve by completing the square to obtain exact solutions. Check your work using a grapher, if appropriate.

21. $x^2 + 6x = 7$
22. $x^2 + 8x = -15$
23. $x^2 = 8x - 9$
24. $x^2 = 22 + 10x$
25. $x^2 + 8x + 25 = 0$
26. $x^2 + 6x + 13 = 0$
27. $3x^2 + 5x - 2 = 0$
28. $2x^2 - 5x - 3 = 0$

Use the quadratic formula to find exact solutions. Check your work using a grapher, if appropriate.

29. $x^2 - 2x = 15$
30. $x^2 + 4x = 5$
31. $5m^2 + 3m = 2$
32. $2y^2 - 3y - 2 = 0$
33. $3x^2 + 6 = 10x$
34. $3t^2 + 8t + 3 = 0$
35. $x^2 + x + 2 = 0$
36. $x^2 + 1 = x$
37. $5t^2 - 8t = 3$
38. $5x^2 + 2 = x$
39. $3x^2 + 4 = 5x$
40. $2t^2 - 5t = 1$

Solve graphically. Round solutions to three decimal places, where appropriate.

41. $x^2 - 8x + 12 = 0$
42. $5x^2 + 42x + 16 = 0$

43. $7x^2 - 43x + 6 = 0$
44. $10x^2 - 23x + 12 = 0$
45. $6x + 1 = 4x^2$
46. $3x^2 + 5x = 3$
47. $2x^2 - 4 = 5x$
48. $4x^2 - 2 = 3x$

Solve.

49. $\dfrac{1}{4} + \dfrac{1}{5} = \dfrac{1}{t}$

50. $\dfrac{1}{3} - \dfrac{5}{6} = \dfrac{1}{x}$

51. $\dfrac{x + 2}{4} - \dfrac{x - 1}{5} = 15$

52. $\dfrac{t + 1}{3} - \dfrac{t - 1}{2} = 1$

53. $\dfrac{1}{2} + \dfrac{2}{x} = \dfrac{1}{3} + \dfrac{3}{x}$

54. $\dfrac{1}{t} + \dfrac{1}{2t} + \dfrac{1}{3t} = 5$

55. $\dfrac{3x}{x + 2} + \dfrac{6}{x} = \dfrac{12}{x^2 + 2x}$

56. $\dfrac{5x}{x - 4} - \dfrac{20}{x} = \dfrac{80}{x^2 - 4x}$

57. $\dfrac{4}{x^2 - 1} - \dfrac{2}{x - 1} = \dfrac{3}{x + 1}$

58. $\dfrac{3y + 5}{y^2 + 5y} + \dfrac{y + 4}{y + 5} = \dfrac{y + 1}{y}$

59. $\dfrac{490}{x^2 - 49} = \dfrac{5x}{x - 7} - \dfrac{35}{x + 7}$

60. $\dfrac{3}{m + 2} + \dfrac{2}{m} = \dfrac{4m - 4}{m^2 - 4}$

61. $\dfrac{1}{x - 6} - \dfrac{1}{x} = \dfrac{6}{x^2 - 6x}$

62. $\dfrac{8}{x^2 - 4} = \dfrac{x}{x - 2} - \dfrac{2}{x + 2}$

63. $\dfrac{8}{x^2 - 2x + 4} = \dfrac{x}{x + 2} + \dfrac{24}{x^3 + 8}$

64. $\dfrac{18}{x^2 - 3x + 9} - \dfrac{x}{x + 3} = \dfrac{81}{x^3 + 27}$

65. $\sqrt{3x - 4} = 1$
66. $\sqrt[3]{2x + 1} = -5$
67. $\sqrt[4]{x^2 - 1} = 1$
68. $\sqrt{m + 1} - 5 = 8$
69. $\sqrt{y - 1} + 4 = 0$
70. $\sqrt[5]{3x + 4} = 2$
71. $\sqrt[3]{6x + 9} + 8 = 5$
72. $\sqrt{6x + 7} = x + 2$

73. $\sqrt{x-3} + \sqrt{x+2} = 5$

74. $\sqrt{x} - \sqrt{x-5} = 1$

75. $\sqrt{3x-5} + \sqrt{2x+3} + 1 = 0$

76. $\sqrt{2m-3} = \sqrt{m+7} - 2$

77. $\sqrt{x} - \sqrt{3x-3} = 1$

78. $\sqrt{2x+1} - \sqrt{x} = 1$

79. $\sqrt{2y-5} - \sqrt{y-3} = 1$

80. $\sqrt{4p+5} + \sqrt{p+5} = 3$

81. $x^{1/3} = -2$

82. $t^{1/5} = 2$

83. $t^{1/4} = 3$

84. $m^{1/2} = -7$

Solve.

85. $|x| = 7$

86. $|x| = 4.5$

87. $|x| = -10.7$

88. $|x| = -\frac{3}{5}$

89. $|x-1| = 4$

90. $|x-7| = 5$

91. $|3x| = 1$

92. $|5x| = 4$

93. $|x| = 0$

94. $|6x| = 0$

95. $|3x+2| = 1$

96. $|7x-4| = 8$

97. $\left|\frac{1}{2}x - 5\right| = 17$

98. $\left|\frac{1}{3}x - 4\right| = 13$

99. $|x-1| + 3 = 6$

100. $|x+2| - 5 = 9$

Solve graphically.

101. $x + y = 2,$
$3x + y = 0$

102. $x + y = 1,$
$3x + y = 7$

103. $\frac{1}{2}x + y = \frac{3}{2},$
$2x + 8y = 13$

104. $x + 2y = -\frac{2}{3},$
$\frac{1}{4}x - \frac{1}{2}y = 1$

105. $y + 1 = 2x,$
$y - 1 = 2x$

106. $2x - y = 1,$
$3y = 6x - 3$

Solve using the substitution method. Use a grapher to check your answer.

107. $x + y = 9,$
$2x - 3y = -2$

108. $3x - y = 5,$
$x + y = \frac{1}{2}$

109. $x - 2y = 7,$
$x = y + 4$

110. $x + 4y = 6,$
$x = -3y + 3$

111. $y = 2x - 6,$
$5x - 3y = 16$

112. $3x + 5y = 2,$
$2x - y = -3$

113. $x - 5y = 4,$
$y = 7 - 2x$

114. $5x + 3y = -1,$
$x + y = 1$

115. $2x - 3y = 5,$
$5x + 4y = 1$

116. $3x + 4y = 6,$
$2x + 3y = 5$

Solve using the elimination method. Also determine whether each system is consistent or inconsistent and dependent or independent. Use a grapher to check your answer.

117. $x + 2y = 7,$
$x - 2y = -5$

118. $3x + 4y = -2,$
$-3x - 5y = 1$

119. $x - 3y = 2,$
$6x + 5y = -34$

120. $x + 3y = 0,$
$20x - 15y = 75$

121. $0.3x - 0.2y = -0.9,$
$0.2x - 0.3y = -0.6$

122. $0.2x - 0.3y = 0.3,$
$0.4x + 0.6y = -0.2$

123. $3x - 12y = 6,$
$2x - 8y = 4$

124. $2x + 6y = 7,$
$3x + 9y = 10$

125. $\frac{1}{5}x + \frac{1}{2}y = 6,$
$\frac{3}{5}x - \frac{1}{2}y = 2$

126. $\frac{2}{3}x + \frac{3}{5}y = -17,$
$\frac{1}{2}x - \frac{1}{3}y = -1$

127. $2x = 5 - 3y,$
$4x = 11 - 7y$

128. $7(x - y) = 14,$
$2x = y + 5$

Solve.

129. $A = \frac{1}{2}bh,$ for b
(Area of a triangle)

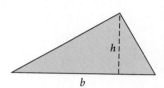

130. $A = \pi r^2,$ for π
(Area of a circle)

131. $P = 2l + 2w,$ for w
(Perimeter of a rectangle)

132. $A = P + Prt,$ for P
(Simple interest)

133. $A = \frac{1}{2}h(b_1 + b_2),$ for h
(Area of a trapezoid)

134. $A = \frac{1}{2}h(b_1 + b_2),$ for b_2

135. $V = \frac{4}{3}\pi r^3,$ for π
(Volume of a sphere)

136. $V = \frac{4}{3}\pi r^3,$ for r^3

137. $F = \frac{9}{5}C + 32,$ for C
(Temperature conversion)

138. $Ax + By = C,$ for y
(Standard linear equation)

139. $\dfrac{P_1 V_1}{T_1} = \dfrac{P_2 V_2}{T_2}$, for T_1
(A chemistry formula for gases)

140. $\dfrac{1}{F} = \dfrac{1}{m} + \dfrac{1}{p}$, for F
(A formula from optics)

141. $\dfrac{1}{R} = \dfrac{1}{R_1} + \dfrac{1}{R_2}$, for R_2
(Resistance)

142. $A = P(1 + i)^2$, for i
(Compound interest)

143. $\dfrac{1}{F} = \dfrac{1}{m} + \dfrac{1}{p}$, for p
(A formula from optics)

Discussion and Writing

144. Use a graphical argument to explain why the equation $x^2 - 6x + 9 = 1 - x^4$ has no solution.

145. Explain why it is necessary to check the possible solutions of a rational equation.

146. Explain in your own words why it is necessary to check the possible solutions when the principle of powers is used to solve an equation.

Synthesis

Solve.

147. $2x - \{x - [3x - (6x + 5)]\} = 4x - 1$

148. $(x - 2)^3 = x^3 - 2$

149. $(x + 1)^3 = (x - 1)^3 + 26$

150. $\dfrac{x + 3}{x + 2} - \dfrac{x + 4}{x + 3} = \dfrac{x + 5}{x + 4} - \dfrac{x + 6}{x + 5}$

151. $(x - 3)^{2/3} = 2$

152. $\sqrt{15 + \sqrt{2x + 80}} = 5$

153. $\sqrt{x + 5} + 1 = \dfrac{6}{\sqrt{x + 5}}$

154. $x^{2/3} = x + 1$

Distance, Midpoints, and Circles
B

- Find the distance between two points in the plane and find the midpoint of a segment.
- Find an equation of a circle with a given center and radius, and given an equation of a circle, find the center and the radius.
- Graph equations of circles.

In carpentry, surveying, engineering, and other fields, it is often necessary to determine distances and midpoints and to produce accurately drawn circles.

The Distance Formula

Suppose that a conservationist needs to determine the distance across an irregularly shaped pond. One way in which the conservationist might proceed is to measure two legs of a right triangle that is situated as shown below. The Pythagorean theorem, $a^2 + b^2 = c^2$, can then be used to find the length of the hypotenuse, which is the distance across the pond.

A similar strategy is used to find the distance between two points in a plane. For two points (x_1, y_1) and (x_2, y_2), we can draw a right triangle in which the legs have lengths $|x_2 - x_1|$ and $|y_2 - y_1|$.

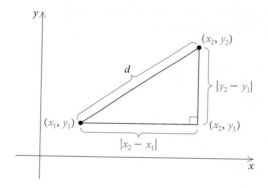

Using the Pythagorean theorem, we have

$$d^2 = |x_2 - x_1|^2 + |y_2 - y_1|^2.$$

Because we are squaring, parentheses can replace the absolute-value symbols:

$$d^2 = (x_2 - x_1)^2 + (y_2 - y_1)^2.$$

Taking the principal square root, we obtain the distance formula.

The Distance Formula

The **distance** d between any two points (x_1, y_1) and (x_2, y_2) is given by

$$d = \sqrt{(x_2 - x_1)^2 + (y_2 - y_1)^2}.$$

The subtraction of the x-coordinates can be done in any order, as can the subtraction of the y-coordinates. Although we derived the distance formula by considering two points not on a horizontal or a vertical line, the distance formula holds for *any* two points.

EXAMPLE 1 The point $(-2, 5)$ is on a circle that has $(3, -1)$ as its center. Find the length of the radius of the circle.

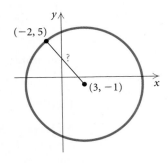

Solution Since the length of the radius is the distance from the center to a point on the circle, we substitute into the distance formula:

$$r = \sqrt{[3 - (-2)]^2 + [-1 - 5]^2}$$ Either point can serve as (x_1, y_1).

$$= \sqrt{5^2 + (-6)^2} = \sqrt{61} \approx 7.8.$$ Rounded to the nearest thousandth

The radius of the circle is approximately 7.8. ▬

Midpoints of Segments

The distance formula can be used to develop a way of determining the *midpoint* of a segment when the endpoints are known. We state the formula and leave its proof to the exercises.

The Midpoint Formula

If the endpoints of a segment are (x_1, y_1) and (x_2, y_2), then the coordinates of the **midpoint** are

$$\left(\frac{x_1 + x_2}{2}, \frac{y_1 + y_2}{2} \right).$$

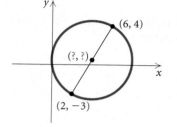

Note that we obtain the coordinates of the midpoint by averaging the coordinates of the endpoints. This is an easy way to remember the midpoint formula.

EXAMPLE 2 The diameter of a circle connects two points $(2, -3)$ and $(6, 4)$ on the circle. Find the coordinates of the center of the circle.

Solution Using the midpoint formula, we obtain

$$\left(\frac{2 + 6}{2}, \frac{-3 + 4}{2}\right), \quad \text{or} \quad \left(\frac{8}{2}, \frac{1}{2}\right), \quad \text{or} \quad \left(4, \frac{1}{2}\right).$$

The coordinates of the center are $\left(4, \frac{1}{2}\right)$.

Circles

A **circle** is the set of all points in a plane that are a fixed distance r from a *center* (h, k). Thus if a point (x, y) is to be r units from the center, we must have

$$r = \sqrt{(x - h)^2 + (y - k)^2}. \qquad \text{Using the distance formula}$$

Squaring both sides gives an equation of a circle. The distance r is the length of a *radius* of the circle.

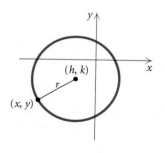

The Equation of a Circle

The equation, in standard form, of a circle with center (h, k) and radius r is

$$(x - h)^2 + (y - k)^2 = r^2.$$

EXAMPLE 3 Find an equation of the circle having radius 5 and center $(3, -7)$.

Solution Using the standard form, we have

$$[x - 3]^2 + [y - (-7)]^2 = 5^2 \qquad \text{Substituting}$$
$$(x - 3)^2 + (y + 7)^2 = 25.$$

EXAMPLE 4 Graph the circle $(x + 5)^2 + (y - 2)^2 = 16$.

Solution We write the equation in standard form to determine the center and the radius:

$$[x - (-5)]^2 + [y - 2]^2 = 4^2.$$

The center is $(-5, 2)$ and the radius is 4. We locate the center and draw the circle using a compass.

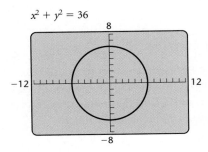

FIGURE 1

FIGURE 2

Circles can also be graphed using a grapher. We show one method of doing so here. Another method was discussed in Section 5.2.

When we graph a circle, we select a viewing window in which the distance between units is visually the same on both axes. This procedure is called **squaring the viewing window.** We do this so that the graph will not be distorted. A graph of the circle $x^2 + y^2 = 36$ in a nonsquared window is shown in Fig. 1.

On many graphers, the distance between units on the y-axis is about $\frac{2}{3}$ the distance between units on the x-axis. When we choose a window in which the length of the y-axis is $\frac{2}{3}$ the length of the x-axis, the window will be squared. The windows with dimensions $[-6, 6, -4, 4]$, $[-9, 9, -6, 6]$, and $[-12, 12, -8, 8]$ are examples of squared windows. A graph of the circle $x^2 + y^2 = 36$ in a squared window is shown in Fig. 2.

Many graphers have an option on the ZOOM menu that squares the window automatically.

EXAMPLE 5 Graph the circle $(x - 2)^2 + (y + 1)^2 = 16$.

Solution The circle $(x - 2)^2 + (y + 1)^2 = 16$ has center $(2, -1)$ and radius 4, so the viewing window $[-9, 9, -6, 6]$ is a good choice in this case. After entering the window dimensions, we select the CIRCLE feature from the DRAW menu and enter the coordinates of the center and the radius to draw the graph of the circle.

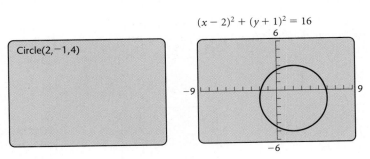

Exercise Set B

Find the distance between the pair of points. Give an exact answer and, where appropriate, an approximation to three decimal places.

1. $(4, 6)$ and $(5, 9)$

2. $(-3, 7)$ and $(2, 11)$

3. $(6, -1)$ and $(9, 5)$

4. $(-4, -7)$ and $(-1, 3)$

5. $(\sqrt{3}, -\sqrt{5})$ and $(-\sqrt{6}, 0)$

6. $(-\sqrt{2}, 1)$ and $(0, \sqrt{7})$

7. The points $(-3, -1)$ and $(9, 4)$ are the endpoints of the diameter of a circle. Find the length of the radius of the circle.

8. The point $(0, 1)$ is on a circle that has center $(-3, 5)$. Find the length of the diameter of the circle.

Use the distance formula and the Pythagorean theorem to determine whether the set of points could be vertices of a right triangle.

9. $(-4, 5)$, $(6, 1)$, and $(-8, -5)$

10. $(-3, 1)$, $(2, -1)$, and $(6, 9)$

11. The points $(-3, 4)$, $(0, 5)$, and $(3, -4)$ are all on the circle $x^2 + y^2 = 25$. Show that these three points are vertices of a right triangle.

12. The points $(-3, 4)$, $(2, -1)$, $(5, 2)$, and $(0, 7)$ are vertices of a quadrilateral. Show that the quadrilateral is a rectangle. (*Hint:* Show that the quadrilateral's opposite sides are the same length and that the two diagonals are the same length.)

13.–18. Find the midpoint of each segment having the endpoints given in Exercises 1–6, respectively.

19. Graph the rectangle described in Exercise 12. Then determine the coordinates of the midpoint for each of the four sides. Are the midpoints vertices of a rectangle?

20. Graph the square with vertices $(-5, -1)$, $(7, -6)$, $(12, 6)$, and $(0, 11)$. Then determine the midpoint for each of the four sides. Are the midpoints vertices of a square?

21. The points $(\sqrt{7}, -4)$ and $(\sqrt{2}, 3)$ are endpoints of the diameter of a circle. Determine the center of the circle.

22. The points $(-3, \sqrt{5})$ and $(1, \sqrt{2})$ are endpoints of the diagonal of a square. Determine the center of the square.

In Exercises 23 and 24, how would you change the window so that the circle is not distorted? Answers may vary.

23.

$(x + 3)^2 + (y - 2)^2 = 36$

24.

$(x - 4)^2 + (y + 5)^2 = 49$

Find an equation for a circle satisfying the given conditions.

25. Center $(2, 3)$, radius of length $\frac{5}{3}$

26. Center $(4, 5)$, diameter of length 8.2

27. Center $(-1, 4)$, passes through $(3, 7)$

28. Center $(6, -5)$, passes through $(1, 7)$

29. The points $(7, 13)$ and $(-3, -11)$ are at either end of a diameter.

30. The points $(-9, 4)$, $(-2, 5)$, $(-8, -3)$, and $(-1, -2)$ are vertices of an inscribed square.

31. Center $(-2, 3)$, tangent (touching at one point) to the y-axis

32. Center $(4, -5)$, tangent to the x-axis

Find the center and the radius of the circle. Then graph the circle using a square viewing window.

33. $x^2 + y^2 = 4$

34. $x^2 + y^2 = 81$

35. $x^2 + (y - 3)^2 = 16$

36. $(x + 2)^2 + y^2 = 100$

37. $(x - 1)^2 + (y - 5)^2 = 36$

38. $(x - 7)^2 + (y + 2)^2 = 25$

39. $(x + 4)^2 + (y + 5)^2 = 9$

40. $(x + 1)^2 + (y - 2)^2 = 64$

Discussion and Writing

41. Explain how the Pythagorean theorem is used to develop the equation of a circle in standard form.

42. Explain how you could find the coordinates of a point $\frac{7}{8}$ of the way from point A to point B.

Synthesis

Find the distance between the pair of points and find the midpoint of the segment having the given points as endpoints.

43. $(a, \sqrt{a})$ and $(a + h, \sqrt{a + h})$

44. $\left(a, \dfrac{1}{a}\right)$ and $\left(a + h, \dfrac{1}{a + h}\right)$

Find an equation of a circle satisfying the given conditions.

45. Center $(2, -7)$ with an area of 36π square units

46. Center $(-5, 8)$ with a circumference of 10π units

47. *Swimming Pool.* A swimming pool is being constructed in the corner of a yard, as shown. Before installation, the contractor needs to know measurements a_1 and a_2. Find them.

48. *An Arch of a Circle in Carpentry.* Ace Carpentry needs to cut an arch for the top of an entranceway. The arch needs to be 8 ft wide and 2 ft high. To draw the arch, the carpenters will use a stretched string with chalk attached at an end as a compass.

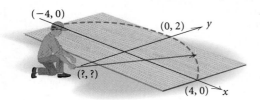

$(-4, 0)$
$(0, 2)$ y
$(?, ?)$
$(4, 0)$ x

a) Using a coordinate system, locate the center of the circle.
b) What radius should the carpenters use to draw the arch?

*Determine whether each of the following points lies on the **unit circle,** $x^2 + y^2 = 1$.*

49. $\left(\dfrac{\sqrt{3}}{2}, -\dfrac{1}{2} \right)$

50. $(0, -1)$

51. $\left(-\dfrac{\sqrt{2}}{2}, \dfrac{\sqrt{2}}{2} \right)$

52. $\left(\dfrac{1}{2}, -\dfrac{\sqrt{3}}{2} \right)$

53. Find the point on the y-axis that is equidistant from the points $(-2, 0)$ and $(4, 6)$.

54. Consider any right triangle with base b and height h, situated as shown. Show that the midpoint of the hypotenuse P is equidistant from the three vertices of the triangle.

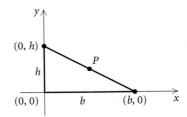

y
$(0, h)$
P
h
$(0, 0)$ b $(b, 0)$ x

55. Prove the midpoint formula by showing that:

a) $\left(\dfrac{x_1 + x_2}{2}, \dfrac{y_1 + y_2}{2} \right)$ is equidistant from the points (x_1, y_1) and (x_2, y_2); and

b) the distance from (x_1, y_1) to the midpoint plus the distance from (x_2, y_2) to the midpoint equals the distance from (x_1, y_1) to (x_2, y_2).

Data Analysis, Curve Fitting, and Regression
C

- *Analyze a set of data to determine whether it can be modeled by a linear, quadratic, cubic, or quartic function.*
- *Use regression to fit a curve to a set of data; then use this model to make predictions.*

Mathematical Models

When a real-world problem can be described in mathematical language, we have a **mathematical model.** For example, the natural numbers constitute a mathematical model for situations in which counting is essential. Situations in which algebra can be brought to bear often require the use of functions as models.

Mathematical models are abstracted from real-world situations. Procedures within the mathematical model give results that allow one to predict what will happen in that real-world situation. If the predictions are inaccurate or the results of experimentation do not conform to the model, the model needs to be changed or discarded.

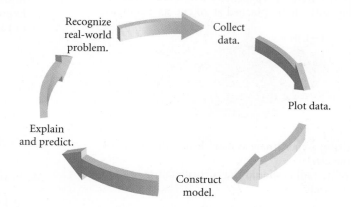

Mathematical modeling can be an ongoing process. For example, finding a mathematical model that will enable an accurate prediction of population growth is not a simple problem. Any population model that one might devise would need to be reshaped as further information is acquired.

Curve Fitting

We will develop and use many kinds of mathematical models in this text. In this chapter, we have seen some types of functions that can be used as models. Let's look more closely at three of them. The quadratic and cubic functions are nonlinear.

Linear function: Quadratic function: Cubic function:
$y = mx + b$ $y = ax^2 + bx + c, a > 0$ $y = ax^3 + bx^2 + cx + d, a > 0$

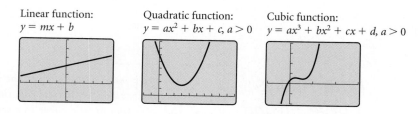

In general, we try to find a function that fits, as well as possible, observations (data), theoretical reasoning, and common sense. We call this **curve fitting;** it is one aspect of mathematical modeling.

Let's first look at some data and related graphs or scatterplots and determine whether any of the above functions, linear or nonlinear, seem to fit either set of data.

Number of Apartment Households in the United States

YEAR, x	NUMBER OF APARTMENT HOUSEHOLDS, y (IN MILLIONS)	SCATTERPLOT	
1970, 0	8.5		It appears that the data points can be represented or fitted by a straight line.
1975, 5	9.9		
1980, 10	10.8		
1985, 15	12.9		The graph is linear.
1990, 20	14.2		
1997, 27	14.5		

Sources: U.S. Bureau of the Census; National Multi-Housing Council

Number of Cellular Phone Subscribers

YEAR, x	NUMBER OF CELLULAR PHONE SUBSCRIBERS, y (IN MILLIONS)	SCATTERPLOT	
1985, 0	0.3		It appears that the data points cannot be represented by a straight line.
1986, 1	0.7		
1987, 2	1.2		The graph is nonlinear.
1988, 3	2.1		
1989, 4	3.5		
1990, 5	5.3		
1991, 6	7.6		
1992, 7	11.0		
1993, 8	16.0		
1994, 9	24.1		
1995, 10	33.8		
1996, 11	44.0		
1997, 12	55.3		

Source: Cellular Telecommunications Industry Association

Looking at the scatterplots, we see that the apartment households data seem to be rising in a manner to suggest that a linear function might fit, although a "perfect" straight line cannot be drawn through the data points. A linear function does not seem to fit the cellular phone data.

The Regression Line

We now consider **linear regression,** a procedure that can be used to model a set of data using a linear function. Although discussion leading to a complete understanding of this method belongs in a statistics course, we present the procedure here because we can carry it out easily using technology. The grapher gives us the powerful capability to find linear models and to make predictions using them.

EXAMPLE 1   *Number of Apartment Households in the United States.* Consider the data just presented on the number of apartment households in the United States.

a) Fit a regression line to the data using the linear regression feature on a grapher.

b) Use the linear model to predict the number of apartment households in 2003.

Solution

a) First, we enter the data in lists on a grapher. We enter the values of the independent variable x in list L1 and the corresponding values of the dependent variable y in L2 (see Fig. 1). The grapher can then create a scatterplot of the data, as shown in Fig. 2.

L1	L2	L3
0	8.5	------
5	9.9	
10	10.8	
15	12.9	
20	14.2	
27	14.5	

L2(7) =

FIGURE 1

FIGURE 2

Consider the data points (0, 8.5), (5, 9.9), (10, 10.8), (15, 12.9), (20, 14.2), and (27, 14.5) on the graph. We want to fit a **regression line,**

$$y = mx + b,$$

to these data points. We use the LINEAR REGRESSION feature to find the regression line.

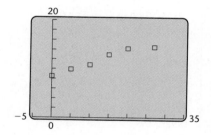

EDIT **CALC** TESTS
1:1–Var Stats
2:2–Var Stats
3:Med–Med
4:LinReg(ax+b)
5:QuadReg
6:CubicReg
7↓QuartReg

LinReg
y=ax+b
a=.2402037351
b=8.717385399

The result is

$$y = 0.2402037351x + 8.717385399. \qquad \text{Regression line}$$

We can then graph the regression line on the same graph as the scatterplot.

b) To predict the number of apartment households in 2003, we evaluate the function found in part (a) for $x = 33$. (In 1970, $x = 0$; in 1990, $x = 20$. Thus in 2003, $x = 33$.)

We see from the grapher that the number of apartment households in 2003 is predicted to be about 16.6 million. ▬

The Correlation Coefficient

On some graphers with DIAGNOSTICS turned on, a constant r between -1 and 1, called the **coefficient of linear correlation,** appears with the equation of the regression line. Though we cannot develop a formula for calculating r in this text, keep in mind that it is used to describe the strength of the linear relationship between x and y. The closer $|r|$ is to 1, the better the correlation. A positive value of r also indicates that the regression line has a positive slope and a negative value of r indicates that the regression line has a negative slope. For the apartment household data just discussed, $r = 0.9755068364$, which indicates a very good linear correlation. The following scatterplots summarize the interpretation of a correlation coefficient.

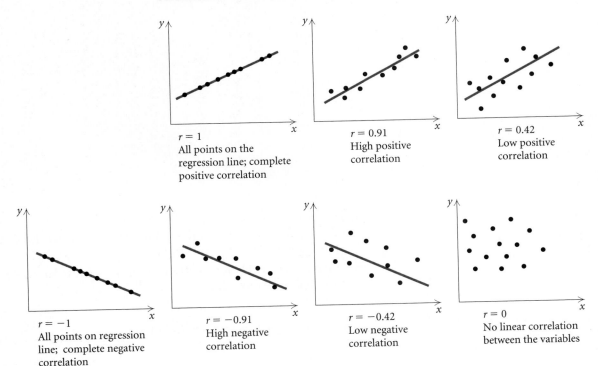

Keep in mind that a high linear correlation coefficient does not necessarily indicate a "cause-and-effect" connection between the variables. We might be able to calculate a high positive correlation between stock prices and rainfall in India, but before we place our life savings in the market, further analysis and common sense should be applied!

EXAMPLE 2   *Garbage Production.* America creates more garbage than any other nation. According to Denis Hayes, president of Seattle's nonprofit Bullitt Foundation and one of the founders of Earth Day in 1970, "we need to be an Heirloom Society instead of a Throw-Away Society." The Environmental Protection Agency estimates that, on average, we each produce 4.4 lb of garbage a day (*Source*: "Take Out the Trash, and Put It—Where?", by Bernard Bavzer, *Parade Magazine*, June 13, 1999). The following graph illustrates the annual production of municipal solid waste in the United States for certain years.

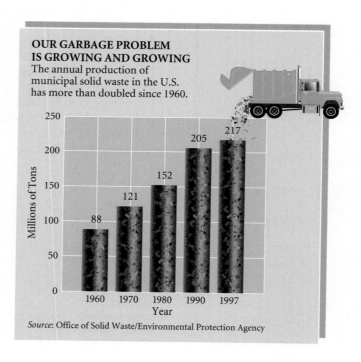

OUR GARBAGE PROBLEM IS GROWING AND GROWING
The annual production of municipal solid waste in the U.S. has more than doubled since 1960.

Source: Office of Solid Waste/Environmental Protection Agency

a) Fit a regression line to the data and use it to predict the amount of municipal solid waste in the United States in 2005. Enter the data as (1960, 88), (1970, 121), and so on.

b) What is the correlation coefficient for the regression line? How close a fit is the regression line?

Solution

a) Using a grapher, we get the following.

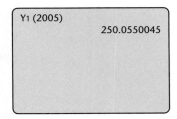

```
LinReg
  y=ax+b
  a=3.650586114
  b=−7069.370153
  r²=.9865098876
  r=.9932320412
```

```
Y₁ (2005)
                    250.0550045
```

The regression line is

$$y = 3.650586114x - 7069.370153.$$

To predict the amount of municipal solid waste in 2005, we find the value of the regression function when $x = 2005$. Thus the amount of solid waste will be about 250 million tons in 2005.

b) The correlation coefficient for the regression line is about 0.9932. Since the closer $|r|$ is to 1 the better the correlation, this equation fits the data closely.

As we have moved through this text, we have developed a "library" of functions that can serve as models for many applications. Let's now add quadratic, cubic, and quartic functions to this library.

Linear Function:
$y_1 = mx + b$

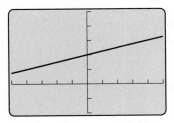

Quadratic Function:
$y_2 = ax^2 + bx + c, a > 0$

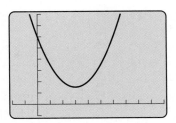

Quadratic Function:
$y_3 = ax^2 + bx + c, a < 0$

Cubic Function:
$y_4 = ax^3 + bx^2 + cx + d$

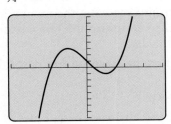

Quartic Function:
$y_5 = ax^4 + bx^3 + cx^2 + dx + e$

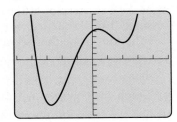

Given a set of data, how can we determine if it can be modeled with one of these functions? One way is to graph the data. For example, consider the scatterplots shown below.

FIGURE 1

FIGURE 2

FIGURE 3

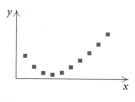

FIGURE 4

The data points in Fig. 1 rise and then fall, suggesting that a quadratic function $f(x) = ax^2 + bx + c$, $a < 0$, might fit the data. In Fig. 2, the data points fall at a steady rate, so a linear function $f(x) = mx + b$ could be used. The data points in Fig. 3 rise, then fall, and then rise again, so a cubic function could be used. And, finally, the data points in Fig. 4 fall and then rise, suggesting that a quadratic function $f(x) = ax^2 + bx + c$, $a > 0$, might fit the data.

QUADRATIC MODELS

EXAMPLE 1 *Leisure Time.* The following table shows the median number of hours of leisure time that Americans had each week in various years.

YEAR, x	MEDIAN NUMBER OF LEISURE HOURS PER WEEK
1973, 0	26.2
1980, 7	19.2
1987, 14	16.6
1993, 20	18.8
1997, 24	19.5

Source: Louis Harris and Associates

a) Make a scatterplot of the data, letting x represent the number of years since 1973, and determine whether a linear function, a quadratic function, or neither seems to fit the data.

b) Use a grapher to fit the type of function determined in part (a) to the data.

c) Graph the equation with the scatterplot.

d) Use the function found in part (b) to estimate the number of leisure hours per week in 1978, 1990, and 2005.

Solution

a) We enter the data in a grapher, letting x represent the number of years since 1973. That is, 1973 corresponds to $x = 0$. Then, since $1980 - 1973 = 7$, we know that 1980 corresponds to $x = 7$. Similarly, 1987 corresponds to $x = 14$, 1993 to $x = 20$, and 1997 to $x = 24$. Some graphers have a ZOOM option that automatically selects a window that includes all the data points. We use this option to obtain the window and the scatterplot shown below.

```
WINDOW
 Xmin=-2.4
 Xmax=26.4
 Xscl=1
 Ymin=14.968
 Ymax=27.832
 Yscl=1
 Xres=1
```

The data points fall and then rise, so a quadratic function $f(x) = ax^2 + bx + c$, $a > 0$, could be used to model the data.

b) Using the quadratic regression feature on a grapher, we find a quadratic function:

$$f(x) = 0.0402302792x^2 - 1.214631021x + 26.02596908.$$

```
QuadReg
 y=ax²+bx+c
 a=.0402302792
 b=-1.214631021
 c=26.02596908
 R²=.9720817074
```

c) We can store this equation on the equation-editor screen at the same time that it is found or immediately after it is found. Then we can graph the regression equation with the scatterplot.

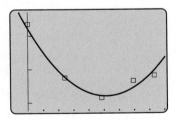

d) We use a table set in ASK mode to find the desired function values. Note that since $1978 - 1973 = 5$, the year 1978 corresponds to $x = 5$. Similarly, 1990 corresponds to $x = 17$ and 2005 to $x = 32$.

X	Y₁
5	20.959
17	17.004
32	28.354

X =

We estimate that Americans had about 21.0 leisure hours per week in 1978 and about 17.0 hours in 1990, and we predict that they will have about 28.4 leisure hours per week in 2005. ▬

CUBIC AND QUARTIC MODELS We continue our discussion of curve fitting with examples of cubic and quartic regression.

EXAMPLE 5   *Declining Number of Farms in the United States.* **Today** U.S. farm acreage is about the same as it was in the early part of the twentieth century, but the number of farms has shrunk.

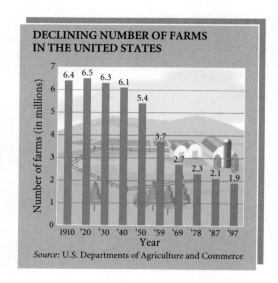

DECLINING NUMBER OF FARMS IN THE UNITED STATES

Source: U.S. Departments of Agriculture and Commerce

Looking at the graph above, we note that the data could be modeled with a cubic or a quartic function.

a) Model the data with both cubic and quartic functions. Let the first coordinate of each data point be the number of years after 1900. That is, enter the data as (10, 6.4), (20, 6.5), and so on. Then using R^2, the **coefficient of determination,** decide which function is the better fit. The R^2-value gives an indication of how well the function fits the data. The closer R^2 is to 1, the better the fit.

b) Graph the function with the scatterplot of the data.

c) Use the answer to part (a) to estimate the number of farms in 1900, 1975, and 2003.

Solution

a) Using the REGRESSION feature with DIAGNOSTIC turned on, we get the following.

```
CubicReg
 y=ax³+bx²+cx+d
 a=2.9566515E−5
 b=−.0049844076
 c=.1742288914
 d=4.983981559
 R²=.9828373098
```

```
QuarticReg
 y=ax⁴+bx³+...+e
 a=1.5229297E−7
 b=−3.092184E−6
 c=−.0026500648
 d=.1117312855
↓e=5.463775105
```

```
QuarticReg
 y=ax⁴+bx³+...+e
↑b=−3.092184E−6
 c=−.0026500648
 d=.1117312855
 e=5.463775105
 R²=.9836750763
```

Since the R^2-value for the quartic function is closer to 1 than that for the cubic function, the quartic function is the better fit. Note that a and b are given in scientific notation on the grapher, but we convert to decimal notation when we write the function.

$$f(x) = 0.00000015229297x^4 - 0.000003092184x^3$$
$$- 0.0026500648x^2 + 0.1117312855x + 5.463775105$$

b) The graph is shown below.

c) We evaluate the function found in part (a).

```
Y₁(0)
                    5.463775105
Y₁(75)
                    2.451136695
Y₁(103)
                    2.61935514
```

Thus we estimate that there were about 5.5 million farms in 1900 and about 2.5 million in 1975. Looking at the bar graph shown on the preceding page, we see that these estimates appear to be fairly accurate.

Since the number of farms will probably continue to decrease, the estimate of 2.6 million for 2003 is probably not a good prediction. Both the cubic and the quartic models have high correlation coefficients over the domain of the data, but this correlation does not reflect the degree of accuracy for extended values. It is always important when using regression to evaluate predictions with common sense and knowledge of current trends. ■

As useful as regression models might be in certain situations, we must keep in mind that almost every model has its limitations.

Exercise Set C

In Exercises 1–4, determine whether the graph might be modeled by a linear function.

1.

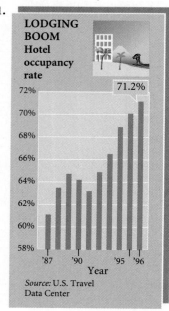

LODGING BOOM
Hotel occupancy rate

71.2%

Source: U.S. Travel Data Center

2.

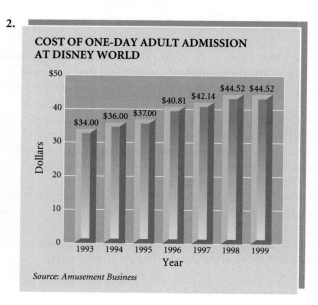

COST OF ONE-DAY ADULT ADMISSION AT DISNEY WORLD

$34.00 $36.00 $37.00 $40.81 $42.14 $44.52 $44.52

Dollars

1993 1994 1995 1996 1997 1998 1999
Year

Source: Amusement Business

3.

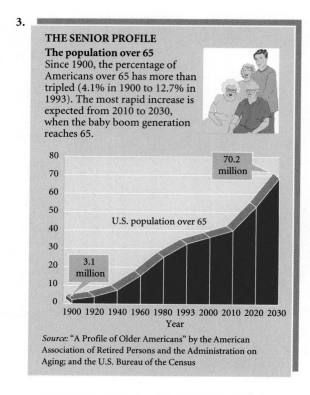

THE SENIOR PROFILE
The population over 65
Since 1900, the percentage of Americans over 65 has more than tripled (4.1% in 1900 to 12.7% in 1993). The most rapid increase is expected from 2010 to 2030, when the baby boom generation reaches 65.

70.2 million

U.S. population over 65

3.1 million

1900 1920 1940 1960 1980 1993 2000 2010 2020 2030
Year

Source: "A Profile of Older Americans" by the American Association of Retired Persons and the Administration on Aging; and the U.S. Bureau of the Census

4.

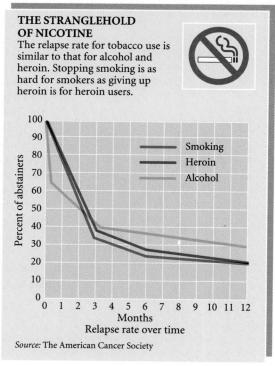

THE STRANGLEHOLD OF NICOTINE
The relapse rate for tobacco use is similar to that for alcohol and heroin. Stopping smoking is as hard for smokers as giving up heroin is for heroin users.

Percent of abstainers

Smoking
Heroin
Alcohol

0 1 2 3 4 5 6 7 8 9 10 11 12
Months
Relapse rate over time

Source: The American Cancer Society

In Exercises 5–8, determine whether the scatterplot might be fit by a linear model.

5.

6.

7.

8.

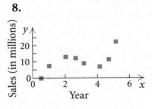

Exercises 9–14 involve using a grapher and linear regression to model data.

9.

Number of Shopping Centers

YEAR, x	NUMBER OF SHOPPING CENTERS, y	SCATTERPLOT
1980, 0	22,050	
1984, 4	25,508	
1988, 8	32,563	
1991, 11	37,975	
1992, 12	38,966	
1993, 13	39,633	
1994, 14	40,274	
1995, 15	41,145	
1996, 16	42,045	
1997, 17	42,874	

Source: International Council of Shopping Centers

a) Use regression to model the data with a linear function and use it to predict the number of shopping centers in 2005.
b) What is the correlation coefficient for the regression line? How close a fit is the regression line?

10.
Cases of Skin Cancers in the United States

YEAR, x	NUMBER OF CASES, y (IN THOUSANDS)	SCATTERPLOT
1988, 0	5800	
1989, 1	6000	
1990, 2	6300	
1991, 3	6500	
1992, 4	6700	

Source: American Cancer Society

a) Use regression to model the data with a linear function and use it to estimate the number of cases of skin cancer in the United States in 1998 and 2000. Also use the function to predict the number of cases in 2010.
b) What is the correlation coefficient for the regression line? How close a fit is the regression line?

11. *The Cost of Tuition at State Universities*

COLLEGE YEAR, x	ESTIMATED TUITION, y
1997–1998, 0	$ 9,838
1998–1999, 1	10,424
1999–2000, 2	11,054
2000–2001, 3	11,717
2001–2002, 4	12,420

Source: College Board, Senate Labor Committee

a) Use regression to model the data with a linear function and use it to predict the cost of college tuition in 2004–2005, 2006–2007, and 2010–2011.
b) What is the correlation coefficient for the regression line? How close a fit is the regression line?
c) A college president asserts that in the college year 2006–2007, the cost of tuition will be $16,619. How well does your estimate compare?

12. *Book Buying Growth in the United States*

YEAR, x	BOOK SALES, y (IN BILLIONS)
1992, 0	$21
1993, 1	23
1994, 2	24
1995, 3	25
1996, 4	26

Source: Book Industry Trends 1995

a) Use regression to model the data with a linear function and use it to predict book sales in 2000, 2005, and 2010.

b) What is the correlation coefficient for the regression line? How close a fit is the regression line?

c) A bookstore owner asserts that in 2000, book sales will be $45 billion. How well does your estimate stack up to this?

13. *Maximum Heart Rate.*   A person who is exercising should not exceed his or her maximum heart rate, which is determined on the basis of that person's sex, age, and resting heart rate. The following table relates resting heart rate and maximum heart rate for a 20-yr-old man.

RESTING HEART RATE, H (IN BEATS PER MINUTE)	MAXIMUM HEART RATE, M (IN BEATS PER MINUTE)
50	166
60	168
70	170
80	172

Source: American Heart Association

a) Use regression to model the data with a linear function.

b) Predict a maximum heart rate if the resting heart rate is 40, 65, 76, and 84.

c) What is the correlation coefficient? How confident are you about using the regression line as a predictor?

14. *Study Time versus Grades.*   A math instructor asked her students to keep track of how much time each spent studying a chapter on functions in her algebra–trigonometry course. She collected the information together with test scores from that chapter's test. The data are listed in the table at the top of the next column.

STUDY TIME, x (IN HOURS)	TEST GRADE, y (IN PERCENT)
23	81
15	85
17	80
9	75
21	86
13	80
16	85
11	93

a) Use regression to model the data with a linear function.

b) Predict a student's score if he or she studies 24 hr, 6 hr, and 18 hr.

c) What is the correlation coefficient? How confident are you about using the regression line as a predictor?

For the scatterplots and graphs in Exercises 15–20, determine which, if any, of the following functions might be used as a model for the data.

a) *Linear,* $f(x) = mx + b$
b) *Quadratic,* $f(x) = ax^2 + bx + c,\ a > 0$
c) *Quadratic,* $f(x) = ax^2 + bx + c,\ a < 0$
d) *Polynomial, not quadratic or linear*

15.

16.

17.

18.

19.

20.

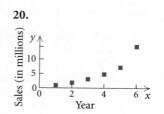

21. *Morning Neswpapers.* The number of morning newspapers in the United States in various years is shown in the following table.

YEAR	NUMBER OF MORNING NEWSPAPERS
1920	437
1940	380
1960	312
1980	387
1990	559
1997	705

Source: *Editor & Publisher*

a) Make a scatterplot of the data, letting x represent the number of years since 1920.
b) Determine whether a linear function, a quadratic function, or neither seems to fit the data.
c) Use a grapher to fit the type of function determined in part (b) to the data.
d) Graph the equation with the scatterplot.
e) Use the function found in part (c) to predict the number of morning newspapers in 2004 and 2007.

22. *Home Education.* The number of children who were home-educated in the United States in various years is shown in the following table.

YEAR	NUMBER OF HOME-EDUCATED CHILDREN (IN THOUSANDS)
1983	92.5
1988	225
1992	703
1995	1060
1997	1347

Source: National Home Education Research Institute

a) Make a scatterplot of the data, letting x represent the number of years since 1983.
b) Determine whether a linear function, a quadratic function, or neither seems to fit the data.
c) Use a grapher to fit the type of function determined in part (b) to the data.
d) Graph the equation with the scatterplot.
e) Use the function found in part (c) to predict the number of children who will be home-educated in 2005 and 2010.

23. *Medicare Assets.* The end-of-year assets of the U.S. Hospital Insurance program, also known as Medicare Part A, for several recent years are shown in the following table.

YEAR	ASSETS (IN MILLIONS)
1990	$ 95,631
1993	126,131
1995	129,520
1998	110,584
2000	100,852

Source: U.S. Health Care Financing Administration

a) Make a scatterplot of the data, letting x represent the number of years since 1990.
b) Determine whether a linear function, a quadratic function, or neither seems to fit the data.
c) Use a grapher to fit the type of function determined in part (b) to the data.
d) Graph the equation with the scatterplot.
e) Use the function found in part (c) to predict the assets of the Hospital Insurance program in 2003 and 2005.

24. *Social Security Assets.* The assets of the U.S. Social Security Trust Fund for several recent years are shown in the following table.

YEAR	ASSETS (IN BILLIONS)
1995	$339.8
1996	353.6
1997	369.1
1998	382.9
1999	396.3
2000	413.4

Source: Social Security Administration

a) Make a scatterplot of the data, letting x represent the number of years since 1995.
b) Determine whether a linear function, a quadratic function, or neither seems to fit the data.
c) Use a grapher to fit the type of function determined in part (b) to the data.
d) Graph the equation with the scatterplot.
e) Use the function found in part (c) to predict the assets of the Social Security Trust Fund in 2004 and 2010.

25. *Mothers in the Labor Force.* The following table shows the percentage of mothers in the labor force who have children under 6 yr old.

YEAR	PERCENTAGE OF MOTHERS IN THE LABOR FORCE
1975	39.0%
1985	53.5
1997	65.0

Source: U.S. Bureau of Labor Statistics

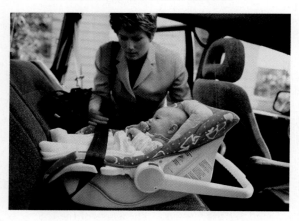

a) Make a scatterplot of the data, letting *x* represent the number of years since 1975.
b) Determine whether a linear function, a quadratic function, or neither seems to fit the data.
c) Use a grapher to fit the type of function determined in part (b) to the data.
d) Graph the equation with the scatterplot.
e) Use the function found in part (c) to predict the percentage of mothers with children under 6 yr old who will be in the labor force in 2003 and 2010.

26. *Senior Citizens.* The following table shows the number of Americans age 65 and over in various years.

YEAR	AMERICANS AGE 65 AND OVER (IN MILLIONS)
1900	3.1
1920	4.9
1940	9.0
1960	16.7
1980	25.5
2000	35.3

Source: U.S. Bureau of the Census

a) Make a scatterplot of the data, letting *x* represent the number of years since 1900.
b) Determine whether a linear function, a quadratic function, or neither seems to fit the data.
c) Use a grapher to fit the type of function determined in part (b) to the data.
d) Graph the equation with the scatterplot.
e) Use the function found in part (c) to predict the number of Americans age 65 and over in 2006 and 2011.

27. *Hours of Sleep versus Death Rate.* The following table shows the relationship between the average number of hours that a group of males slept per day and their death rates.

AVERAGE NUMBER OF HOURS OF SLEEP, x	DEATH RATE PER 100,000 MALES, y
5	1121
6	805
7	626
8	813
9	967

Source: "Hiding in the Hammond Report," *Hospital Practice*, by Harold J. Morowitz

a) Make a scatterplot of the data, letting *x* represent the average number of hours of sleep.
b) Determine whether a linear function, a quadratic function, or neither seems to fit the data.
c) Use a grapher to fit the type of function determined in part (b) to the data.
d) Graph the equation with the scatterplot.
e) Use the function found in part (c) to estimate the death rate of males who slept for an average of 7.5 hr per day and 10 hr per day.

28. *Funeral Costs.* Since the Federal Trade Commission began regulating funeral directors in 1984, the average cost for the funeral of an adult has greatly increased, as shown in the following table.

YEAR	AVERAGE COST OF FUNERAL
1980	$1809
1985	2737
1991	3742
1995	4624
1996	4782
1998	5020

Source: National Funeral Directors Association

a) Make a scatterplot of the data, letting x represent the number of years since 1980.

b) Use a grapher to fit quadratic, cubic, and quartic functions to the data. By comparing the values of R^2, determine the function that best fits the data.

c) Graph the function of best fit with the scatterplot.

d) With each function found in part (b), estimate the cost of a funeral in 2000 and 2010 and determine which function gives the most realistic estimates.

Discussion and Writing

29. Why does a correlation coefficient of 1 not guarantee the existence of a cause-and-effect relationship?

30. Find a cubic function that fits the data shown in the graph, noting that $R^2 = 1$. Then, using this model, estimate the number of gunfire deaths in 2000 and discuss the limitations of the model.

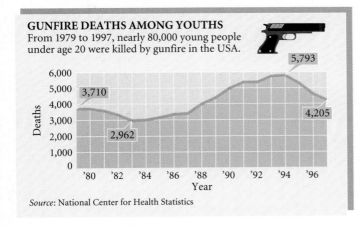

GUNFIRE DEATHS AMONG YOUTHS
From 1979 to 1997, nearly 80,000 young people under age 20 were killed by gunfire in the USA.

Source: National Center for Health Statistics

Synthesis

31. *The Effect of Advertising on Movie Revenue.* The following table shows data about the box office revenues and advertising budgets of several movies.

MOVIE	ADVERTISING BUDGET (IN MILLIONS)	BOX OFFICE REVENUE (IN MILLIONS)
The Lion King	$23.3	$300.4
Forrest Gump	25.0	298.5
True Lies	20.5	146.3
The Santa Clause	19.4	137.8
The Flintstones	10.6	130.6
Clear & Present Danger	17.9	121.8
Speed	17.2	121.2
The Mask	11.2	118.8
Maverick	13.8	101.6
Interview with a Vampire	15.4	100.7

Source: The Hollywood Reporter

a) Make a scatterplot of the data, letting x represent advertising budget and y represent box office revenue.

b) Determine whether a linear function, a quadratic function, or neither seems to fit the data.

c) Use a grapher to fit the type of function determined in part (b) to the data.

d) Graph the equation with the scatterplot.

e) Suppose a producer wants to make a movie that will have a box office revenue of $180,000,000. What amount should be allotted for the advertising budget of the movie?

Answers

Introduction to Graphs and the Graphing Calculator

1.

3.

5.

7.

9. Yes; no   **11.** Yes; no   **13.** No; yes   **15.** No; yes

17.

19.

21.

23.

25.

27.

29.

31.

33.

 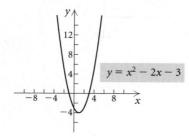

35. (b)　**37.** (a)

39.

41.

43.

45.

47.

49.

51.

53. $[-25, 25, -25, 25]$　**55.** Standard window
57. $[-1, 1, -0.3, 0.3]$　**59.** $(1, 5)$　**61.** $\left(\frac{5}{3}, \frac{7}{3}\right)$
63. $\left(\frac{11}{6}, \frac{1}{3}\right)$　**65.** $(-2, 4), (2, 4)$
67. $(-0.8284271, 3.1715729), (4.8284271, 8.8284271)$

Chapter 1

Exercise Set 1.1

1. Yes　**3.** Yes　**5.** No　**7.** Yes　**9.** Yes　**11.** Yes
13. No　**15.** Function; domain: $\{2, 3, 4\}$;
range: $\{10, 15, 20\}$　**17.** Not a function;
domain: $\{-7, -2, 0\}$; range: $\{3, 1, 4, 7\}$
19. Function; domain: $\{-2, 0, 2, 4, -3\}$; range: $\{1\}$

21. $f(-1) = 2$; $f(0) = 0$; $f(1) = -2$

23. (a) 1; **(b)** 6; **(c)** 22; **(d)** $3x^2 + 2x + 1$; **(e)** $3t^2 - 4t + 2$

25. (a) 8; **(b)** -8; **(c)** $-x^3$; **(d)** $27y^3$; **(e)** $8 + 12h + 6h^2 + h^3$

27. (a) $\frac{1}{8}$; **(b)** 0; **(c)** does not exist; **(d)** $\frac{81}{53}$, or

approximately 1.5283; **(e)** $\dfrac{x + h - 4}{x + h + 3}$

29. 0; does not exist; does not exist as a real number; $\dfrac{1}{\sqrt{3}}$,

or $\dfrac{\sqrt{3}}{3}$ **31.** All real numbers, or $(-\infty, \infty)$

33. $\{x \mid x \neq 0\}$, or $(-\infty, 0) \cup (0, \infty)$ **35.** $\{x \mid x \neq 2\}$, or
$(-\infty, 2) \cup (2, \infty)$ **37.** $\{x \mid x \neq -1 \ and \ x \neq 5\}$, or
$(-\infty, -1) \cup (-1, 5) \cup (5, \infty)$ **39.** Domain: all real
numbers; range: $[0, \infty)$ **41.** Domain: $[-3, 3]$; range: $[0, 3]$
43. Domain: all real numbers; range: all real numbers
45. Domain: $(-\infty, 7]$; range: $[0, \infty)$
47. Domain: all real numbers; range: $(-\infty, 3]$ **49.** No
51. Yes **53.** Yes **55.** No **57.** Function; domain: $[0, 5]$;
range: $[0, 3]$ **59.** Function; domain: $[-2\pi, 2\pi]$;
range: $[-1, 1]$ **61.** 645 m; 0 m **63.** 0.4 acre; 20.4 acres;
50.6 acres; 416.9 acres; 1033.6 acres **65.** Discussion and
Writing **67.** $f(x) = x$, $g(x) = x + 1$ **69.** 35

Exercise Set 1.2

1. (a) Yes; **(b)** yes; **(c)** yes **3. (a)** Yes; **(b)** no; **(c)** no
5. $-\frac{3}{5}$ **7.** 0 **9.** $\frac{1}{5}$ **11.** $-\frac{5}{3}$ **13.** Not defined
15. $2a + h$ **17.** 6.7% grade; $y = 0.067x$
19. \$100 per year **21.** $\frac{1}{6}$ km per minute **23.** $\frac{3}{5}$; $(0, -7)$
25. $-\frac{1}{2}$; $(0, 5)$ **27.** $-\frac{3}{2}$; $(0, 5)$ **29.** $\frac{4}{3}$; $(0, -7)$
31. $y = \frac{2}{9}x + 4$ **33.** $y = -4x - 7$ **35.** $y = -4.2x + \frac{3}{4}$
37. $y = \frac{2}{9}x + \frac{19}{3}$ **39.** $y = 3x - 5$ **41.** $y = -\frac{3}{5}x - \frac{17}{5}$
43. $y = -3x + 2$ **45.** $y = -\frac{1}{2}x + \frac{7}{2}$ **47.** Parallel
49. Perpendicular **51.** $y = \frac{2}{7}x + \frac{29}{7}$; $y = -\frac{7}{2}x + \frac{31}{2}$
53. $y = -0.3x - 2.1$; $y = \frac{10}{3}x + \frac{70}{3}$
55. $y = -\frac{3}{4}x + \frac{1}{4}$; $y = \frac{4}{3}x - 6$ **57.** $x = 3$; $y = -3$
59. (a) $W(h) = 3.5h - 110$;
(b)

$y = 3.5x - 110$

(c) 107 lb; **(d)** $\{h \mid h > 31.43\}$, or $(31.43, \infty)$

61. (a)

$y = 2x + 115$

(b) 115, 75, 135, 179;
(c) Below $-57.5°$, stopping
distance is negative; above
$32°$, ice doesn't form.

63. (a) $\frac{11}{10}$. For each mile per hour faster that the car
travels, it takes $\frac{11}{10}$ ft longer to stop;
(b)
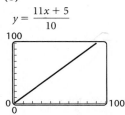
$y = \dfrac{11x + 5}{10}$

(c) 6, 11.5, 22.5, 55.5, 72; **(d)** $\{r \mid r > 0\}$, or $(0, \infty)$. If r is
allowed to be 0, the function says that a stopped car
travels $\frac{1}{2}$ ft before stopping.

65. $\dfrac{\text{Height of corn (in feet)}}{\text{Number of weeks after planting}}$

67. $C(t) = 60 + 40t$; $y = 60 + 40x$
$C(6) = \$300$

69. $C(x) = 800 + 3x$; $y = 800 + 3x$
$C(75) = \$1025$

71. Discussion and Writing **73.** [1.1] 40 **74.** [1.1] 10
75. [1.1] $a^2 + 3a$ **76.** [1.1] $a^2 + 2ah + h^2 - 3a - 3h$
77. -7.75 **79.** $F(C) = \frac{9}{5}C + 32$; $C(F) = \frac{5}{9}(F - 32)$
81. False **83.** False **85.** $f(x) = x + b$

Exercise Set 1.3

1. (a) $(-5, 1)$; **(b)** $(3, 5)$; **(c)** $(1, 3)$
3. (a) $(-3, -1)$, $(3, 5)$; **(b)** $(1, 3)$; **(c)** $(-5, -3)$
5. Increasing: $(0, \infty)$; decreasing: $(-\infty, 0)$; relative
minimum: 0 at $x = 0$

7. Increasing: $(-\infty, 0)$; decreasing: $(0, \infty)$; relative maximum: 5 at $x = 0$
9. Increasing: $(3, \infty)$; decreasing: $(-\infty, 3)$; relative minimum: 1 at $x = 3$
11. Increasing: $(1, 3)$; decreasing: $(-\infty, 1)$, $(3, \infty)$; relative maximum: -4 at $x = 3$; relative minimum: -8 at $x = 1$
13. Increasing: $(-1.552, 0)$, $(1.552, \infty)$; decreasing: $(-\infty, -1.552)$, $(0, 1.552)$; relative maximum: 4.07 at $x = 0$; relative minima: -2.314 at $x = -1.552$, -2.314 at $x = 1.552$
15. (a)

(b) 102.2; **(c)** 6 days after the illness began, the temperature is 102.2°F.
17. Increasing: $(-1, 1)$; decreasing: $(-\infty, -1)$, $(1, \infty)$
19. Increasing: $(-1.414, 1.414)$; decreasing: $(-2, -1.414)$, $(1.414, 2)$ **21.** $d(t) = \sqrt{(120t)^2 + (400)^2}$
23. $A(x) = 24x - x^2$ **25.** $A(w) = 10w - \dfrac{w^2}{2}$
27. $d(s) = \dfrac{14}{s}$ **29. (a)** $A(x) = x(30 - x)$, or $30x - x^2$;
(b) $\{x \mid 0 < x < 30\}$;
(c)

(d) 15 ft by 15 ft
31. (a) $V(x) = x(12 - 2x)(12 - 2x)$, or $4x(6 - x)^2$;
(b) $\{x \mid 0 < x < 6\}$;
(c)

(d) 8 cm by 8 cm by 2 cm
33. (a) $A(x) = x\sqrt{256 - x^2}$; **(b)** $\{x \mid 0 < x < 16\}$;

(c)

(d) 11.314 ft by 11.314 ft
35.

37.

39.

41.

43.

45.

47.
$$y = \begin{cases} \sqrt[3]{x}, & \text{for } x \leq -1, \\ x^2 - 3x, & \text{for } -1 < x < 4, \\ \sqrt{x - 4}, & \text{for } x \geq 4 \end{cases}$$

49. 33   **51.** −1   **53.** 5   **55.** 2   **57.** 0   **59.** 0
61. (a) Domain of f, all real numbers; domain of g,
$[-4, \infty)$; domain of $f + g$, $f - g$, and fg, $[-4, \infty)$; domain
of ff, all real numbers; domain of f/g, $(-4, \infty)$; domain of
g/f, $[-4, 3) \cup (3, \infty)$; **(b)** $(f + g)(x) = x - 3 + \sqrt{x + 4}$;
$(f - g)(x) = x - 3 - \sqrt{x + 4}$;
$(fg)(x) = (x - 3)\sqrt{x + 4}$; $(ff)(x) = x^2 - 6x + 9$;
$(f/g)(x) = (x - 3)/\sqrt{x + 4}$; $(g/f)(x) = \sqrt{x + 4}/(x - 3)$
63. (a) Domain of f, g, $f + g$, $f - g$, fg, and ff, all real
numbers; domain of f/g, all real numbers except -3 and $\frac{1}{2}$;
domain of g/f, all real numbers except 0;
(b) $(f + g)(x) = x^3 + 2x^2 + 5x - 3$;
$(f - g)(x) = x^3 - 2x^2 - 5x + 3$;
$(fg)(x) = 2x^5 + 5x^4 - 3x^3$; $(ff)(x) = x^6$;
$(f/g)(x) = x^3/(2x^2 + 5x - 3)$;
$(g/f)(x) = (2x^2 + 5x - 3)/x^3$
65. $(f + g)(x) = 2x^2 - 2$; $(f - g)(x) = -6$;
$(fg)(x) = x^4 - 2x^2 - 8$; $(f/g)(x) = (x^2 - 4)/(x^2 + 2)$
67. $(f + g)(x) = \sqrt{x - 7} + x^2 - 25$;
$(f - g)(x) = \sqrt{x - 7} - x^2 + 25$;
$(fg)(x) = \sqrt{x - 7}(x^2 - 25)$;
$(f/g)(x) = \sqrt{x - 7}/(x^2 - 25)$
69. Domain of F: $[0, 9]$; domain of G: $[3, 10]$; domain of
$F + G$: $[3, 9]$ **71.** $[3, 6) \cup (6, 8) \cup (8, 9]$
73.

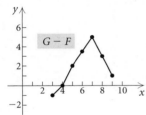

75. (a) $P(x) = -0.4x^2 + 57x - 13$; **(b)** $R(100) = 2000$;
$C(100) = 313$; $P(100) = 1687$; **(c)** left to the student
77. $6x + 3h - 2$ **79.** $3x^2 + 3xh + h^2$
81. $\dfrac{7}{(x + h + 3)(x + 3)}$ **83.** Discussion and Writing
85. [1.1] All real numbers, or $(-\infty, \infty)$
86. [1.1] $\{x \mid x \neq 3\}$ **87.** [1.1] $\{x \mid x \neq -5\}$
88. [1.1] All real numbers, or $(-\infty, \infty)$
89. Increasing: $(-5, -2)$, $(4, \infty)$; decreasing: $(-\infty, -5)$,
$(-2, 4)$; relative maximum: 560 at $x = -2$; relative
minima: 425 at $x = -5$, -304 at $x = 4$
91. (a)

(b) $C(t) = 2(\llbracket t \rrbracket + 1)$, $t > 0$
93. $\{x \mid -5 \le x < -4 \text{ or } 5 \le x < 6\}$
95. (a) $h(r) = \dfrac{30 - 5r}{3}$; **(b)** $V(r) = \pi r^2\left(\dfrac{30 - 5r}{3}\right)$;
(c) $V(h) = \pi h\left(\dfrac{30 - 3h}{5}\right)^2$

Exercise Set 1.4

1. x-axis, no; y-axis, yes; origin, no
3. x-axis, yes; y-axis, no; origin, no
5. x-axis, no; y-axis, no; origin, yes
7. x-axis, no; y-axis, yes; origin, no
9. x-axis, no; y-axis, no; origin, no
11. x-axis, no; y-axis, yes; origin, no
13. x-axis, no; y-axis, no; origin, yes
15. x-axis, no; y-axis, no; origin, yes
17. x-axis, yes; y-axis, yes; origin, yes
19. x-axis, no; y-axis, yes; origin, no
21. x-axis, yes; y-axis, yes; origin, yes
23. x-axis, no; y-axis, no; origin, no
25. x-axis, no; y-axis, no; origin, yes
27. Even   **29.** Odd   **31.** Neither   **33.** Odd   **35.** Even
37. Odd   **39.** Even   **41.** Even
43. Start with the graph
of $g(x) = x^2$. Shift it up
1 unit.
45. Start with the graph
of $f(x) = x^3$. Shift it left
5 units.

$y = x^2 + 1$

$y = (x + 5)^3$

47. Start with the graph of
$g(x) = x^2$. Reflect it across
the x-axis.
49. Start with the graph of
$g(x) = \sqrt{x}$. Stretch it
vertically by multiplying
each y-coordinate by 3,
then shift it down 5 units.

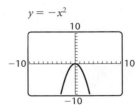

$y = -x^2$

$y = 3\sqrt{x} - 5$

51. Start with the graph of $f(x) = |x|$. Stretch it horizontally by multiplying each x-coordinate by 3, then shift it down 4 units.

$y = \left|\frac{1}{3}x\right| - 4$

53. Start with the graph of $g(x) = x^2$. Shift it left 5 units, then down 4 units.

$y = (x + 5)^2 - 4$

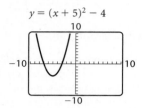

55. Start with the graph of $g(x) = x^2$. Shift it right 5 units, shrink it vertically by multiplying each y-coordinate by $\frac{1}{4}$, and reflect it across the x-axis.

$y = -\frac{1}{4}(x - 5)^2$

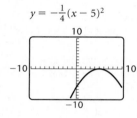

57. Start with the graph of $g(x) = 1/x$. Shift it left 3 units, then up 2 units. Use DOT mode.

$y = \frac{1}{x + 3} + 2$

59. $f(x) = -(x - 8)^2$ **61.** $f(x) = |x + 7| + 2$

63. $f(x) = \frac{1}{2x} - 3$ **65.** $f(x) = -(x - 3)^2 + 4$

67. $f(x) = \sqrt{-(x + 2)} - 1$ **69.** $f(x) = 0.83(x + 4)^3$

71.

73.

75.

77. (f)   **79.** (f)   **81.** (d)   **83.** (c)
85. $f(-x) = 2(-x)^4 - 35(-x)^3 + 3(-x) - 5 = 2x^4 + 35x^3 - 3x - 5 = g(x)$ **87.** $g(x) = x^3 - 3x^2 + 2$
89. $k(x) = (x + 1)^3 - 3(x + 1)^2$ **91.** Discussion and Writing   **93.** Discussion and Writing
94. [1.1] (a) 22; (b) -22; (c) $4a^3 - 5a$; (d) $-4a^3 + 5a$
95. [1.1] (a) 38; (b) 38; (c) $5a^2 - 7$; (d) $5a^2 - 7$
96. [1.2] Slope is $\frac{2}{9}$; y-intercept is $\left(0, \frac{1}{9}\right)$.
97. [1.2] $y = -\frac{1}{8}x + \frac{7}{8}$ **99.** Odd
101. x-axis, yes; y-axis, yes; origin, yes
103. x-axis, yes; y-axis, no; origin, no
105. Start with the graph of $g(x) = [\![x]\!]$. Shift it right $\frac{1}{2}$ unit. Domain: all real numbers; range: all integers.
107. $(3, 8)$; $(3, 6)$; $\left(\frac{3}{2}, 4\right)$ **109.** 5   **111.** True
113. $E(-x) = \dfrac{f(-x) + f(-(-x))}{2} = \dfrac{f(-x) + f(x)}{2} = E(x)$
115. (a) $E(x) + O(x) = \dfrac{f(x) + f(-x)}{2} + \dfrac{f(x) - f(-x)}{2} = \dfrac{2f(x)}{2} = f(x)$; (b) $f(x) = \dfrac{-22x^2 + \sqrt{x} + \sqrt{-x} - 20}{2} + \dfrac{8x^3 + \sqrt{x} - \sqrt{-x}}{2}$

Exercise Set 1.5

1. (a) $(-0.5, -2.25)$; (b) $x = -0.5$; (c) minimum: -2.25
3. (a) $(4, -4)$; (b) $x = 4$; (c) minimum: -4
5. (a) $\left(\frac{7}{2}, -\frac{1}{4}\right)$; (b) $x = \frac{7}{2}$; (c) minimum: $-\frac{1}{4}$
7. (a) $\left(-\frac{3}{2}, \frac{7}{2}\right)$; (b) $x = -\frac{3}{2}$; (c) minimum: $\frac{7}{2}$
9. (a) $\left(\frac{1}{2}, \frac{3}{2}\right)$; (b) $x = \frac{1}{2}$; (c) maximum: $\frac{3}{2}$
11. (f)   **13.** (b)   **15.** (h)   **17.** (c)
19. (a) $(3, -4)$; (b) minimum: -4; (c) $[-4, \infty)$; (d) increasing: $(3, \infty)$; decreasing: $(-\infty, 3)$
21. (a) $(-1, -18)$; (b) minimum: -18; (c) $[-18, \infty)$; (d) increasing: $(-1, \infty)$; decreasing: $(-\infty, -1)$
23. (a) $\left(5, \frac{9}{2}\right)$; (b) maximum: $\frac{9}{2}$; (c) $\left(-\infty, \frac{9}{2}\right]$; (d) increasing: $(-\infty, 5)$; decreasing: $(5, \infty)$
25. 4.5 in.   **27.** Base: 10 cm; height: 10 cm   **29.** 350
31. $797; 40   **33.** 4800 yd^2

35. Discussion and Writing

37. [1.3] 3

35.

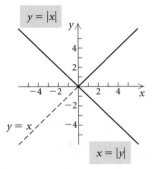

38. [1.3] $4x + 2h - 1$

39. [1.4]

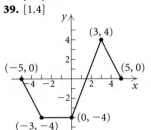

$g(x) = -2f(x)$

40. [1.4]

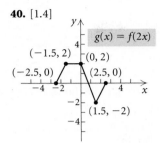

$g(x) = f(2x)$

41. -236.25

43.

$y = (|x| - 5)^2 - 3$

45. Pieces should be $\dfrac{24\pi}{4 + \pi}$ in. and $\dfrac{96}{4 + \pi}$ in.

37.

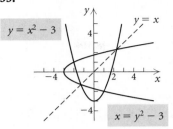

$y = |x|$

$y = x$

$x = |y|$

39. Yes   **41.** No   **43.** No   **45.** Yes   **47.** Yes

49. No   **51.** No   **53.** Yes

55.
$y_1 = 0.8x + 1.7,$
$y_2 = \dfrac{x - 1.7}{0.8}$

Domain and range of both f and f^{-1}: all real numbers

57.
$y_1 = \dfrac{1}{2}x - 4,$
$y_2 = 2x + 8$

Domain and range of both f and f^{-1}: all real numbers

59.
$y_1 = \sqrt{x - 3},$
$y_2 = x^2 + 3, x \geqslant 0$

Domain of f: $[3, \infty]$, range of f: $[0, \infty)$; domain of f^{-1}: $[0, \infty)$, range of f^{-1}: $[3, \infty)$

Exercise Set 1.6

1. $(f \circ g)(x) = (g \circ f)(x) = x$
3. $(f \circ g)(x) = (g \circ f)(x) = x$
5. $(f \circ g)(x) = 20; (g \circ f)(x) = 0.05$
7. $(f \circ g)(x) = |x|; (g \circ f)(x) = x$
9. $(f \circ g)(x) = (g \circ f)(x) = x$
11. $(f \circ g)(x) = x^3 - 2x^2 - 4x + 6;$
$(g \circ f)(x) = x^3 - 5x^2 + 3x + 8$
13. $f(x) = x^5; g(x) = 4 + 3x$
15. $f(x) = \dfrac{1}{x}; g(x) = (x - 2)^4$
17. $f(x) = \dfrac{x - 1}{x + 1}; g(x) = x^3$
19. $f(x) = x^6; g(x) = \dfrac{2 + x^3}{2 - x^3}$
21. $f(x) = \sqrt{x}; g(x) = \dfrac{x - 5}{x + 2}$
23. $f(x) = x^3 - 5x^2 + 3x - 1; g(x) = x + 2$
25. $f(x) = 2(x - 20)$
27. $\{(8, 7), (8, -2), (-4, 3), (-8, 8)\}$
29. $\{(-1, -1), (4, -3)\}$ **31.** $x = 4y - 5$
33. $y^3x = -5$

61.

$$y_1 = x^2 - 4, x \geq 0; \quad y_2 = \sqrt{4 + x}$$

Domain of f: $[0, \infty)$, range of f: $[-4, \infty)$; domain of f^{-1}: $[-4, \infty)$, range of f^{-1}: $[0, \infty)$

63.

$$y_1 = (3x - 9)^3, \quad y_2 = \frac{\sqrt[3]{x} + 9}{3}$$

Domain and range of both f and f^{-1}: all real numbers

65. (a) One-to-one; **(b)** $f^{-1}(x) = x - 4$

67. (a) One-to-one; **(b)** $f^{-1}(x) = \dfrac{x + 1}{2}$

69. (a) One-to-one; **(b)** $f^{-1}(x) = \dfrac{4}{x} - 7$

71. (a) One-to-one; **(b)** $f^{-1}(x) = \dfrac{3x + 4}{x - 1}$

73. (a) One-to-one; **(b)** $f^{-1}(x) = \sqrt[3]{x} + 1$

75. (a) Not one-to-one; **(b)** does not have an inverse that is a function

77. (a) One-to-one; **(b)** $f^{-1}(x) = \sqrt{\dfrac{x + 2}{5}}, x \geq -2$

79. (a) One-to-one; **(b)** $f^{-1}(x) = x^2 - 1, x \geq 0$

81. $\frac{1}{3}x$ **83.** $-x$ **85.** $x^3 + 5$

87. $(f^{-1} \circ f)(x) = f^{-1}(f(x))$
$$= f^{-1}\left(\tfrac{7}{8}x\right)$$
$$= \tfrac{8}{7}\left(\tfrac{7}{8}x\right) = x;$$
$(f \circ f^{-1})(x) = f(f^{-1}(x))$
$$= f\left(\tfrac{8}{7}x\right)$$
$$= \tfrac{7}{8}\left(\tfrac{8}{7}x\right) = x$$

89. $(f^{-1} \circ f)(x) = f^{-1}\left(\dfrac{1 - x}{x}\right)$

$$= \dfrac{1}{\dfrac{1 - x}{x} + 1} = \dfrac{1}{\dfrac{1}{x}} = x;$$

$$(f \circ f^{-1})(x) = f\left(\frac{1}{x + 1}\right)$$

$$= \frac{1 - \dfrac{1}{x + 1}}{\dfrac{1}{x + 1}} = \frac{\dfrac{x + 1 - 1}{x + 1}}{\dfrac{1}{x + 1}} = x$$

91. 5; a **93. (a)** 36, 40, 44, 52, 60; **(b)** $g^{-1}(x) = \dfrac{x}{2} - 12$;

(c) 6, 8, 10, 14, 18

95. (a) 0.5, 11.5, 22.5, 55.5, 72;
(b), (d)

$$y_1 = \frac{11x + 5}{10}, \quad y_2 = \frac{10x - 5}{11}$$

(c) $D^{-1}(r) = \dfrac{10r - 5}{11}$; the speed, in miles per hour, that the car is traveling when the reaction distance is r feet

97. Discussion and Writing
99. [1.2]

$$y = \tfrac{3}{2}x - 4$$

100. [1.5]

$$y = -x^2 + x - 4$$

101. [1.2] $y = \frac{1}{3}x - \frac{11}{3}$

102. [1.2] Slope: $-\frac{17}{8}$; y-intercept: $\left(0, \frac{3}{4}\right)$
103. Yes   **105.** No   **107.** $f(x) = x^2 - 3$, for inputs $x \geq 0$; $f^{-1}(x) = \sqrt{x + 3}$, for inputs $x \geq -3$
109. Answers may vary. $f(x) = 3/x, f(x) = 1 - x, f(x) = x$

Review Exercises, Chapter 1

1. [1.1] Not a function; domain: {3, 5, 7}; range: {1, 3, 5, 7}
2. [1.1] Function; domain: {−2, 0, 1, 2, 7}; range: {−7, −4, −2, 2, 7}   **3.** [1.1] No   **4.** [1.1] Yes   **5.** [1.1] No
6. [1.1] Yes   **7.** [1.1] All real numbers
8. [1.1] $\{x \mid x \neq 0\}$ **9.** [1.1] $\{x \mid x \neq 5 \text{ and } x \neq 1\}$
10. [1.1] $\{x \mid x \neq -4 \text{ and } x \neq 4\}$
11. [1.1] Domain: [−5, 3]; range: [0, 4]
12. [1.1] Domain: all real numbers; range: [−5, ∞)
13. [1.1] Domain: all real numbers; range: all real numbers
14. [1.1] Domain: all real numbers; range: [−19, ∞)
15. [1.1] **(a)** −3; **(b)** 9; **(c)** $a^2 - 3a - 1$; **(d)** $2x + h - 1$
16. [1.2] **(a)** Yes; **(b)** no; **(c)** no, strictly speaking, but data might be modeled by a linear regression function.
17. [1.2] **(a)** Yes; **(b)** yes; **(c)** yes   **18.** [1.2] −2; (0, −7)
19. [1.2] $y + 4 = -\frac{2}{3}(x - 0)$
20. [1.2] $y + 1 = 3(x + 2)$
21. [1.2] $y - 1 = \frac{1}{3}(x - 4)$, or $y + 1 = \frac{1}{3}(x + 2)$
22. [1.2] Parallel   **23.** [1.2] Neither
24. [1.2] Perpendicular   **25.** [1.2] $y = -\frac{2}{3}x - \frac{1}{3}$
26. [1.2] $y = \frac{3}{2}x - \frac{5}{2}$
27. [1.2] $C(t) = 25 + 20t$;   **28.** [1.2] **(a)** 70°C, 220°C,
$145   10,020°C;
(b)

$y = 25 + 20x$

$y = 10x + 20$

(c) [0, 5600]

29. [1.3]

30. [1.3]

31. [1.3]

32. [1.3]

33. [1.3] **(a)** Increasing: (−∞, −0.860), (6.339, ∞); decreasing: (−0.860, 6.339); **(b)** relative maximum: (−0.860, 710.094); relative minimum: (6.339, 504.836)
34. [1.3] **(a)** Increasing: (−∞, −4); decreasing: (−4, ∞); **(b)** relative maximum: (−4, 15)
35. [1.3] **(a)** Increasing: (−3.589, 2.089); decreasing: (−5, −3.589), (2.089, 3); **(b)** relative maximum: (2.089, 4.247); relative minimum: (−3.589, −8.755)
36. [1.3] $A(x) = 2x\sqrt{4 - x^2}$

37. [1.3] **(a)** $A(x) = x^2 + \dfrac{432}{x}$; **(b)** (0, ∞);

(c)

$y = x^2 + \dfrac{432}{x}$

(d) $x = 6$ in.; height = 3 in.

38. [1.4] x-axis, yes; y-axis, yes; origin, yes
39. [1.4] x-axis, yes; y-axis, yes; origin, yes
40. [1.4] x-axis, no; y-axis, no; origin, no
41. [1.4] x-axis, no; y-axis, yes; origin, no
42. [1.4] x-axis, no; y-axis, no; origin, yes
43. [1.4] x-axis, no; y-axis, yes; origin, no
44. [1.4] Even   **45.** [1.4] Even   **46.** [1.4] Odd
47. [1.4] Even   **48.** [1.4] Even   **49.** [1.4] Neither
50. [1.4] Odd   **51.** [1.4] Even   **52.** [1.4] Even
53. [1.4] Odd   **54.** [1.4] $f(x) = (x + 3)^2$
55. [1.4] $f(x) = -\sqrt{x - 3} + 4$
56. [1.4] $f(x) = 2|x - 3|$
57. [1.4]

$y = f(x - 1)$

58. [1.4]

$y = f(2x)$

59. [1.4]

60. [1.4]

61. [1.3] **(a)** Domain of f, $\{x \mid x \neq 0\}$; domain of g, all real numbers; domain of $f + g$, $f - g$, and fg, $\{x \mid x \neq 0\}$;

domain of f/g, $\left\{ x \mid x \neq 0 \text{ and } x \neq \dfrac{3}{2} \right\}$.

(b) $(f + g)(x) = \dfrac{4}{x^2} + 3 - 2x$; $(f - g)(x) = \dfrac{4}{x^2} - 3 + 2x$;

$fg(x) = \dfrac{12}{x^2} - \dfrac{8}{x}$; $(f/g)(x) = \dfrac{4}{x^2(3 - 2x)}$

62. [1.3] **(a)** Domain of f, g, $f + g$, $f - g$, and fg, all real

numbers; domain of f/g, $\left\{ x \mid x \neq \dfrac{1}{2} \right\}$;

(b) $(f + g)(x) = 3x^2 + 6x - 1$;
$(f - g)(x) = 3x^2 + 2x + 1$; $fg(x) = 6x^3 + 5x^2 - 4x$;

$(f/g)(x) = \dfrac{3x^2 + 4x}{2x - 1}$

63. [1.3] $P(x) = -0.5x^2 + 105x - 6$

64. [1.5] **(a)** $\left(\dfrac{3}{8}, -\dfrac{7}{16}\right)$; **(b)** $x = \dfrac{3}{8}$; **(c)** maximum: $-\dfrac{7}{16}$

65. [1.5] **(a)** $(1, -2)$; **(b)** $x = 1$; **(c)** minimum: -2

66. [1.5] (d)   **67.** [1.5] (c)   **68.** [1.5] (b)   **69.** [1.5] (a)

70. [1.5] 6 ft by 6 ft

71. [1.5] $\dfrac{15 - \sqrt{115}}{2}$ cm, or about 2.1 cm

72. [1.5] 2.1 cm

73. [1.6] **(a)** Domain of $f \circ g$, $\left\{ x \mid x \neq \dfrac{3}{2} \right\}$; domain of $g \circ f$,

$\{x \mid x \neq 0\}$; **(b)** $(f \circ g)(x) = \dfrac{4}{(3 - 2x)^2}$; $(g \circ f)(x) = 3 - \dfrac{8}{x^2}$

74. [1.6] **(a)** Domain of $f \circ g$ and $g \circ f$, all real numbers;
(b) $(f \circ g)(x) = 12x^2 - 4x - 1$; $(g \circ f)(x) = 6x^2 + 8x - 1$

75. [1.6] $f(x) = \sqrt{x}$, $g(x) = 5x + 2$. Answers may vary.

76. [1.6] $f(x) = 4x^2 + 9$, $g(x) = 5x - 1$. Answers may vary.

77. [1.6] $\{(-2.7, 1.3), (-3, 8), (3, -5), (-3, 6), (-5, 7)\}$

78. [1.6] **(a)** $x = 3y^2 + 2y - 1$; **(b)** $0.8y^3 - 5.4x^2 = 3y$

79. [1.6] **(a)** Yes; **(b)** $f^{-1}(x) = x^2 + 6$, $x \geq 0$

80. [1.6] **(a)** Yes; **(b)** $f^{-1}(x) = \sqrt[3]{x + 8}$ **81.** [1.6] **(a)** No

82. [1.6] **(a)** Yes; **(b)** $f^{-1}(x) = \ln x$ **83.** [1.6] 657

84. Discussion and Writing [1.4] In the graph of
$y = f(cx)$, the constant c stretches or shrinks the graph of
$y = f(x)$ horizontally. The constant c in $y = cf(x)$
stretches or shrinks the graph of $y = f(x)$ vertically. For
$y = f(cx)$, the x-coordinates of $y = f(x)$ are divided by c;

for $y = cf(x)$, the y-coordinates of $y = f(x)$ are multiplied
by c.

85. Discussion and Writing [1.4] **(a)** To draw the graph of
y_2 from the graph of y_1, reflect across the x-axis the
portions of the graph for which the y-coordinates are
negative. **(b)** To draw the graph of y_2 from the graph of y_1,
draw the portion of the graph of y_1 to the right of the
y-axis; then draw its reflection across the y-axis.

86. Discussion and Writing [1.5] You can conclude that
$|a_1| = |a_2|$ since these constants determine how wide the
parabolas are. Nothing can be concluded about the h's and
the k's.

87. Discussion and Writing [1.6] The inverse of a function

$f(x)$ is written $f^{-1}(x)$, whereas $[f(x)]^{-1}$ means $\dfrac{1}{f(x)}$.

88. [1.1] $\{x \mid x < 0\}$

89. [1.1] $\{x \mid x \neq -3 \text{ and } x \neq 0 \text{ and } x \neq 3\}$

90. [1.3], [1.4] Let $f(x)$ and $g(x)$ be odd functions. Then by
definition, $f(-x) = -f(x)$, or $f(x) = -f(-x)$, and
$g(-x) = -g(x)$, or $g(x) = -g(-x)$. Thus,
$(f + g)(x) = f(x) + g(x) = -f(-x) + [-g(-x)] =$
$-[f(-x) + g(-x)] = -(f + g)(-x)$ and $f + g$ is odd.

91. [1.4] Reflect the graph of $y = f(x)$ across the x-axis
and then across the y-axis.   **92.** [1.5] ± 6 **93.** [1.6] No

Chapter 2

Exercise Set 2.1

1. $\sin \phi = \dfrac{15}{17}$, $\cos \phi = \dfrac{8}{17}$, $\tan \phi = \dfrac{15}{8}$, $\csc \phi = \dfrac{17}{15}$,

$\sec \phi = \dfrac{17}{8}$, $\cot \phi = \dfrac{8}{15}$

3. $\sin \alpha = \dfrac{\sqrt{3}}{2}$, $\cos \alpha = \dfrac{1}{2}$, $\tan \alpha = \sqrt{3}$, $\csc \alpha = \dfrac{2\sqrt{3}}{3}$,

$\sec \alpha = 2$, $\cot \alpha = \dfrac{\sqrt{3}}{3}$

5. $\sin \phi = \dfrac{7\sqrt{65}}{65}$, $\cos \phi = \dfrac{4\sqrt{65}}{65}$, $\tan \phi = \dfrac{7}{4}$,

$\csc \phi = \dfrac{\sqrt{65}}{7}$, $\sec \phi = \dfrac{\sqrt{65}}{4}$, $\cot \phi = \dfrac{4}{7}$

7. $\cos \theta = \dfrac{7}{25}$, $\tan \theta = \dfrac{24}{7}$, $\csc \theta = \dfrac{25}{24}$, $\sec \theta = \dfrac{25}{7}$,

$\cot \theta = \dfrac{7}{24}$

9. $\sin \phi = \dfrac{2\sqrt{5}}{5}$, $\cos \phi = \dfrac{\sqrt{5}}{5}$, $\csc \phi = \dfrac{\sqrt{5}}{2}$, $\sec \phi = \sqrt{5}$,

$\cot \phi = \dfrac{1}{2}$

11. $\sin \theta = \dfrac{2}{3}$, $\cos \theta = \dfrac{\sqrt{5}}{3}$, $\tan \theta = \dfrac{2\sqrt{5}}{5}$,
$\sec \theta = \dfrac{3\sqrt{5}}{5}$, $\cot \theta = \dfrac{\sqrt{5}}{2}$

13. $\dfrac{\sqrt{2}}{2}$ **15.** 2   **17.** $\dfrac{\sqrt{3}}{3}$ **19.** $\dfrac{1}{2}$ **21.** 127.3 ft

23. 9.72°   **25.** 35.01°   **27.** 3.03°   **29.** 49.65°
31. 17°36′   **33.** 83°1′30″   **35.** 11°45′   **37.** 47°49′36″
39. 0.6293   **41.** 0.0737   **43.** 1.2765   **45.** 0.7621
47. 0.9336   **49.** 12.4288   **51.** 1.0000   **53.** 1.7032
55. 30.8°   **57.** 12.5°   **59.** 64.4°   **61.** 46.5°   **63.** 25.2°
65. 38.6°   **67.** 45°   **69.** 60°   **71.** 45°

73. $\cos 20° = \sin 70° = \dfrac{1}{\sec 20°}$

75. $\tan 52° = \cot 38° = \dfrac{1}{\cot 52°}$

77. $\sin 25° \approx 0.4226$, $\cos 25° \approx 0.9063$, $\tan 25° \approx 0.4663$, $\csc 25° \approx 2.366$, $\sec 25° \approx 1.103$, $\cot 25° \approx 2.145$

79. $\sin 8° = q$, $\cos 8° = p$, $\tan 8° = \dfrac{1}{r}$, $\csc 8° = \dfrac{1}{q}$, $\sec 8° = \dfrac{1}{p}$, $\cot 8° = r$ **81.** Discussion and Writing

82. [1.3]

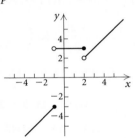

83. [1.4] *x*-axis, no; *y*-axis, yes; origin, no
84. [1.4] *x*-axis, yes; *y*-axis, no; origin, no
85. [1.4] *x*-axis, no; *y*-axis, no; origin, yes
87. 0.6534
89. Area $= \frac{1}{2}ab$. But $a = c \sin A$, so Area $= \frac{1}{2}bc \sin A$.

Exercise Set 2.2

1. $F = 60°$, $d = 3$, $f \approx 5.2$
3. $A = 22.7°$, $a \approx 52.7$, $c \approx 136.6$
5. $P = 47°38′$, $n \approx 34.4$, $p \approx 25.4$
7. $B = 2°17′$, $b \approx 0.39$, $c = 9.74$
9. $A \approx 77.2°$, $B \approx 12.8°$, $a \approx 439$
11. $B = 42.42°$, $a \approx 35.7$, $b \approx 32.6$
13. $B = 55°$, $a \approx 28.0$, $c \approx 48.8$
15. $A \approx 62.4°$, $B \approx 27.6°$, $a \approx 3.56$
17. About 62.2 ft   **19.** About 2.5 ft   **21.** About 606 ft
23. About 96.7 cm   **25.** About 599 ft   **27.** About 8 km

29. About 275 ft   **31.** About 24 km   **33.** Discussion and Writing   **35.** [1.5] **(a)** $(4, -6)$; **(b)** $x = 4$; **(c)** minimum: -6 at $x = 4$ **36.** [1.5] **(a)** $(1, -4)$; **(b)** $x = 1$; **(c)** minimum: -4 at $x = 1$ **37.** [1.3] $x^2 + 2x - 6$
38. [1.3] 12   **39.** 3.3   **41.** Cut so that $\theta = 79.38°$
43. $\theta \approx 27°$

Exercise Set 2.3

1. III   **3.** III   **5.** I   **7.** III   **9.** II   **11.** II
13. 434°, 794°, −286°, −646°
15. 475.3°, 835.3°, −244.7°, −604.7°
17. 180°, 540°, −540°, −900°
19. 72.89°, 162.89°   **21.** 77°56′46″, 167°56′46″
23. 44.8°, 134.8°

25. $\sin \beta = \dfrac{5}{13}$, $\cos \beta = -\dfrac{12}{13}$, $\tan \beta = -\dfrac{5}{12}$, $\csc \beta = \dfrac{13}{5}$, $\sec \beta = -\dfrac{13}{12}$, $\cot \beta = -\dfrac{12}{5}$

27. $\sin \phi = -\dfrac{2\sqrt{7}}{7}$, $\cos \phi = -\dfrac{\sqrt{21}}{7}$, $\tan \phi = \dfrac{2\sqrt{3}}{3}$, $\csc \phi = -\dfrac{\sqrt{7}}{2}$, $\sec \phi = -\dfrac{\sqrt{21}}{3}$, $\cot \phi = \dfrac{\sqrt{3}}{2}$

29. $\sin \theta = -\dfrac{2\sqrt{13}}{13}$, $\cos \theta = \dfrac{3\sqrt{13}}{13}$, $\tan \theta = -\dfrac{2}{3}$

31. $\sin \theta = \dfrac{5\sqrt{41}}{41}$, $\cos \theta = \dfrac{4\sqrt{41}}{41}$, $\tan \theta = \dfrac{5}{4}$

33. $\cos \theta = -\dfrac{2\sqrt{2}}{3}$, $\tan \theta = \dfrac{\sqrt{2}}{4}$, $\csc \theta = -3$, $\sec \theta = -\dfrac{3\sqrt{2}}{4}$, $\cot \theta = 2\sqrt{2}$

35. $\sin \theta = -\dfrac{\sqrt{5}}{5}$, $\cos \theta = \dfrac{2\sqrt{5}}{5}$, $\tan \theta = -\dfrac{1}{2}$, $\csc \theta = -\sqrt{5}$, $\sec \theta = \dfrac{\sqrt{5}}{2}$

37. $\sin \phi = -\dfrac{4}{5}$, $\tan \phi = -\dfrac{4}{3}$, $\csc \phi = -\dfrac{5}{4}$, $\sec \phi = \dfrac{5}{3}$, $\cot \phi = -\dfrac{3}{4}$

39. 30°; $-\dfrac{\sqrt{3}}{2}$ **41.** 45°; 1   **43.** 0   **45.** 45°; $-\dfrac{\sqrt{2}}{2}$
47. 30°; 2   **49.** 30°; $\sqrt{3}$ **51.** 30°; $-\dfrac{\sqrt{3}}{3}$
53. Undefined   **55.** −1   **57.** 60°; $\sqrt{3}$ **59.** 45°; $\dfrac{\sqrt{2}}{2}$
61. 45°; $-\sqrt{2}$ **63.** 1   **65.** 0   **67.** 0   **69.** 0
71. Positive: cos, sec; negative: sin, csc, tan, cot
73. Positive: tan, cot; negative: sin, csc, cos, sec

75. Positive: sin, csc; negative: cos, sec, tan, cot
77. Positive: all
79. $\sin 319° = -0.6561$, $\cos 319° = 0.7547$,
$\tan 319° = -0.8693$, $\csc 319° \approx -1.5242$,
$\sec 319° \approx 1.3250$, $\cot 319° \approx -1.1504$
81. $\sin 115° = 0.9063$, $\cos 115° = -0.4226$,
$\tan 115° = -2.1445$, $\csc 115° \approx 1.1034$,
$\sec 115° \approx -2.3663$, $\cot 115° \approx -0.4663$
83. East: about 130 km; south: 75 km
85. About 223 km   **87.** -1.1585 **89.** -1.4910
91. 0.8771   **93.** 0.4352   **95.** 0.9563   **97.** 2.9238
99. 275.4°   **101.** 200.1°   **103.** 288.1°   **105.** 72.6°
107. Discussion and Writing
109. [1.2] -1 **110.** [1.2] Undefined
111. [1.1] Domain: $\{x \mid x \neq -2\}$; range: $\{x \mid x \neq 1\}$
112. [1.1] Domain: $\left\{x \mid x \neq -\frac{3}{2} \text{ and } x \neq 5\right\}$; range:
all real numbers   **113.** 19.625 in.

Exercise Set 2.4

1.
3.

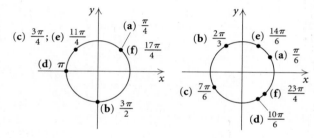

5. $M: \frac{2\pi}{3}, -\frac{4\pi}{3}$; $N: \frac{3\pi}{2}, -\frac{\pi}{2}$; $P: \frac{5\pi}{4}, -\frac{3\pi}{4}$; $Q: \frac{11\pi}{6}, -\frac{\pi}{6}$

7.

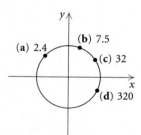

9. $\frac{9\pi}{4}, -\frac{7\pi}{4}$ **11.** $\frac{19\pi}{6}, -\frac{5\pi}{6}$

13. Complement: $\frac{\pi}{6}$; supplement: $\frac{2\pi}{3}$

15. Complement: $\frac{\pi}{8}$; supplement: $\frac{5\pi}{8}$ **17.** $\frac{5\pi}{12}$ **19.** $\frac{10\pi}{9}$

21. $-\frac{214.6\pi}{180}$ **23.** $-\pi$ **25.** 4.19   **27.** -1.05
29. 2.06   **31.** 0.02   **33.** $-135°$ **35.** 1440°   **37.** 57.30°
39. 134.47°
41. $0° = 0$ radian, $30° = \frac{\pi}{6}$, $45° = \frac{\pi}{4}$, $60° = \frac{\pi}{3}$,
$90° = \frac{\pi}{2}$, $135° = \frac{3\pi}{4}$, $180° = \pi$, $225° = \frac{5\pi}{4}$, $270° = \frac{3\pi}{2}$,
$315° = \frac{7\pi}{4}$, $360° = 2\pi$ **43.** 1.1, 63°   **45.** 3.2 yd
47. $\frac{5\pi}{3}$, or about 5.24   **49.** $3150 \dfrac{\text{cm}}{\text{min}}$
51. About 18,852 revolutions per hour   **53.** 1047 mph
55. 10 mph   **57.** About 202   **59.** Discussion and
Writing   **61.** [2.1] 37.3°   **62.** [2.1] 5.01°
63. [2.1] 81°12′36″   **64.** [2.1] 3°6′
65. (a) 5°37′30″; (b) 19°41′15″   **67.** 1.676 radians/sec
69. 1.46 nautical miles

Exercise Set 2.5

1. (a) $\left(-\frac{3}{4}, -\frac{\sqrt{7}}{4}\right)$; (b) $\left(\frac{3}{4}, \frac{\sqrt{7}}{4}\right)$; (c) $\left(\frac{3}{4}, -\frac{\sqrt{7}}{4}\right)$
3. (a) $\left(\frac{2}{5}, \frac{\sqrt{21}}{5}\right)$; (b) $\left(-\frac{2}{5}, -\frac{\sqrt{21}}{5}\right)$; (c) $\left(-\frac{2}{5}, \frac{\sqrt{21}}{5}\right)$
5. $\left(\frac{\sqrt{2}}{2}, -\frac{\sqrt{2}}{2}\right)$ **7.** 0   **9.** $\sqrt{3}$ **11.** 0   **13.** $-\frac{\sqrt{3}}{2}$
15. 1   **17.** $\frac{\sqrt{3}}{2}$ **19.** $-\frac{\sqrt{2}}{2}$ **21.** 0   **23.** 0   **25.** 0.4816
27. 1.3065   **29.** -2.1599 **31.** 1   **33.** -1.1747 **35.** -1
37. -0.7071 **39.** 0   **41.** 0.8391
43. (a)

(b)

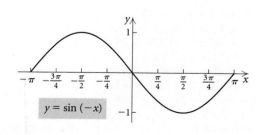

(c) same as (b); (d) the same

45. (a) See Exercise 43(a);
(b)

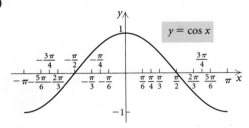

$$y = \sin(x + \pi)$$

(c) same as (b); **(d)** the same
47. (a)

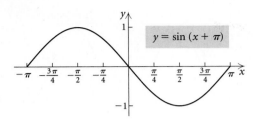

$$y = \cos x$$

(b)

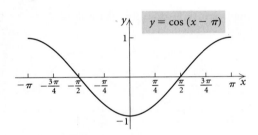

$$y = \cos(x - \pi)$$

(c) same as (b); **(d)** the same
49. Even: cosine, secant; odd: sine, tangent, cosecant, cotangent   **51.** Positive: I, III; negative: II, IV
53. Positive: I, IV; negative: II, III
55. Discussion and Writing
57. [1.4]

$$y_1 = x^2, \quad y_2 = 2x^2 - 3$$

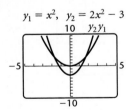

Stretch the graph of f vertically, then shift it down 3 units.

58. [1.4]

$$y_1 = x^2, \quad y_2 = (x - 2)^2$$

Shift the graph of f right 2 units.
59. [1.4]

$$y_1 = |x|, \quad y_2 = \tfrac{1}{2}|x - 4| + 1$$

Shift the graph of f to the right 4 units, shrink it vertically, then shift it up 1 unit.
60. [1.4]

$$y_1 = x^3, \quad y_2 = -x^3$$

Reflect the graph of f across the x-axis.
61. [1.4] $y = -(x - 2)^3 - 1$ **62.** [1.4] $y = \dfrac{1}{4x} + 3$
63. $\cos x$ **65.** $\sin x$ **67.** $\sin x$ **69.** $-\cos x$
71. $-\sin x$
73. (a) $\dfrac{\pi}{2} + 2k\pi, \ k \in \mathbb{Z}$; **(b)** $\pi + 2k\pi, \ k \in \mathbb{Z}$;

(c) $k\pi, \ k \in \mathbb{Z}$; **(d)** $\left(-\dfrac{3\pi}{4} + 2k\pi, \dfrac{\pi}{4} + 2k\pi\right), \ k \in \mathbb{Z}$
75. Domain: $(-\infty, \infty)$; range: $[0, 1]$; period: π;

amplitude: $\dfrac{1}{2}$ **77.** $\left[-\dfrac{\pi}{2} + 2k\pi, \dfrac{\pi}{2} + 2k\pi\right], \ k \in \mathbb{Z}$

79. $\left\{x \mid x \neq \dfrac{\pi}{2} + k\pi, \ k \in \mathbb{Z}\right\}$ **81.** 1

83.

$$y = 3 \sin x$$

85.

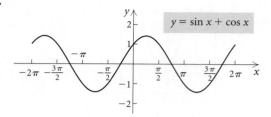

$y = \sin x + \cos x$

Exercise Set 2.6

1. Amplitude: 1; period: 2π; phase shift: 0

$y = \sin x + 1$

3. Amplitude: 3; period: 2π; phase shift: 0

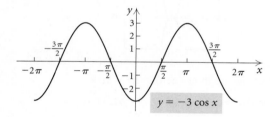

$y = -3 \cos x$

5. Amplitude: 2; period: 4π; phase shift: 0

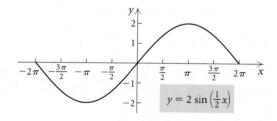

$y = 2 \sin \left(\frac{1}{2} x\right)$

7. Amplitude: $\frac{1}{2}$; period: 2π; phase shift: $-\frac{\pi}{2}$

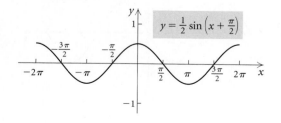

$y = \frac{1}{2} \sin \left(x + \frac{\pi}{2}\right)$

9. Amplitude: 3; period: 2π; phase shift: π

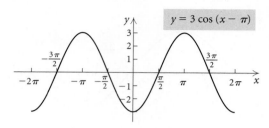

$y = 3 \cos (x - \pi)$

11. Amplitude: $\frac{1}{3}$; period: 2π; phase shift: 0

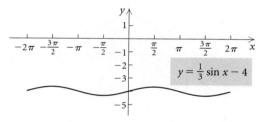

$y = \frac{1}{3} \sin x - 4$

13. Amplitude: 1; period: 2π; phase shift: 0

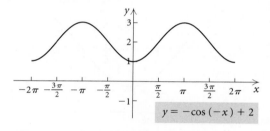

$y = -\cos (-x) + 2$

15. Amplitude: 2; period: 4π; phase shift: π

$y = 2 \cos \left(\frac{1}{2} x - \frac{\pi}{2}\right)$

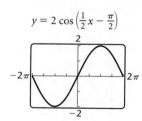

17. Amplitude: $\frac{1}{2}$; period: π; phase shift: $-\frac{\pi}{4}$

$y = -\frac{1}{2} \sin \left(2x + \frac{\pi}{2}\right)$

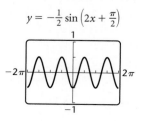

19. Amplitude: 3; period: 2; phase shift: $\dfrac{3}{\pi}$

$$y = 2 + 3\cos(\pi x - 3)$$

21. Amplitude: $\dfrac{1}{2}$; period: 1; phase shift: 0

$$y = -\dfrac{1}{2}\cos(2\pi x) + 2$$

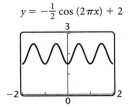

23. Amplitude: 1; period: 4π; phase shift: π

$$y = -\sin\left(\dfrac{1}{2}x - \dfrac{\pi}{2}\right) + \dfrac{1}{2}$$

25. Amplitude: 1; period: 1; phase shift: 0

$$y = \cos(-2\pi x) + 2$$

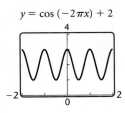

27. Amplitude: $\dfrac{1}{4}$; period: 2; phase shift: $\dfrac{4}{\pi}$

$$y = -\dfrac{1}{4}\cos(\pi x - 4)$$

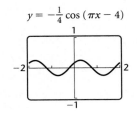

29. (b)   **31.** (h)   **33.** (a)   **35.** (f)

37. $y = \dfrac{1}{2}\cos x + 1$ **39.** $y = \cos\left(x + \dfrac{\pi}{2}\right) - 2$

41.

$y = 2\cos x + \cos 2x$

43.

$y = \sin x + \cos 2x$

45.

$y = \sin x - \cos x$

47.

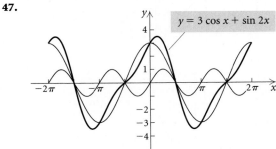

$y = 3\cos x + \sin 2x$

49.

$$y = x + \sin x$$

51.

$$y = \cos x - x$$

53.

$y = \cos 2x + 2x$

55.

$y = 4 \cos 2x - 2 \sin x$

57. Discussion and Writing
59. [2.3] 16°54′50″; 106°54′50″   **60.** [2.3] 66.9°; 156.9°

61. [2.4] $-\dfrac{\pi}{3}$ **62.** [2.4] $\dfrac{59\pi}{90}$

63.

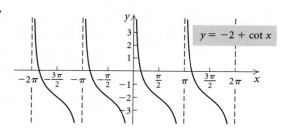

$y = -\tan x$

65.

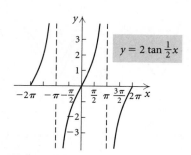

$y = -2 + \cot x$

67.

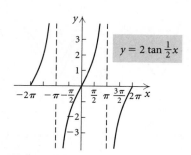

$y = 2 \tan \tfrac{1}{2}x$

69.

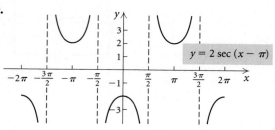

$y = 2 \sec (x - \pi)$

71.

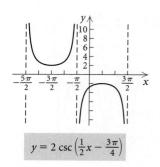

$y = 2 \csc \left(\tfrac{1}{2}x - \tfrac{3\pi}{4}\right)$

73. −9.42, −6.28, −3.14, 3.14, 6.28, 9.42
75. −3.14, 0, 3.14
77. (a)

$y = 101.6 + 3 \sin \left(\tfrac{\pi}{8}x\right)$

(b) 104.6°, 98.6°
79. (a)

$y = 3000 \left[\cos \tfrac{\pi}{45}(x - 10)\right]$

(b) amplitude: 3000; period: 90; phase shift: 10
81. (a)

$y = 6e^{-0.8x} \cos (6\pi x) + 4$

(b) 4 in.

Review Exercises, Chapter 2

1. [2.1] $\sin \theta = \dfrac{3\sqrt{73}}{73}$, $\cos \theta = \dfrac{8\sqrt{73}}{73}$, $\tan \theta = \dfrac{3}{8}$,

$\csc \theta = \dfrac{\sqrt{73}}{3}$, $\sec \theta = \dfrac{\sqrt{73}}{8}$, $\cot \theta = \dfrac{8}{3}$ **2.** [2.1] $\dfrac{\sqrt{2}}{2}$

3. [2.1] $\dfrac{\sqrt{3}}{3}$ **4.** [2.3] $-\dfrac{\sqrt{2}}{2}$ **5.** [2.3] $\dfrac{1}{2}$

6. [2.3] Undefined   **7.** [2.3] $-\sqrt{3}$ **8.** [2.1] 22°16′12″

9. [2.1] 47.56°   **10.** [2.3] 0.4452   **11.** [2.3] 1.1315
12. [2.3] 0.9498   **13.** [2.3] −0.9092   **14.** [2.3] −1.5282
15. [2.3] −0.2778   **16.** [2.3] 205.3°   **17.** [2.3] 47.2°
18. [2.1] 60°   **19.** [2.1] 60°
20. [2.1] sin 30.9° ≈ 0.5135, cos 30.9° ≈ 0.8581,
tan 30.9° ≈ 0.5985, csc 30.9° ≈ 1.9474, sec 30.9° ≈ 1.1654,
cot 30.9° ≈ 1.6709
21. [2.2] $b ≈ 4.5$, $A ≈ 58.1°$, $B ≈ 31.9°$
22. [2.2] $A = 38.83°$, $b ≈ 37.9$, $c ≈ 48.6$ **23.** [2.2] 1748 m

24. [2.2] 14 ft   **25.** [2.3] 425°, −295°   **26.** [2.4] $\dfrac{\pi}{3}$, $-\dfrac{5\pi}{3}$

27. [2.3] Complement: 76.6°; supplement: 166.6°

28. [2.4] Complement: $\dfrac{\pi}{3}$; supplement: $\dfrac{5\pi}{6}$

29. [2.3] $\sin \theta = \dfrac{3\sqrt{13}}{13}$, $\cos \theta = \dfrac{-2\sqrt{13}}{13}$, $\tan \theta = -\dfrac{3}{2}$,

$\csc \theta = \dfrac{\sqrt{13}}{3}$, $\sec \theta = -\dfrac{\sqrt{13}}{2}$, $\cot \theta = -\dfrac{2}{3}$

30. [2.3] $\sin \theta = -\dfrac{2}{3}$, $\cos \theta = -\dfrac{\sqrt{5}}{3}$, $\cot \theta = \dfrac{\sqrt{5}}{2}$,

$\sec \theta = -\dfrac{3\sqrt{5}}{5}$, $\csc \theta = -\dfrac{3}{2}$ **31.** [2.3] About 1743 mi

32. [2.4]

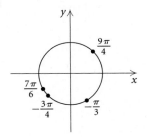

33. [2.4] II, $\dfrac{121}{150}$ π, 2.534   **34.** [2.4] IV, $-\dfrac{\pi}{6}$, −0.524

35. [2.4] 270°   **36.** [2.4] 171.89°   **37.** [2.4] $\dfrac{7\pi}{4}$, or 5.5 cm

38. [2.4] 2.25, 129°   **39.** [2.4] About 37.9 ft/min
40. [2.4] 497,829 radians/hr

41. [2.5] $\left(\dfrac{3}{5}, \dfrac{4}{5}\right)$, $\left(-\dfrac{3}{5}, -\dfrac{4}{5}\right)$, $\left(-\dfrac{3}{5}, \dfrac{4}{5}\right)$

42. [2.5] −1   **43.** [2.5] 1   **44.** [2.5] $-\dfrac{\sqrt{3}}{2}$ **45.** [2.5] $\dfrac{1}{2}$

46. [2.5] $\dfrac{\sqrt{3}}{3}$ **47.** [2.5] −1   **48.** [2.5] −0.9056

49. [2.5] 0.9218   **50.** [2.5] Undefined   **51.** [2.5] 4.3813
52. [2.5] −6.1685   **53.** [2.5] 0.8090

54. [2.5]

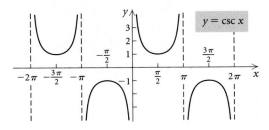

55. [2.5] Period of sin, cos, sec, csc: 2π; period of tan, cot: π

56. [2.5]

FUNCTION	DOMAIN	RANGE
Sine	$(-\infty, \infty)$	$[-1, 1]$
Cosine	$(-\infty, \infty)$	$[-1, 1]$
Tangent	$\left\{x \mid x \neq \dfrac{\pi}{2} + k\pi,\ k \in \mathbb{Z}\right\}$	$(-\infty, \infty)$

57. [2.3]

FUNCTION	I	II	III	IV
Sine	+	+	−	−
Cosine	+	−	−	+
Tangent	+	−	+	−

58. [2.6] Amplitude: 1; period: 2π; phase shift: $-\dfrac{\pi}{2}$

59. [2.6] Amplitude: $\dfrac{1}{2}$; period: π; phase shift: $\dfrac{\pi}{4}$

60. [2.6] (d)   **61.** [2.6] (a)   **62.** [2.6] (c)   **63.** [2.6] (b)
64. [2.6]

$y = 3 \cos x + \sin x$

65. Discussion and Writing [2.1], [2.4] Both degrees and radians are units of angle measure. A degree is defined to be $\frac{1}{360}$ of one complete positive revolution. Degree notation has been in use since Babylonian times. Radians are defined in terms of intercepted arc length on a circle, with one radian being the measure of the angle for which the arc length equals the radius. There are 2π radians in one complete revolution.
66. Discussion and Writing [2.5] The graph of the cosine function is shaped like a continuous wave, with "high" points at $y = 1$ and "low" points at $y = -1$. The maximum value of the cosine function is 1, and it occurs at all points where $x = 2k\pi$, $k \in \mathbb{Z}$.
67. Discussion and Writing [2.6] When x is very large or very small, the amplitude of the function becomes small. The dimensions of the window must be adjusted to be able to see the shape of the graph. Also, when x is 0, the function is undefined, but this may not be obvious from the graph.
68. Discussion and Writing [2.5] No; $\sin x$ is never greater than 1.   **69.** [2.5] All values
70. [2.6] Domain $(-\infty, \infty)$; range $[-3, 3]$; period 4π

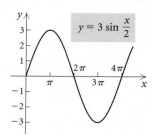

71. [2.6] $y_2 = 2 \sin\left(x + \dfrac{\pi}{2}\right) - 2$
72. [2.6] The domain consists of the intervals $\left(-\dfrac{\pi}{2} + 2k\pi,\ \dfrac{\pi}{2} + 2k\pi\right)$, $k \in \mathbb{Z}$.
73. [2.3] $\cos x = -0.7890$, $\tan x = -0.7787$, $\cot x = -1.2842$, $\sec x = -1.2674$, $\csc x = 1.6276$

Chapter 3

Exercise Set 3.1

1. $\sin^2 x - \cos^2 x$ **3.** $\sin y + \cos y$
5. $1 - 2 \sin\phi \cos\phi$ **7.** $\sin^3 x + \csc^3 x$
9. $\cos x (\sin x + \cos x)$
11. $(\sin x + \cos x)(\sin x - \cos x)$
13. $(2 \cos x + 3)(\cos x - 1)$
15. $(\sin x + 3)(\sin^2 x - 3 \sin x + 9)$ **17.** $\tan x$
19. $\sin x + 1$ **21.** $\dfrac{2 \tan t + 1}{3 \tan t + 1}$ **23.** 1

25. $\dfrac{5\cot\phi}{\sin\phi+\cos\phi}$ **27.** $\dfrac{1+2\sin s+2\cos s}{\sin^2 s-\cos^2 s}$

29. $\dfrac{5(\sin\theta-3)}{3}$ **31.** $\sin x\cos x$

33. $\sqrt{\cos\alpha}\,(\sin\alpha-\cos\alpha)$ **35.** $1-\sin y$

37. $\dfrac{\sqrt{\sin x\cos x}}{\cos x}$ **39.** $\dfrac{\sqrt{2}\cot y}{2}$

41. $\dfrac{\cos x}{\sqrt{\sin x\cos x}}$ **43.** $\dfrac{1+\sin y}{\cos y}$

45. $\cos\theta=\dfrac{\sqrt{a^2-x^2}}{a},\ \tan\theta=\dfrac{x}{\sqrt{a^2-x^2}}$

47. $\sin\theta=\dfrac{\sqrt{x^2-9}}{x},\ \cos\theta=\dfrac{3}{x}$

49. $\sin\theta\tan\theta$ **51.** $\dfrac{\sqrt{6}-\sqrt{2}}{4}$

53. $\dfrac{\sqrt{3}+1}{1-\sqrt{3}}$, or $-2-\sqrt{3}$ **55.** $\dfrac{\sqrt{6}+\sqrt{2}}{4}$

57. $\sin 59°\approx0.8572$ **59.** $\tan 52°\approx1.2799$ **61.** 0

63. $-\dfrac{7}{25}$ **65.** -1.5789 **67.** $2\sin\alpha\cos\beta$

69. $\cos u$ **71.** Discussion and Writing   **73.** [2.5] $\dfrac{\sqrt{2}}{2}$

74. [2.5] 2   **75.** [2.1] 1.9417   **76.** [2.1] 1.6645

77. 0°; the lines are parallel   **79.** $\dfrac{3\pi}{4}$, or 135°

81. 22.83°

83. $\dfrac{\cos(x+h)-\cos x}{h}$

$$=\dfrac{\cos x\cos h-\sin x\sin h-\cos x}{h}$$

$$=\dfrac{\cos x\cos h-\cos x}{h}-\dfrac{\sin x\sin h}{h}$$

$$=\cos x\left(\dfrac{\cos h-1}{h}\right)-\sin x\left(\dfrac{\sin h}{h}\right)$$

85. Let $x=\dfrac{\pi}{5}$. Then $\dfrac{\sin 5x}{x}=\dfrac{\sin\pi}{\pi/5}=0\neq\sin 5$. Answers may vary.

87. Let $\alpha=\dfrac{\pi}{4}$. Then $\cos(2\alpha)=\cos\dfrac{\pi}{2}=0$, but $2\cos\alpha=2\cos\dfrac{\pi}{4}=\sqrt{2}$. Answer may vary.

89. Let $x=\dfrac{\pi}{6}$. Then $\dfrac{\cos 6x}{\cos x}=\dfrac{\cos\pi}{\cos\dfrac{\pi}{6}}=\dfrac{-1}{\sqrt{3}/2}\neq6$. Answer may vary.

91. $\dfrac{6-3\sqrt{3}}{9+2\sqrt{3}}\approx0.0645$ **93.** 168.7°

95. $\cos 2\theta=\cos^2\theta-\sin^2\theta$, or $1-2\sin^2\theta$, or $2\cos^2\theta-1$

97. $\tan\left(x+\dfrac{\pi}{4}\right)=\dfrac{\tan x+\tan\dfrac{\pi}{4}}{1-\tan x\tan\dfrac{\pi}{4}}=\dfrac{1+\tan x}{1-\tan x}$

99. $\sin(\alpha+\beta)+\sin(\alpha-\beta)=\sin\alpha\cos\beta+\cos\alpha\sin\beta+\sin\alpha\cos\beta-\cos\alpha\sin\beta=2\sin\alpha\cos\beta$

Exercise Set 3.2

1. (a) $\tan\dfrac{3\pi}{10}\approx1.3763$, $\csc\dfrac{3\pi}{10}\approx1.2361$, $\sec\dfrac{3\pi}{10}\approx1.7013$, $\cot\dfrac{3\pi}{10}\approx0.7266$; **(b)** $\sin\dfrac{\pi}{5}\approx0.5878$, $\cos\dfrac{\pi}{5}\approx0.8090$, $\tan\dfrac{\pi}{5}\approx0.7266$, $\csc\dfrac{\pi}{5}\approx1.7013$, $\sec\dfrac{\pi}{5}\approx1.2361$, $\cot\dfrac{\pi}{5}\approx1.3763$

3. (a) $\cos\theta=-\dfrac{2\sqrt{2}}{3}$, $\tan\theta=-\dfrac{\sqrt{2}}{4}$, $\csc\theta=3$, $\sec\theta=-\dfrac{3\sqrt{2}}{4}$, $\cot\theta=-2\sqrt{2}$; **(b)** $\sin\left(\dfrac{\pi}{2}-\theta\right)=-\dfrac{2\sqrt{2}}{3}$, $\cos\left(\dfrac{\pi}{2}-\theta\right)=\dfrac{1}{3}$, $\tan\left(\dfrac{\pi}{2}-\theta\right)=-2\sqrt{2}$, $\csc\left(\dfrac{\pi}{2}-\theta\right)=-\dfrac{3\sqrt{2}}{4}$, $\sec\left(\dfrac{\pi}{2}-\theta\right)=3$, $\cot\left(\dfrac{\pi}{2}-\theta\right)=-\dfrac{\sqrt{2}}{4}$; **(c)** $\sin\left(\theta-\dfrac{\pi}{2}\right)=\dfrac{2\sqrt{2}}{3}$, $\cos\left(\theta-\dfrac{\pi}{2}\right)=\dfrac{1}{3}$, $\tan\left(\theta-\dfrac{\pi}{2}\right)=2\sqrt{2}$, $\csc\left(\theta-\dfrac{\pi}{2}\right)=\dfrac{3\sqrt{2}}{4}$, $\sec\left(\theta-\dfrac{\pi}{2}\right)=3$, $\cot\left(\theta-\dfrac{\pi}{2}\right)=\dfrac{\sqrt{2}}{4}$ **5.** $\sec\left(x+\dfrac{\pi}{2}\right)=-\csc x$

7. $\tan\left(x-\dfrac{\pi}{2}\right)=-\cot x$

9. $\sin 2\theta=\dfrac{24}{25}$, $\cos 2\theta=-\dfrac{7}{25}$, $\tan 2\theta=-\dfrac{24}{7}$; II

11. $\sin 2\theta=\dfrac{24}{25}$, $\cos 2\theta=-\dfrac{7}{25}$, $\tan 2\theta=-\dfrac{24}{7}$; II

13. $\sin 2\theta=-\dfrac{120}{169}$, $\cos 2\theta=\dfrac{119}{169}$, $\tan 2\theta=-\dfrac{120}{119}$; IV

15. $\cos 4x=1-8\sin^2 x\cos^2 x$, or $\cos^4 x-6\sin^2 x\cos^2 x+\sin^4 x$, or $8\cos^4 x-8\cos^2 x+1$

17. $\dfrac{\sqrt{2+\sqrt{3}}}{2}$ **19.** $\dfrac{\sqrt{2+\sqrt{2}}}{2}$ **21.** $2+\sqrt{3}$

23. 0.6421   **25.** 0.1735

27. (d); $\dfrac{\cos 2x}{\cos x - \sin x} = \dfrac{\cos^2 x - \sin^2 x}{\cos x - \sin x}$

$\qquad = \dfrac{(\cos x + \sin x)(\cos x - \sin x)}{\cos x - \sin x}$

$\qquad = \cos x + \sin x$

$\qquad = \dfrac{\sin x}{\sin x}(\cos x + \sin x)$

$\qquad = \sin x \left(\dfrac{\cos x}{\sin x} + \dfrac{\sin x}{\sin x} \right)$

$\qquad = \sin x \,(\cot x + 1)$

29. (d); $\dfrac{\sin 2x}{2 \cos x} = \dfrac{2 \sin x \cos x}{2 \cos x} = \sin x$ **31.** $\cos x$

33. 1   **35.** $\cos 2x$ **37.** 8   **39.** Discussion and Writing

41. [3.1] $\sin^2 x$ **42.** [3.1] 1   **43.** [3.1] $-\cos^2 x$

44. [3.1] $\csc^2 x$

45. $\sin 141° \approx 0.6293$, $\cos 141° \approx -0.7772$,
$\tan 141° \approx -0.8097$, $\csc 141° \approx 1.5891$,
$\sec 141° \approx -1.2867$, $\cot 141° \approx -1.2350$

47. $-\cos x(1 + \cot x)$ **49.** $\cot^2 y$

51. $\sin \theta = -\dfrac{15}{17}$, $\cos \theta = -\dfrac{8}{17}$, $\tan \theta = \dfrac{15}{8}$

53. (a) 9.80359 m/sec^2; (b) 9.80180 m/sec^2;
(c) $g = 9.78049(1 + 0.005264 \sin^2 \phi + 0.000024 \sin^4 \phi)$;
(d) g is greatest at 90°N and 90°S (at the poles); g is least at 0° (at the equator).

Exercise Set 3.3

1.

$\sec x - \sin x \tan x$	$\cos x$
$\dfrac{1}{\cos x} - \sin x \cdot \dfrac{\sin x}{\cos x}$	
$\dfrac{1 - \sin^2 x}{\cos x}$	
$\dfrac{\cos^2 x}{\cos x}$	
$\cos x$	

3.

$\dfrac{1 - \cos x}{\sin x}$	$\dfrac{\sin x}{1 + \cos x}$
	$\dfrac{\sin x}{1 + \cos x} \cdot \dfrac{1 - \cos x}{1 - \cos x}$
	$\dfrac{\sin x\,(1 - \cos x)}{1 - \cos^2 x}$
	$\dfrac{\sin x\,(1 - \cos x)}{\sin^2 x}$
	$\dfrac{1 - \cos x}{\sin x}$

5.

$\dfrac{1 + \tan \theta}{1 - \tan \theta} + \dfrac{1 + \cot \theta}{1 - \cot \theta}$	0
$\dfrac{1 + \dfrac{\sin \theta}{\cos \theta}}{1 - \dfrac{\sin \theta}{\cos \theta}} + \dfrac{1 + \dfrac{\cos \theta}{\sin \theta}}{1 - \dfrac{\cos \theta}{\sin \theta}}$	
$\dfrac{\dfrac{\cos \theta + \sin \theta}{\cos \theta}}{\dfrac{\cos \theta - \sin \theta}{\cos \theta}} + \dfrac{\dfrac{\sin \theta + \cos \theta}{\sin \theta}}{\dfrac{\sin \theta - \cos \theta}{\sin \theta}}$	
$\dfrac{\cos \theta + \sin \theta}{\cos \theta} \cdot \dfrac{\cos \theta}{\cos \theta - \sin \theta} +$	
$\dfrac{\sin \theta + \cos \theta}{\sin \theta} \cdot \dfrac{\sin \theta}{\sin \theta - \cos \theta}$	
$\dfrac{\cos \theta + \sin \theta}{\cos \theta - \sin \theta} + \dfrac{\sin \theta + \cos \theta}{\sin \theta - \cos \theta}$	
$\dfrac{\cos \theta + \sin \theta}{\cos \theta - \sin \theta} - \dfrac{\cos \theta + \sin \theta}{\cos \theta - \sin \theta}$	
0	

7.

$\dfrac{\cos^2 \alpha + \cot \alpha}{\cos^2 \alpha - \cot \alpha}$	$\dfrac{\cos^2 \alpha \tan \alpha + 1}{\cos^2 \alpha \tan \alpha - 1}$
$\dfrac{\cos^2 \alpha + \dfrac{\cos \alpha}{\sin \alpha}}{\cos^2 \alpha - \dfrac{\cos \alpha}{\sin \alpha}}$	$\dfrac{\cos^2 \alpha \dfrac{\sin \alpha}{\cos \alpha} + 1}{\cos^2 \alpha \dfrac{\sin \alpha}{\cos \alpha} - 1}$
$\dfrac{\cos \alpha \left(\cos \alpha + \dfrac{1}{\sin \alpha} \right)}{\cos \alpha \left(\cos \alpha - \dfrac{1}{\sin \alpha} \right)}$	$\dfrac{\sin \alpha \cos \alpha + 1}{\sin \alpha \cos \alpha - 1}$
$\dfrac{\cos \alpha + \dfrac{1}{\sin \alpha}}{\cos \alpha - \dfrac{1}{\sin \alpha}}$	
$\dfrac{\dfrac{\sin \alpha \cos \alpha + 1}{\sin \alpha}}{\dfrac{\sin \alpha \cos \alpha - 1}{\sin \alpha}}$	
$\dfrac{\sin \alpha \cos \alpha + 1}{\sin \alpha \cos \alpha - 1}$	

9.

$$\begin{array}{c|c}
\dfrac{2\tan\theta}{1+\tan^2\theta} & \sin 2\theta \\[2ex]
\dfrac{2\tan\theta}{\sec^2\theta} & 2\sin\theta\cos\theta \\[2ex]
\dfrac{2\sin\theta}{\cos\theta}\cdot\dfrac{\cos^2\theta}{1} & \\[2ex]
2\sin\theta\cos\theta &
\end{array}$$

11.

$$\begin{array}{c|c}
1-\cos 5\theta\cos 3\theta-\sin 5\theta\sin 3\theta & 2\sin^2\theta \\
1-[\cos 5\theta\cos 3\theta+\sin 5\theta\sin 3\theta] & 1-\cos 2\theta \\
1-\cos(5\theta-3\theta) & \\
1-\cos 2\theta &
\end{array}$$

13.

$$\begin{array}{c|c}
2\sin\theta\cos^3\theta+2\sin^3\theta\cos\theta & \sin 2\theta \\
2\sin\theta\cos\theta(\cos^2\theta+\sin^2\theta) & 2\sin\theta\cos\theta \\
2\sin\theta\cos\theta &
\end{array}$$

15.

$$\begin{array}{c|c}
\dfrac{\tan x-\sin x}{2\tan x} & \sin^2\dfrac{x}{2} \\[3ex]
\dfrac{1}{2}\left[\dfrac{\dfrac{\sin x}{\cos x}-\sin x}{\dfrac{\sin x}{\cos x}}\right] & \dfrac{1-\cos x}{2} \\[5ex]
\dfrac{1}{2}\dfrac{\sin x-\sin x\cos x}{\cos x}\cdot\dfrac{\cos x}{\sin x} & \\[3ex]
\dfrac{1-\cos x}{2} &
\end{array}$$

17.

$$\begin{array}{c|c}
\sin(\alpha+\beta)\sin(\alpha-\beta) & \sin^2\alpha-\sin^2\beta \\
\left(\begin{array}{c}\sin\alpha\cos\beta+\\\cos\alpha\sin\beta\end{array}\right)\left(\begin{array}{c}\sin\alpha\cos\beta-\\\cos\alpha\sin\beta\end{array}\right) & \begin{array}{l}1-\cos^2\alpha-\\(1-\cos^2\beta)\end{array} \\
\sin^2\alpha\cos^2\beta-\cos^2\alpha\sin^2\beta & \cos^2\beta-\cos^2\alpha \\
\begin{array}{c}\cos^2\beta(1-\cos^2\alpha)-\\\cos^2\alpha(1-\cos^2\beta)\end{array} & \\
\begin{array}{c}\cos^2\beta-\cos^2\alpha\cos^2\beta-\\\cos^2\alpha+\cos^2\alpha\cos^2\beta\end{array} & \\
\cos^2\beta-\cos^2\alpha &
\end{array}$$

19.

$$\begin{array}{c|c}
\tan\theta(\tan\theta+\cot\theta) & \sec^2\theta \\
\tan^2\theta+\tan\theta\cot\theta & \\
\tan^2\theta+1 & \\
\sec^2\theta &
\end{array}$$

21.

$$\begin{array}{c|c}
\dfrac{1+\cos^2 x}{\sin^2 x} & 2\csc^2 x-1 \\[2ex]
\dfrac{1}{\sin^2 x}+\dfrac{\cos^2 x}{\sin^2 x} & \\[2ex]
\csc^2 x+\cot^2 x & \\
\csc^2 x+\csc^2 x-1 & \\
2\csc^2 x-1 &
\end{array}$$

23.

$$\begin{array}{c|c}
\dfrac{1+\sin x}{1-\sin x}+\dfrac{\sin x-1}{1+\sin x} & 4\sec x\tan x \\[2ex]
\dfrac{(1+\sin x)^2-(1-\sin x)^2}{1-\sin^2 x} & 4\cdot\dfrac{1}{\cos x}\cdot\dfrac{\sin x}{\cos x} \\[2ex]
\dfrac{(1+2\sin x+\sin^2 x)-(1-2\sin x+\sin^2 x)}{\cos^2 x} & \dfrac{4\sin x}{\cos^2 x} \\[2ex]
\dfrac{4\sin x}{\cos^2 x} &
\end{array}$$

25.

$$
\begin{array}{c|c}
\cos^2 \alpha \cot^2 \alpha & \cot^2 \alpha - \cos^2 \alpha \\
(1 - \sin^2 \alpha) \cot^2 \alpha & \\
\cot^2 \alpha - \sin^2 \alpha \cdot \dfrac{\cos^2 \alpha}{\sin^2 \alpha} & \\
\cot^2 \alpha - \cos^2 \alpha &
\end{array}
$$

27.

$$
\begin{array}{c|c}
2 \sin^2 \theta \cos^2 \theta + \cos^4 \theta & 1 - \sin^4 \theta \\
\cos^2 \theta \,(2 \sin^2 \theta + \cos^2 \theta) & (1 + \sin^2 \theta)(1 - \sin^2 \theta) \\
\cos^2 \theta \,(\sin^2 \theta + \sin^2 \theta + \cos^2 \theta) & (1 + \sin^2 \theta)(\cos^2 \theta) \\
\cos^2 \theta \,(\sin^2 \theta + 1) &
\end{array}
$$

29.

$$
\begin{array}{c|c}
\dfrac{1 + \sin x}{1 - \sin x} & (\sec x + \tan x)^2 \\[2mm]
\dfrac{1 + \sin x}{1 - \sin x} \cdot \dfrac{1 + \sin x}{1 + \sin x} & \left(\dfrac{1}{\cos x} + \dfrac{\sin x}{\cos x}\right)^2 \\[2mm]
\dfrac{(1 + \sin x)^2}{1 - \sin^2 x} & \dfrac{(1 + \sin x)^2}{\cos^2 x} \\[2mm]
\dfrac{(1 + \sin x)^2}{\cos^2 x} &
\end{array}
$$

31. B;

$$
\begin{array}{c|c}
\dfrac{\cos x + \cot x}{1 + \csc x} & \cos x \\[2mm]
\dfrac{\dfrac{\cos x}{1} + \dfrac{\cos x}{\sin x}}{1 + \dfrac{1}{\sin x}} & \\[4mm]
\dfrac{\sin x \cos x + \cos x}{\sin x} \cdot \dfrac{\sin x}{\sin x + 1} & \\[2mm]
\dfrac{\cos x\,(\sin x + 1)}{\sin x + 1} & \\[2mm]
\cos x &
\end{array}
$$

33. A;

$$
\begin{array}{c|c}
\sin x \cos x + 1 & \dfrac{\sin^3 x - \cos^3 x}{\sin x - \cos x} \\[2mm]
& \dfrac{(\sin x - \cos x)(\sin^2 x + \sin x \cos x + \cos^2 x)}{\sin x - \cos x} \\[2mm]
& \sin^2 x + \sin x \cos x + \cos^2 x \\[2mm]
& \sin x \cos x + 1
\end{array}
$$

35. C;

$$
\begin{array}{c|c}
\dfrac{1}{\cot x \sin^2 x} & \tan x + \cot x \\[2mm]
\dfrac{1}{\dfrac{\cos x}{\sin x} \cdot \sin^2 x} & \dfrac{\sin x}{\cos x} + \dfrac{\cos x}{\sin x} \\[4mm]
\dfrac{1}{\cos x \sin x} & \dfrac{\sin^2 x + \cos^2 x}{\cos x \sin x} \\[2mm]
& \dfrac{1}{\cos x \sin x}
\end{array}
$$

37. Discussion and Writing

39. [1.6] **(a)**, **(d)**

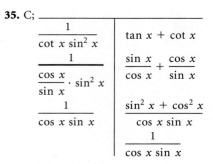

(b) yes; **(c)** $f^{-1}(x) = \dfrac{x + 2}{3}$

40. [1.6] **(a)**, **(d)**

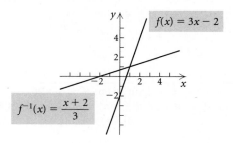

(b) yes; **(c)** $f^{-1}(x) = \sqrt[3]{x - 1}$

41. [1.6] **(a)**, **(d)**

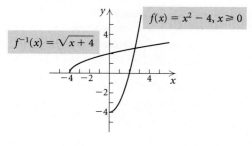

(b) yes; **(c)** $f^{-1}(x) = \sqrt{x + 4}$

42. [1.6] **(a)**, **(d)**

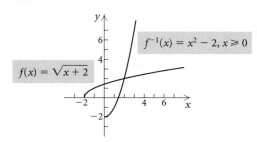

$f^{-1}(x) = x^2 - 2, x \geq 0$

$f(x) = \sqrt{x + 2}$

(b) yes; **(c)** $f^{-1}(x) = x^2 - 2, x \geq 0$

43.

$\ln|\tan x|$	$-\ln|\cot x|$
$\ln\left|\dfrac{1}{\cot x}\right|$	
$\ln|1| - \ln|\cot x|$	
$0 - \ln|\cot x|$	
$-\ln|\cot x|$	

45. $\log(\cos x - \sin x) + \log(\cos x + \sin x)$
$= \log[(\cos x - \sin x)(\cos x + \sin x)]$
$= \log(\cos^2 x - \sin^2 x) = \log\cos 2x$

47. $\dfrac{1}{\omega C(\tan\theta + \tan\phi)} = \dfrac{1}{\omega C\left(\dfrac{\sin\theta}{\cos\theta} + \dfrac{\sin\phi}{\cos\phi}\right)}$

$= \dfrac{1}{\omega C\left(\dfrac{\sin\theta\cos\phi + \sin\phi\cos\theta}{\cos\theta\cos\phi}\right)}$

$= \dfrac{\cos\theta\cos\phi}{\omega C\sin(\theta + \phi)}$

Exercise Set 3.4

1. $-\dfrac{\pi}{3}, -60°$ **3.** $\dfrac{\pi}{4}, 45°$ **5.** $\dfrac{\pi}{4}, 45°$ **7.** $0, 0°$

9. $\dfrac{\pi}{6}, 30°$ **11.** $\dfrac{\pi}{6}, 30°$ **13.** $-\dfrac{\pi}{6}, -30°$

15. $-\dfrac{\pi}{6}, -30°$ **17.** $\dfrac{\pi}{2}, 90°$ **19.** $\dfrac{\pi}{3}, 60°$

21. $0.352, 20.2°$ **23.** $1.292, 74.0°$ **25.** $2.946, 168.8°$
27. $-0.160, -9.2°$ **29.** $0.829, 47.5°$ **31.** $-0.960, -55.0°$
33. $\sin^{-1}: [-1, 1];\ \cos^{-1}: [-1, 1];\ \tan^{-1}: (-\infty, \infty)$

35. $\theta = \sin^{-1}\left(\dfrac{2000}{d}\right)$ **37.** 0.3 **39.** $\dfrac{\pi}{4}$ **41.** $\dfrac{\pi}{5}$

43. $-\dfrac{\pi}{3}$ **45.** $\dfrac{1}{2}$ **47.** 1 **49.** $\dfrac{\pi}{3}$ **51.** $\dfrac{\sqrt{11}}{33}$

53. $\dfrac{a}{\sqrt{a^2 + 9}}$ **55.** $\dfrac{\sqrt{q^2 - p^2}}{p}$ **57.** $\dfrac{p}{3}$ **59.** $\dfrac{\sqrt{3}}{2}$

61. $-\dfrac{\sqrt{2}}{10}$ **63.** $xy + \sqrt{(1 - x^2)(1 - y^2)}$ **65.** 0.9861

67. Discussion and Writing   **69.** Discussion and Writing
70. [1.4] x-axis: no; y-axis: yes; origin: no
71. [1.4] x-axis: yes; y-axis: yes; origin: yes
72. [1.4] Neither   **73.** [1.4] Odd
75.

$\sin^{-1} x + \cos^{-1} x$	$\dfrac{\pi}{2}$
$\sin(\sin^{-1} x + \cos^{-1} x)$	$\sin\dfrac{\pi}{2}$
$[\sin(\sin^{-1} x)][\cos(\cos^{-1} x)] +$ $[\cos(\sin^{-1} x)][\sin(\cos^{-1} x)]$	1
$x \cdot x + \sqrt{1 - x^2} \cdot \sqrt{1 - x^2}$	
$x^2 + 1 - x^2$	
1	

77.

$\sin^{-1} x$	$\tan^{-1}\dfrac{x}{\sqrt{1 - x^2}}$
$\sin(\sin^{-1} x)$	$\sin\left(\tan^{-1}\dfrac{x}{\sqrt{1 - x^2}}\right)$
x	x

79.

$\arcsin x$	$\arccos\sqrt{1 - x^2}$
$\sin(\arcsin x)$	$\sin(\arccos\sqrt{1 - x^2})$
x	x

81. $\theta = \arctan\dfrac{y + h}{x} - \arctan\dfrac{y}{x}; 38.7°$

Exercise Set 3.5

1. $\dfrac{\pi}{6} + 2k\pi, \dfrac{11\pi}{6} + 2k\pi$, or $30° + k \cdot 360°, 330° + k \cdot 360°$

3. $\dfrac{2\pi}{3} + k\pi$, or $120° + k \cdot 180°$ **5.** $98.09°, 261.91°$

7. $\dfrac{4\pi}{3}, \dfrac{5\pi}{3}$ **9.** $\dfrac{\pi}{4}, \dfrac{3\pi}{4}, \dfrac{5\pi}{4}, \dfrac{7\pi}{4}$ **11.** $\dfrac{\pi}{6}, \dfrac{5\pi}{6}, \dfrac{3\pi}{2}$

13. $\dfrac{\pi}{6}, \dfrac{\pi}{2}, \dfrac{3\pi}{2}, \dfrac{11\pi}{6}$ **15.** $109.47°, 120°, 240°, 250.53°$

17. $0, \dfrac{\pi}{4}, \dfrac{3\pi}{4}, \pi, \dfrac{5\pi}{4}, \dfrac{7\pi}{4}$ **19.** $139.81°, 220.19°$

21. $37.22°, 169.35°, 217.22°, 349.35°$ **23.** $0, \pi, \dfrac{7\pi}{6}, \dfrac{11\pi}{6}$

25. $0, \dfrac{\pi}{2}, \pi, \dfrac{3\pi}{2}$ **27.** $0, \pi$ **29.** $\dfrac{3\pi}{4}, \dfrac{7\pi}{4}$

31. $\dfrac{2\pi}{3}, \dfrac{4\pi}{3}, \dfrac{3\pi}{2}$ **33.** $\dfrac{\pi}{4}, \dfrac{3\pi}{4}, \dfrac{5\pi}{4}, \dfrac{7\pi}{4}$ **35.** $\dfrac{\pi}{12}, \dfrac{5\pi}{12}$

37. $0.967, 1.853, 4.109, 4.994$ **39.** $\dfrac{2\pi}{3}, \dfrac{4\pi}{3}$

41. $1.114, 2.773$ **43.** 0.515 **45.** $0.422, 1.756$
47. Discussion and Writing   **49.** [2.2] $B = 35°$,
$b \approx 140.7, c \approx 245.4$ **50.** [2.2] $R \approx 15.5°, T \approx 74.5°$,
$t \approx 13.7$

51. [1.2] $y = \frac{3}{4}x - 5$ **52.** [1.2] $y = -10$
53. $\frac{\pi}{3}, \frac{2\pi}{3}, \frac{4\pi}{3}, \frac{5\pi}{3}$ **55.** $\frac{\pi}{3}, \frac{4\pi}{3}$ **57.** 0
59. $e^{3\pi/2 + 2k\pi}$, where k (an integer) ≤ -1
61. (a) $y = 7\sin(-2.6180x + 0.5236) + 7$;
(b) \$10,500, \$13,062
63. 1.24 days, 6.76 days   **65.** 16.5°N   **67.** 1   **69.** 0.1923

Review Exercises, Chapter 3

1. [3.1] $\csc^2 x$ **2.** [3.1] 1   **3.** [3.1] $\tan^2 y - \cot^2 y$
4. [3.1] $\dfrac{(\cos^2 x + 1)^2}{\cos^2 x}$ **5.** [3.1] $\csc x(\sec x - \csc x)$
6. [3.1] $(3\sin y + 5)(\sin y - 4)$
7. [3.1] $(10 - \cos u)(100 + 10\cos u + \cos^2 u)$ **8.** [3.1] 1
9. [3.1] $\dfrac{1}{2}\sec x$ **10.** [3.1] $\dfrac{3\tan x}{\sin x - \cos x}$
11. [3.1] $\dfrac{3\cos y + 3\sin y + 2}{\cos^2 y - \sin^2 y}$ **12.** [3.1] 1
13. [3.1] $\dfrac{1}{4}\cot x$ **14.** [3.1] $\sin x + \cos x$
15. [3.1] $\dfrac{\cos x}{1 - \sin x}$ **16.** [3.1] $\dfrac{\cos x}{\sqrt{\sin x}}$ **17.** [3.1] $3\sec\theta$
18. [3.1] $\cos x \cos\dfrac{3\pi}{2} - \sin x \sin\dfrac{3\pi}{2}$
19. [3.1] $\dfrac{\tan 45° - \tan 30°}{1 + \tan 45° \tan 30°}$
20. [3.1] $\cos(27° - 16°)$, or $\cos 11°$ **21.** [3.1] $\dfrac{-\sqrt{6} - \sqrt{2}}{4}$
22. [3.1] $2 - \sqrt{3}$ **23.** [3.1] -0.3745 **24.** [3.2] $-\sin x$
25. [3.2] $\sin x$ **26.** [3.2] $-\cos x$
27. [3.2] **(a)** $\sin\alpha = -\dfrac{4}{5}$, $\tan\alpha = \dfrac{4}{3}$, $\cot\alpha = \dfrac{3}{4}$,
$\sec\alpha = -\dfrac{5}{3}$, $\csc\alpha = -\dfrac{5}{4}$; **(b)** $\sin\left(\dfrac{\pi}{2} - \alpha\right) = -\dfrac{3}{5}$,
$\cos\left(\dfrac{\pi}{2} - \alpha\right) = -\dfrac{4}{5}$, $\tan\left(\dfrac{\pi}{2} - \alpha\right) = \dfrac{3}{4}$,
$\cot\left(\dfrac{\pi}{2} - \alpha\right) = \dfrac{4}{3}$, $\sec\left(\dfrac{\pi}{2} - \alpha\right) = -\dfrac{5}{4}$,
$\csc\left(\dfrac{\pi}{2} - \alpha\right) = -\dfrac{5}{3}$; **(c)** $\sin\left(\alpha + \dfrac{\pi}{2}\right) = -\dfrac{3}{5}$,
$\cos\left(\alpha + \dfrac{\pi}{2}\right) = \dfrac{4}{5}$, $\tan\left(\alpha + \dfrac{\pi}{2}\right) = -\dfrac{3}{4}$,
$\cot\left(\alpha + \dfrac{\pi}{2}\right) = -\dfrac{4}{3}$, $\sec\left(\alpha + \dfrac{\pi}{2}\right) = \dfrac{5}{4}$,
$\csc\left(\alpha + \dfrac{\pi}{2}\right) = -\dfrac{5}{3}$ **28.** [3.2] $-\sec x$

29. [3.2] $\tan 2\theta = \dfrac{24}{7}$, $\cos 2\theta = \dfrac{7}{25}$, $\sin 2\theta = \dfrac{24}{25}$; I
30. [3.2] $\dfrac{\sqrt{2 - \sqrt{2}}}{2}$
31. [3.2] $\sin 2\beta = 0.4261$, $\cos\dfrac{\beta}{2} = 0.9940$,
$\cos 4\beta = 0.6369$ **32.** [3.2] $\cos x$ **33.** [3.2] 1
34. [3.2] $\sin 2x$ **35.** [3.2] $\tan 2x$

36. [3.3]

$\dfrac{1 - \sin x}{\cos x}$	$\dfrac{\cos x}{1 + \sin x}$
$\dfrac{1 - \sin x}{\cos x} \cdot \dfrac{\cos x}{\cos x}$	$\dfrac{\cos x}{1 + \sin x} \cdot \dfrac{1 - \sin x}{1 - \sin x}$
$\dfrac{\cos x - \sin x \cos x}{\cos^2 x}$	$\dfrac{\cos x - \sin x \cos x}{1 - \sin^2 x}$
	$\dfrac{\cos x - \sin x \cos x}{\cos^2 x}$

37. [3.3]

$\dfrac{1 + \cos 2\theta}{\sin 2\theta}$	$\cot\theta$
$\dfrac{1 + 2\cos^2\theta - 1}{2\sin\theta\cos\theta}$	$\dfrac{\cos\theta}{\sin\theta}$
$\dfrac{\cos\theta}{\sin\theta}$	

38. [3.3]

$\dfrac{\tan y + \sin y}{2\tan y}$	$\cos^2\dfrac{y}{2}$
$\dfrac{1}{2}\left[\dfrac{\dfrac{\sin y + \sin y \cos y}{\cos y}}{\dfrac{\sin y}{\cos y}}\right]$	$\dfrac{1 + \cos y}{2}$
$\dfrac{1}{2}\left[\dfrac{\sin y(1 + \cos y)}{\cos y} \cdot \dfrac{\cos y}{\sin y}\right]$	
$\dfrac{1 + \cos y}{2}$	

39. [3.3]

$\dfrac{\sin x - \cos x}{\cos^2 x}$	$\dfrac{\tan^2 x - 1}{\sin x + \cos x}$
	$\dfrac{\dfrac{\sin^2 x}{\cos^2 x} - 1}{\sin x + \cos x}$
	$\dfrac{\sin^2 x - \cos^2 x}{\cos^2 x} \cdot \dfrac{1}{\sin x + \cos x}$
	$\dfrac{\sin x - \cos x}{\cos^2 x}$

40. [3.3] B;
$$\begin{array}{c|c}
\csc x - \cos x \cot x & \sin x \\
\dfrac{1}{\sin x} - \cos x\,\dfrac{\cos x}{\sin x} & \\
\dfrac{1 - \cos^2 x}{\sin x} & \\
\dfrac{\sin^2 x}{\sin x} & \\
\sin x &
\end{array}$$

41. [3.3] D;
$$\begin{array}{c|c}
\dfrac{1}{\sin x \cos x} - \dfrac{\cos x}{\sin x} & \dfrac{\sin x \cos x}{1 - \sin^2 x} \\
\dfrac{1}{\sin x \cos x} - \dfrac{\cos^2 x}{\sin x \cos x} & \dfrac{\sin x \cos x}{\cos^2 x} \\
\dfrac{1 - \cos^2 x}{\sin x \cos x} & \dfrac{\sin x}{\cos x} \\
\dfrac{\sin^2 x}{\sin x \cos x} & \\
\dfrac{\sin x}{\cos x} &
\end{array}$$

42. [3.3] A;
$$\begin{array}{c|c}
\dfrac{\cot x - 1}{1 - \tan x} & \dfrac{\csc x}{\sec x} \\
\dfrac{\frac{\cos x}{\sin x} - \frac{\sin x}{\sin x}}{\frac{\cos x}{\cos x} - \frac{\sin x}{\cos x}} & \dfrac{\frac{1}{\sin x}}{\frac{1}{\cos x}} \\
\dfrac{\cos x - \sin x}{\sin x} \cdot \dfrac{\cos x}{\cos x - \sin x} & \dfrac{1}{\sin x} \cdot \dfrac{\cos x}{1} \\
\dfrac{\cos x}{\sin x} & \dfrac{\cos x}{\sin x}
\end{array}$$

43. [3.3] C;
$$\begin{array}{c|c}
\dfrac{\cos x + 1}{\sin x} + \dfrac{\sin x}{\cos x + 1} & \dfrac{2}{\sin x} \\
\dfrac{(\cos x + 1)^2 + \sin^2 x}{\sin x (\cos x + 1)} & \\
\dfrac{\cos^2 x + 2\cos x + 1 + \sin^2 x}{\sin x (\cos x + 1)} & \\
\dfrac{2\cos x + 2}{\sin x (\cos x + 1)} & \\
\dfrac{2(\cos x + 1)}{\sin x (\cos x + 1)} & \\
\dfrac{2}{\sin x} &
\end{array}$$

44. [3.4] $-\dfrac{\pi}{6}$, $-30°$ **45.** [3.4] $\dfrac{\pi}{6}$, $30°$

46. [3.4] $\dfrac{\pi}{4}$, $45°$ **47.** [3.4] 0, 0°   **48.** [3.4] 1.792, 102.7°

49. [3.4] 0.398, 22.8°   **50.** [3.4] $\dfrac{1}{2}$ **51.** [3.4] $\dfrac{\sqrt{3}}{3}$

52. [3.4] $\dfrac{\pi}{7}$ **53.** [3.4] $\dfrac{\sqrt{2}}{2}$ **54.** [3.4] $\dfrac{3}{\sqrt{b^2 + 9}}$

55. [3.4] $-\dfrac{7}{25}$

56. [3.5] $\dfrac{3\pi}{4} + 2k\pi$, $\dfrac{5\pi}{4} + 2k\pi$, or $135° + k \cdot 360°$,

$225° + k \cdot 360°$ **57.** [3.5] $\dfrac{\pi}{3} + k\pi$, or $60° + k \cdot 180°$

58. [3.5] $\dfrac{\pi}{6}, \dfrac{5\pi}{6}, \dfrac{7\pi}{6}, \dfrac{11\pi}{6}$

59. [3.5] $\dfrac{\pi}{4}, \dfrac{\pi}{2}, \dfrac{3\pi}{4}, \dfrac{5\pi}{4}, \dfrac{3\pi}{2}, \dfrac{7\pi}{4}$

60. [3.5] $\dfrac{2\pi}{3}, \pi, \dfrac{4\pi}{3}$

61. [3.5] 0, π **62.** [3.5] $\dfrac{\pi}{4}, \dfrac{3\pi}{4}, \dfrac{5\pi}{4}, \dfrac{7\pi}{4}$

63. [3.5] 0, $\dfrac{\pi}{2}, \pi, \dfrac{3\pi}{2}$ **64.** [3.5] $\dfrac{7\pi}{12}, \dfrac{23\pi}{12}$

65. [3.5] 0.864, 2.972, 4.006, 6.114
66. [3.5] 4.917   **67.** [3.5] No solution in $[0, 2\pi)$
68. Discussion and Writing [3.3]
(a) $2 \cos^2 x - 1 = \cos 2x = \cos^2 x - \sin^2 x$
$= 1 \cdot (\cos^2 x - \sin^2 x)$
$= (\cos^2 x + \sin^2 x)(\cos^2 x - \sin^2 x)$
$= \cos^4 x - \sin^4 x;$
(b) $\cos^4 x - \sin^4 x = (\cos^2 x + \sin^2 x)(\cos^2 x - \sin^2 x)$
$= 1 \cdot (\cos^2 x - \sin^2 x)$
$= \cos^2 x - \sin^2 x = \cos 2x$
$= 2 \cos^2 x - 1;$
(c)
$$\begin{array}{c|l}
2 \cos^2 x - 1 & \cos^4 x - \sin^4 x \\
\cos 2x & (\cos^2 x + \sin^2 x)(\cos^2 x - \sin^2 x) \\
 & 1 \cdot (\cos^2 x - \sin^2 x) \\
 & \cos^2 x - \sin^2 x \\
 & \cos 2x
\end{array}$$
Answer may vary. Method 2 may be the more efficient because it involves straightforward factorization and simplification. Method 1(a) requires a "trick" such as multiplying by a particular expression equivalent to 1.
69. Discussion and Writing [3.4] The ranges of the inverse trigonometric functions are restricted in order that they might be functions.
70. [3.1] 108.4°
71. [3.1]
$\cos (u + v) = \cos u \cos v - \sin u \sin v$
$= \cos u \cos v - \cos\left(\dfrac{\pi}{2} - u\right) \cos\left(\dfrac{\pi}{2} - v\right)$

72. [3.2] $\cos^2 x$

73. [3.2] $\sin\theta = \sqrt{\dfrac{1}{2}+\dfrac{\sqrt{6}}{5}}$; $\cos\theta = \sqrt{\dfrac{1}{2}-\dfrac{\sqrt{6}}{5}}$;

$\tan\theta = \sqrt{\dfrac{5+2\sqrt{6}}{5-2\sqrt{6}}}$

74. [3.3] $\ln e^{\sin t} = \log_e e^{\sin t} = \sin t$

75. [3.4]

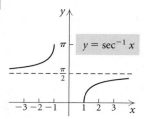

76. [3.4] Let $x = \dfrac{\sqrt{2}}{2}$. Then $\tan^{-1}\dfrac{\sqrt{2}}{2} \approx 0.6155$ and

$\dfrac{\sin^{-1}\dfrac{\sqrt{2}}{2}}{\cos^{-1}\dfrac{\sqrt{2}}{2}} = \dfrac{\dfrac{\pi}{4}}{\dfrac{\pi}{4}} = 1.$ **77.** [3.5] $\dfrac{\pi}{2},\dfrac{3\pi}{2}$

Chapter 4

Exercise Set 4.1

1. $A = 121°$, $a \approx 33$, $c \approx 14$ **3.** $B \approx 57.4°$, $C \approx 86.1°$, $c \approx 40$, or $B \approx 122.6°$, $C \approx 20.9°$, $c \approx 14$
5. $B \approx 44°24'$, $A \approx 74°26'$, $a \approx 33.3$
7. $A = 110.36°$, $a \approx 5$ mi, $b \approx 3$ mi
9. $B \approx 83.78°$, $A \approx 12.44°$, $a \approx 12.30$ yd
11. $B \approx 14.7°$, $C \approx 135.0°$, $c \approx 28.04$ cm
13. No solution   **15.** $B = 125.27°$, $b \approx 302$ m, $c \approx 138$ m
17. 8.2 ft^2 **19.** 12 yd^2 **21.** 596.98 ft^2 **23.** 787 ft^2
25. About 12.86 ft, or 12 ft, 10 in.   **27.** About 51 ft
29. From A: about 35 mi; from B: about 66 mi
31. About 60 mi   **33.** Discussion and Writing
35. [2.1] 1.348, $77.2°$ **36.** [2.1] No angle
37. [2.1] $18.24°$ **38.** [2.1] $125.06°$
39. Use the formula for the area of a triangle and the law of sines.

$$K = \frac{1}{2}bc\sin A \quad\text{and}\quad b = \frac{c\sin B}{\sin C},$$

$$\text{so}\quad K = \frac{c^2\sin A\sin B}{2\sin C}.$$

$$K = \frac{1}{2}ab\sin C \quad\text{and}\quad b = \frac{a\sin B}{\sin A},$$

$$\text{so}\quad K = \frac{a^2\sin B\sin C}{2\sin A}.$$

$$K = \frac{1}{2}bc\sin A \quad\text{and}\quad c = \frac{b\sin C}{\sin B},$$

$$\text{so}\quad K = \frac{b^2\sin A\sin C}{2\sin B}.$$

41.

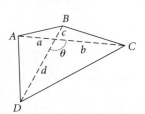

For the quadrilateral $ABCD$, we have

$$\text{Area} = \frac{1}{2}bd\sin\theta + \frac{1}{2}ac\sin\theta$$

$$+ \frac{1}{2}ad(\sin 180° - \theta) + \frac{1}{2}bc\sin(180° - \theta)$$

Note: $\sin\theta = \sin(180° - \theta)$.

$$= \frac{1}{2}(bd + ac + ad + bc)\sin\theta$$

$$= \frac{1}{2}(a+b)(c+d)\sin\theta$$

$$= \frac{1}{2}d_1 d_2\sin\theta,$$

where $d_1 = a + b$ and $d_2 = c + d$.

Exercise Set 4.2

1. $a \approx 15$, $B \approx 24°$, $C \approx 126°$
3. $A \approx 36.18°$, $B \approx 43.53°$, $C \approx 100.29°$
5. $b \approx 75$ m, $A \approx 94°51'$, $C \approx 12°29'$
7. $A \approx 24.15°$, $B \approx 30.75°$, $C \approx 125.10°$
9. No solution
11. $A \approx 79.93°$, $B \approx 53.55°$, $C \approx 46.52°$
13. $c \approx 45.17$ mi, $A \approx 89.3°$, $B \approx 42.0°$
15. $a \approx 13.9$ in., $B \approx 36.127°$, $C \approx 90.417°$
17. Law of sines; $C = 98°$, $a \approx 96.7$, $c \approx 101.9$
19. Law of cosines; $A \approx 73.71°$, $B \approx 51.75°$, $C \approx 54.54°$
21. Cannot be solved
23. Law of cosines; $A \approx 33.71°$, $B \approx 107.08°$, $C \approx 39.21°$
25. About 367 ft   **27.** About 1.5 mi
29. About 37 nautical mi   **31.** About 912 km
33. (a) About 16 ft; (b) about 122 ft^2 **35.** About 4.7 cm
37. Discussion and Writing
39. [2.3] $\dfrac{\sqrt{2}}{2}$ **40.** [2.3] $\dfrac{\sqrt{3}}{2}$ **41.** [2.3] $-\dfrac{1}{2}$

42. [2.3] $-\dfrac{\sqrt{3}}{2}$ **43.** About 9386 ft

45. $A = \dfrac{1}{2}a^2 \sin\theta$; when $\theta = 90°$

Exercise Set 4.3

1. $2 + 11i$ **3.** $5 - 12i$ **5.** $5 + 9i$ **7.** $5 + 4i$
9. $5 + 7i$ **11.** $13 - i$ **13.** $-11 + 16i$ **15.** $35 + 14i$
17. $-14 + 23i$ **19.** 41   **21.** $12 + 16i$ **23.** $-45 - 28i$
25. $\dfrac{-4\sqrt{3} + 10}{41} + \dfrac{5\sqrt{3} + 8}{41}i$ **27.** $-\dfrac{14}{13} + \dfrac{5}{13}i$
29. $\frac{15}{146} + \frac{33}{146}i$ **31.** $-\frac{1}{2} + \frac{1}{2}i$ **33.** $-\frac{1}{2} - \frac{13}{2}i$ **35.** $-i$
37. $-i$ **39.** 1   **41.** i **43.** 625   **45.** Discussion and
Writing   **47.** [1.2] $y = \frac{1}{2}x - \frac{13}{2}$ **48.** [1.3] $x^2 - 3x - 1$
49. [1.3] $\frac{8}{11}$ **50.** [1.3] $2x + h - 3$ **51.** True   **53.** True
55. $a^2 + b^2$

Exercise Set 4.4

1. 5;

Imaginary / Real axes graph showing point $4 + 3i$

3. 1;

Imaginary / Real axes graph showing point i

5. $\sqrt{17}$;

Imaginary / Real axes graph showing point $4 - i$

7. 3;

Imaginary / Real axes graph showing point 3

9. $3 - 3i$; $3\sqrt{2}\left(\cos\dfrac{7\pi}{4} + i\sin\dfrac{7\pi}{4}\right)$, or
$3\sqrt{2}(\cos 315° + i\sin 315°)$

11. $4i$; $4\left(\cos\dfrac{\pi}{2} + i\sin\dfrac{\pi}{2}\right)$, or $4(\cos 90° + i\sin 90°)$

13. $\sqrt{2}\left(\cos\dfrac{7\pi}{4} + i\sin\dfrac{7\pi}{4}\right)$, or $\sqrt{2}(\cos 315° + i\sin 315°)$

15. $3\left(\cos\dfrac{3\pi}{2} + i\sin\dfrac{3\pi}{2}\right)$, or $3(\cos 270° + i\sin 270°)$

17. $2\left(\cos\dfrac{\pi}{6} + i\sin\dfrac{\pi}{6}\right)$, or $2(\cos 30° + i\sin 30°)$

19. $\dfrac{2}{5}(\cos 0 + i\sin 0)$, or $\dfrac{2}{5}(\cos 0° + i\sin 0°)$

21. $\dfrac{3\sqrt{3}}{2} + \dfrac{3}{2}i$ **23.** $-10i$ **25.** $2 + 2i$
27. $2i$ **29.** $4(\cos 42° + i\sin 42°)$
31. $11.25(\cos 56° + i\sin 56°)$ **33.** 4   **35.** $-i$
37. $6 + 6\sqrt{3}i$ **39.** $-2i$ **41.** $8(\cos\pi + i\sin\pi)$
43. $8\left(\cos\dfrac{3\pi}{2} + i\sin\dfrac{3\pi}{2}\right)$ **45.** $\dfrac{27}{2} + \dfrac{27\sqrt{3}}{2}i$
47. $-4 + 4i$ **49.** -1 **51.** $-\dfrac{\sqrt{2}}{2} + \dfrac{\sqrt{2}}{2}i, \dfrac{\sqrt{2}}{2} - \dfrac{\sqrt{2}}{2}i$
53. $2(\cos 157.5° + i\sin 157.5°), 2(\cos 337.5° + i\sin 337.5°)$
55. $\dfrac{\sqrt{3}}{2} + \dfrac{1}{2}i, -\dfrac{\sqrt{3}}{2} + \dfrac{1}{2}i, -i$
57. $\sqrt[3]{4}(\cos 110° + i\sin 110°), \sqrt[3]{4}(\cos 230° + i\sin 230°),$
$\sqrt[3]{4}(\cos 350° + i\sin 350°)$
59. $2, 2i, -2, -2i$;

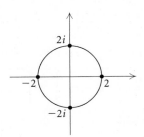

61. cos 36° + *i* sin 36°, cos 108° + *i* sin 108°, −1,
cos 252° + *i* sin 252°, cos 324° + *i* sin 324°;

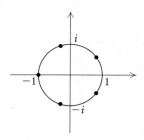

63. $\sqrt[10]{8}$, $\sqrt[10]{8}$(cos 36° + *i* sin 36°), $\sqrt[10]{8}$(cos 72° + *i* sin 72°),
$\sqrt[10]{8}$(cos 108° + *i* sin 108°), $\sqrt[10]{8}$(cos 144° + *i* sin 144°),
−$\sqrt[10]{8}$, $\sqrt[10]{8}$(cos 216° + *i* sin 216°),
$\sqrt[10]{8}$(cos 252° + *i* sin 252°), $\sqrt[10]{8}$(cos 288° + *i* sin 288°),
$\sqrt[10]{8}$(cos 324° + *i* sin 324°)

65. $\dfrac{\sqrt{3}}{2} + \dfrac{1}{2}i$, *i*, $-\dfrac{\sqrt{3}}{2} + \dfrac{1}{2}i$, $-\dfrac{\sqrt{3}}{2} - \dfrac{1}{2}i$, $-i$, $\dfrac{\sqrt{3}}{2} - \dfrac{1}{2}i$

67. 1, $-\dfrac{1}{2} + \dfrac{\sqrt{3}}{2}i$, $-\dfrac{1}{2} - \dfrac{\sqrt{3}}{2}i$

69. cos 67.5° + *i* sin 67.5°, cos 157.5° + *i* sin 157.5°,
cos 247.5° + *i* sin 247.5°, cos 337.5° + *i* sin 337.5°

71. $\sqrt{3} + i$, 2*i*, $-\sqrt{3} + i$, $-\sqrt{3} - i$, $-2i$, $\sqrt{3} - i$

73. Discussion and Writing   **75.** [2.4] 15°

76. [2.4] 540°   **77.** [2.4] $\dfrac{11\pi}{6}$ **78.** [2.4] $-\dfrac{5\pi}{4}$

79. [1.2] Slope: $-\dfrac{1}{5}$; *y*-intercept: (0, −15)

80. [1.2] Slope: 0; *y*-intercept: (0, 7)

81. $-\dfrac{1 + \sqrt{3}}{2} + \dfrac{1 + \sqrt{3}}{2}i$, $\dfrac{1 - \sqrt{3}}{2} + \dfrac{1 - \sqrt{3}}{2}i$

83. cos *θ* − *i* sin *θ*

85. $z = a + bi$, $|z| = \sqrt{a^2 + b^2}$; $\bar{z} = a - bi$,
$|\bar{z}| = \sqrt{a^2 + (-b)^2} = \sqrt{a^2 + b^2}$, ∴ $|z| = |\bar{z}|$

87. $|(a + bi)^2| = |a^2 - b^2 + 2abi|$
$= \sqrt{(a^2 - b^2)^2 + 4a^2b^2}$
$= \sqrt{a^4 + 2a^2b^2 + b^4} = a^2 + b^2$,
$|a + bi|^2 = (\sqrt{a^2 + b^2})^2 = a^2 + b^2$

89. $\dfrac{z}{w} = \dfrac{r_1(\cos \theta_1 + i \sin \theta_1)}{r_2(\cos \theta_2 + i \sin \theta_2)}$

$= \dfrac{r_1}{r_2}(\cos (\theta_1 - \theta_2) + i \sin (\theta_1 - \theta_2))$,

$\left|\dfrac{z}{w}\right| = \sqrt{\left[\dfrac{r_1}{r_2} \cos (\theta_1 - \theta_2)\right]^2 + \left[\dfrac{r_1}{r_2} \sin (\theta_1 - \theta_2)\right]^2}$

$= \sqrt{\dfrac{r_1^2}{r_2^2}} = \dfrac{|r_1|}{|r_2|}$;

$|z| = \sqrt{(r_1 \cos \theta_1)^2 + (r_1 \sin \theta_1)^2}$
$= \sqrt{r_1^2} = |r_1|$;
$|w| = \sqrt{(r_2 \cos \theta_2)^2 + (r_2 \sin \theta_2)^2}$
$= \sqrt{r_2^2} = |r_2|$;

Then $\left|\dfrac{z}{w}\right| = \dfrac{|r_1|}{|r_2|} = \dfrac{|z|}{|w|}$.

91.

Exercise Set 4.5

1. Yes   **3.** No   **5.** Yes   **7.** No   **9.** No   **11.** Yes
13. 55 N, 55°   **15.** 929 N, 19°   **17.** 57.0, 38°
19. 18.4, 37°   **21.** 20.9, 58°   **23.** 68.3, 18°
25. 11 ft/sec, 63°   **27.** 174 nautical mi, S15°E   **29.** 60°
31. 70.7 east; 70.7 south
33. Horizontal: 215.17 mph forward; vertical: 65.78 mph up
35. Horizontal: 390 lb forward; vertical: 675.5 lb up
37. Northerly: 115 km/h; westerly: 164 km/h
39. Perpendicular: 90.6 lb; parallel: 42.3 lb   **41.** 48.1 lb

43. Discussion and Writing   **45.** [2.5] $\dfrac{\sqrt{3}}{2}$

46. [2.5] $\dfrac{\sqrt{3}}{2}$ **47.** [2.5] $\dfrac{\sqrt{2}}{2}$ **48.** [2.5] $\dfrac{1}{2}$

49. (a) (4.950, 4.950); (b) (0.950, −1.978)

Exercise Set 4.6

1. ⟨−9, 5⟩; $\sqrt{106}$ **3.** ⟨−3, 6⟩; $3\sqrt{5}$ **5.** ⟨4, 0⟩; 4
7. $\sqrt{37}$ **9.** ⟨4, −5⟩   **11.** $\sqrt{257}$ **13.** ⟨−9, 9⟩
15. ⟨41, −38⟩   **17.** $\sqrt{261} - \sqrt{65}$ **19.** ⟨−1, −1⟩
21. ⟨−8, 14⟩   **23.** 1   **25.** −34
27.   **29.**

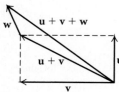

31. (a) w = u + v; (b) v = w − u

33. $\left\langle -\dfrac{5}{13}, \dfrac{12}{13}\right\rangle$ **35.** $\left\langle \dfrac{1}{\sqrt{101}}, -\dfrac{10}{\sqrt{101}}\right\rangle$

37. $\left\langle -\dfrac{1}{\sqrt{17}}, -\dfrac{4}{\sqrt{17}}\right\rangle$ **39.** w = −4i + 6j

41. $\mathbf{s} = 2\mathbf{i} + 5\mathbf{j}$ **43.** $-7\mathbf{i} + 5\mathbf{j}$ **45. (a)** $3\mathbf{i} + 29\mathbf{j}$;
(b) $\langle 3, 29 \rangle$ **47. (a)** $4\mathbf{i} + 16\mathbf{j}$; **(b)** $\langle 4, 16 \rangle$ **49.** $\mathbf{j}$, or $\langle 0, 1 \rangle$
51. $-\dfrac{1}{2}\mathbf{i} - \dfrac{\sqrt{3}}{2}\mathbf{j}$, or $\left\langle -\dfrac{1}{2}, -\dfrac{\sqrt{3}}{2} \right\rangle$ **53.** $248°$
55. $63°$ **57.** $50°$ **59.** $|\mathbf{u}| = 3$; $\theta = 45°$ **61.** 1; $120°$
63. $144.2°$ **65.** $14.0°$ **67.** $101.3°$
69.

71. $\mathbf{u} = -\dfrac{\sqrt{2}}{2}\mathbf{i} - \dfrac{\sqrt{2}}{2}\mathbf{j}$ **73.** $\mathbf{u} = -\dfrac{\sqrt{10}}{10}\mathbf{i} + \dfrac{3\sqrt{10}}{10}\mathbf{j}$
75. $\sqrt{13}\left(\dfrac{2\sqrt{13}}{13}\mathbf{i} - \dfrac{3\sqrt{13}}{13}\mathbf{j} \right)$

77.

79. 174 nautical mi, S17°E   **81.** $60°$
83. 500 lb on left, 866 lb on right
85. Cable: 224-lb tension, boom: 167-lb compression
87. $\mathbf{u} + \mathbf{v} = \langle u_1, u_2 \rangle + \langle v_1, v_2 \rangle$
   $= \langle u_1 + v_1, u_2 + v_2 \rangle$
   $= \langle v_1 + u_1, v_2 + u_2 \rangle$
   $= \langle v_1, v_2 \rangle + \langle u_1, u_2 \rangle$
   $= \mathbf{v} + \mathbf{u}$
89. Discussion and Writing   **91.** [1.1] -1
92. [1.1] Does not exist   **93.** [1.1] 0
94. [1.1] $(-\infty, 1) \cup (1, \infty)$
95. (a) $\cos \theta = \dfrac{\mathbf{u} \cdot \mathbf{v}}{|\mathbf{u}||\mathbf{v}|} = \dfrac{0}{|\mathbf{u}||\mathbf{v}|}$, $\therefore \cos \theta = 0$ and $\theta = 90°$.
(b) Answers may vary. $\mathbf{u} = \langle 2, -3 \rangle$ and $\mathbf{v} = \langle -3, -2 \rangle$;
$\mathbf{u} \cdot \mathbf{v} = 2(-3) + (-3)(-2) = 0$
97. $\dfrac{3}{5}\mathbf{i} - \dfrac{4}{5}\mathbf{j}$, $-\dfrac{3}{5}\mathbf{i} + \dfrac{4}{5}\mathbf{j}$ **99.** $(5, 8)$

Review Exercises, Chapter 4

1. [4.2] $A \approx 153°$, $B \approx 18°$, $C \approx 9°$

2. [4.1] $A = 118°$, $a \approx 37$ in., $c \approx 24$ in.
3. [4.1] $B = 14°50'$, $a \approx 2523$ m, $c \approx 1827$ m
4. [4.1] No solution   **5.** [4.1] 33 m^2 **6.** [4.1] 13.72 ft^2
7. [4.1] 63 ft^2 **8.** [4.2] $92°$, $33°$, $55°$
9. [4.1] 419 ft   **10.** [4.2] About 650 km
11. [4.3] $-2\sqrt{10}\,i$ **12.** [4.3] $-4\sqrt{15}$ **13.** [4.3] $-\dfrac{7}{8}$
14. [4.3] $-18 - 26i$ **15.** [4.3] $\dfrac{11}{10} + \dfrac{3}{10}i$ **16.** [4.3] $1 - 4i$
17. [4.3] $2 - i$ **18.** [4.3] $-i$ **19.** [4.3] 3^{28}
20. [4.4] $\sqrt{29}$;   **21.** [4.4] 4;

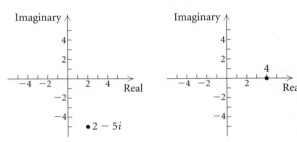

22. [4.4] 2;   **23.** [4.4] $\sqrt{10}$;

24. [4.4] $\sqrt{2}\left(\cos \dfrac{\pi}{4} + i \sin \dfrac{\pi}{4} \right)$, or
$\sqrt{2}(\cos 45° + i \sin 45°)$
25. [4.4] $4\left(\cos \dfrac{3\pi}{2} + i \sin \dfrac{3\pi}{2} \right)$, or $4(\cos 270° + i \sin 270°)$
26. [4.4] $10\left(\cos \dfrac{5\pi}{6} + i \sin \dfrac{5\pi}{6} \right)$, or $10(\cos 150° + i \sin 150°)$
27. [4.4] $\dfrac{3}{4}(\cos 0 + i \sin 0)$, or $\dfrac{3}{4}(\cos 0° + i \sin 0°)$
28. [4.4] $2 + 2\sqrt{3}i$ **29.** [4.4] 7 **30.** [4.4] $-\dfrac{5}{2} + \dfrac{5\sqrt{3}}{2}i$
31. [4.4] $1 - \sqrt{3}i$ **32.** [4.4] $1 + \sqrt{3} + (-1 + \sqrt{3})i$
33. [4.4] $-i$ **34.** [4.4] $2i$ **35.** [4.4] $3\sqrt{3} + 3i$
36. [4.4] $8(\cos 180° + i \sin 180°)$
37. [4.4] $4(\cos 7\pi + i \sin 7\pi)$ **38.** [4.4] $-8i$
39. [4.4] $-\dfrac{1}{2} - \dfrac{\sqrt{3}}{2}i$
40. [4.4] $\sqrt[4]{2}\left(\cos \dfrac{3\pi}{8} + i \sin \dfrac{3\pi}{8} \right)$,
$\sqrt[4]{2}\left(\cos \dfrac{11\pi}{8} + i \sin \dfrac{11\pi}{8} \right)$

41. [4.4] $\sqrt[3]{6}(\cos 110° + i \sin 110°)$,
$\sqrt[3]{6}(\cos 230° + i \sin 230°)$, $\sqrt[3]{6}(\cos 350° + i \sin 350°)$
42. [4.4] 3, 3i, −3, −3i

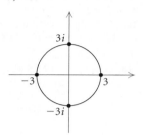

43. [4.4] 1, $\cos 72° + i \sin 72°$, $\cos 144° + i \sin 144°$,
$\cos 216° + i \sin 216°$, $\cos 288° + i \sin 288°$

44. [4.4] $\cos 22.5° + i \sin 22.5°$, $\cos 112.5° + i \sin 112.5°$,
$\cos 202.5° + i \sin 202.5°$, $\cos 292.5° + i \sin 292.5°$
45. [4.4] $\dfrac{1}{2} + \dfrac{\sqrt{3}}{2}i$, −1, $\dfrac{1}{2} - \dfrac{\sqrt{3}}{2}i$
46. [4.5] 13.7, 71°   **47.** [4.5] 98.7, 15°
48. [4.5]   **49.** [4.5]

50. [4.5] 666.7 N, 36°   **51.** [4.5] 29 km/h, 329°
52. [4.5] 102.4 nautical mi, S43°E   **53.** [4.6] ⟨−4, 3⟩
54. [4.6] ⟨2, −6⟩   **55.** [4.6] $\sqrt{61}$ **56.** [4.6] ⟨10, −21⟩
57. [4.6] ⟨14, −64⟩   **58.** [4.6] 5 + $\sqrt{116}$ **59.** [4.6] 14
60. [4.6] $\left\langle -\dfrac{3}{\sqrt{10}}, -\dfrac{1}{\sqrt{10}} \right\rangle$ **61.** [4.6] −9**i** + 4**j**
62. [4.6] 194.0°   **63.** [4.6] $\sqrt{34}$; $\theta = 211.0°$
64. [4.6] 111.8°   **65.** [4.6] 85.1°   **66.** [4.6] 34**i** − 55**j**
67. [4.6] **i** − 12**j**   **68.** [4.6] 5$\sqrt{2}$
69. [4.6] 3$\sqrt{65}$ + $\sqrt{109}$ **70.** [4.6] −5**i** + 5**j**

71. [4.6]   **72.** [4.6]

73. [4.6] $\sqrt{10}\left(\dfrac{3\sqrt{10}}{10}\mathbf{i} - \dfrac{\sqrt{10}}{10}\mathbf{j} \right)$

74. Discussion and Writing [4.1] When only the lengths of the sides of a triangle are known, the law of sines cannot be used to find the first angle because there will be two unknowns in the equation. To use the law of sines, you must know the measure of at least one angle and the length of the side opposite it.
75. Discussion and Writing [4.1], [4.2] A triangle has no solution when a sine or cosine value found is less than −1 or greater than 1. A triangle also has no solution if the sum of the angle measures calculated is greater than 180°. A triangle has only one solution if only one possible answer is found, or if one of the possible answers has an angle sum greater than 180°. A triangle has two solutions when two possible answers are found and neither results in an angle sum greater than 180°.
76. Discussion and Writing [4.2] For 150-yd shot, about 13.1 yd; for 300-yd shot, about 26.2 yd
77. [4.6] $\frac{36}{13}\mathbf{i} + \frac{15}{13}\mathbf{j}$ **78.** [4.1] 50.52°, 129.48°

Chapter 5

Exercise Set 5.1

1. (f)   **3.** (b)   **5.** (d)
7. V: (0, 0);   **9.** V: (0, 0); F: $\left(-\frac{3}{2}, 0\right)$;
F: (0, 5); D: $y = -5$ D: $x = \frac{3}{2}$

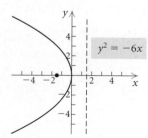

11. V: $(0, 0)$;
F: $(0, 1)$; D: $y = -1$

$$x^2 - 4y = 0$$

13. V: $(0, 0)$; F: $\left(\frac{1}{8}, 0\right)$;
D: $x = -\frac{1}{8}$

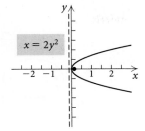

$$x = 2y^2$$

15. $y^2 = 16x$ **17.** $x^2 = -4\pi y$
19. $(y - 2)^2 = 14\left(x + \frac{1}{2}\right)$
21. V: $(-2, 1)$; F: $\left(-2, -\frac{1}{2}\right)$; D: $y = \frac{5}{2}$

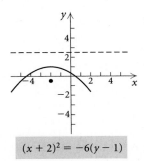

$$(x + 2)^2 = -6(y - 1)$$

23. V: $(-1, -3)$; F: $\left(-1, -\frac{7}{2}\right)$; D: $y = -\frac{5}{2}$

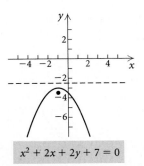

$$x^2 + 2x + 2y + 7 = 0$$

25. V: $(0, -2)$; F: $\left(0, -1\frac{3}{4}\right)$; D: $y = -2\frac{1}{4}$

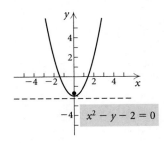

$$x^2 - y - 2 = 0$$

27. V: $(-2, -1)$;
F: $\left(-2, -\frac{3}{4}\right)$;
D: $y = -1\frac{1}{4}$

$$y = x^2 + 4x + 3$$

29. V: $\left(5\frac{3}{4}, \frac{1}{2}\right)$;
F: $\left(6, \frac{1}{2}\right)$;
D: $x = 5\frac{1}{2}$

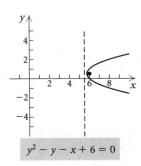

$$y^2 - y - x + 6 = 0$$

31. **(a)** $y^2 = 16x$; **(b)** $3\frac{33}{64}$ ft   **33.** $\frac{2}{3}$ ft, or 8 in.

35. Discussion and Writing   **37.** [3.4] $-\frac{\pi}{4}$, $-45°$

38. [3.4] $\frac{\pi}{3}$, $60°$ **39.** [3.4] $\frac{\pi}{6}$ **40.** [3.4] $\frac{1}{2}$
41. $(x + 1)^2 = -4(y - 2)$
43. V: $(0.867, 0.348)$; F: $(0.867, -0.191)$; D: $y = 0.887$
45. 10 ft, 11.6 ft, 16.4 ft, 24.4 ft, 35.6 ft, 50 ft

Exercise Set 5.2

1. (b)   **3.** (d)   **5.** (a)
7. $(7, -2)$; 8   **9.** $(-2, 3)$; 5

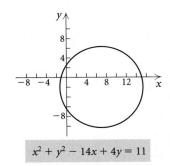

$$x^2 + y^2 - 14x + 4y = 11$$

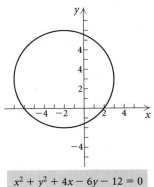

$$x^2 + y^2 + 4x - 6y - 12 = 0$$

11. $(-3, 5)$; $\sqrt{34}$

$x^2 + y^2 + 6x - 10y = 0$

13. $\left(\dfrac{9}{2}, -2\right)$; $\dfrac{5\sqrt{5}}{2}$

$x^2 + y^2 - 9x = 7 - 4y$

15. (c)   **17.** (d)

19. V: $(2, 0)$, $(-2, 0)$;
F: $(\sqrt{3}, 0)$, $(-\sqrt{3}, 0)$

$\dfrac{x^2}{4} + \dfrac{y^2}{1} = 1$

21. V: $(0, 4)$, $(0, -4)$;
F: $(0, \sqrt{7})$, $(0, -\sqrt{7})$

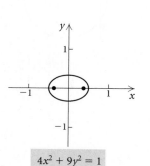

$16x^2 + 9y^2 = 144$

23. V: $(-\sqrt{3}, 0)$, $(\sqrt{3}, 0)$;
F: $(-1, 0)$, $(1, 0)$

$2x^2 + 3y^2 = 6$

25. V: $\left(-\dfrac{1}{2}, 0\right)$, $\left(\dfrac{1}{2}, 0\right)$;
F: $\left(-\dfrac{\sqrt{5}}{6}, 0\right)$, $\left(\dfrac{\sqrt{5}}{6}, 0\right)$

$4x^2 + 9y^2 = 1$

27. $\dfrac{x^2}{49} + \dfrac{y^2}{40} = 1$ **29.** $\dfrac{x^2}{25} + \dfrac{y^2}{64} = 1$

31. $\dfrac{x^2}{9} + \dfrac{y^2}{5} = 1$

33. C: $(1, 2)$; V: $(4, 2)$, $(-2, 2)$; F: $(1 + \sqrt{5}, 2)$, $(1 - \sqrt{5}, 2)$

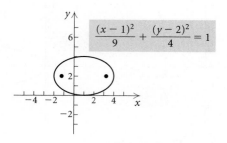

$\dfrac{(x-1)^2}{9} + \dfrac{(y-2)^2}{4} = 1$

35. C: $(-3, 5)$; V: $(-3, 11)$, $(-3, -1)$; F: $(-3, 5 + \sqrt{11})$, $(-3, 5 - \sqrt{11})$

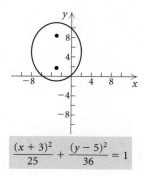

$\dfrac{(x+3)^2}{25} + \dfrac{(y-5)^2}{36} = 1$

37. C: $(-2, 1)$; V: $(-10, 1)$, $(6, 1)$; F: $(-6, 1)$, $(2, 1)$

$3(x+2)^2 + 4(y-1)^2 = 192$

39. C: $(2, -1)$; V: $(-1, -1)$, $(5, -1)$; F: $(2 + \sqrt{5}, -1)$, $(2 - \sqrt{5}, -1)$

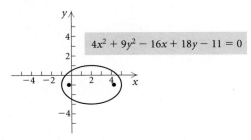

$$4x^2 + 9y^2 - 16x + 18y - 11 = 0$$

41. C: $(1, 1)$; V: $(1, 3)$, $(1, -1)$; F: $(1, 1 + \sqrt{3})$, $(1, 1 - \sqrt{3})$

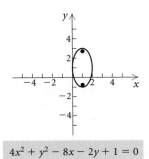

$$4x^2 + y^2 - 8x - 2y + 1 = 0$$

43. Example 2; $\dfrac{3}{5} < \dfrac{\sqrt{12}}{4}$ **45.** $\dfrac{x^2}{15} + \dfrac{y^2}{16} = 1$

47. $\dfrac{x^2}{2500} + \dfrac{y^2}{144} = 1$ **49.** 2×10^6 mi

51. Discussion and Writing

53. [1.6] **(a)** Yes; **(b)** $f^{-1}(x) = \dfrac{x + 3}{2}$

54. [1.6] **(a)** Yes; **(b)** $f^{-1}(x) = \sqrt[3]{x - 2}$

55. [1.6] **(a)** Yes; **(b)** $f^{-1}(x) = \dfrac{5}{x} + 1$, or $\dfrac{5 + x}{x}$

56. [1.6] **(a)** Yes; **(b)** $f^{-1}(x) = x^2 - 4$, $x \geq 0$

57. $\dfrac{(x - 3)^2}{4} + \dfrac{(y - 1)^2}{25} = 1$ **59.** $\dfrac{x^2}{9} + \dfrac{y^2}{484/5} = 1$

61. C: $(2.003, -1.005)$; V: $(-1.017, -1.005)$, $(5.023, -1.005)$ **63.** About 9.1 ft

Exercise Set 5.3

1. (b)   **3.** (c)   **5.** (a)   **7.** $\dfrac{y^2}{9} - \dfrac{x^2}{16} = 1$

9. $\dfrac{x^2}{4} - \dfrac{y^2}{9} = 1$

11. C: $(0, 0)$; V: $(2, 0)$, $(-2, 0)$; F: $(2\sqrt{2}, 0)$, $(-2\sqrt{2}, 0)$; A: $y = x$, $y = -x$

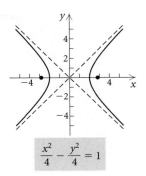

$$\dfrac{x^2}{4} - \dfrac{y^2}{4} = 1$$

13. C: $(2, -5)$; V: $(-1, -5)$, $(5, -5)$; F: $(2 - \sqrt{10}, -5)$, $(2 + \sqrt{10}, -5)$; A: $y = -\dfrac{x}{3} - \dfrac{13}{3}$, $y = \dfrac{x}{3} - \dfrac{17}{3}$

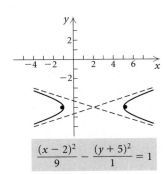

$$\dfrac{(x - 2)^2}{9} - \dfrac{(y + 5)^2}{1} = 1$$

15. C: $(-1, -3)$; V: $(-1, -1)$, $(-1, -5)$; F: $(-1, -3 + 2\sqrt{5})$, $(-1, -3 - 2\sqrt{5})$; A: $y = \frac{1}{2}x - \frac{5}{2}$, $y = -\frac{1}{2}x - \frac{7}{2}$

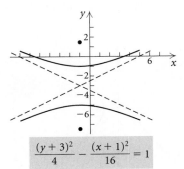

$$\dfrac{(y + 3)^2}{4} - \dfrac{(x + 1)^2}{16} = 1$$

17. C: $(0, 0)$; V: $(-2, 0)$, $(2, 0)$; F: $(-\sqrt{5}, 0)$, $(\sqrt{5}, 0)$;
A: $y = -\frac{1}{2}x$, $y = \frac{1}{2}x$

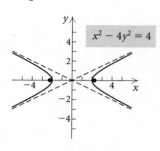

19. C: $(0, 0)$; V: $(0, -3)$, $(0, 3)$; F: $(0, -3\sqrt{10})$, $(0, 3\sqrt{10})$;
A: $y = \frac{1}{3}x$, $y = -\frac{1}{3}x$

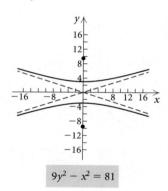

21. C: $(0, 0)$; V: $(-\sqrt{2}, 0)$, $(\sqrt{2}, 0)$; F: $(-2, 0)$, $(2, 0)$;
A: $y = x$, $y = -x$

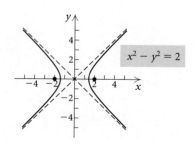

23. C: $(0, 0)$; V: $\left(0, -\frac{1}{2}\right)$, $\left(0, \frac{1}{2}\right)$; F: $\left(0, -\frac{\sqrt{2}}{2}\right)$,
$\left(0, \frac{\sqrt{2}}{2}\right)$; A: $y = x$, $y = -x$

25. C: $(1, -2)$; V: $(0, -2)$, $(2, -2)$; F: $(1 - \sqrt{2}, -2)$,
$(1 + \sqrt{2}, -2)$; A: $y = -x - 1$, $y = x - 3$

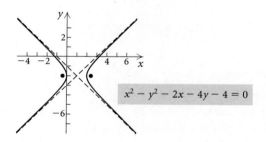

27. C: $\left(\frac{1}{3}, 3\right)$; V: $\left(-\frac{2}{3}, 3\right)$, $\left(\frac{4}{3}, 3\right)$; F: $\left(\frac{1}{3} - \sqrt{37}, 3\right)$,
$\left(\frac{1}{3} + \sqrt{37}, 3\right)$; A: $y = 6x + 1$, $y = -6x + 5$

29. C: $(3, 1)$; V: $(3, 3)$, $(3, -1)$; F: $(3, 1 + \sqrt{13})$, $(3, 1 - \sqrt{13})$; A: $y = \frac{2}{3}x - 1$, $y = -\frac{2}{3}x + 3$

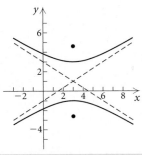

$$9y^2 - 4x^2 - 18y + 24x - 63 = 0$$

31. C: $(1, -2)$; V: $(2, -2)$, $(0, -2)$; F: $(1 + \sqrt{2}, -2)$, $(1 - \sqrt{2}, -2)$; A: $y = x - 3$, $y = -x - 1$

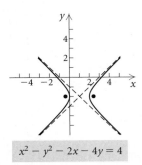

$$x^2 - y^2 - 2x - 4y = 4$$

33. C: $(-3, 4)$; V: $(-3, 10)$, $(-3, -2)$; F: $(-3, 4 + 6\sqrt{2})$, $(-3, 4 - 6\sqrt{2})$; A: $y = x + 7$, $y = -x + 1$

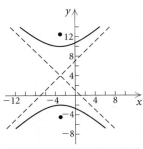

$$y^2 - x^2 - 6x - 8y - 29 = 0$$

35. Example 3; $\dfrac{\sqrt{5}}{1} > \dfrac{5}{4}$ **37.** $\dfrac{x^2}{9} - \dfrac{(y - 7)^2}{16} = 1$

39. $\dfrac{y^2}{25} - \dfrac{x^2}{11} = 1$ **41.** Discussion and Writing

43. [4.1] $B = 24°$, $a \approx 44$, $c \approx 28$

44. [4.2] $b \approx 61$, $A \approx 19.7°$, $C \approx 17.3°$

45. [4.1] About 5.01 yd^2 **46.** [4.1] About 21.75 cm^2

47. $\dfrac{(y + 5)^2}{9} - (x - 3)^2 = 1$

49. C: $(-1.460, -0.957)$; V: $(-2.360, -0.957)$, $(-0.560, -0.957)$; A: $y = -1.429x - 3.043$, $y = 1.429x + 1.129$

51. $\dfrac{x^2}{345.96} - \dfrac{y^2}{22{,}154.04} = 1$

Exercise Set 5.4

1. (e)   **3.** (c)   **5.** (b)   **7.** $(-4, -3)$, $(3, 4)$

9. $(0, 2)$, $(3, 0)$ **11.** $(-5, 0)$, $(4, 3)$, $(4, -3)$

13. $(3, 0)$, $(-3, 0)$ **15.** $(0, -3)$, $(4, 5)$

17. $(-2, 1)$ **19.** $(3, 4)$, $(-3, -4)$, $(4, 3)$, $(-4, -3)$

21. $\left(\dfrac{6\sqrt{21}}{7}, \dfrac{4i\sqrt{35}}{7}\right)$, $\left(\dfrac{6\sqrt{21}}{7}, -\dfrac{4i\sqrt{35}}{7}\right)$, $\left(-\dfrac{6\sqrt{21}}{7}, \dfrac{4i\sqrt{35}}{7}\right)$, $\left(-\dfrac{6\sqrt{21}}{7}, -\dfrac{4i\sqrt{35}}{7}\right)$

23. $(3, 2)$, $\left(4, \dfrac{3}{2}\right)$

25. $\left(\dfrac{5 + \sqrt{70}}{3}, \dfrac{-1 + \sqrt{70}}{3}\right)$, $\left(\dfrac{5 - \sqrt{70}}{3}, \dfrac{-1 - \sqrt{70}}{3}\right)$

27. $(\sqrt{2}, \sqrt{14})$, $(-\sqrt{2}, \sqrt{14})$, $(\sqrt{2}, -\sqrt{14})$, $(-\sqrt{2}, -\sqrt{14})$ **29.** $(1, 2)$, $(-1, -2)$, $(2, 1)$, $(-2, -1)$

31. $\left(\dfrac{15 + \sqrt{561}}{8}, \dfrac{11 - 3\sqrt{561}}{8}\right)$, $\left(\dfrac{15 - \sqrt{561}}{8}, \dfrac{11 + 3\sqrt{561}}{8}\right)$

33. $\left(\dfrac{7 - \sqrt{33}}{2}, \dfrac{7 + \sqrt{33}}{2}\right)$, $\left(\dfrac{7 + \sqrt{33}}{2}, \dfrac{7 - \sqrt{33}}{2}\right)$

35. $(3, 2)$, $(-3, -2)$, $(2, 3)$, $(-2, -3)$

37. $\left(\dfrac{5 - 9\sqrt{15}}{20}, \dfrac{-45 + 3\sqrt{15}}{20}\right)$, $\left(\dfrac{5 + 9\sqrt{15}}{20}, \dfrac{-45 - 3\sqrt{15}}{20}\right)$ **39.** $(3, -5)$, $(-1, 3)$

41. $(8, 5)$, $(-5, -8)$ **43.** $(3, 2)$, $(-3, -2)$

45. $(2, 1)$, $(-2, -1)$, $(1, 2)$, $(-1, -2)$

47. $\left(4 + \dfrac{3i\sqrt{6}}{2}, -4 + \dfrac{3i\sqrt{6}}{2}\right)$, $\left(4 - \dfrac{3i\sqrt{6}}{2}, -4 - \dfrac{3i\sqrt{6}}{2}\right)$

49. $(3, \sqrt{5})$, $(-3, -\sqrt{5})$, $(\sqrt{5}, 3)$, $(-\sqrt{5}, -3)$

51. $\left(\dfrac{8\sqrt{5}}{5}i, \dfrac{3\sqrt{105}}{5}\right)$, $\left(\dfrac{8\sqrt{5}}{5}i, -\dfrac{3\sqrt{105}}{5}\right)$, $\left(-\dfrac{8\sqrt{5}}{5}i, \dfrac{3\sqrt{105}}{5}\right)$, $\left(-\dfrac{8\sqrt{5}}{5}i, -\dfrac{3\sqrt{105}}{5}\right)$

53. $(2, 1)$, $(-2, -1)$, $\left(-i\sqrt{5}, \dfrac{2i\sqrt{5}}{5}\right)$, $\left(i\sqrt{5}, -\dfrac{2i\sqrt{5}}{5}\right)$

55. 6 cm by 8 cm   **57.** 4 in. by 5 in.

59. 1 m by $\sqrt{3}$ m   **61.** 30 yd by 75 yd   **63.** 16 ft, 24 ft

65. (a) $I(x) = 21.15508287x + 198.9151934$;
(b) $E(x) = 197.3380928 \cdot 1.053908223^x$; **(c)** about 25 yr
after 1985   **67.** Discussion and Writing
69. [4.3] $68 + 11i$ **70.** [4.3] $-\frac{1}{10} + \frac{3}{10}i$

71. [4.4] $1 + \sqrt{3}\,i$ **72.** [4.4] $\dfrac{\sqrt{3}}{2} - \dfrac{3}{2}i$

73. $(x - 2)^2 + (y - 3)^2 = 1$ **75.** $\dfrac{x^2}{4} + y^2 = 1$

77. $\left(x + \dfrac{5}{13}\right)^2 + \left(y - \dfrac{32}{13}\right)^2 = \dfrac{5365}{169}$

79. There is no number x such that $\dfrac{x^2}{a^2} - \dfrac{\left(\frac{b}{a}x\right)^2}{b^2} = 1$,

because the left side simplifies to $\dfrac{x^2}{a^2} - \dfrac{x^2}{a^2}$, which is 0.

81. $\left(\frac{1}{2}, \frac{1}{4}\right), \left(\frac{1}{2}, -\frac{1}{4}\right), \left(-\frac{1}{2}, \frac{1}{4}\right), \left(-\frac{1}{2}, -\frac{1}{4}\right)$
83. Factor: $x^3 + y^3 = (x + y)(x^2 - xy + y^2)$. We know
that $x + y = 1$, so $(x + y)^2 = x^2 + 2xy + y^2 = 1$, or
$x^2 + y^2 = 1 - 2xy$. We also know that $xy = 1$, so
$x^2 + y^2 = 1 - 2 \cdot 1 = -1$. Then
$x^3 + y^3 = 1 \cdot (-1 - 1) = -2$.   **85.** $(2, 4), (4, 2)$
87. $(3, -2), (-3, 2), (2, -3), (-2, 3)$
89. $(0.965, 4402.33), (-0.965, -4402.33)$
91. $(2.112, -0.109), (-13.041, -13.337)$
93. $(400, 1.431), (-400, 1.431), (400, -1.431),$
$(-400, -1.431)$

Exercise Set 5.5

1. $(0, -2)$ **3.** $(1, \sqrt{3})$ **5.** $(\sqrt{2}, 0)$ **7.** $(\sqrt{3}, 1)$
9. Ellipse or circle   **11.** Hyperbola   **13.** Parabola
15. Hyperbola   **17.** Ellipse or circle
19.   **21.**

23.

25.

27.

29.

31.

33.

35.

37.

39. Discussion and Writing   **41.** [2.4] $\dfrac{2\pi}{3}$

42. [2.4] $-\dfrac{7\pi}{4}$ **43.** [2.4] $60°$ **44.** [2.4] $135°$

45. $x = x' \cos \theta - y' \sin \theta,\ y = x' \sin \theta + y' \cos \theta$

47. $A' + C' = A\cos^2\theta + B\sin\theta\cos\theta + C\sin^2\theta$
$\qquad\qquad + A\sin^2\theta - B\sin\theta\cos\theta + C\cos^2\theta$
$\qquad = A(\sin^2\theta + \cos^2\theta) + C(\sin^2\theta + \cos^2\theta)$
$\qquad = A + C$

Exercise Set 5.6

13. A: $(4, 30°)$, $(4, 390°)$, $(-4, 210°)$; B: $(5, 300°)$,
$(5, -60°)$, $(-5, 120°)$; C: $(2, 150°)$, $(2, 510°)$, $(-2, 330°)$;
D: $(3, 225°)$, $(3, -135°)$, $(-3, 45°)$; answers may vary

15. $(3, 270°)$, $\left(3, \dfrac{3\pi}{2}\right)$ **17.** $(6, 300°)$, $\left(6, \dfrac{5\pi}{3}\right)$

19. $(8, 330°)$, $\left(8, \dfrac{11\pi}{6}\right)$ **21.** $(2, 225°)$, $\left(2, \dfrac{5\pi}{4}\right)$

23. $(7.616, 66.8°)$, $(7.616, 1.166)$
25. $(4.643, 132.9°)$, $(4.643, 2.320)$

27. $\left(\dfrac{5}{2}, \dfrac{5\sqrt{3}}{2}\right)$ **29.** $\left(-\dfrac{3\sqrt{2}}{2}, -\dfrac{3\sqrt{2}}{2}\right)$

31. $\left(-\dfrac{3}{2}, -\dfrac{3\sqrt{3}}{2}\right)$ **33.** $(-1, \sqrt{3})$

35. $(2.19, -2.05)$ **37.** $(1.30, -3.99)$
39. $r(3\cos\theta + 4\sin\theta) = 5$ **41.** $r\cos\theta = 5$
43. $r = 6$ **45.** $r^2\cos^2\theta = 25r\sin\theta$
47. $r^2\sin^2\theta - 5r\cos\theta - 25 = 0$ **49.** $x^2 + y^2 = 25$
51. $y = 2$ **53.** $y^2 = -6x + 9$
55. $x^2 - 9x + y^2 - 7y = 0$ **57.** $x = 5$
59.   **61.**

$r = \sin\theta$ $r = 4\cos 2\theta$

63. (d)   **65.** (g)   **67.** (j)   **69.** (b)   **71.** (e)   **73.** (k)
75. Discussion and Writing

77. [3.5] $\dfrac{\pi}{3}, \dfrac{2\pi}{3}, \dfrac{4\pi}{3}, \dfrac{5\pi}{3}$, or 60°, 120°, 240°, 300°

78. [3.5] $\dfrac{\pi}{3}, \pi, \dfrac{5\pi}{3}$, or 60°, 180°, 300°

79. [1.2]   **80.** [1.2]

$y = 2x - 5$ $4x - y = 6$

81. [1.2]   **82.** [1.2]

$x = -3$ $y = 0$

83. $y^2 = -4x + 4$
85.   **87.**

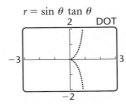

$r = \sin\theta\,\tan\theta$ DOT $r = e^{\theta/10}$

89.   **91.**

$r = \cos 2\theta\,\sec\theta$ DOT $r = \dfrac{1}{4}\tan^2\theta\,\sec\theta$

Exercise Set 5.7

1. (b)   **3.** (a)   **5.** (d)

7. (a) Parabola; **(b)** vertical, 1 unit to the right of the pole;
(c) $\left(\frac{1}{2}, 0\right)$;
(d)

$$r = \frac{1}{1 + \cos\theta}$$

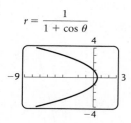

9. (a) Hyperbola; **(b)** horizontal, $\frac{3}{2}$ units below the pole;
(c) $\left(-3, \frac{\pi}{2}\right), \left(1, \frac{3\pi}{2}\right)$;
(d)

$$r = \frac{15}{5 - 10\sin\theta}$$

11. (a) Ellipse; **(b)** vertical, $\frac{8}{3}$ units to the left of the pole;
(c) $\left(\frac{8}{3}, 0\right), \left(\frac{8}{9}, \pi\right)$;
(d)

$$r = \frac{8}{6 - 3\cos\theta}$$

13. (a) Hyperbola; **(b)** horizontal, $\frac{4}{3}$ units above the pole;
(c) $\left(\frac{4}{5}, \frac{\pi}{2}\right), \left(-4, \frac{3\pi}{2}\right)$;
(d)

$$r = \frac{20}{10 + 15\sin\theta}$$

15. (a) Ellipse; **(b)** vertical, 3 units to the right of the pole;
(c) $(1, 0), (3, \pi)$;
(d)

$$r = \frac{9}{6 + 3\cos\theta}$$

17. (a) Parabola; **(b)** horizontal, $\frac{3}{2}$ units below the pole;
(c) $\left(\frac{3}{4}, \frac{3\pi}{2}\right)$;
(d)

$$r = \frac{3}{2 - 2\sin\theta}$$

19. (a) Ellipse; **(b)** vertical, 4 units to the left of the pole;
(c) $(4, 0), \left(\frac{4}{3}, \pi\right)$;
(d)

$$r = \frac{4}{2 - \cos\theta}$$

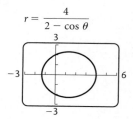

21. (a) Hyperbola; **(b)** horizontal, $\frac{7}{10}$ units above the pole;
(c) $\left(\frac{7}{12}, \frac{\pi}{2}\right), \left(-\frac{7}{8}, \frac{3\pi}{2}\right)$;
(d)

$$r = \frac{7}{2 + 10\sin\theta}$$

23. $y^2 + 2x - 1 = 0$ **25.** $x^2 - 3y^2 - 12y - 9 = 0$
27. $27x^2 + 36y^2 - 48x - 64 = 0$

29. $4x^2 - 5y^2 + 24y - 16 = 0$
31. $3x^2 + 4y^2 + 6x - 9 = 0$
33. $4x^2 - 12y - 9 = 0$ **35.** $3x^2 + 4y^2 - 8x - 16 = 0$
37. $4x^2 - 96y^2 + 140y - 49 = 0$ **39.** $r = \dfrac{6}{1 + 2\sin\theta}$
41. $r = \dfrac{4}{1 + \cos\theta}$ **43.** $r = \dfrac{2}{2 - \cos\theta}$
45. $r = \dfrac{15}{4 + 3\sin\theta}$ **47.** $r = \dfrac{8}{1 - 4\sin\theta}$
49. Discussion and Writing
51. [1.1] $f(t) = (t - 3)^2 + 4$, or $t^2 - 6t + 13$
52. [1.1] $f(2t) = (2t - 3)^2 + 4$, or $4t^2 - 12t + 13$
53. [1.1] $f(t - 1) = (t - 4)^2 + 4$, or $t^2 - 8t + 20$
54. [1.1] $f(t + 2) = (t - 1)^2 + 4$, or $t^2 - 2t + 5$
55. $r = \dfrac{1.5 \times 10^8}{1 + \sin\theta}$

Exercise Set 5.8

1. $y = 12x - 7$, $-\dfrac{1}{2} \le x \le 3$
3. $y = \sqrt[3]{x} - 4$, $-1 \le x \le 1000$ **5.** $x = y^4$, $0 \le x \le 16$
7. $y = \dfrac{1}{x}$, $1 \le x \le 5$ **9.** $y = \dfrac{1}{4}(x + 1)^2$, $-7 \le x \le 5$
11. $y = \dfrac{1}{x}$, $x > 0$ **13.** $x^2 + y^2 = 9$, $-3 \le x \le 3$
15. $x^2 + \dfrac{y^2}{4} = 1$, $-1 \le x \le 1$ **17.** $y = \dfrac{1}{x}$, $x \ge 1$
19. $(x - 1)^2 + (y - 2)^2 = 4$, $-1 \le x \le 3$
21. 0.7071   **23.** −0.2588   **25.** 0.7265   **27.** 0.5460
29. Answers may vary. $x = t$, $y = 4t - 3$; $x = \dfrac{t}{4} + 3$,
$y = t + 9$
31. Answers may vary. $x = t$, $y = (t - 2)^2 - 6t$;
$x = t + 2$, $y = t^2 - 6t - 12$
33. (a) About 36.1 ft; **(b)** about 196.6 ft; **(c)** about 3 sec
35. About 73.1 ft
37.

$x = 2(t - \sin t)$, $y = 2(1 - \cos t)$; $0 \le t \le 4\pi$

39.

$x = t - \sin t$, $y = 1 - \cos t$; $-2\pi \le t \le 2\pi$

41. Discussion and Writing   **43.** [4.6] 301°
44. [4.6] 288°   **45.** [4.6] 36.9°   **46.** [4.6] 81.9°
47. $x = 3\cos t$, $y = -3\sin t$

Review Exercises, Chapter 5

1. [5.1] (d)   **2.** [5.2] (a)   **3.** [5.2] (e)   **4.** [5.3] (g)
5. [5.2] (b)   **6.** [5.2] (f)   **7.** [5.1] (h)   **8.** [5.3] (c)
9. [5.1] $x^2 = -6y$
10. [5.1] F: $(-3, 0)$; V: $(0, 0)$; D: $x = 3$
11. [5.1] V: $(-5, 8)$; F: $\left(-5, \dfrac{15}{2}\right)$; D: $y = \dfrac{17}{2}$
12. [5.2] C: $(2, -1)$; V: $(-3, -1)$, $(7, -1)$; F: $(-1, -1)$,
$(5, -1)$;

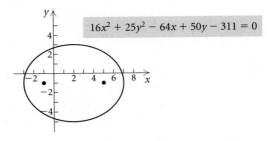

$16x^2 + 25y^2 - 64x + 50y - 311 = 0$

13. [5.2] $\dfrac{x^2}{9} + \dfrac{y^2}{16} = 1$
14. [5.3] C: $\left(-2, \dfrac{1}{4}\right)$; V: $\left(0, \dfrac{1}{4}\right)$ $\left(-4, \dfrac{1}{4}\right)$;
F: $\left(-2 + \sqrt{6}, \dfrac{1}{4}\right)$, $\left(-2 - \sqrt{6}, \dfrac{1}{4}\right)$;
A: $y - \dfrac{1}{4} = \dfrac{\sqrt{2}}{2}(x + 2)$, $y - \dfrac{1}{4} = -\dfrac{\sqrt{2}}{2}(x + 2)$
15. [5.1] 0.167 ft   **16.** [5.4] $(-8\sqrt{2}, 8)$, $(8\sqrt{2}, 8)$
17. [5.4] $\left(3, \dfrac{\sqrt{29}}{2}\right)$, $\left(-3, \dfrac{\sqrt{29}}{2}\right)$, $\left(3, -\dfrac{\sqrt{29}}{2}\right)$,
$\left(-3, -\dfrac{\sqrt{29}}{2}\right)$ **18.** [5.4] $(7, 4)$
19. [5.4] $(2, 2)$, $\left(\dfrac{32}{9}, -\dfrac{10}{9}\right)$ **20.** [5.4] $(0, -3)$, $(2, 1)$
21. [5.4] $(4, 3)$, $(4, -3)$, $(-4, 3)$, $(-4, -3)$
22. [5.4] $(-\sqrt{3}, 0)$, $(\sqrt{3}, 0)$, $(-2, 1)$, $(2, 1)$
23. [5.4] $\left(-\dfrac{3}{5}, \dfrac{21}{5}\right)$, $(3, -3)$

24. [5.4] (6, 8), (6, −8), (−6, 8), (−6, −8)
25. [5.4] (2, 2), (−2, −2), (2√2, √2), (−2√2, −√2)
26. [5.4] 7, 4   **27.** [5.4] 7 m by 12 m   **28.** [5.4] 4, 8
29. [5.4] 32 cm, 20 cm   **30.** [5.4] 11 ft, 3 ft
31. [5.5]

$$\frac{(x')^2}{6} + \frac{(y')^2}{4} = 1$$

32. [5.5]

$$\frac{(x')^2}{3} - \frac{(y')^2}{2} = 1$$

33. [5.5]

$$\frac{(x')^2}{2} - \frac{(y')^2}{4} = 1$$

34. [5.5]

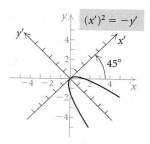

$$(x')^2 = -y'$$

35. [5.6] A: (5, 120°), (5, 480°), (−5, 300°); B: (3, 210°),
(−3, 30°), (−3, 390°); C: (4, 60°), (4, 420°), (−4, 240°);
D: (1, 300°), (1, −60°), (−1, 120°); answers may vary

36. [5.6] (8, 135°), $\left(8, \dfrac{3\pi}{4}\right)$ **37.** [5.6] (5, 270°), $\left(5, \dfrac{3\pi}{2}\right)$
38. [5.6] (5.385, 111.8°), (5.385, 1.951)
39. [5.6] (4.964, 147.8°), (4.964, 2.579)
40. [5.6] $\left(\dfrac{3\sqrt{2}}{2}, \dfrac{3\sqrt{2}}{2}\right)$ **41.** [5.6] (3, 3√3)
42. [5.6] (1.93, −0.52)   **43.** [5.6] (−1.86, −1.35)
44. [5.6] $r(5\cos\theta - 2\sin\theta) = 6$ **45.** [5.6] $r\sin\theta = 3$
46. [5.6] $r = 3$ **47.** [5.6] $r^2\sin^2\theta - 4r\cos\theta - 16 = 0$
48. [5.6] $x^2 + y^2 = 36$ **49.** [5.6] $x^2 + 2y = 1$
50. [5.6] $y^2 - 6x = 9$ **51.** [5.6] $x^2 - 2x + y^2 - 3y = 0$
52. [5.6] (b)   **53.** [5.6] (d)   **54.** [5.6] (a)   **55.** [5.6] (c)
56. [5.7]

$$r = \frac{6}{3 - 3\sin\theta}$$

Horizontal directrix 6 units below the pole;
vertex: $\left(1, \dfrac{3\pi}{2}\right)$

57. [5.7]

$$r = \frac{8}{2 + 4\cos\theta}$$

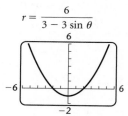

Vertical directrix 2 units to the right of the pole;
vertices: $\left(0, \dfrac{4}{3}\right), (-4, \pi)$

58. [5.7]

$$r = \frac{4}{2 - \cos\theta}$$

Vertical directrix 4 units to the left of the pole;
vertices: (0, 4), $\left(\dfrac{4}{3}, \pi\right)$

59. [5.7]

$$r = \frac{18}{9 + 6 \sin \theta}$$

Horizontal directrix 3 units above the pole;

vertices: $\left(\frac{6}{5}, \frac{\pi}{2}\right)$, $\left(6, \frac{3\pi}{2}\right)$

60. [5.7] $x^2 - 12y - 36 = 0$
61. [5.7] $4x^2 - y^2 - 16x + 16 = 0$
62. [5.7] $3x^2 + 4y^2 - 8x - 16 = 0$
63. [5.7] $9x^2 + 5y^2 + 24y - 36 = 0$
64. [5.7] $r = \dfrac{1}{1 + \frac{1}{2}\cos\theta}$, or $r = \dfrac{2}{2 + \cos\theta}$

65. [5.7] $r = \dfrac{18}{1 - 3\sin\theta}$ **66.** [5.7] $r = \dfrac{4}{1 - \cos\theta}$

67. [5.7] $r = \dfrac{6}{1 + 2\sin\theta}$

68. [5.8]

$$x = t,\ y = 2 + t,\ -3 \leqslant t \leqslant 3$$

$y = 2 + x,\ -3 \leq x \leq 3,\ -1 \leq y \leq 5$
69. [5.8]

$$x = \sqrt{t},\ y = t - 1,\ 0 \leqslant t \leqslant 9$$

$y = x^2 - 1,\ 0 \leq x \leq 3,\ -1 \leq y \leq 8$

70. [5.8]

$$x = 2\cos t,\ y = 2\sin t,\ 0 \leqslant t \leqslant 2\pi$$

$x^2 + y^2 = 1$
71. [5.8]

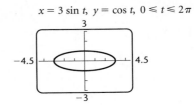

$$x = 3\sin t,\ y = \cos t,\ 0 \leqslant t \leqslant 2\pi$$

$\dfrac{x^2}{9} + y^2 = 1$

72. [5.8] Answers may vary. $x = t$, $y = 2t - 3$; $x = t + 1$, $y = 2t - 1$
73. [5.8] Answers may vary. $x = t$, $y = t^2 + 4$; $x = t - 2$, $y = t^2 - 4t + 8$
74. Discussion and Writing [5.4] An algebraic solution of a nonlinear system of equations may be preferable to a graphical solution if there are complex-number solutions or if any of the equations is difficult to enter on a grapher. However, there are nonlinear systems that are difficult or impossible to solve using algebraic methods. Approximations of the real-number solutions of many of those systems may be found using graphical methods.
75. Discussion and Writing [5.2] The equation of a circle can be written as

$$\frac{(x - h)^2}{a^2} + \frac{(y - k)^2}{b^2} = 1,$$

where $a = b = r$, the radius of the circle. In an ellipse, $a > b$, so a circle is not a special type of ellipse.

76. [5.2] $x^2 + \dfrac{y^2}{9} = 1$ **77.** [5.4] $\dfrac{8}{7}, \dfrac{7}{2}$

78. [5.2], [5.4] $(x - 2)^2 + (y - 1)^2 = 100$

79. [5.3] $\dfrac{x^2}{778.41} - \dfrac{y^2}{39,221.59} = 1$ **80.** [5.6] $y^2 = 4x + 4$

Chapter 6

Exercise Set 6.1

1. 54.5982   **3.** 0.0856

5.

7.

9.

11.

13.

15.

17.

19. Shift the graph of $y = 2^x$ left 1 unit.

$$y = 2^{x+1}$$

21. Shift the graph of $y = 2^x$ down 3 units.

$$y = 2^x - 3$$

23. Reflect the graph of $y = 3^x$ across the y-axis, then across the x-axis, and then shift it up 4 units.

$$y = 4 - 3^{-x}$$

25. Shift the graph of $y = \left(\frac{3}{2}\right)^x$ right 1 unit.

$$y = \left(\frac{3}{2}\right)^{x-1}$$

27. Shift the graph of $y = 2^x$ left 3 units, and then down 5 units.

$$y = 2^{x+3} - 5$$

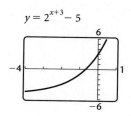

29. Shrink the graph of $y = e^x$ horizontally.

$y = e^{2x}$

31. Shift the graph of $y = e^x$ left 1 unit and reflect it across the y-axis.

$y = e^{-x+1}$

33. Reflect the graph of $y = e^x$ across the y-axis, then across the x-axis, then shift it up 1 unit, and then stretch it vertically.

$y = 2(1 - e^{-x}), x > 0$

35. (a) 799,053; **(b)** 1,363,576;
(c)

$y = 274,390(1.195)^t$

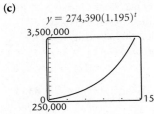

(d) in about 13 yr, or in 2005;
37. (a) 350,000; 233,333; 69,136; 6070;
(b)   **(c)** after about 13 yr

$y = 350,000\left(\frac{2}{3}\right)^x$

39. (a)   ;

$y = 5800(0.8)^x$

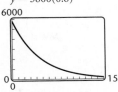

(b) $5800, $4640, $3712, $1900.54, $622.77; **(c)** after 11 yr
41. (a)   ;

$y = 46.6(1.018)^x$

(b) About 65.4 billion ft^3, 78.2 billion ft^3; **(c)** after about 39 yr
43. (a)   ;

$y = 100(1 - e^{-0.04x})$

(b) about 63%; **(c)** after 58 days
45. (c)   **47.** (a)   **49.** (l)   **51.** (g)   **53.** (i)   **55.** (k)
57. (m)   **59.** (1.481, 4.090)   **61.** $(-0.402, -1.662)$,
(1.051, 2.722)   **63.** 4.448   **65.** $(0, \infty)$
67. 2.294, 3.228   **69.** Discussion and Writing
71. Discussion and Writing
72. [3.5] $\frac{\pi}{4}, \frac{3\pi}{4}, \frac{5\pi}{4}, \frac{7\pi}{4}$, or 45°, 135°, 225°, 315°
73. [3.5] $\frac{7\pi}{6}, \frac{11\pi}{6}$, or 210°, 330°   **72.** [3.5] $\frac{\pi}{4}, \frac{5\pi}{4}$,
or 45°, 225°   **75.** [3.5] 1.107, 4.249, or 63.4°, 243.4°
77. $y = 1$
79. (a)

$y = e^{-x^2}$

(b) none; **(c)** relative maximum: 1 at $x = 0$

Exercise Set 6.2

1.

$y = \log_3 x$

3.

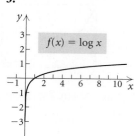

$f(x) = \log x$

5. 4   **7.** 3   **9.** -3 **11.** -2 **13.** 0   **15.** 1
17. $\log_{10} 1000 = 3$ **19.** $\log_8 2 = \frac{1}{3}$ **21.** $\log_e t = 3$
23. $\log_e 7.3891 = 2$
25. $\log_p 3 = k$ **27.** $5^1 = 5$ **29.** $10^{-2} = 0.01$
31. $e^{3.4012} = 30$ **33.** $a^{-x} = M$ **35.** $a^x = T^3$
37. 0.4771   **39.** 2.7259   **41.** -0.2441 **43.** Does not
exist   **45.** 0.6931   **47.** 6.6962   **49.** Does not exist
51. 3.3219   **53.** -0.2614 **55.** 0.7384
57. Shift the graph of $y = \log_2 x$ left 3 units. Domain:
$(-3, \infty)$; vertical asymptote: $x = -3$;

$$y = \frac{\log(x+3)}{\log 2}$$

59. Shift the graph of $y = \log_3 x$ down 1 unit. Domain:
$(0, \infty)$; vertical asymptote: $x = 0$;

$$y = \frac{\log x}{\log 3} - 1$$

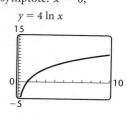

61. Stretch the graph of $y = \ln x$ vertically. Domain:
$(0, \infty)$; vertical asymptote: $x = 0$;

$y = 4 \ln x$

63. Reflect the graph of $y = \ln x$ across the x-axis and
shift it up 2 units. Domain: $(0, \infty)$; vertical asymptote:
$x = 0$;

$y = 2 - \ln x$

65.

$y_1 = 3^x$,
$y_2 = \dfrac{\log x}{\log 3}$

67.

$y_1 = \log x$,
$y_2 = 10^x$

69. (a) 2.3 ft/sec; **(b)** 3.0 ft/sec; **(c)** 2.2 ft/sec; **(d)** 1.9 ft/sec;
(e) 1.7 ft/sec
71. (a) 78%; **(b)** 67.5%, 57%;
(c) $y = 78 - 15 \log(x+1), x \geq 0$;   **(d)** after 73 months

73. (a) 10^{-7}; **(b)** 4.0×10^{-6}; **(c)** 6.3×10^{-4}; **(d)** 1.6×10^{-5}
75. (a) 34 decibels; **(b)** 64 decibels; **(c)** 60 decibels;
(d) 90 decibels   **77.** Discussion and Writing
78. [5.1] $(x-2)^2 = 8(y+10)$, or $x^2 - 4x - 8y - 76 = 0$
79. [5.1] $(y+6)^2 = -12(x-3)$, or $y^2 + 12y + 12x = 0$
80. [5.2] $\dfrac{x^2}{100} + \dfrac{y^2}{36} = 1$ **81.** [5.2] $\dfrac{x^2}{13} + \dfrac{y^2}{49} = 1$ **83.** 3
85. $(0, \infty)$ **87.** $(-\infty, 0) \cup (0, \infty)$ **89.** $\left(-\frac{5}{2}, -2\right)$
91. (d)   **93.** (b)
95. (a)

$y = x \ln x$;

(b) 1; **(c)** relative minimum: -0.368 at $x = 0.368$

97. (a)

(b) 1; **(c)** relative maximum: 0.184 at $x = 1.649$

Exercise Set 6.3

1. $\log_3 81 + \log_3 27$ **3.** $\log_5 5 + \log_5 125$
5. $\log_t 8 + \log_t Y$ **7.** $3 \log_b t$ **9.** $8 \log y$
11. $-6 \log_c K$ **13.** $\log_t M - \log_t 8$
15. $\log_a x - \log_a y$
17. $\log_a 6 + \log_a x + 5 \log_a y + 4 \log_a z$
19. $2 \log_b p + 5 \log_b q - 4 \log_b m - 9$
21. $3 \log_a x - \frac{5}{2} \log_a p - 4 \log_a q$
23. $2 \log_a m + 3 \log_a n - \frac{3}{4} - \frac{5}{4} \log_a b$
25. $\log_a 150$ **27.** $\log 100 = 2$

29. $\log_a x^{-5/2}y^4$, or $\log_a \dfrac{y^4}{x^{5/2}}$ **31.** $\ln x$ **33.** $\ln(x - 2)$

35. $\ln \dfrac{x}{(x^2 - 25)^3}$ **37.** $\ln \dfrac{2^{11/5}x^9}{y^8}$ **39.** -0.5108

41. -1.6094 **43.** $\frac{1}{2}$ **45.** 2.6094 **47.** 4.3174
49. 3   **51.** $|x - 4|$ **53.** $4x$ **55.** w **57.** $8t$
59. Discussion and Writing   **61.** [3.4] $0.986, 56.5°$
62. [3.4] $0.403, 23.1°$ **63.** [3.4] $0.578, 33.1°$
64. [3.4] $2.095, 120.0°$ **65.** 4   **67.** $\log_a(x^3 - y^3)$
69. $\frac{1}{2}\log_a(x - y) - \frac{1}{2}\log_a(x + y)$ **71.** 7   **73.** True
75. True   **77.** True   **79.** -2 **81.** 3
83. $\log_a \left(\dfrac{x + \sqrt{x^2 - 5}}{5} \cdot \dfrac{x - \sqrt{x^2 - 5}}{x - \sqrt{x^2 - 5}} \right)$

$$= \log_a \dfrac{5}{5(x - \sqrt{x^2 - 5})}$$
$$= -\log_a(x - \sqrt{x^2 - 5})$$

Exercise Set 6.4

1. 4   **3.** $\frac{3}{2}$ **5.** 5.044   **7.** $\frac{5}{2}$ **9.** $-3, \frac{1}{2}$ **11.** 0.959
13. 6.908   **15.** 84.191   **17.** -1.710 **19.** 2.844
21. $-1.567, 1.567$ **23.** 0.347   **25.** 625   **27.** 0.0001
29. e **31.** $\dfrac{22}{3}$ **33.** 10   **35.** $\frac{1}{63}$ **37.** 5   **39.** $\frac{21}{8}$
41. 0.367   **43.** 0.621   **45.** -1.532 **47.** 7.062
49. 2.444   **51.** $(4.093, 0.786)$ **53.** $(7.586, 6.684)$
55. Discussion and Writing
56. [1.5] **(a)** $(3, 1)$; **(b)** $x = 3$; **(c)** maximum: 1 when $x = 3$

57. [1.5] **(a)** $(0, -6)$; **(b)** $x = 0$; **(c)** minimum: -6 when
$x = 0$ **58.** [1.5] **(a)** $(2, 4)$; **(b)** $x = 2$; **(c)** minimum: 4
when $x = 2$ **59.** [1.5] **(a)** $(-1, -5)$; **(b)** $x = -1$;
(c) maximum: -5 when $x = -1$ **61.** 10
63. 1, e^4 or 1, 54.598   **65.** $\frac{1}{3}$, 27   **67.** 1, e^2 or 1, 7.389
69. 0, 0.431   **71.** $-9, 9$ **73.** e^{-2}, e^2 or 0.135, 7.389
75. $\frac{7}{4}$ **77.** $(56.598, \infty)$ **79.** 5   **81.** $a = \frac{2}{3}b$ **83.** 88

Exercise Set 6.5

1. (a) $P(t) = 6.0e^{0.013t}$; **(b)** 6.5 billion, 6.9 billion;
(c) in 22.1 yr; **(d)** 53.3 yr
3. (a) 36.5 yr; **(b)** 0.2% per year; **(c)** 21.0 yr; **(d)** 138.6 yr;
(e) 3.4% per year; **(f)** 2.2% per year
5. About 322 yr   **7. (a)** $P(t) = 10,000e^{0.054t}$; **(b)** \$10,555;
\$11,140; \$13,100; \$17,160; **(c)** 12.8 yr   **9.** About 5135 yr
11. (a) 23.1% per minute; **(b)** 3.15% per year; **(c)** 7.2 days;
(d) 11 yr; **(e)** 2.8% per year; **(f)** 0.015% per year;
(g) 0.003% per year   **13. (a)** $k \approx 0.016$; $P(t) = 80e^{-0.016t}$;
(b) about 61 lb per person; **(c)** 86.6 yr
15. (a)

(b) 96; **(c)** 286, 1005, 1719, 1971, 1997
17. (a)

(b) 400, 520, 1214, 2059, 2396, 2478
19. (f)   **21.** (b)   **23.** (c)
25. (a) $y = 4.195491964(1.025306189)^x$, or
$y = 4.195491964e^{0.024991289x}$; since $r \approx 0.9954$, the function
is a good fit.
(b)

(c) 4.8 million, 7.8 million, 51.1 million

27. (a)

;

(b) linear: $y = 1429.214286x - 530.0714286$, $r^2 = 0.9641$; quadratic: $y = 158.0952381x^2 + 480.6428571x + 260.4047619$, $r^2 = 0.9995$; $y = 445.8787388(1.736315606)^x$, $r^2 = 0.9361$; according to r^2, the quadratic has the best fit;

(c)

(d) linear: \$19,479 million, or \$19.479 billion; quadratic: \$37,976 million, or \$37.976 billion; exponential: \$1,009,295 million, or \$1009.295 billion; the revenue from the quadratic model seems most realistic. Answers may vary.

29. Discussion and Writing
31. [4.1] $C \approx 34.8°$, $B \approx 14°$, $b \approx 7.6$
32. [4.2] $A \approx 49.6°$, $B \approx 98.2°$, $C \approx 32.2°$
33. [4.4] $8 + 8i$ **34.** [4.4] $-9 + 46i$
35. \$14,182.70   **37.** \$166.16   **39.** 46.7°F
41. $t = -\dfrac{L}{R}\left[\ln\left(1 - \dfrac{iR}{V}\right)\right]$ **43.** Linear

Review Exercises, Chapter 6

1. [6.1] (c)   **2.** [6.2] (a)   **3.** [6.2] (b)   **4.** [6.1] (f)
5. [6.1] (e)   **6.** [6.2] (d)   **7.** [6.2] $4^2 = x$
8. [6.2] $\log_e 80 = x$ **9.** [6.4] 16   **10.** [6.4] $\frac{1}{5}$
11. [6.4] 4.382   **12.** [6.4] 2   **13.** [6.4] $\frac{1}{2}$
14. [6.4] 5   **15.** [6.4] 4   **16.** [6.4] 9   **17.** [6.4] 1
18. [6.4] 3.912   **19.** [6.3] $\log_b \dfrac{x^3\sqrt{z}}{y^4}$
20. [6.3] $\ln(x^2 - 4)$ **21.** [6.3] $\frac{1}{4}\ln w + \frac{1}{2}\ln r$
22. [6.3] $\frac{2}{3}\log M - \frac{1}{3}\log N$ **23.** [6.3] 0.477
24. [6.3] 1.699   **25.** [6.3] -0.699 **26.** [6.3] 0.233
27. [6.3] $-5k$ **28.** [6.3] $-6t$ **29.** [6.5] 8.1 yr
30. [6.5] 2.3%   **31.** [6.5] About 2623 yr   **32.** [6.2] 5.6
33. [6.2] 30 decibels
34. [6.5] **(a)** $P(t) = 11e^{0.012t}$; **(b)** 11.8 million, 14.3 million; **(c)** in 68.4 yr; **(d)** 57.8 yr

35. [6.5] **(a)** $k = 0.304$; **(b)** $P(t) = 7e^{0.304t}$;
(c)

(d) 292.4 billion, 727.9 billion; **(e)** 2006
36. [6.2] **(a)** 2.7 ft/sec; **(b)** 8,553,143
37. [6.5] **(a)** $y = 11.96557466(1.00877703)^x$;
(b)

(c) 44, 255, 394; **(d)** Comparing the coefficient of correlation values (power: 0.9739; exponential: 0.9955), we see that the exponential function is the better fit.
38. Discussion and Writing [6.3] By the product rule, $\log_2 x + \log_2 5 = \log_2 5x$, not $\log_2(x + 5)$. Also, substituting various numbers for x shows that both sides of the inequality are indeed unequal. You could also graph each side and show that the graphs do not coincide.
39. Discussion and Writing [6.1], [6.2] Reflect the graph of $f(x) = e^x$ across the line $y = x$ and then shift it down 1 unit.
40. [6.4] $\frac{1}{64}$, 64   **41.** [6.4] 1   **42.** [6.4] 16
43. [6.1], [6.2] **(a)**

$$y = 5e^{-x}\ln x$$

(b) relative maximum: 0.486 at $x = 1.763$
44. [6.2] $(1, \infty)$
45. [6.1] $(-0.393, 1.143)$, $(0.827, 1.495)$, $(3.310, 2.000)$

Appendixes

Exercise Set A

1. 4   **3.** 8   **5.** $\frac{4}{5}$ **7.** $-\frac{3}{2}$ **9.** $\frac{2}{3}, \frac{3}{2}$ **11.** $-1, \frac{2}{3}$
13. $-\sqrt{3}, \sqrt{3}$ **15.** 0, 3   **17.** $-\frac{1}{3}, 0, 2$ **19.** $-1, -\frac{1}{7}, 1$

21. $-7, 1$ **23.** $4 \pm \sqrt{7}$ **25.** $-4 \pm 3i$ **27.** $-2, \frac{1}{3}$

29. $-3, 5$ **31.** $-1, \frac{2}{5}$ **33.** $\frac{5 \pm \sqrt{7}}{3}$ **35.** $-\frac{1}{2} \pm \frac{\sqrt{7}}{2}i$

37. $\frac{4 \pm \sqrt{31}}{5}$ **39.** $\frac{5}{6} \pm \frac{\sqrt{23}}{6}i$ **41.** $2, 6$ **43.** $0.143, 6$

45. $-0.151, 1.651$ **47.** $-0.637, 3.137$ **49.** $\frac{20}{9}$ **51.** 286

53. 6 **55.** No solution   **57.** No solution   **59.** No solution   **61.** $\{x \mid x \text{ is a real number } and\ x \neq 0 \text{ and } x \neq 6\}$

63. 2 **65.** $\frac{5}{3}$ **67.** $\pm\sqrt{2}$ **69.** No solution   **71.** -6

73. 7 **75.** No solution   **77.** 1 **79.** $3, 7$ **81.** -8

83. 81 **85.** $-7, 7$ **87.** No solution   **89.** $-3, 5$

91. $-\frac{1}{3}, \frac{1}{3}$ **93.** 0 **95.** $-1, -\frac{1}{3}$ **97.** $-24, 44$

99. $-2, 4$ **101.** $(-1, 3)$ **103.** $(-0.5, 1.75)$

105. No solution   **107.** $(5, 4)$ **109.** $(1, -3)$

111. $(2, -2)$ **113.** $\left(\frac{39}{11}, -\frac{1}{11}\right)$ **115.** $(1, -1)$

117. $(1, 3)$; consistent, independent

119. $(-4, -2)$; consistent, independent

121. $(-3, 0)$; consistent, independent

123. $(4y + 2, y)$ or $\left(x, \frac{1}{4}x - \frac{1}{2}\right)$; consistent, dependent

125. $(10, 8)$; consistent, independent

127. $(1, 1)$; consistent, independent

129. $b = \dfrac{2A}{h}$ **131.** $w = \dfrac{P - 2l}{2}$ **133.** $h = \dfrac{2A}{b_1 + b_2}$

135. $\pi = \dfrac{3V}{4r^3}$ **137.** $C = \frac{5}{9}(F - 32)$ **139.** $T_1 = \dfrac{P_1 V_1 T_2}{P_2 V_2}$

141. $R_2 = \dfrac{RR_1}{R_1 - R}$ **143.** $p = \dfrac{Fm}{m - F}$

145. Discussion and Writing   **147.** $-\frac{2}{3}$

149. $-2, 2$ **151.** $3 \pm 2\sqrt{2}$ **153.** -1

Exercise Set B

1. $\sqrt{10}$, 3.162 **3.** $\sqrt{45}$, 6.708 **5.** $\sqrt{14 + 6\sqrt{2}}$, 4.742

7. 6.5 **9.** Yes

11. The distances between the points are $\sqrt{10}$, 10, and $\sqrt{90}$. Since $(\sqrt{10})^2 + (\sqrt{90})^2 = 10^2$, the three points are vertices of a right triangle.

13. $\left(\dfrac{9}{2}, \dfrac{15}{2}\right)$ **15.** $\left(\dfrac{15}{2}, 2\right)$ **17.** $\left(\dfrac{\sqrt{3} - \sqrt{6}}{2}, -\dfrac{\sqrt{5}}{2}\right)$

19.

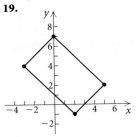

$\left(-\dfrac{1}{2}, \dfrac{3}{2}\right)$, $\left(\dfrac{7}{2}, \dfrac{1}{2}\right)$, $\left(\dfrac{5}{2}, \dfrac{9}{2}\right)$, $\left(-\dfrac{3}{2}, \dfrac{11}{2}\right)$; no

21. $\left(\dfrac{\sqrt{7} + \sqrt{2}}{2}, -\dfrac{1}{2}\right)$

23. Square the window; for example, $[-12, 9, -4, 10]$.

25. $(x - 2)^2 + (y - 3)^2 = \frac{25}{9}$

27. $(x + 1)^2 + (y - 4)^2 = 25$

29. $(x - 2)^2 + (y - 1)^2 = 169$

31. $(x + 2)^2 + (y - 3)^2 = 4$

33. $(0, 0)$; 2;   **35.** $(0, 3)$; 4;

37. $(1, 5)$; 6;   **39.** $(-4, -5)$; 3;

41. Discussion and Writing

43. $\sqrt{h^2 + h + 2a - 2\sqrt{a^2 + ah}}$,

$\left(\dfrac{2a + h}{2}, \dfrac{\sqrt{a} + \sqrt{a + h}}{2}\right)$

45. $(x - 2)^2 + (y + 7)^2 = 36$

47. $a_1 \approx 2.7$ ft, $a_2 \approx 37.3$ ft   **49.** Yes   **51.** Yes

53. $(0, 4)$

55. Let $P_1 = (x_1, y_1)$, $P_2 = (x_2, y_2)$, and

$M = \left(\dfrac{x_1 + x_2}{2}, \dfrac{y_1 + y_2}{2}\right)$. Let $d(AB)$ denote the distance from point A to point B.

(a) $d(P_1 M) = \sqrt{\left(\dfrac{x_1 + x_2}{2} - x_1\right)^2 + \left(\dfrac{y_1 + y_2}{2} - y_1\right)^2}$

$= \dfrac{1}{2}\sqrt{(x_2 - x_1)^2 + (y_2 - y_1)^2}$;

$d(P_2 M) = \sqrt{\left(\dfrac{x_1 + x_2}{2} - x_2\right)^2 + \left(\dfrac{y_1 + y_2}{2} - y_2\right)^2}$

$= \dfrac{1}{2}\sqrt{(x_1 - x_2)^2 + (y_1 - y_2)^2}$

$= \dfrac{1}{2}\sqrt{(x_2 - x_1)^2 + (y_2 - y_1)^2} = d(P_1 M).$

(b) $d(P_1M) + d(P_2M) = \dfrac{1}{2}\sqrt{(x_2 - x_1)^2 + (y_2 - y_1)^2}$

$$+ \frac{1}{2}\sqrt{(x_2 - x_1)^2 + (y_2 - y_1)^2}$$

$$= \sqrt{(x_2 - x_1)^2 + (y_2 - y_1)^2}$$

$$= d(P_1P_2).$$

Exercise Set C

1. Yes **3.** No **5.** No **7.** Yes
9. (a) $y = 1304.844444x + 21950.01111$; 54,571;
(b) $r = 0.9890$, a good fit **11. (a)** $y = 645.7x + 9799.2$;
$14,319.10; \$15,610.50; \$18,193.30;$ **(b)** $r = 0.9994$, a good
fit; **(c)** The president's assertion is about $1000, or 6%,
higher than the estimate from the model.
13. (a) $M = 0.2H + 156$; **(b)** 164, 169, 171, 173; **(c)** $r = 1$;
the regression line fits the data perfectly and should be a
good predictor.   **15.** (b)   **17.** (c)   **19.** (a)
21. (a)   **(b)** quadratic;

(c) $f(x) = 0.1729980538x^2 - 10.70013387x + 465.7917864$;
(d)   **(e)** 788; 844

23. (a)   **(b)** quadratic;

[0, 20, 70,000, 150,000]

(c) $f(x) = -1226.47072x^2 + 12,279.05135x + 97,260.77349$;
(d)   **(e)** $49,615 million;
$5491 million

[0, 20, 70,000, 150,000]

25. (a)   **(b)** Answers may vary;
                                           quadratic;

(c) $f(x) = -0.0223484848x^2 + 1.673484848x + 39$;
(d)   **(e)** 68.3%; 70.2%

27. (a)   **(b)** quadratic;

(c) $f(x) = 93.28571429x^2 - 1336x + 5460.828571$;
(d)   **(e)** 688; 1429

29. Discussion and Writing
31. (a)   **(b)** quadratic;

(c) $f(x) = 2.031904548x^2 - 59.04179598x + 527.2818092$;
(d)

(e) $20.9 million

Index

Index of Applications